Brian Kaye

Chaos & Complexity

Discovering the Surprising Patterns of Science and Technology

Weinheim · New York
Basel · Cambridge · Tokyo

Professor Brian H. Kaye
Laurentian University
Ramsey Lake Road
Sudbury, Ontario P3E 268
Canada

Published jointly by
VCH Verlagsgesellschaft mbH, Weinheim (Federal Republic of Germany)
VCH Publishers Inc., New York, NY (USA)

Editorial Directors: Dr. Peter Gregory and Dr. Ute Anton
Production Manager: Elke Littmann

Cover illustration: The beautiful photograph of a detailed magnification of the Mandelbrot set is taken from a series of Postcards available commercially from Art Matrix, PO 880-P, Ithaca, N.Y., U.S.A., 14851. Used by permission of Art Matrix.

Library of Congress Card No. applied for.

A catalogue record for this book is available from the British Library.

Deutsche Bibliothek Cataloguing-in-Publication Data:
Kaye, Brian H.:
Chaos & complexity : discovering the surprising patterns of science and technology / Brian Kaye. – Weinheim ; New York ; Basel ; Cambridge ; Tokyo : VCH, 1993
ISBN 3-527-29007-9 (Weinheim ...) brosch.
ISBN 1-56081-798-4 (New York) brosch.
ISBN 3-527-29039-7 (Weinheim ...) Gb.
NE: Kaye, Brian H.: Chaos and complexity

Printed on acid-free and chlorine-free paper

Composition, printing and bookbinding: Konrad Triltsch, Graphischer Betrieb, D-97070 Würzburg.
Printed in the Federal Republic of Germany.

This book is dedicated to my children Andrew, Sharon, Alison, and Christopher

Biography

Dr. Brian Kaye was born in Hull, Yorkshire, England, in 1932. He obtained his B.Sc., M.Sc., and Ph.D. degrees from London University after studying at the University College of Hull where he was a George Fredrick Grant Memorial Scholar. After working as a scientific officer at the British Atomic Weapons Research Establishment (Aldermaston) he taught physics at Nottingham Technical College from 1959–1963. He then moved to Chicago where he was a Senior Physicist in the Chemistry Division of the IIT Research Institute (the Research Institute of the Ilinois Institute of Technology). There he studied problems as different as why dirt sticks to the fibers of carpet to the design of better propellants for space rockets.

Since 1968 he has been Professor of Physics at Laurentian University in Sudbury, Ontario. He specializes in powder technology which deals with the manufacture and properties of cosmetics, explosives, powdered metal pigments, drug powders, food powders, and abrasives. He has written a standard text on characterizing powders and authored over 100 scientific papers.

In 1977 his interest in the complex structure of soot involved him in the new subject of fractal geometry, an interest that led to the book, "A Random Walk Through Fractal Dimensions". The philosophical side of science has always interested him and has been complimented by his activities as a methodist local preacher in the Sudbury region of Ontario, Canada. He is just as likely to be found holding a service in a protestant church as he is to be lecturing on fractal geometry and chaos theory at the University.

Acknowledgements

All of the experiments discussed in this book have been carried out by students at Laurentian University and by local highschool students who participated in a series of extra-curricular lectures offered at the university. One of the teachers who participated in this programme for highschool students was Mr. Ron Lewis of Nickle District Secondary School (Sudbury). Mr. Lewis became so entranced by fractals that he took a one year sabbatical to study with the fractal group at Laurentian University. During that year he modified and adapted the experiments described in this book for the curriculum in Ontario high schools. Since his early work with our group Mr. Lewis has gone on to develop a programme for mathematics students in Ontario high schools which is now in its second year of experimental presentation. His students have interacted with students at Laurentian University and all of the students were very willing to pioneer the new experiments which eventually formed the basis of those described in this book. By acting as guinea pigs they enabled me to find the level at which one needed to explain various concepts, and their opinions helped to shape the final form of many of the experiments. The students are too numerous to mention by name but special thanks must go to Garry Clark, my research associate, who prepared the diagrams for the book and pioneered many of the experiments, especially many of the more difficult ones on the Mandelbrot set and the mathematical roots of polynomial equations.

Various secretaries have toiled on the script for this book and a special mention must be made to the efforts of Sharon Kaye, and Alison Kaye (my two daughters), Julie Gratton, and Linda Romas. I wish to thank the Dean of Science at Laurentian University Dr. D. Goldsack who provided funds for some of the preparation of the manuscript. In the early days of the project Remi Trottier gave continuous help and encouragement to the programme.

The initial encouragement to write the first book on fractal geometry came from Dr. Helmut Grünewald who was then a senior official at VCH. He has continued to be interested in the progress of the books on fractal geometry and chaos theory. Dr. Christina Dyllick was a helpful and enthusiastic editor for the Random Walk book and helped me plan this book on Chaos and Complexity. Walter Greulich continued the supportive editorial work until it was taken over by Dr. Peter Gregory and Dr. Ute Anton. I wish to thank them all for their help and constant patience with my Yorkshire grammar and creative spelling!

Discovering The Surprising Patterns of Chaos And Complexity

In response to interest shown in the first book, "*A Random Walk Through Fractal Dimensions*", I started to prepare a set of laboratory scripts to enable students to carry out experiments in applied fractal geometry. I soon found, however, that the students wishing to carry out the experiments lacked a background in the general use of probability relationships which they needed to be able to plot the necessary graphs in their applied fractal experiments. Accordingly, I started to develop a series of experiments aimed at familiarizing students with the standard probability relationships such as the Gaussian distribution, log–normal distribution etc.

The title of this book comes from the fact that as the students carried out their investigations, using a mixture of Monte Carlo routines and physical measurements, they expressed surprise at the way in which patterns in apparently chaotic systems manifest themselves from time to time. Very early in the development of the laboratory projects students in high schools at some distance from the university started to use the scripts as the basis of science fair projects and what are known as enrichment class studies (Ontario high schools have an enriched mathematics programme for gifted students).

The fact that these scripts were being used by students who could not travel to the university necessitated the development of more descriptive material to explain the underlying science of the experiments and we also found that traditional dictionaries lagged well behind the need in explaination of terms such as fractal and Julia sets. Accordingly it was necessary to start developing a vocabulary resource for these students and hence the vocabulary notes sprinkled throughout the text. It was found that students needed a very wide base in the necessary vocabulary if they were to understand the experiments that they were carrying out.

At the same time as the material was being used for high school students, the text of the experiments was found to be useful in workshops for industrial scientists interested in acquiring a background in applied fractals. The term applied fractals is being used in this book to differentiate between the exotic patterns of theoretical fractals, such as the Mandelbrot sets, Julia sets, and the escape maps of estimates of the roots of polynomial equations, and the applications of fractal geometry to materials science problems such as the ruggedness of fractured surfaces.

As I developed these experiments I was surprised to find which experiments were found to be of greatest interest to the students. For example, I had some doubts as to whether it was necessary to put the pixilated rainbow experiment in chapter one but Mr. Ron Lewis of Nickle District Secondary School reported to me that his students found this a very useful precursor to the experiments on the Mandelbrot set.

Probably the most popular experiment in the book is the discovery of the Poisson distribution using candy coated chocolate buttons!

The choice of the material to be covered in theses experiments was based upon the availability of equipment to the average reader so that we have not discussed the measurement of mass fractals using light scattering theory and other much more complex fractal measurements. Some of the topics that were included in this book were deliberately chosen to lay the ground work for a third book I intend to write entitled, "Chaos, Catastrophe, and Spherical Chickens". In this third book of the planned series the impact on applied science of the new subjects of deterministic chaos, catastrophe theory and experimental maths will be explored. The way in which the developments of the these three subjects have pointed out the major failures of simplistic modelling called spherical chicken paradigms will be discussed. In the last decade of the 20th century we will probably see a return to the real world of experimental physics after the scientific community begins to realize that computer modelling will not solve all of its problems.

Most of the experiments described in this book can be carried out with simple pencil and paper but obviously, access to a personal computer would reduce the drudgery of some of them. Hopefully, after exploring the material presented in this book, the readers will gain a new appreciation of the physical significance of the theorems of probability theory and also grasp the significance of fractal structures manifest in the physical world around us.

Chapter Index

Wordfinder XIV

Chapter 1 **Preparing for a Journey Through Deterministic Chaos and Complex Systems** 1

Section 1.1 Deterministic Chaos? 3
Section 1.2 Complexity and the Failure of Reductionism 8
Section 1.3 Pure Mathematics and Crazy Numbers 12
Section 1.4 What is the Point of Geometry? 18
Section 1.5 Painting a Rainbow by Tracking Vanishing Pixels in Linespace 25
Section 1.6 Beware of Averages! 36
Section 1.7 Noisy Telephone Lines and Cantorian Dusts 45
Section 1.8 Monte Carlo Explorations with a Cluster of Drunks 50
Section 1.9 Empirical Facts and General Truths 60
Section 1.10 Escape From Time Space Into Phase Space 64
Exercises 71
References 77

Chapter 2 **Mandelbrot Sets, Julia Lace and Fatou Dusts** 79

Section 2.1 A Three-fold Path to Geometric Understanding 81
Section 2.2 The Mandelbrot Set – A Prison for Restless Pixels 85
Section 2.3 Julia Lace and Fatou Dusts 103
References 105

Chapter 3 **The Normal Distribution of Drunks** 107

Section 3.1 Can Tumbling Coins Tell Us Anything About Staggering Drunks? 109
Section 3.2 Seductive (Often Meaningless) Patterns in Random Time Sequences 112
Section 3.3 Technical Names for the Bell Curve 114
Section 3.4 Mathematical Description of the Structure of the Gaussian Probability Curve 115
Section 3.5 Stretching Data Mathematically with Gaussian Probability Graph Paper 120
Section 3.6 When are we Likely to Encounter Gaussian Probability Distribution in Stochastic Systems? 124
Section 3.7 Gaussianly Scattered Drunks 126
Section 3.8 Gaussian Variations in the Structure of Powder Mixtures 127

Section 3.9 Using The Gaussian Probability Curve to Characterize Uncertainty in Experimental Data 133
Section 3.10 The Power of Self-cancelling Errors to Improve Experimentally Based Estimates of a Physical Quantity 137
Section 3.11 Using Dot Counting to Estimate Irregular Areas 140
Section 3.12 Finding the Percentage of a Given Material in an Ore 145
Section 3.13 Aperture Size Distribution in a Woven Wire Sieve 147
Section 3.14 Characterizing The Weight Variation in a Population of Candy Covered Chocolate Buttons 150
Section 3.15 Buffon's Needle – A Surprising Pattern of Events 155
Section 3.16 Exploiting the Efficiency of Antithetic Variates to Improve the Efficiency of the Estimates of π by Buffon's Needle Experiments 158
Section 3.17 How Much Work in Needed to Improve One's Confidence in an Average Value? 163
Exercises 167
References 175

Chapter 4 The Rare Events of Cooperative Chaos 177

Section 4.1 "Double or Quits" – A Desperate Gamble 179
Section 4.2 The Surprise of a (Half) Lifetime! 184
Section 4.3 Making Things Normal Again with Logarithms 189
Section 4.4 How Long is a Game of Snakes and Ladders? 193
Exercises 201
References 206

Chapter 5 Prussian Horses and Fishy Statistics 207

Section 5.1 Death Rates from Horse Kicks in the Prussian Cavalry 209
Section 5.2 Using Poisson Graph Paper in the Assessment of Dust Deposition Density in Air Pollution Studies 213
Section 5.3 Stingy with the Yellow Buttons? 215
Section 5.4 Using Poisson Trackers to Monitor Chaos in a Powder Mixer 220
Section 5.5 Segregated and Tumbled Jelly Beans 223
Section 5.6 A Cautionary Tale of Tails 226
Exercises 227
References 229

Chapter 6 Rubber Number Logic and the Swinging Mouse 231

Section 6.1 The Dimensions of Reality 233
Section 6.2 Topolgy, Topography & Stretched Relationships 238

Section 6.3 Stokes' Law Versus Galileo 244
Section 6.4 Rubber Number Logic for Studying the Flow of Viscous Fluid Through Pipes 253
Section 6.5 Dimensionless Numbers as Indicators of Similar Structure and Behavior 256
Exercises 262
References 265

Chapter 7 Congregating Drunks, Soot and Other Pigments 267

Section 7.1 Congregating Drunks Create Surprising Patterns! 269
Section 7.2 Characterizing the Structure of Fractal Agglomerates 275
Section 7.3 Fractal Fingers Generated by Electrolytic Deposition 282
Section 7.4 Creating Fractal Fingers of Moving Fluid in a Hele–Shaw Cell 287
Exercises 289
References 304

Chapter 8 Infinite Coastlines and Other Wiggly Lines 305

Section 8.1 Easy Questions and Impossible Answers 307
Section 8.2 Beware! Richardson Plots May Have More Than One Straight Line Data Relationship Lurking in the Scatter of the Data Points 311
Section 8.3 Characterizing Profiles Which Manifest Various Fractal Structures Around Their Perimeter 314
Section 8.4 Estimating Fractal Dimensions by Penny Plating Procedures . 324
Section 8.5 Mosaic Amalgamation – Another Variation of the Minkowski Sausage Method for Characterizing Rugged Curves 331
Section 8.6 Fractal Rabbits and Manitoulin Island 334
Section 8.7 How to Tell a Vulcan from Another Carbon Black 340
Section 8.8 Putting Fractal Dimensions to Work in Applied Science 344
Section 8.9 More Ideas for Putting Fractals to Work 359
Exercises 363
References 383

Chapter 9 Invisible Carpets, Swiss Cheese and a Slice of Bread 387

Section 9.1 Fractalicious Bread and Cheese 389
Section 9.2 Exploring the Fractal Structure of Felts and Filters 403
Section 9.3 Characterizing the Porous Nature of Bone and Sandstone 408
Exercises 411
References 415

Chapter 10 A New Wrinkle on Surface Fractals 417

Section 10.1 Characterizing Canyons and the Effect of Sunshine 419
Section 10.2 Simulating Passoja's Method for Studying the Roughness of Metal Fractures 425
Section 10.3 How Many Islands are there in a Lake? 427
Exercises 434
References 436

Chapter 11 Zipf's Law and the Surprising Patterns of Word Occurrences 437

Section 11.1 Hyperbolic Word Frequencies 439
Section 11.2 The Size Distribution of Population Centers 443
Section 11.3 Beware of Procrustean Thinking 445
Section 11.4 Levy Flights – A Generalized Theory of Brownian Motion .. 450
References 461

Chapter 12 Climbing Fig Trees to Discover Fascinating Numbers ... 463

Section 12.1 Magic Numbers for Communicating with Space Aliens? 465
Section 12.2 Population Ecology – Malthus Modified 466
Section 12.3 Climbing Attractor Fig Trees by Means of Parabolic Cobwebs 477
Exercises 482
References 482

Chapter 13 Coincidences, Clusters and Catastrophes 483

Section 13.1 Strange Coincidences and Significant Clusters 485
Section 13.2 Simulating Significant and Nonsignificant Patterns of Accidents and Diseases 487
Section 13.3 The Importance of Understanding Coincidences and Clustering in Fine Particle Science 499
Section 13.4 The Catastrophic Behavior of Dripping Taps and Tumbling Rocks 515
Section 13.5 Avalanches and Earthquakes 528
Exercises 530
References 532

Chapter 14 Mathematical Watersheds and Rooting Around in Drainage Basins 535

Section 14.1 The Concepts of Fluxions 537
Section 14.2 Using Newton's Method for Discovering the Roots of Equations 543
References 552

Chapter 15 Fourier Analysis, Fractal Dimension and Formation Dynamics 553

Section 15.1 Hot Rocks and Musical Notes 555
Section 15.2 Harmonious Rocks and Fractal Profiles 561
Section 15.3 Fourier Analysis, Fractal Structure and Fortune Hunting 569
References 574

Author Index 575
Subject Index 579

Wordfinder

This wordfinder lists the location of the occurrence in the text where the meaning of a word is set out for the reader. When the reader encounters a word which is vaguely remembered the wordfinder refers the reader back to the place in the text where that word is definded. Thus, if the reader is browsing through Section 6.2 and encounters the word *postulate* the wordfinder will indicate that this word is defined on page 47.

A

absorbed 330
abstruse 90
absurd 53
acceleration 237
acceleration due to gravity 242
accuracy 135
acoustics 559
ad infinitum 35
ad nauseum 26
addendum 233
address of the point 23
adjective 474
adsorbed 330
aerodynamic diameter 344
aerosol 503
agglomerate 274
aggregate 274
Al-Khwarizmi 72
algorithm 71
alloy 363
Alma Mater 400
alumina 378
aluminum oxide 378
amalgam 333
amber 425
amplitude 558
amplitude of the oscillation 64
analysis 9
analytical geometry 81
anion 285
antithetic 160
antithetic variate characterization 160
aperture 148
applied mathematician 12
approximate 245
arbitrary 88
archbishop 375
Archimedes' 249
Archimedes principle 249
archipelago 375
arithmetic probability paper 120
artifact 43
aspect ratio 258
asterisk 238
astrology 238
asymptote 75
asymptotic 75
atom 22
attractor in phase space 70
attractor orbit 70
Auebarch 443
average 36

B

back filling 354
bacteria 510
balance 472

ball mill 347
ballistic growth 286
banker 180
bar graph 57
barometer 345
bell curve 114
Bequerel 184
berserk 38
Berylliosis 244
beryllium powder 244
Bessel 118
Bessel functions 118
Bessel's correction 118
BET method 330
bimodal 443
bimodal distribution 153
biodegradable 513
biography 240
biology 238
black lung 345
black-body radiation 19
blind pore 392
blissymbolics 538
blowing agent 392
blue moon 353
boundary fractal dimension 311
boundary layer 255
brass 363
Briggsian logarithms 119
bronze 363
Brown 56
Brownian motion 56
Brunauer, Emmett and Teller 329
Buffon 155
Buffon's needle 156
butterfly effect 11

C

calculate 52
Cantor 46
Cantorian bar 46
Cantorian dust 46
Cantorian set theory 85
capture trees 503
carbon black 280
Cardano 87
Cartesian coordinate 24
Cartesian product of two sets 84
casein 401
CAT scanning 389
catastrophe 486, 515
cation 285
caustic soda 508
Cayley 547
central limits theorem 126
chance 40
chaos 3
chaotic 7
charge coupled device camera 331
Charybdis 114
Chaye's dot-counting technique 146
chocolate 401
chord 315
chromatography 371
chromium 371
chromosomes 371
cinder 350
cinema 30
circumference 75
claimed accuracy 136
cobweb diagram 473, 477
coefficient of viscosity 247
coin 333
coincidence 485
coincidence errors 500
coins 334
collimated 508
compensating errors 138
complementary pattern 269
complete 269
complex 88
complex numbers 87
complexity 9

concave 256
concentric 275
conic section 429
contour lines 419
convex 256
convex hull 256
convoluted 327
cooperative interaction 180
coordinate geometry 81
coordinates of the point 23
corrosion 308
cosmetic technology 3
cosmetics 3
cosmology 3
cosmos 3
critical velocity 253, 260
critically self-organized systems 528
culinary 400
cumulative data histogram 120
cumulative oversize histogram 120
cumulative undersize histogram 120
cytologist 323
cytology 323

D

damping of a vibration 65
data set 117
deMoivre 115
debris 321
democracy 466
demographers 466
dendritic agglomerates 503
denominator 27
density 249
depth filter 403
derived dimensions 236
Descartes 22
detergent 517
determinism 3
deterministic chaos 5
devil's staircase 48
diagonal 319
diameter 256
differential 542
differential calculus 537
differential equations 543
differential function 542
diffuse 274
diffusion 274
diffusion limited aggregate 274
dilatancy 355
dilate 355
dime 111
dimension 234
dimensional homogeneity 233
discrepancy 303
displacement 237
displacement vector 55
distributed histogram 120
divide 52
DLA 274
Doppler 454
Doppler shift 454
Dow Jones Industrial Index 381
drainage basin 93
dynamic viscosity 260
dynamics 31
dynamite 31

E

ecology 439
ecology of populations 439
eddy 251
electrolyte 285
elute 370
elutriation 370
empirical fact 62
energy 333
envelope 256
epidemic 486
epidemiology 486
epoxy resins 426

equal 471
equation 473
equilibrium 472
equipaced method 314
equipaced polygon estimation technique 314
ergodic sequences 41, 166
erode 308
error 133
estimate 136
Euclid 5
Euclidean geometry 5
examination 472
excipient powder 512
experimental mathematics 14
explicit 19
exponent 27
exponential decay 182
exponential form 27
exponential notation 26
extrapolation 124

F

factorial operation 119
feedback signal 466
Feigenbaum number 466, 477
felt 403
ferrography 322
fiber 258
filter 406
fluorescence 184
fluxions 539
flyash 355
forensic science 371
Fourier 555
Fourier analysis 555
fractal 6
fractal dimension 6, 233
fractal geometry 5
fractal origami 9
fractal rabbits 334
frequency 558
frequency of oscillation 64
fume 280
fundamental dimensions 236

G

Galileo 12
Galton 126
gambler 180
game theory 195
Gamow 50
gangue 442
gas adsorption studies 329
Gauss 115
Gaussian probability curve 114
Gaussian probability paper 120
genus of the system 239
geography 240
geology 238
geometric probability 156
geometric signature waveform 561
geometrical probability 390
geometry 238
globe 274
globule 274
graph 240
"guesstimate" 136

H

half-life 188
hardener 426
harmonic notation 558
harmonic spectrum 559
Heisenberg 4
Heisenberg's Uncertainty Principle 4
herbicides 360
heterogeneous 133
histogram 57
homogeneous 133
hull 256

human ecology 439
hygiene 258
hyperbola 429
hyperbole 429
hyperbolic function 429
hypotenuse 319
hypothesis 151

I

ideal fractal curves 308
image 180
imaginary 88
imaginary numbers 87
implicit 19
indented 326
indentured servant 326
index 256
inert 244
intangible 75
integer 391
integrate 391
inviscid fluid 247
irrational numbers 16
Isles of Scilly 375
isobars 345
isosceles 319
isotherms 345
iterate 26

J

javelin 474
Julia 90
juxtaposition agglomerate 502

K

Karman street 251
kinematic viscosity 260
kinematics 30
kinetic energy 263, 476
Koch islands 71

L

laminar flow 245
Laplace 4
Laplacian determinism 4
laws of chance 111
Leibniz 539
leukocytes 323
Levy 453
Levy flights 453
liberate 401
ligament 363
limiting parameters 242
log cycle 29
log-log graph paper 29
log-normal graph paper 192
log-normally distributed 192
log-probability graph paper 192
logarithm 29
logarithmic decay 181
logarithmic graph paper 29
logarithmic scale 29
Lord Rayleigh 243
leukemia 323

M

Mach number 259
Mach, Ernst 259
Malthus 466
Mandelbrot 5
Mandelbrot set 5
map area 429
Markovian chain 41
mass 237
mathematics 13
matrix 400
May equation 473
mean 36
meander 140
median value 198
mesh 174

metal 333
metallurgy 333
metastasis 459
microtome 390
Minkowski 324
Minkowski sausage 324
mistletoe 246
modal analysis 146
mode 153, 198
modulus of a complex number 87
molasses 246
monochromatic 434
Monte Carlo routines 59
mosaic 14
mosaic amalgamation 334
Murphy's first law 498
Murphy's second law 498
muses 14
museum 14

N

nanometer 350
nanophase material 350
Napier 29, 118
Napierian logarithms 119
natural fractal curves 308
natural logarithms 119
nest of sieves 446
neutrons 508
Newton 3
Newton's method of approximation 537
Newtonian fluids 248
Nirvana 194
non-Newtonian fluids 356
normal probability curve 114
normal probability paper 120
normalization 61
normalization factor 62
nuclear winter 352
nucleating centre 270
nucleus 270
numerator 27

O

observed population lines 212
occlusion 372
odd number 179
odd person 179
odds 180
operational definition 30
operational definition of zero 31
operational dimensionality 236
operative parameters 242
ordered pairs 84
original data 118
orthodontist 321
orthodox 18
orthogonal 321
Oughtred 15
outgassing 330
oversize 120

P

π, pi 14
parabola 429
paradox 18
parameter 241
pastes 248
pendant 321
pendulum 321
percolating 512
perimeter 75
period 242
periscope 75
perpendicular 321
pesticide 360
phase angle of the oscillation 67
phase of the oscillation 67
phase space 65
phase space graph 68
phenomenal 259

phenomenon 259
philology 439
photon 21
pi, π 14
pixel 14
pixoid 22
plagiarism 539
Planck 19
plumb line 320
plumber 320
pneumatic conveying 362
pneumatic drill 362
pneumoconiosis 345
Poincaré 111
point 18
Poise 255
Poiseuille 253
Poiseuille formula 253
Poisson 209
Poisson distribution 209
Poisson probability graph paper 211
polluted 370
polygon 319
pore 392
pores 391
porosity 392
position paper 152
postulate 47
potential energy 263
potential energy 476
precision 135
predator 468
prey 468
primary count loss 500
probability theory 110
probit 123
Procrustean 446
profile 327
projected area 429
projection occlusion 372
prosthesis 322
pseudohomogeneous 133
pulverized 347
pure mathematician 12
putty 355
Pythagoras 319
Pythagoras' theorem 319

Q

quantum 19
quantum geometry 22
quantum theory 20
quench 342

R

radioactivity 184
random 7, 38
random number table 37
random walk theory 59
ratio 15
raw data 118
real 88
reciprocal 27
reciprocating engine 27
reductionism 8
resin 425
resistazone stream counters 499
resolution 308
resolution of inspection 310
resolve 308
rheology 360
Richardson 60
Richardson plot 64, 310
richness of an ore 145
robot 203
rock matrix 401
rock tailings 353, 401
rodent 308
roots of the equation 543
Rosin–Rammler 446
Rosiwal intercept method 391
rubber number logic 233

rule of thumb 338
running average 303

S

sagittal suture 359
saw toothed wave 558
scalar 55
scale of magnification 310
scaling function 44
scanning electron microscope 350
Schumpter 572
Scylla 114
second order equation 471
secondary count gain 501
section 391
self-avoiding random walk 57
self-canceling errors 138
SEM 350
set square 321
shape factors 256
shape index 256
shape indices 256
Shaver's disease 505
shearing stress 248
Sierpinski 393
Sierpinski carpet 393
sieve 174
sigma 117
silhouette 328
silver amalgam 334
simulate 38
sintering 350
Sirens 114
skewed bell curve 192
slurry 248
Smarties® 150
social scientists 442
solar flare 359
soot 277
sound barrier 259
specific gravity 258
speed 55
sponge-type filter 403
square root 53
stable equilibrium 476
standard deviation of the data 117
statistically self-similar 47
stereology 389
Stober Disc Centrifuge 345
stochastic 38
stochastic process 40
stochastic time sequences 114
Stokes 251
Stokes' law 245
strain 248
strange attractors 70
stream counters 499
streamlined 260
stress 247
stretcher 446
stroke 380
structural fractal dimension 313
structural Sierpinski fractal 400
submarine 345
suffix 345
sun spots 359
surd 54
surface filter 508
surface tension 517
symbol 327
symmetry 393

T

tailings ponds 354
tangent 75
tangible 75
TEM 350
termites 393
textural fractal dimension 313
textural Sierpinski fractal 400
Gauss 115
theorem 151

theoretical 151
theory 151
thin layer paper chromatography (TLPC) 371
thixotropic behavior 355
thixotropy 355
Thoria 347
tile 14
tome 390
tomography 390
topical cream 419
topical medication 419
topical products 419
topography 240
topological dimension 239
topology 238
trajectory 474
trajectory of the calculation 474
transformed data 118
transmission electron microscope 350
treacle 246
tremas 393
triangle 319
tribology 322
trigonometry 319
turbid 251
turbulent flow 251

U

uncertainty 137
undersize 120
unstable equilibrium 476
Urania 347

V

vacant 391
vacation 391
variance 118
vector 55
velocity 55
violet catastrophe 19
viscous drag 249
viscous fluid 247
voidage 392
voids 391
volatile 102
volcano 340
von Koch 71

W

wage 180
wager 180
Wallis 30
watershed boundary 377
wavelength 558
wavicle 21
wood pulping 403

Y

yeast 392

Z

Zipf 439
Zipf's law 440

Chapter 1

Preparing for a Journey Through Deterministic Chaos and Complex Systems

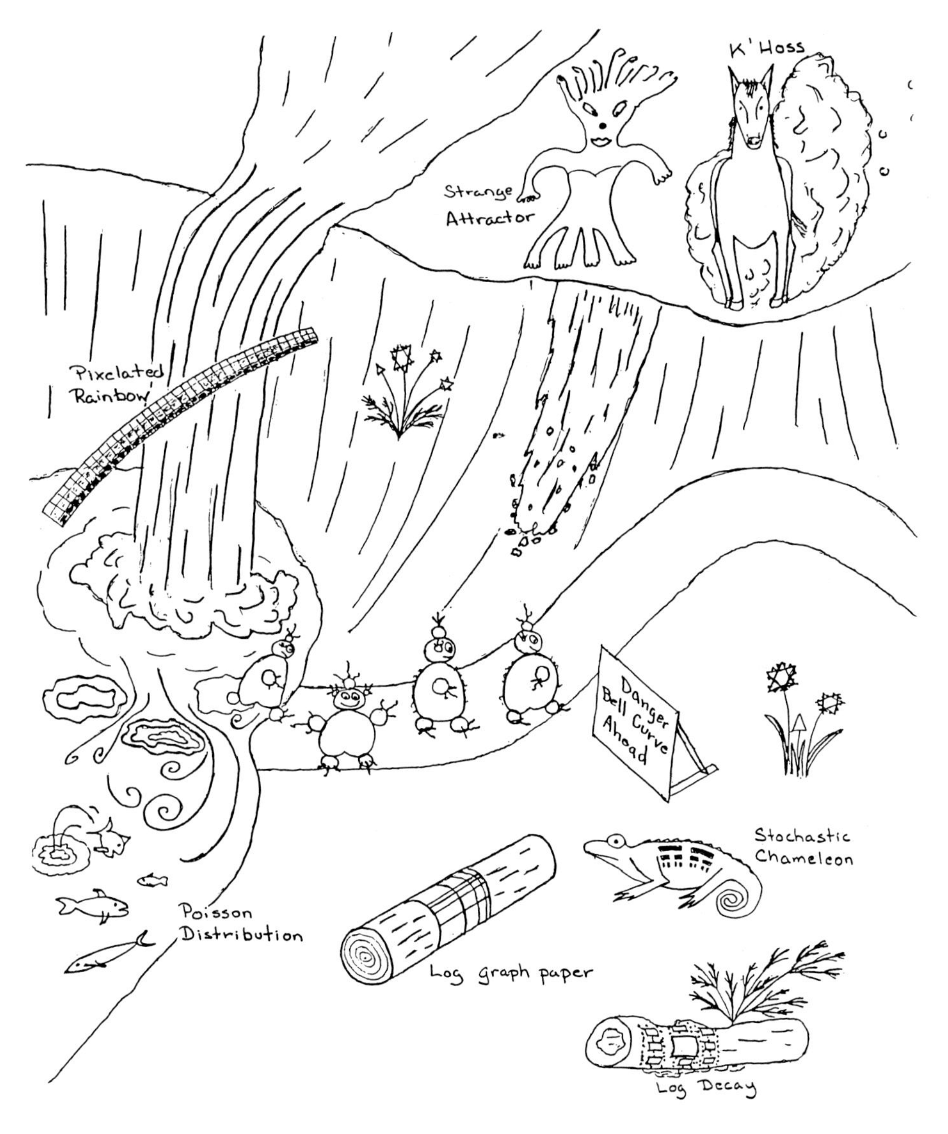

The Mandelbrot set starting their journey of discovery

Chapter 1 Preparing for a Journey Through Deterministic Chaos and Complex Systems 1

Section 1.1 Deterministic Chaos? 3
Section 1.2 Complexity and the Failure of Reductionism 8
Section 1.3 Pure Mathematics and Crazy Numbers 12
Section 1.4 What is the Point of Geometry? 18
Section 1.5 Painting a Rainbow by Tracking Vanishing Pixels in Linespace 25
Section 1.6 Beware of Averages! 36
Section 1.7 Noisy Telephone Lines and Cantorian Dusts 45
Section 1.8 Monte Carlo Explorations with a Cluster of Drunks 50
Section 1.9 Empirical Facts and General Truths 60
Section 1.10 Escape From Time Space Into Phase Space 64
Exercises 71
References 77

Chapter 1

Preparing for a Journey Through Deterministic Chaos and Complex Systems

Section 1.1. Deterministic Chaos?

Deterministic chaos. Why is it worth exploring and what is it? The two words used to describe the subject seem to be self-contradictory. Deterministic implies ordered, structured, predictable. How can the behavior of a chaotic system be predictable? Actually, the title of this section is the name of a new branch of mathematics and physics. Logic is not always a guaranteed aspect of scientific terminology; scientific vocabulary can be as inconsistent as poetry. To understand what is meant by deterministic chaos in modern science, we must dig down into the history of science. To the philosophers of ancient Greece *chaos* represented the mysterious unorganized material which was used by the Gods to fabricate the visible universe. In modern English chaos came to mean a completely unorganized, turbulently unpredictable state. The boiling surf at the base of a cliff, randomly leaping and churning as the waves of the sea smash against the rocks is a typical example of a system which, in everyday English, would be described as chaotic. To describe the ordered universe created by the Gods out of chaos, the Greeks used the word *cosmos*. *Cosmology* is the name given to the study of the origin and structure of the universe. In today's society we define *cosmetic technology* as the art of creating a structured face using the various beauty aids described as *cosmetics* [1] (in the course of the journey of discovery described in this book, many unfamiliar words will be encountered. The first time such a word occurs in the text it will be defined, and the Wordfinder at the front of the book will enable the reader to locate these definitions).

Scientists do not like chaotic systems. The basic drive of the scientist is to study cause and effect relationships and to predict the future behavior of a system. It is difficult to place an exact date on the birth of modern science. For many the intellectual revolution that we call the scientific method began with Galileo, who revolutionized astronomy by examining the moon and the planets with his telescope. Beginning with the pioneering work of Isaac *Newton* (1642–1727) science came to be dominated by a philosophy which has come to be known as *determinism* [2]. By studying the motion of the stars and the planets, and by observing the behavior of objects falling towards the earth, Newton described the ordered behavior of moving objects using theories which are usually described as "Newtons Laws of Motion". Those beginning to study physics learn these laws of motion and use them to predict the behavior of colliding balls on a flat surface. They are taught that, provided they know the initial velocities and pathways of a set of original objects, such as billiard balls approaching each other, they can completely predict the outcome and future behavior of the set of colliding objects. In other words, the future behavior is determined completely by the past state of the objects. In their imagination students, drunk with their success with the billiard balls, dream that by using the three laws of motion they can describe the celestial dance of the whirling galaxies of outer space.

The ability to predict the future movement of objects from observed past behavior became known as determinism. The great expounder of the philosophy of determinism was the French scientist Pierre Simon *Laplace* (1749–1827). Laplace developed the idea that the future of the universe is completely determined by its present state. *Laplacian determinism* claims that when we have sufficient knowledge of the current state of the universe we can completely predict its future. These ideas were put forward in a book entitled "Celestial Mechanics". It is said that Napoleon, on looking through the book, noted that there was no mention of God. "I had no need of that hypothesis", answered Laplace. The spread of determinism as a scientific philosophy resulted in the development of an idea of a clock maker God who had built the universe, wound it up and set it going. This type of God sits at the remote borders of the universe, observes it "ticking" but never interferes with its progress. Scientists generally held the belief that as our ability to describe the universe increased, our knowledge of its future would be complete.

The first jolt to the dominance of determinism as a scientific philosophy came from the impact on science of an idea which came to be known as *Heisenberg's Uncertainty Principle*. This theory, put forward by Karl *Heisenberg* (1927), stated that it was never possible to know both the position and velocity of an object with absolute certainty. Many scientists took the philosophical implication of this idea to mean that the physical universe around us was a manifestation of swirling changes in an underlying soup of uncertainty. Other scientists, including Einstein, refused to accept this as an implication of Heisenberg's Uncertainty Principle [3]. In a key phrase summarizing his rejection of a universe based on uncertainty, Einstein stated "God does not play dice with the Universe". Perhaps by the end of the journeys of discovery suggested in this book, we will be able to hint at the fact that God may play dice with the universe in some situations, but only to create opportunities for humans to exercise their own free will and creativity in a complex universe [4].

Although chastened by the presence of Heisenberg's Uncertainty Principle, modern scientists became increasingly optimistic that they would achieve the highest levels of Laplacian determinism as they used their increasingly complex and massive computers to grasp the details of the universe. From the late 1970s, however,

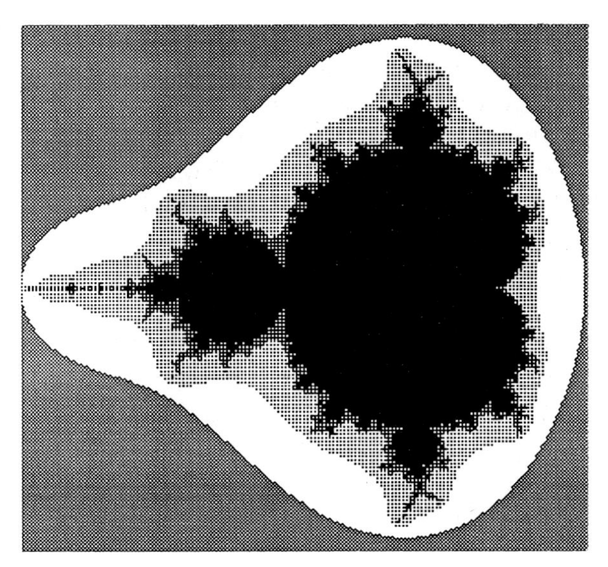

Figure 1.1. The Mandelbrot set and its "fleeing pixel" environment is one of the fractalicious patterns being created in the study of systems falling within the domain of deterministic chaos. Fractals have infinite properties, fractal + delicious = something which is infinitely delicious = fractalicious.

scientists began to undergo a culture shock as the mountain of numbers churned out by the computers began to indicate that there were many physical systems which, although essentially deterministic, were so sensitive to initial conditions that any hope of predicting their future behavior, although a theoretical possibility, would remain forever an impractical dream. As scientists contemplated the surprise predictions of their computers, they christened the new subject concerned with the complex behavior of systems sensitive to initial conditions *deterministic chaos*.

As the subject of deterministic chaos blossomed, scientists working in the field began to publish exotic, complicated graphic summaries of the behavior of dynamic systems. In *Figure 1.1* a typical fascinating picture of the type of graphics generated by students of deterministic chaos is shown. This system is known to mathematicians as the *Mandelbrot set*. Much later in this book we will discover how to generate the Mandelbrot set. For now, however, it is sufficient to notice its fascinating structure and to enjoy its complexity. The Mandelbrot set is named after its discoverer, B.B. *Mandelbrot*, who created a new subject called *fractal geometry*. This geometry of rugged systems is revolutionizing the mathematical description of rough and fractured systems in materials science and is finding many applications in other areas of applied science.

The geometry taught in high school is called *Euclidean geometry* (named after the Greek scholar *Euclid* who was born in 325 BC). Euclidean geometry seeks to reduce the universe to perfect circles, triangles and smooth surfaces. Fractal geometry

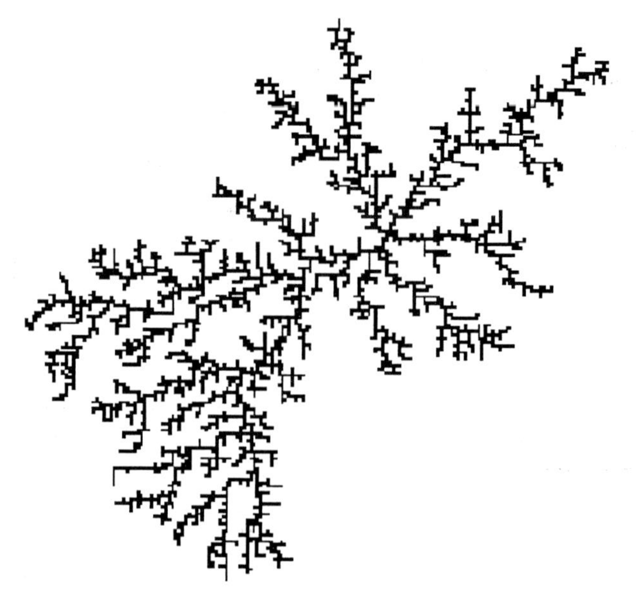

Figure 1.2. The fractal dimensions of structures known as "diffusion limited aggregates" (DLA systems) are being used to characterize the structure of such diverse items as commercially important paint pigments and fine soot particles carrying cancer-causing chemicals into the lung.

tackles the problem of describing the rugged surfaces and structures found in nature. The term *fractal* was coined by Dr. Mandelbrot from the Latin word fractus meaning "broken", because a fracture shows the infinite type of complexity of real rugged systems which can be described using the concepts of fractal geometry. In this book we will discover how to describe complex objects by their *fractal dimension*: a measure of the ruggedness and space-filling ability of a structure.

A simpler but equally surprising pattern encountered in an exploration of fractal geometry is shown in *Figure 1.2.* This structure is known to scientists as a diffusion-limited aggregate [11]. By studying the way in which diffusion-limited aggregates grow, and how other surprising fractal systems evolve, I hope to explain to the reader how deterministic chaos is not an abandonment of determinism but a strategic retreat from full determinism. From the new perspective of fractal geometry and deterministic chaos the behavior of complex systems can no longer be described in a predictable manner, but their probable behavior can be anticipated from the probable interaction of the many causes interacting to produce the final system. The seductive brilliance of fractal imagery and the intriguing patterns of deterministic chaos such as those of Figure 1.1 have attracted the attention of the general public and as a consequence several popular books have invited the average reader to explore the patterns and to enjoy their beauty [2–17].

When a new subject grows explosively scientists are often at a loss for words to describe their endeavors and their discoveries. As they invent the new words they need, they do not always agree on a uniform terminology. The terms they create at the moments of discovery are often more colorful than logical. For example, as we start to look at the swinging behavior of a complex pendulum we will find that mathematicians talk about "strange attractors in phase space". To the uninitiated these objects could be seductive aliens from outer space. Even when one discovers that they are a mathematical concept, the name seems strange to a physicist who expects attractors to have the ability to exert forces on objects, whereas a strange attractor turns out to be nothing more than a probable position or path of a system portrayed in a novel type of graph. The word strange came into the vocabulary of the mathematicians from the reaction of two scientists – the Belgian scientist David Ruell and the Dutch mathematician Floris Takens – to a surprising pattern of behavior observed in a given system. It has now spread into the general vocabulary of deterministic chaos. We will encounter strange attractors in Chapter 13 and discover their ordinariness when the mystery of the vocabulary is stripped away.

When a scientific field of study captures the imagination of the general public, the technical terms being used by the scientist tend to become less precise and convey meanings never intended, as they are used and abused in popular articles on the subject. As the fantastically intricate pictures generated on computers by scientists studying deterministic chaos became widely available to the public through popular articles in magazines such as *New Scientist* and *Scientific American*, the subject became known by the shortened name "chaos". In a very well written and exciting book by James Gleick, the subject is simply called "Chaos" [5]. It would be flying in the face of common usage to attempt to retrain the public to describe the subject as "deterministic chaos". However, in some ways the term "deterministic chaos" for the new subject of complex systems is unfortunate, because it implies to the reader that scientists are able to study completely chaotic systems of the most intricate structure (note that one of the most beautiful books displaying the graphics of deterministic chaos has a subtitle "Images of Complex Dynamical Systems" [12]). The fact that we cannot use so called "chaos theory" to study all chaotic systems is emphasized by the comments of Frances C. Moon, one of the developers of deterministic chaos theory. In his book he wrote:

> *We must distinguish between so called random and chaotic motions. The former term (random) is reserved for problems in which we truly do not know the input forces or we only know some statistical measures of the variables. The term "chaotic" is reserved for those deterministic problems for which there are no random or unpredictable inputs or parameters* [6].

Thus, strictly speaking, a deterministic chaotic system is one in which we know all the contributory causes, but in which the interaction of the causes is so complex that we must study the outcome experimentally rather than theoretically. If we have a randomly chaotic system we are only able to observe the behavior of the system in the absence of a complete knowledge of the contributory causes [7, 8]. In this book we will study the patterns of behavior of systems in which the causes interact chaotically, and strictly speaking the title of this book should be "Deterministic Chaos and Complexity".

The term complexity was added to the title since some of the systems we will study do not seem to be sufficiently disorganized to deserve the adjective chaotic. They do, however, represent systems which are essentially unpredictable in specific cases but which over a long period appear to generate patterns of behavior which, although surprising, occur with amazing regularity. Hopefully, as the reader follows the different experiments and resulting data set out in this book they will find that deterministic chaos is not an abandonment of Laplacian determinism. They should discover for themselves that deterministic chaos represents a retreat from the anticipation of the ability to predict precisely the behavior of systems when they are complex, even though these systems still conform to general patterns of behavior which are infinitely variable within certain limits. The experiments and exercises discussed and suggested will encourage the reader to anticipate possible patterns of behavior in complex systems which will enable them to cope with complexity. It is my opinion that when the intellectual history of the 1980s comes to be written, it will be clear that the evolution of deterministic chaos as a subject will reestablish the respectability of experimental science, which has tended to be overshadowed by the sophistication of often unreal computer models of reality.

One cynical applied scientist was asked to define a successful graduate student of chemical engineering in the early 1980s. He replied that the successful graduate student was the one who could take a real problem from industry and change its structure so that it could be successfully modelled on a computer, even though the results would be completely irrelevant to the industrial situation being studied. (See the discussion of spherical chicken syndrome [18].)

Section 1.2. Complexity and the Failure of Reductionism

Another philosophy of science which was very popular in the mid-20th century is known as *reductionism*. To the uninitiated this philosophy could be called "nothing-butism". The reductionist seeks to reduce a description of a complex system to a simple recipe of ingredients. A reductionist views a human being as "nothing but" so many liters of water and so many grams of calcium, etc. However, the facts set out by a reductionist are of very little use if one needs to understand the workings of the system. From the list of the ingredients constituting the human body, it would be very difficult to predict the intelligence of the human being or its emotions.

When stated in this way, reductionism obviously has severe limitations, but it is surprising how entrenched reductionism has been amongst scientists who see a human being as "nothing but" a computer or "nothing but" a biological organism engaged in the search for survival of the species. The opposite of reductionism is the direct study of complexity, but such a study is often less satisfying to the intellect than the reductionist *analysis* (note that analysis literally means "cut up into parts", and so the description of the human body by its constituent chemicals represents the literal interpretation of the statement "analysis of a human being"). The word *complexity* comes from Latin root words: com, "together", and plectere, "to plait".

Thus, a complex system has many different contributory elements interwoven with each other. The relationship between reductionism and complexity can be illustrated by the two elements of *Figure 1.3*. In Figure 1.3(a), a model of a bird created by folding paper is shown. The art of creating such models by folding paper is known as origami [19]. To describe such a model as nothing but a folded piece of paper misses the point of the three-dimensional ingenuity of the final structure and the creativity put into the model by the origamic sculptor. To reduce the model to its unfolded structure, as shown in Figure 1.3(b) with lines indicating where the paper has been folded, is an exercise in reductionism that makes it almost impossible to predict what the model represents from an examination of the folds.

It is rather interesting that some of the earlier studies of fractal systems involved the patterns created by multiple folding of paper and the extension of such real experiments into multi-folded systems that could only be created in the imagination, since real paper cannot be folded as many times as can be imagined with the mind. We will not be able to review this work on the chaotic and fractal structure of many folded systems, but the interested reader can discover the magic of *fractal origami* in the series of articles by Dekking et al. [20, 21].

The only word in the title of this book that has not been discussed in detail is "surprising". My dictionary has many definitions of the word surprise but the one I like is "To strike with wonder or astonishment". When something surprises us with its unexpected complexity or beauty, we are astonished and filled with wonder. The intricate patterns of Figure 1.1 and the other beautiful patterns of deterministic chaos are sufficiently new to fill us with surprise, but many of the older patterns of behavior are equally surprising if we can forget our familiarity with their known behavior. Sometimes experienced scientists teach the well known facts of science as if they were obvious, although they were actually very surprising to their original discoverers. Indeed, sometimes the discoveries were rejected by the scientific community as contradictory to accepted truths. I hope in the course of this book that the reader will share with me my original surprise when I discovered some of the surprising patterns of complexity, such as the patterns of events which could be expected from an array of drunks staggering away from a lamp post. I hope that I will be able to help the reader enjoy such surprising patterns as those created by randomly falling needles dropping onto a set of parallel lines (Buffon's needle).

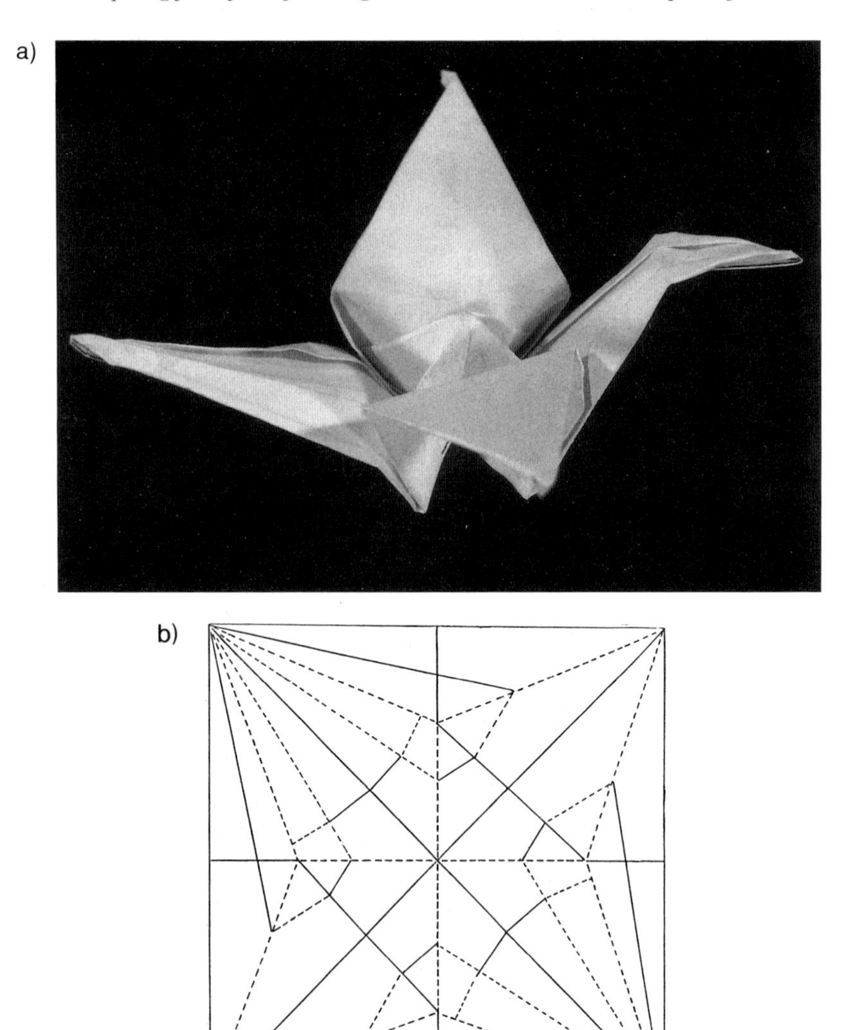

Figure 1.3. The limitations of reductionism as a philosophy can be illustrated by considering what we retain of our knowledge of an origami model when it is reduced to "nothing but" a set of folds on a piece of paper. a) Bird created by paper folding (origami bird model). b) It is not easy to visualize the three-dimensional appearance of the bird of part (a) from its unfolded sheet.

Sometimes the patterns we discover in complex systems are not only surprising but slightly annoying. I know that as a student, I resented the fact that the performances of myself and my colleagues could be fitted with a "bell curve". It seemed to go against human freedom that we should be classified by a mathematical curve which could describe the performance of all students [22].

In this book we shall distinguish between deterministic chaotic systems and complex systems. A system whose behavior involves many known contributory causes but whose behavioral outcome is very sensitive to the initial conditions will be described as deterministically chaotic. Thus attempts at predicting the weather would be an exercise in deterministic chaos. When modelling the weathers evolution from its present state, the smallest uncertainty in the knowledge of any given parameter such as wind strength can result in an enormous variation in the actual weather pattern two or three days into the future. Workers in the field of weather forecasting have coined the phrase *butterfly effect* to describe the chaotic evolution of weather patterns [5]. This phrase refers to the possibility that the fluttering of a butterfly's wings in the Amazon jungle can affect the path of a hurricane in the Gulf of Mexico.

The probable patterns of heads and tails exhibited by 25 coins thrown repeatedly at random onto a table is a study of a complex system. Studies of the probable outcome of events such as heads and tails in a coin throwing experiment is a retreat from Laplacian determinism in that one cannot predict the outcome of any particular throw of the coins. One can only discuss probable patterns from observations or from predictions based on probability theory. As the journeys of exploration suggested in this book are carried out, it will become apparent that probability theory and deterministic chaos are closely related subjects with similar philosophical attitudes in their description of complicated systems. Newtons laws of motion are predictive. Colliding bodies cannot disobey them. The laws of probability are strictly scientific laws in that they are descriptive of probable events and are more properly called relationships rather than laws. Probability relationships only become predictive if they are used to describe probable events manifest in a very large number of events. Even with a large number of the events, ghosts of uncertainty are present in any discussion of the significance of observed data of complex systems. The experiments described in this book are intended to be part of a journey of discovery in which the reader will encounter and grapple with the strange attractors of deterministic chaotic phase space, and learn to recognize and use the amazing patterns which emerge in the behavior of complex multivariable systems. Before setting out on any journey, we must learn how to read maps and plot a pathway. On this journey of discovery one must have more than a nodding acquaintance with logarithms and the exponential notation for writing down large numbers. The remaining sections of this chapter are intended to develop mathematical muscle for the journey of discovery, and to build up a vocabulary so that one can communicate with those who have worked in the areas of intellectual endeavor we will explore. The more experienced reader may wish to skip over these beginning exercises and move ahead to Chapter 2.

Section 1.3. Pure Mathematics and Crazy Numbers

When I arrived at university to start studies leading to a science degree, I had to choose between studying pure and applied mathematics. I wasn't really sure of the difference between the two branches of mathematics, but I decided to study pure mathematics. As I attended classes I was surprised to find that students studying for arts degrees were attending the same lectures. Apparently pure mathematics is artistry with numbers and does not necessarily involve a study of reality.

In the popular image of the scientist, the symbols of mathematics are so firmly associated with the wizardry of applied sciences that it is often assumed that mathematics is a science. Pure mathematicians, however, do not view themselves in this way. G. H. Hardy, a well known pure mathematician at the University of Oxford, England, used to boast that he had never done anything useful in his life [23].

Ian Stewart, in a book on the *Concepts of Modern Mathematics*, makes the statement that

> *The aim of pure mathematics is not practical applications but intellectual satisfaction. The pure mathematician plays with a problem for the fun of it* [24].

Isaac Asimov states that

> *Mathematics is generally not counted amongst the natural sciences, mathematics itself is apart from nature and is a creation of pure mentality being built up step by step by the mathematician with closed eye and thoughtful mind.*

The *applied mathematician* is a practical scientist using the symbolic language created by the *pure mathematician* to organize and solve problems, but the pure mathematician has more in common with the artist than the engineer.

The widely held but inaccurate view of the nature of pure mathematics stems from two sources of confusion. The first is a failure to realize that scientific knowledge, as we know it today, deals only with experimental knowledge, although the record of that knowledge is often written down using the symbols of pure mathematics. The mathematician does not have to be interested in the real world. The second is the fact that most people don't realize that mathematical creations can be pure speculation, completely devoid of any relationship to the universe in which we live. The birth of modern science can be traced to a time when investigators began to avoid speculations about reality, and to explore the structure of the universe by experiment. Scientists, as distinct from philosophers and theologians, based their theories about the universe on observed fact rather than on ideas. For example, some people trace the conflict which developed between the Roman Catholic Church and science, a conflict which continues into the present century, to the work of *Galileo* (1564–1642). It is said that when Galileo invited a priest to look through his new telescope at the craters on the moon, the priest refused saying God would not make a moon with an imperfect surface. Galileo faced facts. The priest avoided experimental facts and preferred to continue to believe in a universe that existed only in his mind. However, although the scientist patiently tracks down reality, the pure mathematician is free to create many varied universes, which may exist only in the imagination. The pattern of numbers created by the mathematician in his equational dreams often turn out to be useful, but that is not the reason for processing and creating patterns with numbers. In the same way, the patterns of sound we call music are created for pleasure. The fact that farmers have found that milk yield can be improved by playing music to cows as they are milked is not a prime reason for creating musical patterns. The link between music and mathematics can even be traced to the original Greek word from which these words are derived. They both come from the Greek root word mathein, "to learn", as outlined in *Figure 1.4*. Originally, songs were described as remembered patterns of sound, just as arithmetic is remembered patterns of numbers.

Many people are often puzzled by the fact that the word *mathematics* is a plural. This is because the word is a short form for what was originally the mathematical

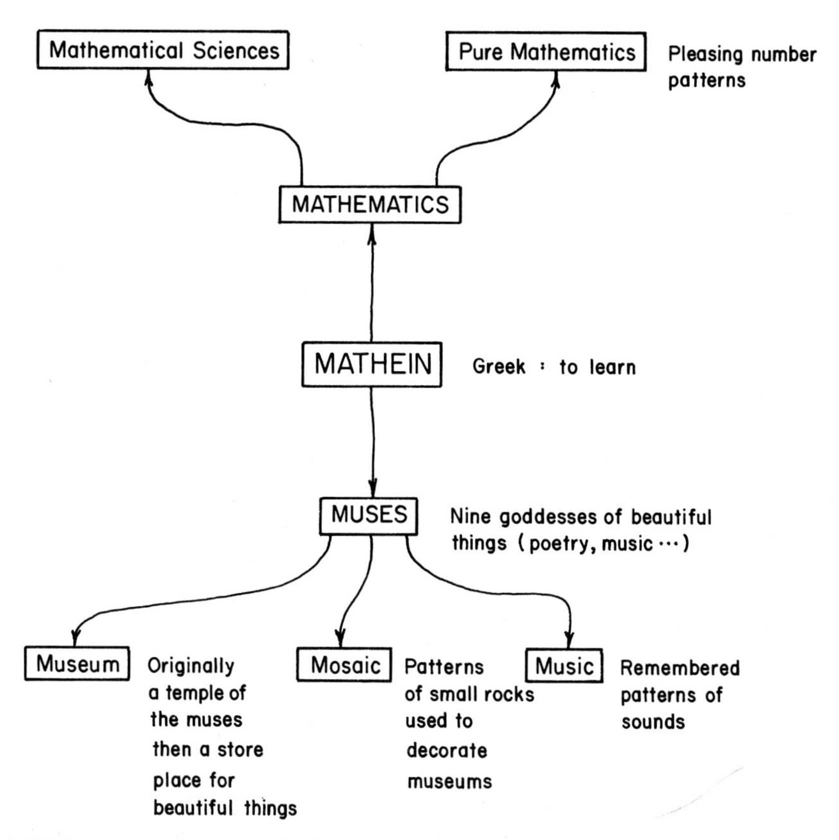

Figure 1.4. Mathematics and music are both remembered patterns of knowledge.

sciences, a group name for arithmetic, algebra, geometry, trigonometry etc. Two other words which are closely related to music and mathematics are the words "museum" and "mosaic". Later in this book, we will discover how the pure mathematicians have learned to create beautiful mosaics on the computer. Therefore, it is useful to have a clear idea of the mathematical meaning of a mosaic pattern. In the stories that the Greeks told of their gods, they described nine goddesses who were responsible for the creation and sponsoring of poetry, history, music, comedy, dance and so on [25]. These goddesses were thought of as being the daughters of Zeus, the supreme god, and the goddess of memory, Mnemosyne. This myth draws attention to the fact that music and mathematics are only pleasures when they can be recalled in the memory. These goddesses were known collectively as the *muses*, a word which is also derived from the Greek word "to learn" (see Figure 1.4). Temples built in honor of these goddesses were known as *museums*. Later this word came to mean any building in which beautiful things are stored. Temples to the muses were often decorated with patterns created with small pieces of colored marble and glass, which were described as *mosaics*. In this type of mosaic, the chips could be of various size and shape and could be oriented in various ways. In a computer generated mosaic, the individual pieces of the mosaic are in a regular mathematical pattern. Sometimes the individual element of a mathematical mosaic is still referred to as a *tile* but a much more widely used word is *pixel*. This word is short for picture element. Pixel will be the preferred term in this book. As the pixels on a computer screen become smaller and smaller, the human brain fuses the appearance of the mosaic into a continuous picture, but if these types of picture are examined with a microscope, the tiny pixels making up the picture are still clearly visible.

How do mathematicians create new ideas? The popular image of brilliant scientists depicts them in their laboratories creating new theories by sheer brilliance of mental ability. In practice the pure mathematician often discovers new ideas by *experimental mathematics* [26]. Consider, for instance, the number π which students encounter at school as they struggle to measure the circumference and area of a circle. I have a very clear memory of the very first time I encountered π. The school teacher wrote on the board that the perimeter of a circle was $2\pi r$. I remember wondering where on earth the teacher had discovered such a fundamental truth. Indeed, I wondered if it was a heavenly revelation that God had handed down to Moses on Mount Sinai. How did the teacher know that this constant was about equal to 3.14? As far as I can remember, my teacher never explained to the class that π is a symbol used to denote the value of the perimeter of a circle divided by the diameter (see *Figure 1.5*). The teacher taught it as a magic recipe for perimeters and areas. It was years later that I discovered that it was defined as the value of the perimeter divided by the diameter.

The discovery of the fact that the perimeter of all circles divided by their diameter is the same value, irrespective of the size of the circle, was one of the great steps forward in the history of mathematics. It is one of the cornerstones of geometry. An

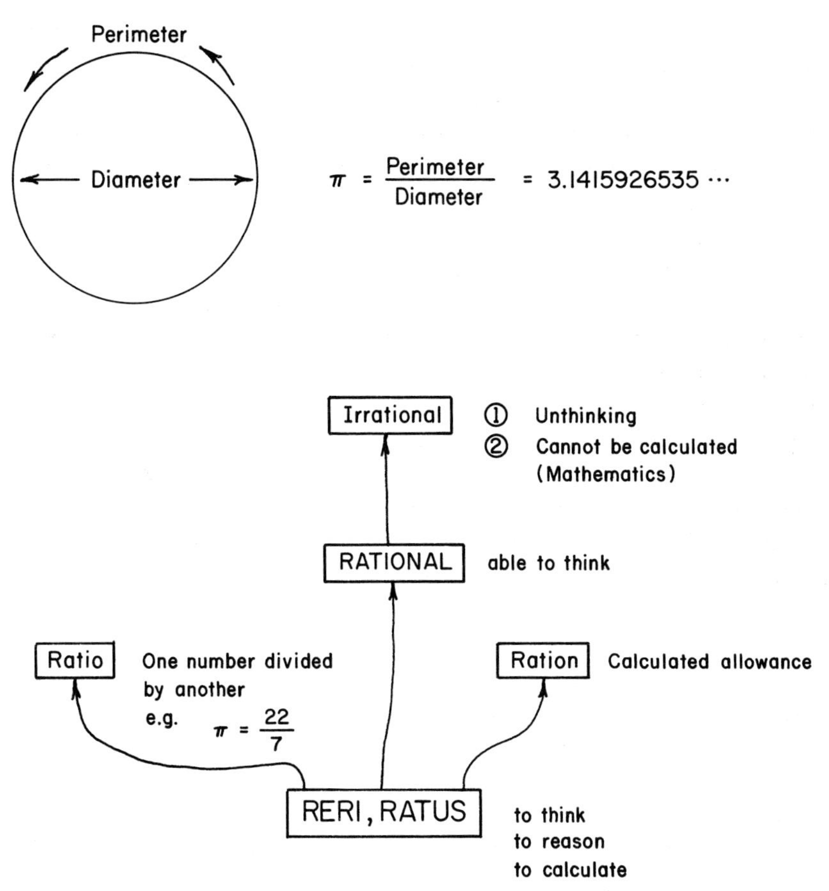

Figure 1.5. Irrational numbers are not crazy; they just cannot be calculated exactly. π is an irrational number.

appreciation of this general relationship was probably based on hundreds of years of experimental mathematics [27]. Those magic recipes for circular perimeters and areas did not leap into the mind of a creative genius. They were the products of centuries of sweat and toil. Generations of engineers digging pits to hold water or building circular granaries discovered experimentally that, whatever size of cylinder they built, the area and the circumference of the circle representing the top of the bin could be described using the ratio of approximately 3.14.

The word *ratio* comes from the Latin word "to think, to reason and to calculate". The Greek letter π is equivalent to our p. It was first used as a symbol for the ratio "perimeter to diameter of a circle" around 1600 by the English mathematician William *Oughtred*. For many purposes one can remember the value of π using the ratio $\frac{22}{7}$ which yields the value of about 3.143. This value is correct to 1 part in 3,000.

In other words, it is accurate within 0.3%. For all geometric work undertaken with a pair of compasses, a ruler and a pencil, this value is more than adequate for the achievable accuracy and precision of the graphical constructions.

As mathematicians developed their skill in handling numbers, they discovered experimentally that the value of the ratio π never reached a finite value, no matter how many decimal places one used to express the calculated value. Pure mathematicians fascinated by π found there was a whole group of numbers which could not be expressed exactly as a simple ratio, or a finite sequence of digits after a decimal point. These types of numbers became known as *irrational numbers*. This name does not come from the fact that these numbers were driving mathematicians crazy as they attempted to calculate finite values for such numbers. It comes from a strict interpretation of the word irrational meaning "not calculable" (see Figure 1.5). By the year 1717 the English mathematician Abraham Sharpe, after working for many years, had calculated π to 72 decimal places, but he was unable to find any indication of a repeating pattern of digits, or an end to the series of digits in the value of π. For the next 200 years pure mathematicians worked long and hard to discover an end to the digits in the sequence of π, all to no avail. This endless search for order in the sequence of digits in the calculated value of π has been described in a book by Beckmann [27]. In *Figure 1.6* the first 207 digits of π are shown. Fink summarizes this search for order in π in the following words:

> *Uncounted hours of human toil have been spent over the past two centuries in computing hundreds of decimal digits and attempting to find some pattern in them that would prove it had a definite value* [28].

It should be noted that the search for the ultimate digit in the sequence of π, although fascinating, has little use other than as an intellectual exercise. It was pursued by pure mathematicians as a problem simply because it was there to be solved. It is an absorbing search for order in a chaotic set of numbers. From the engineers point of view, it can be shown that the value of π to 10 decimal places is sufficient to compute the circumference of the earth to within an inch of its true value. Simon Newcomb, the American astronomer, observed that 30 decimals of π would be enough to specify the circumference of the visible universe to an error too small for the most powerful microscope to detect. Thus, once we pass 30 decimals in the sequence of π, our number cannot correspond to any physical reality in the universe. As we pursue the digits to the first 100 decimal places, and then a second 100 decimal places, our value of π is of no value to the engineer. It is a trail followed by the mathematician seeking a pattern amongst numbers representing a magnitude larger than the known visible universe. The endless patternless sequence of digits of π is a seductive enticement out into insubstantial infinity!

In 1955 a computer was used to calculate the value of π to 10,017 decimal places. In 1989 the number of known digits in π exceeded 1 billion; this number has since been greatly exceeded, and still there is no detectable pattern emerging from the

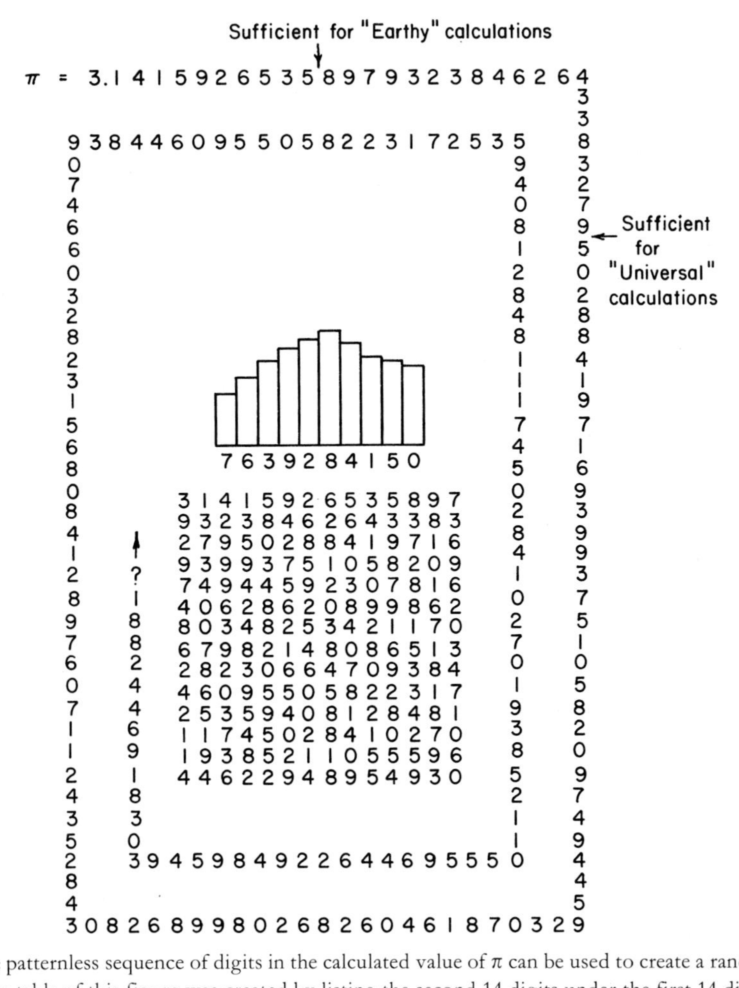

Figure 1.6. The patternless sequence of digits in the calculated value of π can be used to create a random number table (the table of this figure was created by listing the second 14 digits under the first 14 digits, and so on).

string of digits [29]. Although pure mathematicians have proved to their own satisfaction that π is irrational and will never be exactly specified, there are probably still graduate students secretly working on the problem with their big computers hoping to discover experimentally a repeating sequence in the endless cascade of numbers tied to the tail of π. In the next section we too will find ourselves going beyond the limits of the universe as we play with a problem which we shall call "creating a pixel rainbow in line space". We shall also discover that we have to create our own working definitions of zero and infinity.

Section 1.4. What is the Point of Geometry?

At first glance the title of this section appears to crystallize the feelings of many students who are driven into a dry subject littered with obtuse and acute triangles. As they plod through the proofs of Euclidian geometry, students often cannot see any connection between the world in which they live and the fact that "this angle plus that angle equals something else". They really don't care what parallel lines look like at infinity, and would rather study the outcome of a baseball game. Perhaps the boredom of many students with high school geometry comes from the fact that the Greeks were always seeking the underlying perfection beyond observed reality, so that in school geometry the movement is always away from real systems to ideal objects.

When a Greek geometer scratched a line on a wax tablet, he imagined it represented an infinitely thin line which had length but no width and was therefore truly one-dimensional. Their intellectual search for perfection underlying reality started to cause problems when they tried to define a *point*. In Euclidean geometry a point has position but no size. This definition leads to tremendous difficulties when we ask "how many points are there on a line?" In Euclidean geometry the answer is: the same for all lines, independent of their length, an infinite number.

Stewart has pointed out that not only does the classical definition of a point cause difficulty when deciding how many points there are on a line, but the concept of position (the definition of a point) is equally difficult. He states:

> *If you analyze the idea of position it turns out to be as difficult to define as a point, and the two concepts chase each other around in circles* [24].

Trying to manipulate points that have position but no magnitude leads the geometer into the study of paradoxical systems. The word *paradox* has several meanings, but the one relevant to science is: "A statement that is apparently absurd or self contradictory but is, or may be, really true".

Paradox comes from two Greek root words para, meaning "beside", and doxa, meaning "an opinion". Ortho means "correct" in Greek, and anyone who is *orthodox* is considered to have the correct opinions. A paradox is usually two sets of ideas or descriptions of a system which, when put side by side, appear to be self-contradictory. The task of the scientist when faced with a scientific paradox is to explore his ideas to see if he can resolve it. He may be able to resolve the paradox by showing that one set of statements is false or not useful. If he cannot resolve the paradox this way, then he has to live with what appears to be contradictory truths as part of his experience, hoping that future knowledge will help resolve the paradox.

To help understand the role of paradoxical thinking and challenge in science, it is useful to review the history of the paradox which gave birth to the quantum theory of physics. Around the year 1900, physicists faced a crisis in the development of their theories. They were studying the radiant energy given out of a totally black body; a topic known as *black-body radiation*. Their theories predicted an impossible conclusion that a black body should theoretically emit infinite radiation at short wavelengths. This paradoxical conclusion was referred to by physicists as the *violet catastrophe*. The name came from the fact that violet was the shortest wavelength of visible light and hence part of the electromagnetic radiation given out by the black body. In 1900 Max *Planck* (1848–1947) was able to derive a formula which fitted the actual measured radiation spectrum from a black body. He derived his formula by making the very radical proposal that energy was not infinitely subdivisable into smaller and smaller quantities and that there existed a smallest unit of energy which could not be split into smaller proportions. Until then everybody just assumed that energy was infinitely subdivisable.

When people study various scientific problems they usually make several assumptions about the behavior of the system that they are studying. Assumptions about a system come in two categories – implicit and explicit. An *explicit* assumption is one that is clearly stated and therefore open to critical inspection from the very start of the scientific investigation. If one is studying the behavior of a simple pendulum, one often makes the explicit assumption that the weight of the string can be ignored when studying the behavior of the pendulum. An *implicit* assumption is harder to track down and is one that everybody makes when discussing a problem, but which no one has actually clearly stated. Usually an implicit assumption is correct, but occasionally it turns out that an implicit assumption being made by everybody discussing a problem is not correct. The rephrasing of a problem in the light of an implicit assumption which is now made explicit can resolve a set of paradoxical results. Thus, we will find that in a discussion of coastlines it was assumed implicitly for years that there is an answer to the question "How long is the coastline of Great Britain?". It has now been discovered that this apparently simple question has no answer, and can only be answered if one assumes that a certain operational technique is going to be used to measure the length of the coastline. We will find in a later section of this book that a challenge to the implicit assumption underlying the question "How long is the coastline of Great Britain?" was one of the stimuli that led to the development of fractal geometry.

The realization that for generations physicists had implicitly assumed that energy was infinitely subdivisable, and the exploration of the alternative view that, if one assumes explicitly there is a basic small unit of energy which cannot be further subdivided, led to the birth of a new era in physics. In his original work, Planck called the smallest indivisible piece of energy a *quantum* of energy. Some books say that this word comes from the Latin quanta, meaning "how much?". However, in late Latin, the type that would be used by students, the word quantum came to mean "a load" as in "a load of hay". It is probable that Planck was thinking that a quantum of energy was the smallest load that could be carried by electromagnetic waves in the physical universe. The theory of how the universe behaves when one must take into account the "chunkiness" of the energy universe, that it is quantized, became known as the *quantum theory* of physics. The quantum theory of physics is only important if one is dealing with very small quantities of energy. This is like saying that the random structure of a paint film is only important if one is looking at a very small area of the film trying to predict the actual interaction of light and paint pigments. On the much larger scale, looking at the physical properties of paint on a house, one does not need to concern oneself with the actual quantized randomized structure of the pigment in the film. In the same way, if one is looking at the light coming out of a furnace in the foundry, one does not need to worry about the quantum nature of energy. On the other hand, if one is trying to look at the energy arriving on a photoelectric cell from a distant star in a remote galaxy, one has to be concerned with lumps of light energy, called photons, arriving at the photocell.

It should be realized that when Planck discovered that he could derive a formula describing the observed distribution of energy radiated from a black body by treating energy as being quantized, he did not believe that he had stumbled on a fundamental discovery. He half suspected that he had discovered a piece of mathematical juggling needed to get the right answer, and that he was no nearer understanding the real behavior of black body radiation than other scientists. Some people say that Planck noticed that if he took -1 from one part of the formula developed earlier to describe the radiant energy of a black body, and which produced the ultraviolet catastrophe, he could get the workable formula. At this point he was like a student who peeps at the answer at the back of the book and then is faced with the problem of how to work from the answer back to the original question.

Note that in the development of the quantum theory, the reality of the universe came as a surprise to scientists studying the problem, and that the formula they derived seemed to go against common sense (Einstein once said that common sense was the prejudices we acquire before the age of 16!). We are told that Planck struggled for years to find a way around his own discovery, and only became a reluctant convert to his own quantum theory after several years of experimentation.

In the case of the quantum theory of physics, the paradox concerning the difference between the predicted and observed behavior of a black-body radiator was resolved by tackling an implicit assumption which proved to be false. Note, however, that scientists are still not sure why the universe turns out to be quantized. However, they have learned the lesson that, if they wish to describe the universe they observe, then they must learn to live with the experimental fact that energy is not infinitely subdivisable.

When describing the way in which the smallest basic energy unit of light travels across the universe and is absorbed by a surface such as that of a photoelectric cell, the physicist meets another paradox. When studying how the light energy traverses the universe from the stars to the earth, one has to assume that light energy is being propagated by what is known as an electromagnetic wave. One can characterize the motion of the wave with a wavelength parameter that describes the color of the light.

However, when one studies the absorption of the light by a surface, it is essential that one recognizes the quantized nature of the light energy. Einstein received the Nobel prize for discovering and describing the quantized nature of light absorption at a photocell. The smallest package of light energy that can exist is known as a *photon*. When one asks, therefore, what is the nature of light, one receives the paradoxical answer that in some ways it is a wave and in others it is lumpy like a stream of bullets. Sometimes the smallest amount of light energy is referred to as a *wavicle*, but no one knows what a wavicle looks like. It is a name which hides our ignorance. The paradoxical nature of light is something that we have to live with, and we can only hope that as knowledge advances we can look beyond our currently paradoxical description of light energy and grasp a new reality, in which the nature of light is no longer paradoxical.

To appreciate the type of paradoxes that one encounters in Euclidean geometry when one attempts to describe systems with a point which has position and no magnitude, consider the circles shown in *Figure 1.7*. If we ask the question "how many points are there on each of the circles?", the pure mathematician tells us that they all have the same number of points – infinite. Mathematicians prove that each circle has the same number of points by claiming that one can map them in a one to one relationship by drawing a line such as A → B from the centre of the circle to cross each of the lines. It is claimed that one can draw lines through sets of three points on the circles which will eventually link up all the points on the three circles. Students intuitively reject this conclusion. As they grapple with triple infinity they

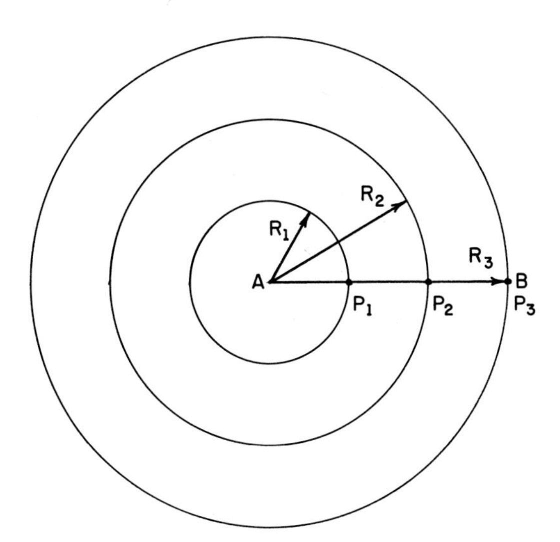

Figure 1.7. From the view point of traditional geometry (in which a point has position but no magnitude) all three circles shown above have the same number of points – infinite. This is "proved" in the theory of Cantorian sets by the fact that these three circles can be regarded as mapped sets of triplets such as P_1, P_2 and P_3.

wonder how they can find their way out of the paradox. They know that in the real world if they tried to make circles with pieces of wire, they would need a lot more wire to make the large circle than the small one. The paradox seems to deepen if one asks the question "how many points lie within the areas of the three circles?". Again the answer is the same. The number of points in the small circle and the large circle is infinite. The pure mathematician enjoys juggling with these paradoxes, but the applied scientist can find his way out of the paradox when dealing with real circles and real areas by adopting the same philosophical position as that of the quantum physicist. Just as the physicist has to realize that there is no such thing as an infinitely small piece of energy, the applied geometer has to treat all diagrams as if they were mosaics based upon patterns generated with finite sized tiles or pixels. The pixel becomes the quantum of area. If the geometer constructs a *quantum geometry* based upon a finite pixel, then paradoxes of infinity are avoided. Using quantum concepts the number of pixels needed to create the three areas of the three circles of Figure 1.7 are in direct proportion to the actual areas of the circles. When we construct a quantum geometry based on a finite pixel of known area we are actually following in the steps of the ancient Greek mathematicians. Vilenken tells us

> *The Greek mathematicians refused to have anything to do with the notion of infinity and excluded it from their mathematical arguments. They assumed that all geometric figures consisted of a finite number of minute indivisible parts which they called atoms* [30].

The word *atom* comes from two Greek words, temnein, meaning "to cut" and the prefix a, meaning "not". Something which could be described as an atom meant that it was pieces of matter which could not be cut into smaller pieces. Vilenken goes on to tell us that when using a quantum approach to geometry, it turns out to be impossible to divide a circle into two equal parts, since the centre would have to belong to one of the two parts. This would contradict the equality of the two halves [30]. Throughout this book we have to assume that, even though we sometimes allow our pixels in our imagination to be smaller than the atoms of the world, we are always studying mosaics made up of finite pixels.

To understand how we define a pixel mathematically, it is necessary to start looking at how we define position in two-dimensional space. In *Figure 1.8*(a) a system for locating a point developed by the French scientist Rene *Descartes* (1596–1650) is shown. Descartes was the first to suggest that we could locate any point in space using three reference lines at right angles to each other.

In the system of Figure 1.8 we are limiting ourselves to two-dimensional space so we only need two axes (the discerning student will realize that if we use the ideas of quantum geometry in three-dimensional space our pixel becomes a very small cube. We need three pieces of information to locate the position of such a *pixoid* in three-dimensional space. However, for now we will restrict our discussion to two-dimensional space).

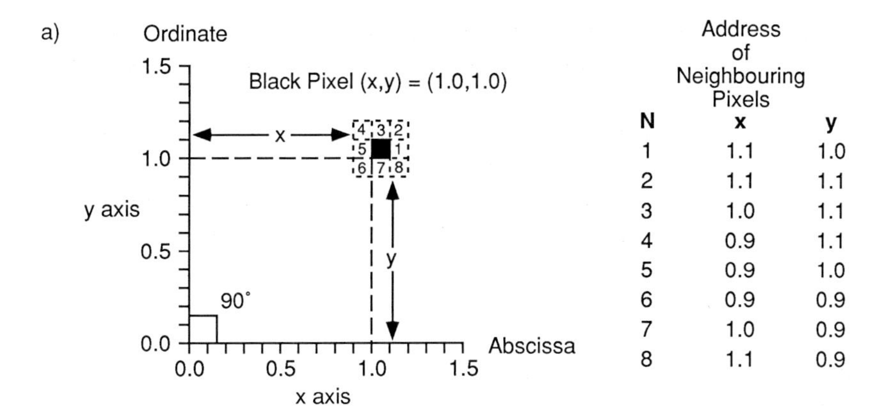

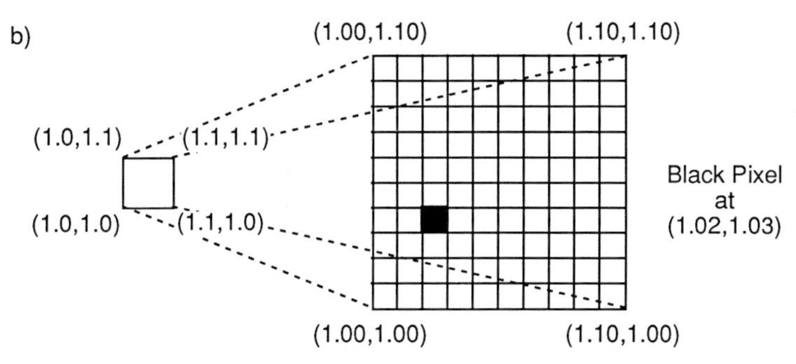

Figure 1.8. In quantum geometry all space is made up of quantized pixels. The size of the quantum space considered in any coordinated space can be varied at will. Only when λ, the side length of a pixel, becomes vanishingly small, does quantized geometry merge with idealized Euclidian geometry. a) In many real geometric problems the size and address of the sets of pixels constituting the fabricated space are useful pieces of information. b) The pixel size of quantized space can be reduced to any level of coordinate inspection.

We are told that Descartes was inspired to invent his system for locating points in space as he was resting in a tent during an army campaign. It is said that, as he watched a fly moving around the inside of a tent, it occurred to him that the fly's position could be specified by giving the distances of the fly from the three reference axes which we now describe as the x, y and z coordinates.

If we restrict our discussion to two-dimensional space, then to locate a point we give the distance that one must move along the abscissa (usually referred to as the x direction) and then the distance that one must move along the y axis from the new position on the x axis. The two distances that one must move along the axes to locate the point are referred to as the *coordinates of the point*. It is also known as the *address of the point*. Thus, the black pixel of Figure 1.8(a) is located at the position given by the coordinates x = 1.0 and y = 1.0. By giving the coordinates to two decimal places, one is stating the accuracy with which information can be plotted on the graph. By stating the accuracy with which a data point is located on a graph, one is implicitly assuming that the graph is made up of squares with side length equal to the magnitude of the accuracy claimed for the data plotting. In this case a data point represented by a pixel located at the address (1.0, 1.0) is assumed to have a pixel the magnitude of which is 0.1×0.1.

Any pattern that appears to be created by plotted data is only a mosaic created with a tile or pixel of size equal to the accuracy of the plotting. When we give the data point we are giving the address of the corner of the pixel. We are imagining that it is surrounded by eight other small tiles whose addresses are given in Figure 1.8(a). In quantum geometry we can state that the coordinates represent a set of instructions for locating tiles on a mosaic. The idea that one can locate a point by using these types of mathematical address is described as a *Cartesian coordinate* system, in honor of Descartes (at the time of Descartes, Latin was the international language of science, and therefore Descartes used to sign his scientific papers with a Latin form of his name – Renatus Cartesius).

When we come to discuss the complicated pictures created by computers, the pixel size corresponds to the resolution of the printer and/or the graphic screen of the computer used to display the picture.

We can always mathematically magnify the structure of a pixel by imagining the basic unit to be itself constituted from an array of pixels. If we increase our resolution of inspection, or portrayal, of our mosaic of Figure 1.8(a) by an order of 10, we take our original pixel and expand it to the pixel system shown in Figure 1.8(b). We now can locate a black pixel in this expanded pixel array with an address such as (1.02, 1.03). If we wished to expand our inspection of this pixel, we could consider it to be constituted of 100 sub-pixels located within the coordinate system defined by 1.020 to 1.029, and 1.030 and 1.039. Mathematically, one can continue to expand the pixel inspection resolution ad infinitum, even though in the real world the smallest pixel size ceases to have any physical reality. It can be much smaller than the piece of pigment we can use to depict it on a page. In quantum geometry, however, the pixel can never be of zero size. Therefore, no matter how high our resolution, we can always quantify the structure of a picture by counting pixels.

The philosophical difference between a point which has position and no magnitude and the address of a pixel which can be very, very small will become clearer as we explore the Mandelbrot set. To introduce several important concepts involved in manipulating pixels in geometric space, a simple exercise leading to the creation of a pixel rainbow in one-dimensional space will be described in the next section.

Section 1.5. Painting a Rainbow by Tracking Vanishing Pixels in Linespace

As discussed in the previous section, an ideal line to a Greek geometer was infinitely thin and existed in one-dimensional space. If we construct a line with pixels, our line always has a width equal to the size of the pixels being used to create the lines. In this section, we will carry out our first exercise in quantum geometry as we study how to make a pixel rainbow out of the behavioral pattern of fugitive pixels. The linespace we shall use in constructing our first pixel rainbow is shown in *Figure 1.9*. It consists of 21 pixels of the size γ, indicated in the drawing. What patterns can a pure mathematician create with the pixels, or more exactly the addresses of the set (collection) of pixels, making up the portion of linespace shown at the top of Figure 1.9?

Let us imagine that our pixels are free to move about in linespace and that we can keep track of the address of any pixel by quoting the number locating the right hand edge of the pixel. The initial address of the pixel nearest zero in the linespace of Figure 1.9 is 0.9. To create a colored line out of the pixels of the line we could simply

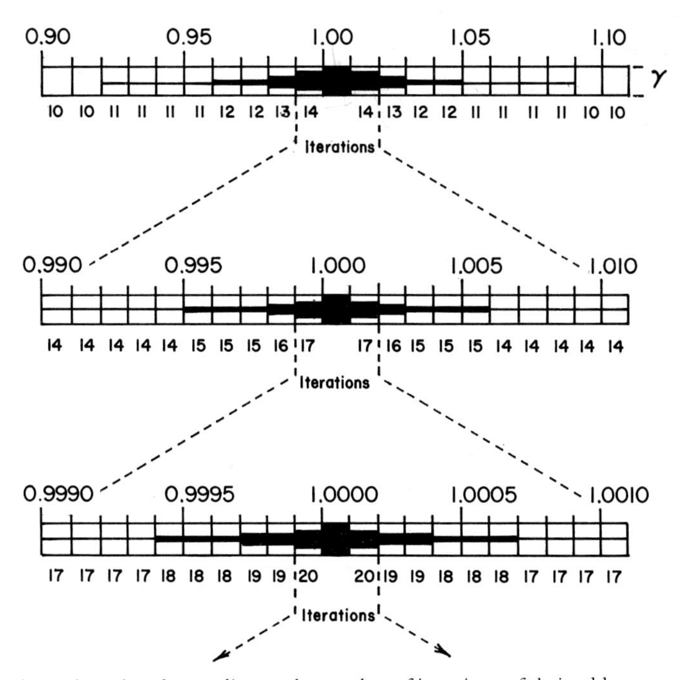

Figure 1.9. Pixels can be colored according to the number of iterations of their address required to make them travel beyond the limits of the universe (in black and white the rainbow effect can be portrayed by the amount of black added to the pixel square).

fill in each pixel with a different color, choosing each color at random. To a pure mathematician that would be too simple and he would prefer to have a mathematical recipe for selecting the colors. One of the things that mathematicians like to do with numbers is see how they change in magnitude when one repeatedly subjects them to a mathematical operation. One way we can create a pixel rainbow from the pixels of the linespace of Figure 1.9 is to color them according to the way in which they move when we perform the operation that we will write symbolically as

$$z_{n+1} = z_n^2 \rightarrow \text{ad nauseum}$$

This symbolic recipe means that we create a new number z_{n+1} by taking the number z_n and multiplying it by itself (that is we square the number). Then, to create the new number z_{n+2}, we multiply the number z_{n+1} by itself. The $\rightarrow$ in the recipe tells us to keep on repeating the process until we are sick and tired of doing the operation (*ad nauseum* is a Latin phrase meaning "until you are sick". A dictionary of Latin phrases defines it "to the point of disgust"!). In the language of the mathematician this process is called "iteration of the operation square the number"

Iterate comes from a Latin word meaning "again". To be able to carry out the mathematical operation "square the number" on the addresses of the pixels of the linespace of Figure 1.9, it is necessary to develop a working knowledge of a technique for writing down and manipulating large numbers which is known as *exponential notation.*

Thus, if we were to carry out this operation on the number 2, we would generate the sequence of numbers summarized in column 2 of *Table 1.1*. In the next column we have written the result of the operation using decimal notation. The following column contains the numbers in exponential notation, using powers of ten. To create this notation, mathematicians have agreed to interpret the small number written above a 10, such as 10^2, as the number of times that 10 is used as a multiplying factor

Table 1.1. Large numbers generated by an iterative process can be written down conveniently using exponential notation. This table demonstrates the result of iterating "square the number" starting with the number 2.

Iteration	z	z^2	Exponential Notation	Computer Representation
1	2	4	4×10^0	4
2	4	16	1.6×10^1	16
3	16	256	2.56×10^2	256
4	256	65536	6.5536×10^4	6.5536E + 04
5	65536	4294967296	4.2949×10^9	4.2949E + 09
6			1.8446×10^{19}	1.8446E + 19
7	Numbers too large		3.4028×10^{38}	3.4028E + 38
8	to be shown here.		1.1579×10^{77}	1.1579E + 77

on the number 1

$$10^2 = 10 \times 10 = 100 \quad \text{and}$$
$$10^3 = 10 \times 10 \times 10 = 1000$$

This small number in the air is described as an *exponent*. A number such as $100 = 10^2$ is said to be written in the *exponential form*. Mathematicians have also agreed to interpret the symbol 10^0 as meaning 1, and 10^{-3} as

$$10^{-3} = \frac{1}{10^3} = \frac{1}{1000} = 0.001$$

That is, a negative exponential is the *reciprocal* of the corresponding positive exponential number. The word reciprocal is one of those words which has such a different everyday meaning from its scientific meaning that it can cause real confusion when first encountered by the student beginning a study of science. On the radio we may hear that two countries have established reciprocal relationships and will exchange trade information on commercially important reciprocating engines. The word reciprocal comes from two Latin words recos, "turning backward" and procos, "turning forward". In a *reciprocating engine* the piston first goes forward and then reverses its motion. In reciprocal relationships each side gives to the other the same type of information; that is, information goes back and forth. It is not clear how a number such as 3/2 came to be called the reciprocal of 2/3. One explanation can be arrived at by considering a ratio of numbers in general such as a/b. In such a number, "a" is known as the *numerator*, and "b" the *denominator*. When we create the reciprocal number b/a, we reverse the roles of the original numerator and denominator. That is, the two parts of the number go back and forth in the ratio and its reciprocal.

Mathematicians have shown that when multiplying numbers written in exponential form, one adds the exponents. Thus:

$$10^2 \times 10^3 = 10^5$$

In *Figure 1.10* the range of magnitudes encountered in the real universe are expressed in exponential notation based on 1 cm [31]. If we had been dealing with magnitudes expressed in centimeters then the operation "square the number" iterated 7 times starting with 2 cm would result in a number representing a distance larger than the diameter of the universe. However, from a pure mathematician's point of view, the fact that we have run out of space in the universe does not stop us from continuing with the iteration process. Thus:

$$(z^2) \text{ iterated 8 times} = 1.16 \times 10^{77} \text{ when } z = 2$$

Only our patience limits how far we can go with the iteration process when we are using exponential notation. When computers are used to carry out an operational iteration, they print out large numbers using exponential notation in the form shown

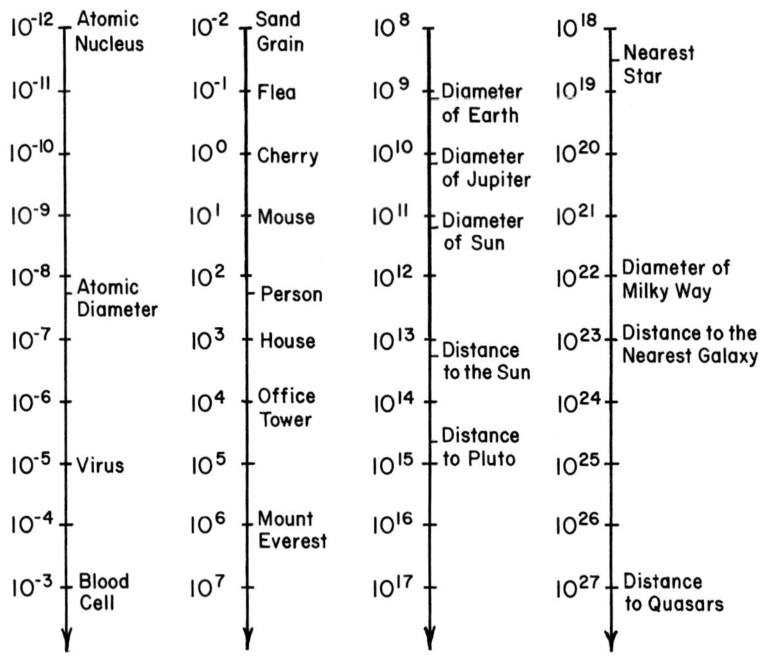

Figure 1.10. Exponential numbers are a convenient way of expressing large variations in magnitude. All values are in centimeters.

Table 1.2. Iteration of the operation "square the result of squaring starting at 5" ad nauseum soon generates gigantic numbers.

Iteration	z	z^2	Exponential Notation	Computer Representation
1	5	25	2.5×10^1	25
2	25	625	6.25×10^2	625
3	625	390625	3.9062×10^5	3.9062E + 05
4			1.5259×10^{11}	1.5259E + 11
5	Numbers too large		2.3283×10^{22}	2.3283E + 22
6	to be shown here.		5.4210×10^{44}	5.4210E + 44
7			2.9387×10^{89}	2.9387E + 89

in column 4 of Table 1.1. Thus 1.024×10^3 is written as 1.0240 E03. In *Table 1.2* the data generated by the iteration of the operation "square the number, starting with 5" is shown.

In *Figure 1.11* two different ways of depicting rapid growth of the iterated numbers are shown. In Figure 1.11(a) traditional graph paper, with straightforward number scales, is used to summarize the growth in the operational sequences of

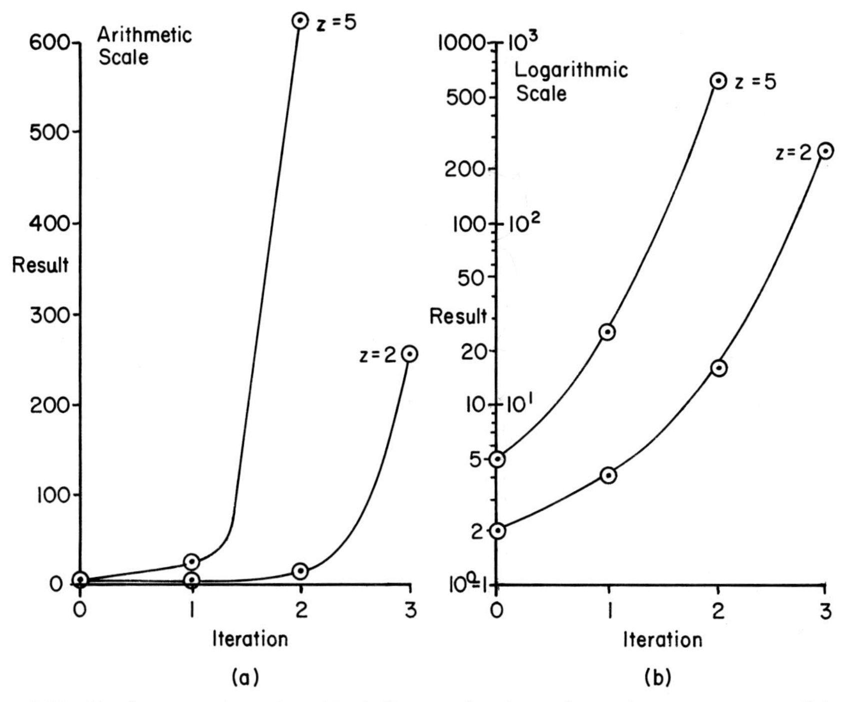

Figure 1.11. Graph paper using a logarithmically spaced scale can be used to portray more of the data generated by an iterated operation than ordinary graph paper using linearly (arithmetic) spaced scales. a) Iterations plotted on linear scales. b) Iterations plotted using a logarithmic scale on the y axis.

Table 1.1 and Table 1.2. As can be seen from the graph, the data soon outgrows the ordinary graph paper. To accommodate more of the data of the type of Table 1.1 one can use graph paper with logarithmic scales.

The word *logarithm* means "proportional number". The word was originally coined by a Scottish mathematician, John *Napier* (1550–1617), to describe a new way of expressing numbers. The root words in logarithm are logos, which in Greek can mean "to calculate" and arithmos, meaning "a number". On a logarithmic scale equal distances represent equal increases in proportion. If we look at the *logarithmic scale* of the graph in Figure 1.11 (b), we see that the same distance is occupied by the increase from 10 to 100 (a ratio of 10) as is occupied by the range from 100 to 1,000, which is again a ratio of 10. When an increase of a given ratio such as 10 as in Figure 1.11 (b) is completed, it is said that 1 cycle of the logarithmic scale has been completed. On the scale of Figure 1.11 (b), there are 3 logarithmic cycles completed. This term is often abbreviated to *log cycle*. Graph paper with the scales shown in Figure 1.11 (b) is referred to as 3 cycle *logarithmic graph paper*. When both scales are logarithmic the paper is described as *log-log graph paper*. When the data for the iterative operation "keep squaring the number" starting with 2 and 5 are plotted on logarithmic graph paper, it can be seen that a much wider range in data can be accommodated.

One of the things that we have avoided in our discussion so far is the fact that we cannot define a magnitude for infinity. As we have demonstrated, we can keep on squaring a number in an endless sequence. To say that we stop when we are sick of it (ad nauseum) is not very scientific. If we wish to make any statement about the difference between the sequence generated by operating on 2 and on 5, we have to decide when we will agree that the number we have generated is so large that it might as well be infinity.

A decision to stop the calculation can be as simple as deciding that the multiplication has exceeded a given limit. In our short form table given above, we can decide that, to all intents and purposes we have reached infinity when our calculation exceeds one million. This may seem a long way from infinity, but if you set yourself the task of counting a million grains of rice you would certainly reach ad nauseum before you finished the counting! In a more practical sense, if one is using a small calculator, multiplying a million by itself has a product which is larger than the display system of the calculator. If we decide that for all practical purposes, the sequence has reached infinity when the next iteration would cause the product to exceed one million, then we can see that the difference between the sequence generated by operating on 2 and the iteration of the operation on 5, is that the sequence based on 2 reaches what we have defined as infinity after 5 iterations, and the sequence based on 5 reaches infinity after 4 iterations. Note that an *operational definition* is one in which a definition is established based upon a specified operation.

Thus, the operation "examine the product to see if it exceeds one million" establishes a working definition of what constitutes infinity and this constitutes our operational definition of infinity in this specific problem. [Note that the symbol ∞ is the symbol used to denote infinity. It was invented by an English mathematician, John *Wallis* (1616–1703)]. By setting a practical limit to what we will regard as infinity, we have managed to find a difference between the two series we are studying, both of which go to infinity. One sequence reaches our operational limit of infinity more quickly than the other. This difference in the rate at which a number operated on by an iterative process reaches an operational definition of infinity is the basis of a procedure we are going to set up to color our pixel rainbow in linespace.

Returning to our linespace as shown in Figure 1.9, we could decide to iterate the address of a given pixel and find out how many times we had to carry out the operation until either the pixel reaches 0 or vanishes into infinity. Note that the mathematician does not have to worry about any forces required to move a pixel or to accelerate it in its journey towards infinity. When scientists study movement without looking at any necessary forces they are said to be studying *kinematics*. This term comes from the Greek word kinema, meaning "motion". The same word has given the English language the word *cinema*, a place where we watch moving pictures. When mathematicians consider forces causing motion the subject is called *dynamics* from the Greek word dynamis, "power" (a word to be found in *dynamite* – a powerful explosive).

In *Table 1.3* the sequence of numbers generated by iterating the calculation "square the number and then square the result" is shown. It can be seen that if one writes this number using the exponential format, the pixel with the address 1.10 reaches our operational infinity limit after 8 iterations.

When we come to look at the movement of the pixel with the address 0.90, we need an *operational definition of zero.* In an arbitrary manner, we can decide that the pixel can be considered as having reached zero when it is less than one millionth of a pixel away from zero. Using this operational definition of zero the data of Table 1.3(b) shows that at the 7th iteration the pixel has reached zero. Notice that, again, we could keep moving closer and closer to zero and never reach it by creating an endless series of negative exponential numbers which approach zero but are always a very, very small distance from zero. If we interpret out exponential numbers with respect to the physical world we note from Figure 1.10 that at 10^{-3} cm (10 microns) the paper we are using to draw our pixels on is made from fibers 2×10^{-3} cm in width and the clay used to make glossy paper is made of fineparticles 5×10^{-5} cm in diameter. At 10^{-7} cm we are looking at distances of the same magnitude as molecular sizes. At 10^{-8} cm we are down to the size of atoms, and so on as shown in Figure 1.10. With our operational definition of zero as being 1×10^{-10} cm, we are considering distances from zero of less than a millionth of an atom. When we come to explore the Mandelbrot set in the next chapter, it becomes very important to remember that zero and infinity are ideals, and that in any real world one must set operational limits to the numbers that we generate when moving towards these two limiting concepts of zero and infinity.

In *Table 1.4* a complete study of the movement of all of the pixels of the linespace of Figure 1.9 is shown. On the computer we used it was convenient to set the operational limit of infinity as being reached whenever the next iteration would take the number being iterated to an exponential value in excess of 20. In the same way we defined having reached zero when the moving pixel had reached a value such that the next iteration would require an exponential smaller than -20 to describe its position. In Table 1.4 the address of the pixel is listed in the first column, the number of iterations is shown in the centre column and the magnitude reached is listed in the last column. It can be seen that the number of iterations to reach zero increases as one moves nearer to the value 1.00. The number of iterations to escape to our pragmatic value of infinity decreases as one considers pixels further and further away from 1.00. Although it may be trivial for linespace, it is useful to start describing the number 1 as the boundary between those pixels whose addresses when iterated move to infinity and those which when iterated with the operation "square the number" move to zero. Note that in linespace, this boundary between escaping and permanently contained pixels is a sharp infinitely thin line at the point 1.0000000 . . . → zeros ad nauseum. Pixels to the left of 1.000 can never escape from linespace, no

Table 1.3. Iteration of the operation z^2 on the number 1.10 soon results in a number larger than the known universe, just as the same operation on 0.90 reaches what we have defined as zero by the 7th iteration. a) Iteration of the operation $z_{n+1} = z_n^2$ for $z_1 = 1.10$. b) Iteration of the operation $z_{n+1} = z_n^2$ for $z_1 = 0.90$.

a)

Iteration	z	z^2	Exponential Notation	Computer Representation
1	1.10	1.21	1.21×10^0	1.21
2	1.21	1.4641	1.4641×10^0	1.4641
3	1.4641	2.1435	2.1435×10^0	2.1435
4	2.1435	4.5950	4.5959×10^0	4.5950
5	4.5950	21.1138	2.1114×10^1	21.114
6	21.111	445.791	4.4579×10^2	445.791
7	445.791	198730	1.9873×10^5	1.9873E + 5
8			3.9494×10^{10}	3.9494E + 10
9	Numbers too large		1.5597×10^{21}	1.5579E + 21
10	to be shown here.		2.4328×10^{42}	2.4328E + 42

b)

Iteration	z	z^2	Exponential Notation	Computer Representation
1	0.9	0.81	8.1×10^{-1}	0.81
2	0.81	0.6561	6.561×10^{-1}	0.6561
3	0.6561	0.430467	4.30467×10^{-1}	0.430467
4	0.430467	0.185302	1.85302×10^{-1}	0.185302
5	0.185302	0.034336	3.43368×10^{-2}	0.034336
6	0.034336	0.001179	1.17901×10^{-3}	1.1790E – 03
7	0.001179	0.000001	1.39008×10^{-6}	1.3900E – 06
8			1.93233×10^{-12}	1.9323E – 12
9	Numbers too small		3.73392×10^{-24}	3.7339E – 24
10	to be shown here.		1.39421×10^{-47}	1.3942E – 47
11			1.94383×10^{-94}	1.9438E – 94

Table 1.4. Iterations of the pixel positions in line space spaced 0.01 apart.

Number	Iterations	Result	Number	Iterations	Result
0.90	10	1.93233509E – 12	1.01	14	5.01588497E + 17
0.91	10	3.27036705E – 11	1.02	13	4.10348189E + 17
0.92	11	2.87983554E – 19	1.03	12	1.39739239E + 13
0.93	11	7.29910714E – 17	1.04	12	2.76783095E + 17
0.94	11	1.74365733E – 14	1.05	11	7.06189398E + 10
0.95	11	3.93084627E – 12	1.06	11	9.04905476E + 12
0.96	12	7.01040023E – 19	1.07	11	1.10788421E + 15
0.97	12	2.84613901E – 14	1.08	11	1.29706844E + 17
0.98	13	1.07399824E – 18	1.09	10	3.81226896E + 09
0.99	14	1.32360557E – 18	1.10	10	3.94936699E + 10

matter how many times they are iterated, and those to the right of this reference point always escape to infinity. The difference in behavior of the pixels to the left and to the right can, however, be summarized with respect to their relative swiftness of movement to infinity or zero by looking at the number of iterations required to reach the operationally defined zero or infinity.

To aid comprehension of masses of data, mathematicians have started to create color-coded pictures with their computers, to help us absorb all the information generated by the computer. We can use this strategy to display the data of Table 1.4 on the linespace of Figure 1.9. If we decide to let the number 14 be represented by red, 13 by yellow, 12 by green, 11 by blue and 10 by violet, then we can color the pixels of Figure 1.9 to generate a rainbow. Once we are given the interpretive code, the rainbow tells us in a very vivid way about the behavior of pixel addresses when subject to the specific iterative operation. The construction of the colored version of Figure 1.9 is left to the creative instincts of the reader.

There is no limit to the intensity with which we can inspect the structure of our pixel rainbow. For example, we can decide to increase our inspection resolution (that is, the detail that we can observe in our model) by considering the 20 pixels that inhabit the linespace in the strip identified by the addresses 0.990 to 1.010. The calculations for the movement towards infinity or zero of the addresses of our more highly resolved set of pixels in our chosen linespace is summarized in *Table 1.5*. One can now set up the linespace in Figure 1.9 in which the resolved pixels are expanded in size to occupy the same real space as the set of pixels in the less resolved set of calculations given in Table 1.4. In each of the pixels, the number of iterations required to move to our operationally defined zero or to escape to operationally defined infinity are listed. In *Tables 1.6* through to *1.10* more and more highly resolved sets of pixels in our linespace are studied, and the number of iterations required to reach zero or infinity are summarized in the tables. It is left as a creative exercise for the student to color the squares in a kind of "paint by number" sequence

Table 1.5. Iterations of the pixel positions in line space spaced 0.001 apart.

Number	Iterations	Result	Number	Iterations	Result
0.990	14	1.32360557E − 18	1.001	17	1.67439743E + 14
0.991	14	8.27351337E − 17	1.002	16	1.64722790E + 14
0.992	14	5.15003559E − 15	1.003	15	4.54198914E + 10
0.993	14	3.19244477E − 13	1.004	15	1.59433118E + 14
0.994	14	1.97074888E − 11	1.005	15	5.55109512E + 17
0.995	15	1.46784338E − 18	1.006	14	4.37854122E + 10
0.996	15	5.50178183E − 15	1.007	14	2.56278902E + 12
0.997	15	2.04520985E − 11	1.008	14	1.49396767E + 14
0.998	16	5.68581573E − 15	1.009	14	8.67399054E + 15
0.999	17	5.77998988E − 15	1.010	14	5.01588497E + 17

Table 1.6. Iterations of the pixel positions in line space spaced 0.0001 apart.

Number	Iterations	Result	Number	Iterations	Result
0.9990	17	5.77998988 E − 15	1.0001	20	2.42235686 E + 11
0.9991	17	1.53593799 E − 13	1.0002	19	2.41943669 E + 11
0.9992	17	4.08017439 E − 12	1.0003	19	1.18750436 E + 17
0.9993	17	1.08354860 E − 10	1.0004	18	2.41273665 E + 11
0.9994	18	8.27423247 E − 18	1.0005	18	1.68820049 E + 14
0.9995	18	5.82751522 E − 15	1.0006	18	1.18044186 E + 17
0.9996	18	4.10171475 E − 12	1.0007	17	9.08254082 E + 09
0.9997	19	8.32318587 E − 18	1.0008	17	2.40013002 E + 11
0.9998	19	4.11264715 E − 12	1.0009	17	6.34040269 E + 12
0.9999	20	4.11817481 E − 12	1.0010	17	1.67439743 E + 14

Table 1.7. Iterations of the pixel positions in line space spaced 0.00001 apart.

Number	Iterations	Result	Number	Iterations	Result
0.99990	20	4.11817481 E − 12	1.00001	24	1.65151808 E + 18
0.99991	20	5.66627709 E − 11	1.00002	23	1.64426689 E + 18
0.99992	21	6.07844409 E − 19	1.00003	22	4.59429841 E + 13
0.99993	21	1.15031023 E − 16	1.00004	22	1.64258030 E + 18
0.99994	21	2.17719758 E − 14	1.00005	21	2.42408706 E + 11
0.99995	21	4.12116404 E − 12	1.00006	21	4.58645380 E + 13
0.99996	22	6.08414290 E − 19	1.00007	21	8.67571526 E + 15
0.99997	22	2.17892682 E − 14	1.00008	21	1.64146927 E + 18
0.99998	23	6.08955992 E − 19	1.00009	20	1.76156565 E + 10
0.99999	24	6.08955992 E − 19	1.00010	20	2.42235686 E + 11

Table 1.8. Iterations of the pixel positions in line space spaced 0.000001 apart.

Number	Iterations	Result	Number	Iterations	Result
0.999990	24	6.08955992 E − 19	1.000001	27	3.85334168 E + 14
0.999991	24	4.04059147 E − 17	1.000002	26	3.79352161 E + 14
0.999992	24	2.67849066 E − 15	1.000003	25	8.54424369 E + 10
0.999993	24	1.77847752 E − 13	1.000004	25	3.76412914 E + 14
0.999994	24	1.17702519 E − 11	1.000005	25	1.65705838 E + 18
0.999995	25	6.11374731 E − 19	1.000006	24	8.54424369 E + 10
0.999996	25	2.67849066 E − 15	1.000007	24	5.65017393 E + 12
0.999997	25	1.17830412 E − 11	1.000008	24	3.74937877 E + 14
0.999998	26	2.68848296 E − 15	1.000009	24	2.48310103 E + 16
0.999999	27	2.69836091 E − 15	1.000010	24	1.65151808 E + 18

Table 1.9. Iterations of the pixel positions in line space spaced 0.0000001 apart.

Number	Iterations	Result	Number	Iterations	Result
0.9999990	27	2.69836091 E − 15	1.0000001	30	6.61615746 E + 11
0.9999991	27	7.77525463 E − 14	1.0000002	29	5.30478308 E + 11
0.9999992	27	2.22859474 E − 12	1.0000003	29	3.42236571 E + 17
0.9999993	27	6.36066414 E − 11	1.0000004	28	4.96408563 E + 11
0.9999994	28	3.31188450 E − 18	1.0000005	28	4.00662440 E + 14
0.9999995	28	2.79088928 E − 15	1.0000006	28	3.11454634 E + 17
0.9999996	28	2.22859474 E − 12	1.0000007	27	1.65646521 E + 10
0.9999997	29	3.43715099 E − 18	1.0000008	27	4.67369011 E + 11
0.9999998	29	2.36077255 E − 12	1.0000009	27	1.34944663 E + 13
0.9999999	30	2.49327989 E − 12	1.0000010	27	3.85334168 E + 14

Table 1.10. Iterations of the pixel positions in line space spaced 0.00000001 apart.

Number	Iterations	Result	Number	Iterations	Result
0.99999990	30	2.49327989 E − 12	1.00000001	33	2.64693811 E + 10
0.99999991	30	3.54831185 E − 11	1.00000002	32	9.73912936 E + 09
0.99999992	31	2.41287703 E − 19	1.00000003	32	1.92256746 E + 14
0.99999993	31	5.23127845 E − 17	1.00000004	31	3.52746654 E + 09
0.99999994	31	1.06038679 E − 14	1.00000005	31	8.23176554 E + 11
0.99999995	31	2.49327989 E − 12	1.00000006	31	1.92256746 E + 14
0.99999996	32	3.45150588 E − 19	1.00000007	31	2.44049399 E + 16
0.99999997	32	1.42956830 E − 14	1.00000008	31	6.01798995 E + 18
0.99999998	33	3.45150588 E − 19	1.00000009	30	3.97561590 E + 10
0.99999999	34	7.68364596 E − 19	1.00000010	30	6.61615746 E + 11

to create personal rainbows to summarize what is known about the kinematic behavior of the pixels when subjected to the specified iterative operation.

To summarize what we have discovered in this section; first of all, when carrying out iterative operations on numbers, we have to set operational limits to the magnitude of infinity and zero. Secondly, any given system in linespace can be examined with higher and higher resolution *ad infinitum*, (a Latin phrase meaning "until we reach infinity"). In fact we note that as the size of the pixel γ decreases, the number of pixels in the linespace tends to infinity. At the resolution represented by the data in Table 1.10 there are already 100,000,000 pixels in the linespace stretching from 0 to 1.

We have also discovered that color coding of linespace helps us to summarize in a vivid and perhaps beautiful way the difference in kinematic behavior of pixels located at various points in linespace when we carry out our iterative operation on the address of the pixel. To the reader who finds the construction of a pixel rainbow a trivial exercise, I would hint that the pixel manipulative skills learned in this section will turn out to be invaluable when we explore the Mandelbrot set.

Section 1.6. Beware of Averages!

The word *average* has leaked out of science into everyday language. We hear daily in the news about average salaries, average consumption rates, etc., but in reality many people have no idea how useless and even dangerous an average statistic can be. Most people distrust statistics, but are often unaware of how seductive an average can be when describing a system. What is an average? My etymological dictionary tells me that an average is

> *The mean value of a number of quantities as obtained by dividing the sum of the quantities by their number. The word first appears, in various forms, about 1500 in connection with the Mediterranean sea trade.*

The dictionary further defines *mean* as "intermediate", "average", from the Latin word for "middle". An example of how many people do not fully understand what is meant by average is illustrated by the fact that a politician is said to have asked of the Prime Minister

> *Did the Prime Minister know that half the children in our educational system have below average reading skills, and what was he going to do about it?*

I know from class discussions that at first even intelligent students find this political question not unreasonable until they realize that, by definition, half the performance level for any task has to be below average!

Another example of how one can be misled by averages is the story of the photographer who received a letter from a young lady claiming to have the same average statistics as Miss Universe. The statistics used to describe Miss Universe were 39-27-36. This gives an average dimension of 34. When the photographer invited the average young lady for a modeling session, he was dismayed to find that her dimensions were 27-36-39. These dimensions did indeed give the same average of 34. However, the visual impression created by the structure of Miss Average was somewhat different to that of Miss Universe.

When I was a student we used to tell a story about the sad fate of a statistician who, when he came to the bank of a river, was advised that the river was on average 4 feet deep. Although he could not swim, we are told that the statistician set off at an average pace only to drown in the middle of the river which was 8 feet deep. The aim of this section is to introduce the reader to the advantages and disadvantages of using an average value to describe a set of data, and to try to prevent the reader from inadvertently drowning in averages.

Readers unfamiliar with the theory of statistics will find that they can gain a good background in the subject by reading an introductory text by Moroney [32]. In this book Moroney tells us that everybody using an average should ask themselves the question "What conclusions will be drawn from this average that I am about to calculate, and will it create a false impression?"

In our definition of average given above it was hinted that the term developed as a trading term amongst merchants in the Mediterranean. Moroney elaborates on the history of the word and tells us that the mathematical use of the word appears to have developed in connection with assessing the value of a cargo left on a ship when it reached its final destination. In the middle ages many ships went to sea overloaded with cargo. As a consequence, when the ship was caught in a storm it was necessary to lighten the load of the ship to prevent it from sinking by throwing some of the cargo overboard. The damage to the overall value of the cargo of the ship was known as "havori". Havori eventually came to be the word for the amount of money to be paid by those whose cargo reached its destination to those whose cargo had to be abandoned to save the ship. An average is what we are left with after we have thrown much of the data originally collected "overboard".

To illustrate how misleading the term average can be, and to introduce ourselves to some important concepts in fractal geometry and deterministic chaos, let us consider a simple gambling game to be played between two players, who we will call Fred and Freda. Let us imagine that they are going to toss a coin with the agreement that if the coin falls with the head showing, then Freda will receive a dollar. If, on the other hand, the coin falls with the tail showing, then Fred will receive a dollar. If one were to discuss this game with many people, they would quickly state that on average neither Freda nor Fred will make a profit at the game, and that in the long run the average winnings will be zero. However, just how long one has to play this game to have a zero win situation is not fully appreciated by people who have limited experience of random systems such as the frequency with which heads and tails appear when one flips a coin. It would be quite wearisome to play this game for a long time by flipping a coin, so one can simulate the game on a computer by using what is known as a *random number table.*

It is quite difficult to pick numbers at random, and the problem has received a great deal of attention from mathematicians and from people who must decide winning numbers in lottery games (for a detailed discussion of this problem see the discussion of random numbers in Reference 10). One of the ways that we can make ourselves a table of digits chosen at random is to take the digits from the value of π. We have already stated that the digits of π are without recognizable pattern in their sequence. This is another way of saying that they are varying at random. The random number table prepared using the first 200 digits of π is shown in Figure 1.6. We can use this random number table to simulate the gambling game being played by Fred and Freda.

Thus far in our development of the vocabulary of *stochastic* systems we have avoided giving a strict definition of randomness. The dictionary tells us that the word *random* comes from an old French word randir meaning "to gallop". The idea was that in the warfare of mediaeval times a knight on horseback would move hither and thither without any obvious plan of attack. A strategy probably aided by the fact that the knight had drunk a large quantity of ale or other alcoholic drink before he had the courage to face the enemy. This pattern of behavior was similar to that practiced by Saxon warriors who used to go into battle wearing a bearskin and fortified with large quantities of mead, an alcoholic drink made from honey. The subsequent behavior of this warrior has given us the term *berserk* (derived from bearskin) to describe highly irregular movement and behavior. It is only by accident of history that we refer to the patternless behavior of the digits in the number π as being similar to that of a drunken horseman. Instead of calling them random we could have called them berserk numbers to conjure up the image of our highly disorganized Saxon warrior.

Simulate is another word which has a different meaning in everyday English and in mathematics. A dictionary describes simulate as "to have or assume a false appearance of". In other words, the aim of simulation in everyday life is to create a false impression of reality. In mathematics the word simulate means "to build a mathematical model" which behaves in every important aspect the same way as the real system. If we make a computer program which models the real gambling game, "Fred" becomes a memory element receiving donations or paying out donations which are represented by the numbers 1 and -1. We make the choice between 1 and -1 as an addition or debit from Fred's to Freda's store by selecting numbers at random from our number table and deciding to interpret an even number as "heads" and an odd number as "tails" (note that for this game 0 is interpreted as an even number). In this way we eliminate the physical tossing of the coin. We use a random number table to simulate the coin tossing. We can play the game on the computer with lightning speed, completing the equivalent of a days gambling in a fraction of a second. In *Figure 1.12* a simplified random number table generated to facilitate playing the game of Freda versus Fred on a computer is shown. We can move through this table in any way we choose to select heads or tails in the game of chance being pursued by Freda and Fred. If we were to start at the top left hand corner of the random number table and select digits by moving down the column the first 50 digits would be as shown in *Figure 1.13*. This would be converted to heads and tails to give the sequence listed in Figure 1.13(a). Using this sequence we find that in the first 50 tosses of the coin the progress of Fred versus Freda would be as shown in Figure 1.13(b). To be able to summarize the progress of the game we can focus on the number of tosses of the coin between a zero sum situation. That is for the set of data shown in Figure 1.13(b) the lengths designated λ_1, λ_2, λ_3 etc. on the figure.

12121211211222211111121112112222112221122221122221
12221221111122212211121111211222221221212212111112
22211222111212112212122221222221211112111121211211
22111212212121122111122111211111222221122221122221
22112211222122112112112112222112122222112211112122
22222212221222221122112222111222222111222111112122
11221121211222122122212222212122222211111211211221
12112111122212111121121221111211211222221111122122
12112212222211222111111121211122112111212222221111
11222211122211221221121122122221112112211112122212
11121111121112112122211122121112221211221122122222
22221211211122212121211211112221111112222121212111
21112112121122112121221222212122122111212221121111
11212211212212121111121112212221111111222111212211
21211221211111221221111221222222121112212121211122
21221111111211222212212211122122221121121112222221
121222221122221112212212212121222222211122122121211
12112221111212122112211112121211211221112211122221
11222122121212122222122112211122211221222222122121
21112212212221111122121112122222211112112222211122
22221222222222222221112111122112111121211111111112
21112211211121212111112222222222111211122111111222
11212111112212221112111221112111211211122112221222
122212222112211222122121212221212111211112221121211
12222112221222111121211212111122212221221221212112
12212121211211112111221112111221122121222111111121
11211221222222121121111211222222221221211112121222
22222111112212222211112111211222212222122222212212
11111212111121111122121121212222211111212222121121
12212112222212212211122121212222121111221211222121
12211222111211222112211111212121121121122112212112
21111221211211222112122121121211212112121212212222
21112211211222221222211111212121121122111212212111
22212212122121222121221122121211122221212222121122221
22121112112212212212122122112112122122122111212222
22211212222221211221122211112221112111112211212222
121112112221222222122221221112111122221212211221121
2112211121111211221222211112121212112121122221112221
21121211211112121121221122222212122112121121111121
22222221222112111122112221112122221122221112211211
22222211212111112212122221122121221111121212221212
11222112112222112211212121212221221212122111222122
222122111112122212222211112211111122222111211222212
221222121211112221121211111212222222212112212221211
21211112112112111111221212221212122121211221121121
2121122221211222211112221121211222222212222111212212
22211121122112222112112112222222122222221221112112
22112121211212221212112221112221122121222212212122
22112121112112212122122211211212211222212111222212
11121211222222121121221221121111111121111112211221

Figure 1.12. Simplified random numbers for playing the Freda-Fred gambling game.

When we look at the state of the winnings for either Fred or Freda at any one time, the movement of the dollar between the two players at the next toss of the coin is independent of who is winning at the time of the toss of the coin. When the outcome of an event in a sequence, such as the record of the winnings of Fred or Freda, is independent of what has gone before, it is said to be determined entirely by what we

a)

Number Sequence
from
Random Number Table

11 2222 11111 22 1 22 111 222 11111 2 111 22222 1 2222 1 2222222 1

Head and Tail Sequence

TT HHHH TTTTT HH T HH TTT HHH TTTTT H TTT HHHHH T HHHH T HHHHHHH T

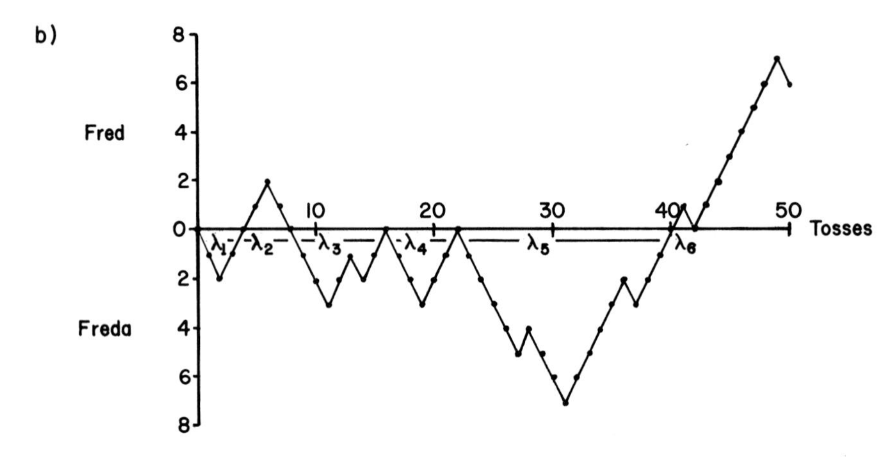

Figure 1.13. In the first 50 tosses Freda at first appears to be on a winning stretch, until Fred suddenly has a change of fortune. a) A record of the heads and tails sequence as generated from a random number table. b) Graphical summary of the data of (a).

call chance. *Chance* is defined in a dictionary as "events which happen without assignable cause or an unexpected event". It comes from the Latin word cadere, which means "to fall". The meaning of unpredictable behavior in the word chance can probably be traced back to the fact that witch doctors and their relatives in the Roman empire used to pretend to predict what was going to happen in the future by letting several objects, such as chicken bones or sticks, fall onto the floor and interpreting the pattern created by the falling objects.

The mathematician describes a process in which a series of events is determined by chance, independently of what has happened previously, as a *stochastic process*. This comes from a Greek word meaning "to guess". The idea is that if one is looking at a stochastic variable such as the sequence of heads and tails displayed by a flipped coin, then one can never predict the outcome of any one particular flip. One can only guess which number will show. In essence, a stochastic variable is one which we have to guess. A sequence of events in which a new event is independent of the previous event, but in which the chain of events generates a final physical quantity such as accumulated gambling gains is known as a *Markovian chain.* The name comes from that of a Russian mathematician, Markov (1856–1922) who helped to develop the theory of the study of stochastic events happening in a sequence such as that in a gambling game.

If we look at the values of λ_1, λ_2, the number of simulated coin flips between zero sum situations between Freda and Fred, we can ask ourselves the question "Is it reasonable to expect the values of λ, as recorded in a very long sequence of coin tossing experiments, to be describable by any mathematical function?"

If a group of students are asked this type of question, they will usually express the opinions that the values of λ will vary at random, or as one student from Great Britain picturesquely put it "The values of λ will be higgledy-piggledy". A British dictionary defines higgledy-piggledy as "haphazard, in confusion, origin obscure".

Experimentally, we will now explore the possibility that there is a mathematical pattern in the variable λ. The first thing we will discover when we look for a possible pattern in the sequence of zero state winnings between Fred and Freda is that λ can have much larger values than most people imagine, and that we have to play the game for much longer than most people would predict before a pattern emerges. Mathematicians have a word for very long, time-consuming sequences of events. They describe them as *ergodic sequences.* The word ergodic is derived from a Greek word meaning "tiresome" or "weary". In *Table 1.11* the number of moves between zero states, for a game played until the number of zero-state events which occurred reached twenty five thousand times, are summarized. The data has been summarized as giving the values of λ greater than or equal to a given number of moves. This data is summarized graphically on a special type of graph paper shown in *Figure 1.14.* This graph paper has both the ordinate and the abscissa marked off in logarithmic cycles. When the data of Table 1.11 is plotted on this graph paper, we have the surprising result that the data fits a straight line relationship. The data is not scattered higgledy-piggledy, but can be described by a mathematical equation

$$\log N_0 = k + m \log \lambda$$

where "log" means take the logarithm of the number, k is a constant, m is the slope of the line and N_0 is the number of coin flips between zero-sum states equal to or greater than a given value of λ.

The fact that students find that the number of coin tosses between zero state situations in the Freda versus Fred game is surprising is attested to by the comment made by the students when reporting the data of Figure 1.14.

> *It was found that one player in the game could stay ahead for over 10,000,000 flips of the coin. This means that if we assume that we could flip a coin every 5 seconds then we would have to flip a coin 24 hours a day for over a year and a half before the winning changed hands – the two players would commit suicide long before this happened.*

Table 1.11. Toss sequences between changes of fortune for the Freda-Fred game as simulated on a computer (program written by Ian Robb of Laurentian University).

Chord Length λ	Frequency f	Cumulative Frequency $\Sigma f \geq \lambda(N_0)$
2	12430	25000
4	3135	12570
6	1540	9435
8	1006	7895
10	704	6889
20	1724	6185
40	1277	4458
60	595	3181
80	347	2586
100	239	2239
200	598	2000
400	394	1402
600	200	1008
800	92	808
1000	58	700
2000	191	642
4000	128	451
6000	63	323
8000	39	260
10000	27	221
20000	52	194
40000	47	142
60000	15	95
80000	14	80
100000	10	66
200000	19	56
400000	15	37
600000	8	22
800000	3	14
1000000	2	11
2000000	3	9
4000000	4	6
6000000	0	2
8000000	0	2
10000000	2	2

The students' solution to ending the ergodic sequence of the Markovian chain of events in the coin flipping experiments would seem to be rather drastic. Surely they could have agreed to let one of them win and go for a cup of coffee!

If we calculate the average number of coin flips between zero-sum states for the data of Figure 1.14, we find that the average is 4.37. This cannot correspond to

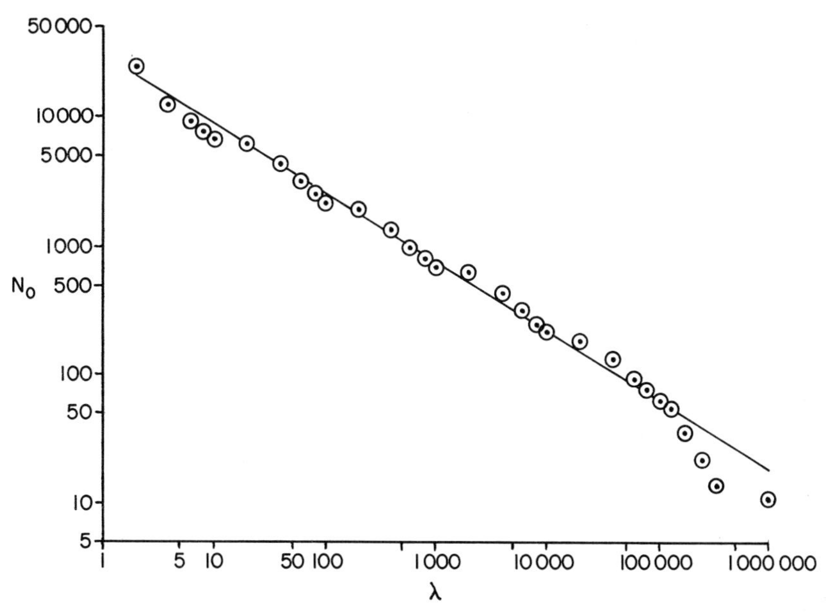

Figure 1.14. The distribution of the number of tosses of the coin between back-to-zero situations (no one winning) is a scaling function, as shown by the straight-line relationship between the data from Table 1.11. N_0: Cumulative number of lengths greater than or equal to the stated length λ. λ: Distance between zero states.

anything in a real game, because one cannot have a fractional coin flip. In our usual numbering system in which we use numbers 0–9 (this system is known as the decimal system from the Latin deci meaning "ten") the numbers 0, 1, 2, 3, etc., are known as digits. This word comes from the Latin word digitus, meaning "a finger". In the language of mathematics it is said that the variable "coin flips" can only have digital values. We are so familiar with the convenience of numbers such as the average number of coin flips being 4.37, that we sometimes overlook the lack of correspondence between the mathematical description and physical reality. The fractional average is only of use to us if we are interested in answering questions such as, "If the Freda-Fred game continues until 20 zero sums have been achieved, on average how many coin flips will constitute a 20 zero-sum sequence?"

The average is of no use in predicting the length of any particular flip number between two consecutive zero-sum states. A concept such as a fractional quantity of a digitized variable is described by the mathematician as a convenient *artifact*. An artifact is defined in a dictionary as "something made by a human being; not necessarily corresponding to anything in the real world". When we begin our discussion of fractal dimensions later in this book we will find that the subject is littered with convenient artifacts which do not correspond readily to images we try to create in the mind when presented with mathematical relationships.

When a physical phenomena or a random variable can be described by the type of relationship shown above, the scientist has a special word for this type of relationship. They describe it as a *scaling function*. Historically, this term came into use because of the fact that the behavior of this type of variable when examined at any scale of magnitude is the same. When we come to discuss fractal geometry, we will discover that an ideal fractal system is a scaling function which displays the same basic variability at any inspection magnification.

When presented with the data of Figure 1.14, students ask "Why does such a mathematical relationship describe the zero sum separation data?" The answer is that this data pattern was discovered experimentally, and the fact that the equation can be used to describe the data is a surprising pattern of complexity. As far as I am

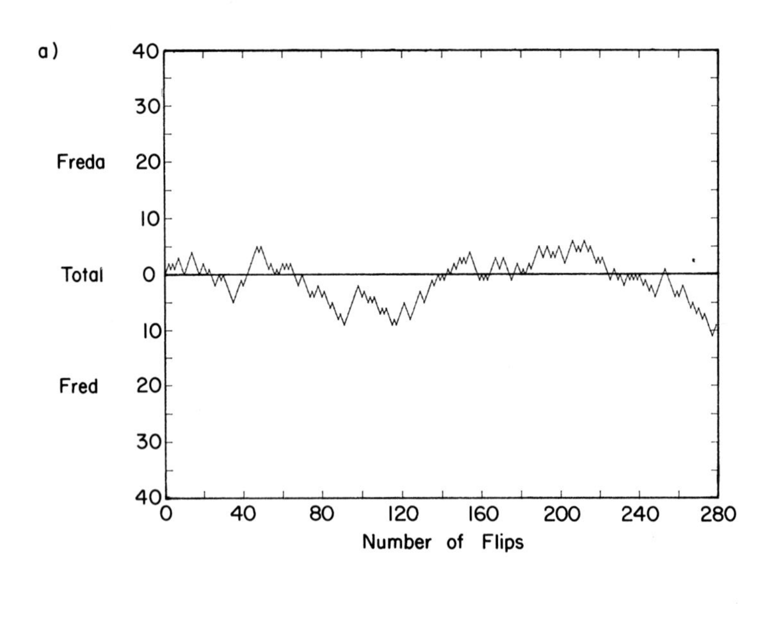

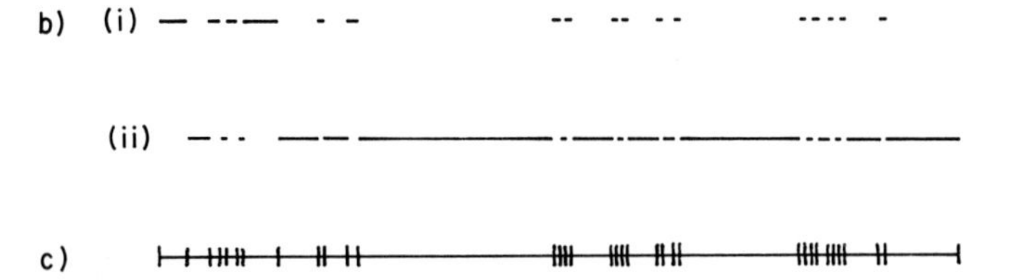

Figure 1.15. Alternative ways of summarizing the length of coin tossing between zero states in the Freda-Fred gambling game. a) A Freda-Fred game consisting of 280 tosses of the coin. b) Alternate lengths between zero-sum events. i) One possible set of chords generated by zero sum events. ii) Complimentary set of chords to part (i). c) Zero-sum events in time-line space.

aware, no one can predict from theory that this type of data pattern would exist in the Freda-Fred gambling game. How did I know which graph paper to plot the data to find a straight line relationship? I tried many different kinds of graph paper, and discovered the useful graph paper by experiment, not from theoretical considerations.

Another way of summarizing the occurrence of zero-sum situations in a time sequence of the Freda-Fred gambling game is to plot the number of flips between zero state situations on a line as shown in *Figure 1.15*. In Figure 1.15(a) a Freda-Fred game consisting of 280 flips of the coin is shown. In Figure 1.15(b) alternate lengths between zero sum situations have been colored black. The reason for choosing this format for displaying the data will become apparent later in this discussion. In Figure 1.15(c) the data is summarized in linespace by making a short mark to represent the zero sum situations. The surprising feature of the data when displayed in this way is the apparent tendency of the zero situations to cluster in the linear time sequence. As we shall discover in the next section, this tendency of random events to appear to cluster was one of the aspects of noise in telephone lines that stimulated Mandelbrot to create fractal geometry.

Section 1.7. Noisy Telephone Lines and Cantorian Dusts

Benoit Mandelbrot, the inventor of fractal geometry, tells us that an important stimulus in his development of fractal geometry was his study of a problem involving noise events on a telephone line. He tells us that during his investigation of this problem, he was once looking at a recorded pattern of noise, which looked essentially the same as the graph of the zero state situations in Figure 1.15(c) for the simple gambling game that we have described in the previous section as the Fred-Freda gambling game. The clustering of noise events recalled to his memory some data presented in a book on statistics by W. Feller [33]. Feller had been investigating the Markovian chain of events created by flipping a coin in a game which he called pitch and toss, but which we have called the Freda-Fred gambling game. The fact that the type of data we have summarized in Figure 1.15 was surprising to the first scientists who studied this kind of Markovian chain experimentally is endorsed by Feller's comments on his own coin flipping data. When discussing a set of data shown in his book, similar to our Figure 1.15, Feller said that the data was

> *Especially selected from amongst those he had generated since many of the graphs of the gamblers progress looked too wild to be believable.*

It is worth quoting his statement regarding the surprising nature of his data at length.

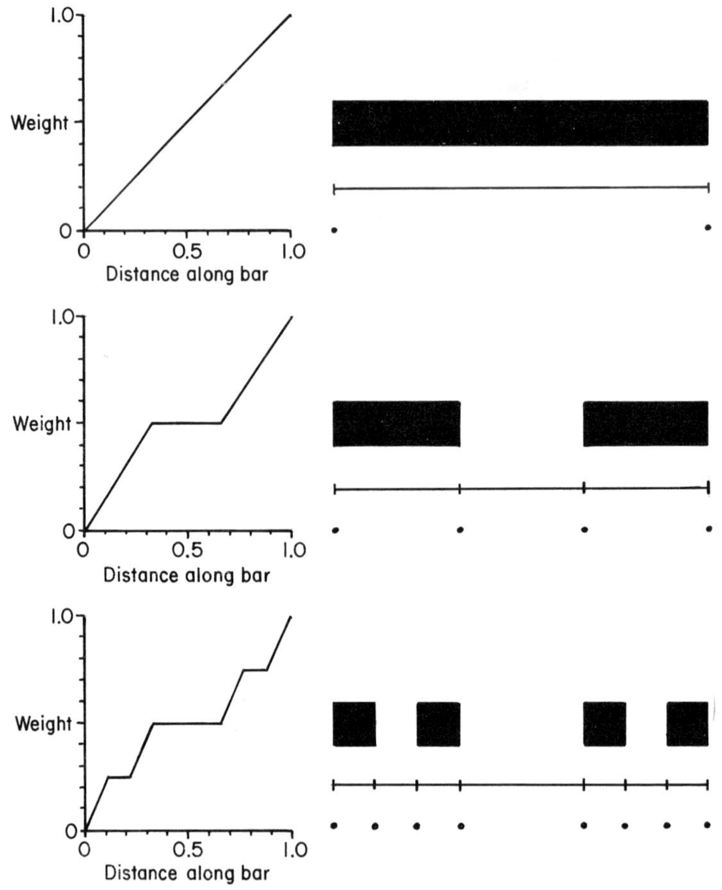

Figure 1.16. Segmented bars lead to Cantorian dusts and devil's staircases.

> *If a modern educator or psychologist were to study the long run of case histories of individual coin tossing games that I recorded, he would classify the majority of coins as maladjusted. In a surprisingly large proportion of cases one player is in the lead almost all of the time, and in very few cases will the lead change sides and fluctuate in a manner that is generally expected of a well behaved coin* [33].

Mandelbrot tells us that not only did the sequence of noise events on the telephone line cause him to remember Feller's data, but that the sequence of chord length between events when depicted as shown in Figure 1.15(b) brought to mind a similar pattern of black and white lines generated when studying a system known as a *Cantorian bar* or *Cantorian dust*. *Cantor* (1845–1918) was a German mathematician who studied the problems of infinity encountered in mathematics and geometry. To help us understand why Feller's data stimulated Mandelbrot to study the clustering of noise events in a telephone line in terms of the mathematical ideas of Cantor, consider the thin metal bar shown at the top of *Figure 1.16*. Let us assume that the bar is cut in the middle and then hammered in such a way that all of the mass of the original bar is concentrated in the two new bars, each of length one third of the original bar. Below each of the beaten bars shown in Figure 1.16, another line is shown in which the beginning and end of the residual bars are shown as dots in linespace.

The next stage in the transformation of a Cantorian bar into a Cantorian dust involves a further cutting of each of the smaller bars in the middle, and subsequent hammering to concentrate the material of the original bar in the 4 slugs of material spaced as shown in the figure. On the left-hand side of the segmented Cantorian bar, a series of graphs demonstrate how the mass of the bar is distributed in space after each segmentation plus concentration by hammering. The graph of the distribution of mass in the 4 slugs begins to look like a staircase, and the pattern of dots in linespace showing the location of the beginning and end of the slugs starts to look like a dust. Furthermore, the patterns of lines marking the location of the slugs in the divided Cantorian bar, and the dots on the corresponding space line begin to look somewhat like the graphical summary of zero-sum situations in our Freda-Fred gambling game of Figure 1.15(c).

Of course, the correspondence between the map of a segmented Cantorian bar and the record of zero-sum states in our Freda-Fred game is not exact. However, it is not unreasonable to suggest that the zero-sum pattern of Figure 1.15(c) is similar to the spacing of the Cantorian slugs when the latter have been shuffled in one-dimensional space. When two systems look as if they might be of the same pattern but with one of them being shuffled at random in the appropriate space, mathematicians say that the two systems are *statistically self-similar*.

On several occasions in this book we will describe a system as being a "statistically self-similar system" version of an ideal system. It is important to appreciate not only the usefulness of such a concept, but also its limitations. For example, Miss Average is a statistically self-similar version of Miss Universe, but the properties they have in common constitute only a limited description of their spatial structures. When Mandelbrot was looking at the pattern of time sequences between noise events, his imagination made a creative leap that resulted in his postulating that the noise events constituted a statistically self-similar version of a Cantorian dust (a dictionary defines a *postulate* as a position assumed as self-evident or something whose possibility of existence is assumed. When we postulate a possible theory this becomes the beginning of a discussion in which we seek to prove the truth of that postulate).

It is interesting to consider the possible reasons why Mandelbrot was able to make the leap in his imagination, bringing together the noisy telephone line pattern and the events in a simple gambling game, to create a theory for describing the noise events using the concepts of Cantorian mathematics. Mandelbrot was able to complete the creative act because of the wide and varied experience he had acquired in many fields of knowledge. If he had been a narrow specialist in an obscure area of mathematics, he could not have made the creative connection between such apparently dissimilar systems. In North America the current philosophy of education seems to be to beat science students into a submissive state with a mass of problems.

An alternative theory of education seeks to present the student with a very wide range of knowledge, so that he can creatively bring together the various aspects of his education. As a student I found problems involving massless springs on frictionless plains boring to the point of nausea. I have always felt that physics problems ad nauseam destroy the ability to think creatively. Obviously, there must be some student exercises in the application of math or physics to the real world in a course of study, but the current balance of educational strategies in many educational institutions seems to exhaust the student rather than build up his intellect. One of my motives in writing this book is to expose student readers to a multitude of exciting ideas to stimulate their creativity before they give up physics because they are suffocated in a swelter of tiresome problems.

We could have described the creation of a Cantorian dust without discussing the Cantorian bar, but the Cantorian bar discussion has the advantage that as the bar is sectioned and beaten, the mass is conserved making it possible to plot the staircase graphs shown in the series of diagrams of Figure 1.16. If we carry out the subdivision of the bar an infinite number of times, we reach a situation in which the original bar is divided into a infinite number of small slugs, each one being of infinite density (the reader should note that in this juggling of infinities we are implicitly assuming that we are using the concepts of Euclidean geometry. If we were using the concepts of quantum geometry, our bar subdivision would have to cease when the slug size was the same as our ultimate pixel size). Even with an infinite number of slugs, one can still plot the distribution of material across the array of infinitely small slugs to generate the graph similar to that shown at the bottom of the staircase sequence. This graph is known as the *devils staircase*. It has an infinite number of steps, within a finite distance. Just the kind of fiendish system created by the devil to torture the minds of humans! The devils staircase of a distributed quantity has turned out to be an important system in deterministic chaos. For example, some scientists have shown that the possible number of frequencies of an oscillator undergoing chaotic variations can be plotted as a devil's staircase [9].

If we consider the number of points forming the Cantorian dust in Figure 1.16 from the perspective of Euclidean geometry, then after we have divided the bar an infinite number of times we have an infinite number of infinitely small points in the original linespace. However, these points are obviously more sparsely distributed in space than the infinity of points forming a continuous line. When studying systems which we now describe as Cantorian dusts, Cantor developed the idea that we can describe a continuous infinite number of points as constituting one-dimensional space. He then went on to describe how the infinite number of points created in our Cantorian bar division forms an infinite number of points in linespace which can be described by a dimension less than one. In fact, for theoretical reasons, the Cantorian dust of the system shown in Figure 1.16 is described by mathematicians as being of dimension 0.6.

In an earlier section, we discussed the problems of quoting an average for a digitized system when the magnitude of the fractional part of the average did not correspond to any physical reality. At that time it was pointed out that such a fractional average quantity was a useful contrivance created by the mathematician to describe his system in a useful manner. When discussing the dimensions of an object, most people feel that they have a good grasp of what is meant by one, two and three dimensions. In their mind they can visualize a line representing one-dimensional space, a sheet of paper representing two-dimensional space and a cube of solid material representing three-dimensional space (see, however, the discussion of the operational definition of dimension in Chapter 6).

When mathematicians describe the dots in linespace as having a dimension less than one, they are creating a concept which does not correspond to any simple system that can be easily visualized. A dimension being allocated to a set of dots occupying but not filling linespace is a similar concept to that of calculating an average for a quantity which can only have digital, that is quantized, existence in real systems. To help understand what the mathematicians are describing when they use fractional dimensions between one and zero to describe the population density of points in space, consider the system shown in *Figure 1.17*. This shows another Cantorian bar being transformed into an infinite set of slugs, along with the corresponding dots showing space occupied by the ends and beginnings of each slug of material. For this system the bar is cut into four parts in each stage of segmentation. Obviously the system again generates a Cantorian dust with an infinite number of points. If, however, we compare the number of dots created at any given stage of subdivision, the Cantorian dust corresponding to the division depicted in Figure 1.17 obviously creates a "dust" of points which appear to occupy space more efficiently than the Cantorian dust of Figure 1.16.

By allocating a fractional dimension to the two sets of dust which differ in their ability to populate linespace, the mathematician is able to characterize and compare the structure of the two sets of infinite points. The reader should note that the essential step in being able to distinguish between the two Cantorian dusts is to look at the efficiency with which they fill space, rather than to count the dots constituting the dust. We see that the fractional dimension allocated to the dust in linespace is a mathematical convenience not directly related to the simple concept of dimensionality which we use in everyday speech to describe the objects we encounter around us.

A full discussion of various types of Cantorian dust which can be constructed by the mathematician is beyond the scope of this book. The interested reader should pursue the topic in Mandelbrot's books on fractal geometry [9, 10]. Mandelbrot showed that the array of points on the line representing the occurrence of noise events on a noisy telephone line had a Cantorian dust dimension of 0.3. For reasons which will be discussed later in this book, the fractional dimension of a Cantorian

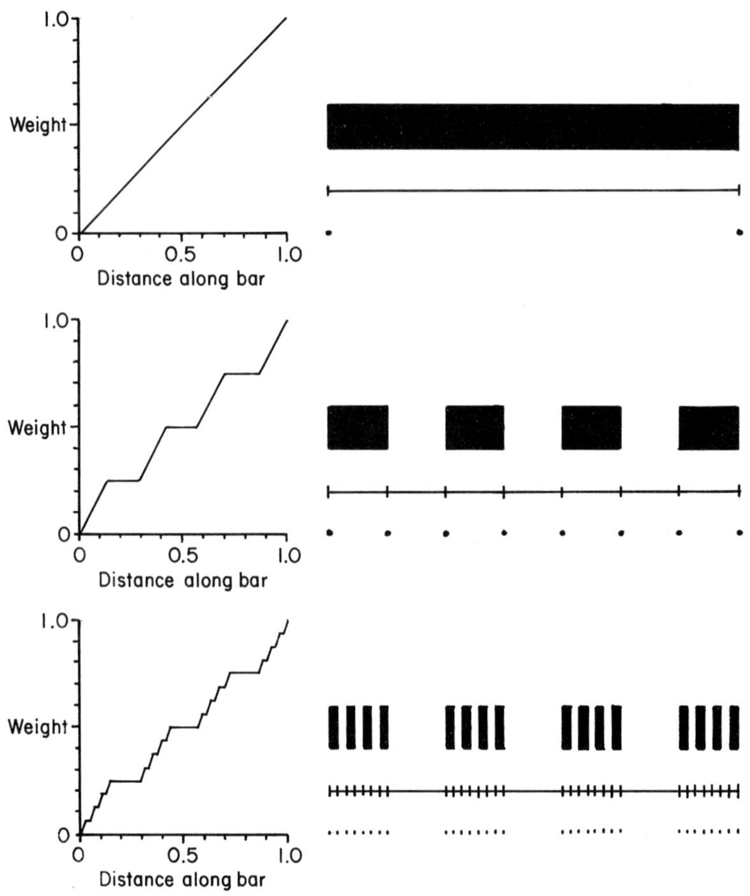

Figure 1.17. The devil's staircase can be generated in many fiendishly different ways.

dust is actually a fractal dimension in one-dimensional space. Fractal dimensions in one-dimensional space are important for interpreting data from an experiment in which the structure of an object is explored by means of a line search. This aspect of applied fractal geometry will be discussed in greater detail later in the book.

Section 1.8. Monte Carlo Explorations with a Cluster of Drunks

George *Gamow* (1904–1968), who was one of the pioneers of nuclear physics, was the first to suggest that because of nuclear reactions occurring within the sun, the

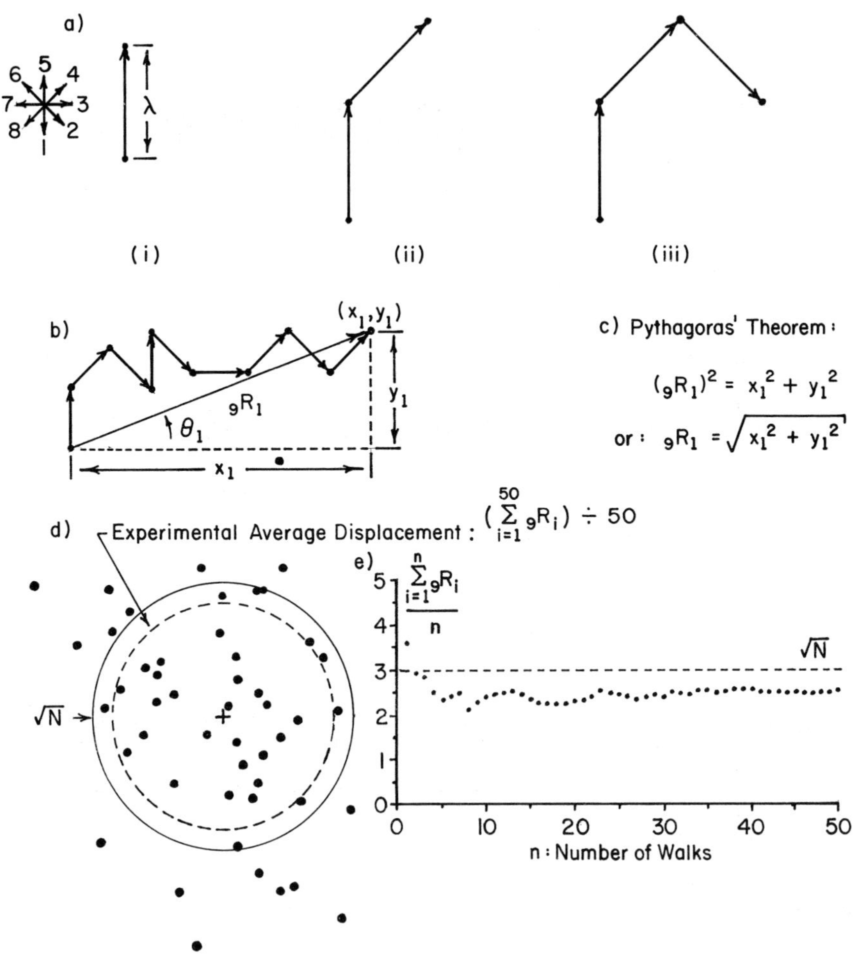

Figure 1.18. The dispersal of a set of drunks as they leave a lamp post can be modeled using a random number table to choose directions. a) The first three steps in our simulated progress of a staggering drunk. b) The distance from the lamp post of our drunk after nine steps is called "the displacement vector" $({}_9R_1, \theta_1)$. c) The magnitude of the displacement can be calculated from Pythagoras's theorem. d) Dispersion pattern after 50 attempts of 9 steps each to move away from the lamp post. e) Many random walks are required before the measured average distance converges on the theoretically predicted value of $L\sqrt{N}$ (for this set of walks, $L = 1$ and $N = 9$, therefore $R_{av} = 3$).

sun was heating up and that human life on earth could be destroyed some day not by freezing because the sun ran out of energy, but by frying conditions generated by the overheating of an energetic sun. Gamow was very good at explaining difficult ideas in everyday language. In 1946 he wrote a book "One, Two, Three, ... Infinity" which has been widely read [34]. Although this book is dated in parts it is still an exciting book to read. It was in this book that I encountered what I considered to be two very surprising facts. The first concerned a problem known as "Buffon's needle", which we will discuss in depth in Chapter 3. The other was a formula for the average progress of a drunk staggering away from a lamp post. The situation studied in such a problem is illustrated in *Figure 1.18*(a). It is assumed that a drunk takes a series of steps away from the lamp post with the direction of each step being entirely at random and independent of the distance from the lamp post. Figure 1.18(d) shows the positions of 50 drunks, who are assumed to have been at a party and have continued their frivolities leaning on the same lamp post until it was time to go home. After the drunks have taken the same number of steps, how far on average are they from the lamp post? Surprisingly, the scattering of the drunks about their lamp post is not chaotic. The average distance of the drunks from the lamp post turns out to be describable by a mathematical equation. Even in their drunken stupor, it appears that the drunks must move within limits describable by a stochastic relationship. Before we can present the reader with the formula describing the surprising patterns of dispersal of the drunks, it is necessary to explore the concept of the square root of a number.

As pointed out in an earlier section, geometry began as a practical subject concerned with the building of pyramids, the measuring of fields and the location of boundaries to establish ownership of fields and the limits of cities. A difficult task facing the early surveyors was the calculation of the area of a field from the measured length of a side of the field. An important step in the development of applied mathematics for surveyors was the creation of techniques for manipulating the data collected by measuring the dimensions of a field or building. One technique which helped the surveyor was the move away from the specific counting of defined objects, such as apples, to a general idea of counting of unspecified things. They probably did this originally by using little stones to represent things such as the number of steps taken when measuring the length of the side of a field. This possible origin of number representation of objects by small stones is hinted at by the fact that the word to *calculate* comes from a Latin word calculus, "a small stone". Another word that hints at the use of small stones in the early history of mathematics is the fact that the word *divide* means "to see separately". If one were dividing 6 by 3, one could take the 6 stones and arrange them into groups of 3 which could be "seen to be separate", that is, divided. The symbol for division also illustrates this early practice for calculations with small stones. Thus, the — of ÷ is the line drawn between the two dots so that they can be viewed separately.

An important task facing the early surveyors was the calculation of the length of the side of a field of a known area. As he manipulated his stones on a surface, the ancient mathematician probably discovered that certain numbers of pebbles could be arranged into square patterns. As shown below, 16 pebbles can be arranged into a square made of four sets of four pebbles. From one point of view, this square was being created by repeating the set of stones present in one edge of the square.

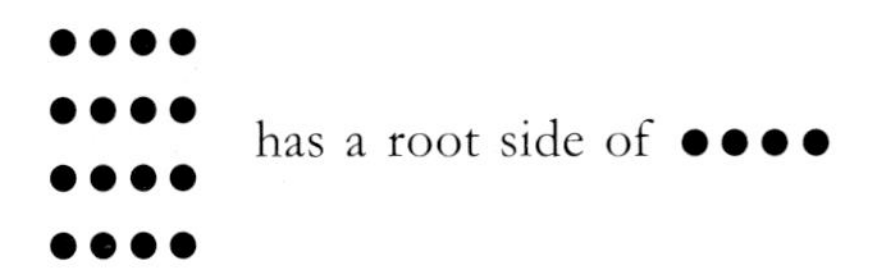

The number of stones on one side of such a square array came to be described as the "root of the square". In common usage this soon became known as the *square root* of the number. Mathematicians invented a symbol that looked like a root pulled out of the ground to represent the square root of a number. Using this symbol, the fact that 16 stones can be formed into a square with four stones along one side is written symbolically as

$$4 = \sqrt{16}$$

Using exponential relationships, mathematicians have agreed to write the square root operation as a fraction. Therefore the relationship is written $4 = (16)^{1/2}$.

Mathematicians have a habit of extending concrete ideas represented by the patterns such as that made by the small stones shown above to create difficult abstract ideas which can no longer be represented in a simple manner by a physical system. Mathematicians soon started to ask the question "What is the square root of numbers such as 2 and 3?". One cannot arrange 3 pebbles into a square pattern. However, one can imagine that there was a field which had sides of length α which had an area of 3 units of area. One could then write for this field that

$$\sqrt{3} = \alpha$$

The only problem is that if one attempts to calculate $\sqrt{3}$, one ends up with an endless series of digits in the same way that we found π to have an endless number of random digits. You can try this for yourself if you have a calculator that gives "square roots". If you put 16 on the display and press the square root sign, $\sqrt{\ }$, the number 4 appears in the display. If you put 3 on the keyboard and press the square root sign you immediately use up all the digits in the display. My calculator gave 1.7320508 for the square root of 3. If you use bigger and bigger calculators, there is no end to the number of digits. $\sqrt{3}$ is an irrational number in the same way that π is an irrational number.

Over the years, the word irrational was used in everyday speech to describe someone who was "crazy and unreasonable". In medieaval times a copyist, when writing down a mathematical text, made an interesting substitution for the word irrational when applied to irrational numbers. He described numbers such as $\sqrt{3}$ as *absurd*, since absurd also meant crazy and irrational. Did he make the change to absurd because it is shorter than irrational? When you have had lots of copying to do by hand, the difference reduces the effort required. The prejudices of ancient society against handicapped people show in this word, since absurd comes from a Latin word surdus meaning "deaf or unintelligent". Even absurd soon became shortened by the copyist to *surd*. Generations of students have learned to call numbers such as $\sqrt{3}$ and $\sqrt{2}$ surds.

The reader who is encountering irrational, absurd and abstract square roots for the first time may share the emotions of a mathematician from Trinity College, Cambridge, England which are summarized in a limerick given by Gamow in his book "One, Two, Three, ... Infinity" (a limerick is a form of humorous verse in a five-line jingle, named after an original poem referring to Limerick in Ireland).

There was a young fellow from Trinity,
Who took the square root of infinity,
But the number of digits
Gave him the fidgets;
He dropped Math and took up Divinity!

Now that we know what is meant by a square root, and the way in which the quantity is written, we can return to the problem of summarizing the progress of a drunk staggering away from a lamp post, as discussed by George Gamow.

When reading Gamow's account of the progress of the drunk, I was surprised to find that the average distance travelled by a drunk staggering away from his lamp post is given by the relationship

$$d_a = \lambda\sqrt{N}$$

Where d_a = average distance of the drunks from the lamp post and N = the number of steps of size λ taken by the drunks.

Gamow gives the proof of this result in his book, but we can test the truth of the relationship for ourselves by simulating the progress of the drunk using a piece of graph paper and a random number table. First of all, we have to draw a probability star for use in interpreting random numbers as directions in space. In Figure 1.18(a) this probability star for 8 directions is shown. In the simple experiment we will discuss here, we will assume that each step taken by the drunk as he moves away from the lamp post has the same size, λ. To choose a direction in space for the first step, we will enter the random number table and choose a digit between 1 and 8 (that means that in moving through a random number table to choose digits we will ignore zero and nine). If we enter our random number table (*Fig. 4.6.*) at the top right hand corner, the first digit that we encounter is 5, and so we mark off a step λ in the direction indicated by this number in the direction star of Figure 1.18(a) (i).

To simulate the second step taken by our drunk, we choose the direction of the second step from our random number table. We then draw a line of length equal to λ in this direction from the tip of the arrow representing the first step as shown in Figure 18(a) (ii). The direction of the third step is again chosen at random from the random number table. The third arrow representing the third step is drawn from the position reached after the second step, as illustrated in Figure 1.18(a) (iii). In Figure 1.18(b) the result of nine steps taken this way are shown. We can measure the displacement of the drunk directly by drawing the line denoted by ${}_9R_1$. Alternatively, if the whole experiment is being conducted on a computer, we could keep track of the x, y coordinates of the point reached after each step, so that we could use Pythagoras's theorem regarding the properties of a right angled triangle to calculate the value of ${}_9R_1$, as shown in Figure 1.18(c). Thus for our first simulated staggering progress, the distance reached by the drunk as illustrated in Figure 1.18(b) is $\sqrt{(x_1^2 + y_1^2)}$. In Figure 1.18(d) the dispersal pattern created by 50 drunks attempting to leave the lamp post are shown. From the relationship of Gamow we know that the drunks on average should reach a distance of the square root of nine, which equals three, steps from the lamp post (we will assume that each step is of unit length, that is, $\lambda = 1$). The actual average distance calculated from the simulated attempts of the drunk to leave the lamp post is shown as a circle in Figure 1.18(d). The theoretical average distance is shown in the same diagram. Again we note that a knowledge of what the average distance should be is of no use in predicting the actual distance reached in any one attempt to stagger away from the lamp post. The average distance even after 50 attempts is not exactly the same as the value predicted by the formula. One has to make many more attempts before the measured average distance of the drunk starts to converge on the theoretical value as shown by the data of Figure 1.18(e).

In the language of the mathematician, the distance from the lamp post to the point reached by the drunk is referred to as the *displacement vector* of the drunk. When discussing various quantities the mathematician uses the two terms *vector* and *scalar* to refer to two basically different quantities. When we know the distance of the drunk from the lamp post without knowing the direction, the magnitude of the distance is known as a scalar quantity from the Latin word for "a ladder", scala. A ladder, until it is used, has magnitude but not direction. When a ladder is placed upon a wall the height that the ladder reaches is related to its length but is fixed essentially by the direction of the ladder. If we were to look at our sketch in Figure 1.18(b), the ladder of ${}_9R_1$ only reaches the height y_1 because of the angle at which the ladder has been placed. When we are interested in a quantity which has direction and magnitude, it is called a vector quantity. The word vector comes from the Latin root word vehere, meaning "to carry". The same word gives us the English word "vehicle" to describe a type of cart in which we carry objects. Different mathematical symbols are used to indicate vector and scalar quantities. In most books, a small arrow placed above the quantity is used to indicate that we are dealing with a vector rather than a scalar quantity. In everyday life we use the term *speed* and *velocity* as interchangeable terms. To the scientist a speed is a scalar quantity giving no information on direction, whereas a velocity has to have the direction of travel specified as well as the magnitude of the velocity.

If a drunk was the only person or item that moved away from a reference point by a series of steps taken at random, then the surprising pattern illustrated in Figure 1.18(d) would remain an academic curiosity. However, the movement of tiny

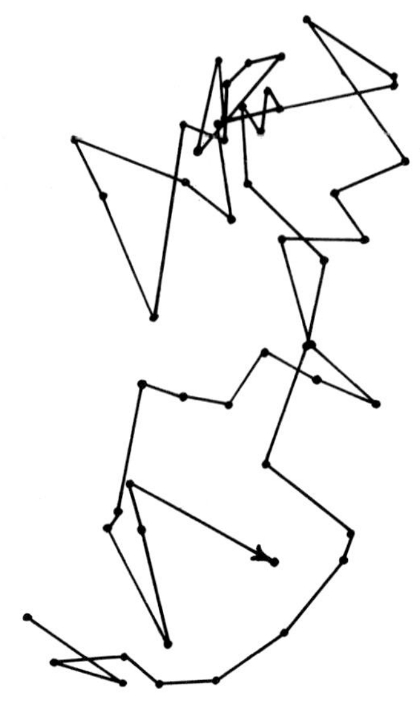

Figure 1.19. A stochastic wandering of a drunk, viewed by a giant who could barely see the drunk, would simulate Brownian motion.

objects in a fluid or a gas which are of the same order of size as the molecules in the fluid behave in a manner exactly the same as our drunk, and so the theory of random dispersion from a given starting point is very important in physical science. Let us imagine that we take the data of Figure 1.18(d) and create a Markovian chain of events in which we allow the various drunkard sets of nine steps to be taken in sequence. This involves drawing a sequence of arrows in which the resultant vector of nine steps is added to the previous nine steps to create the pattern of events shown in *Figure 1.19*. Imagine too that a creature from outer space had been looking down on our almost invisible drunk, and was unaware that the displacement of the drunk was taking place as a consequence of nine unobserved steps. Then all that this being from outerspace would observe would be the stochastic pattern of dispersal. It would be unaware of what was actually causing the movement of the drunk.

In 1827, a Scottish botanist Robert *Brown* was looking at a suspension of pollen in water under the microscope (remember, in 1827 the atomic or molecular theory of matter had not been developed). As he looked at his pollen in the water, he noted that the individual pollen grains moved about irregularly. At the time he thought this was the result of the life hidden within the pollen grains. However, when he studied dye particles suspended in water he observed the same erratic motion. This type of erratic motion of tiny objects in liquids and air is now known as *Brownian motion* in honor of its discoverer. We now know that a small object such as a pollen grain moves around in the fluid because of the fact that at any one instant in time the random bombardment of the pollen grain by molecules of water is never exactly balanced. Therefore, at any one instance the pollen grain experiences a net force in one direction, and then a fraction of a second later it experiences a push in another direction completely at random. The direction of the "net push" varies and Brownian motion is a Markovian chain created by the stochastic variable "net force" or "push"! It is interesting to note that when Einstein was developing his theories of the atomic and quantized structure of the universe, he predicted Brownian motion and was unaware of the fact that it had actually been observed experimentally half a century before it was predicted theoretically. As we will discover in a subsequent section, Brownian motion played an important part in the evolution of the theories of fractal geometry developed by Mandelbrot.

Another area where random dispersal of energy from a point is an important subject is the movement of items such as an electron in a crystal lattice. A full discussion of this topic would be beyond the scope of this book, but we note, however, that in some important problems scientists have to develop a variation of our drunkards walk which is known as a self-avoiding random walk. We carry out this type of walk on a square lattice as illustrated in *Figure 1.20*.

On a *self-avoiding random walk* in two-dimensional space, we specify that the point moving on the grid cannot return to a point which it has already visited. It is relatively easy to play this game in two-dimensional space allowing a random number between 1 and 4 to direct the movement along the grid. Playing the game visually we can soon recognize forbidden moves, and a typical set of movements in a self-avoiding random walk is shown in Figure 1.20. If you play this game experimentally you will soon discover something which is quite surprising to many people who start to play the game. By experiment one finds that many of the random walks quickly end in a dead-end trap where it is not possible to move away from the point of arrival. No one has ever worked out the theory for predicting the distance travelled from the original point before becoming trapped in two-dimensional space. However, one can simulate the self-avoiding random walk and soon generate data such as that of Figure 1.20. This data is summarized in a type of graph known as a *histogram*. A more common name for this type of graph is *bar graph*. To construct the histogram one erects a bar representing the number of events of a given magnitude forming the base of the bar. The name histogram comes from the Greek word for a loom on which woollen blankets and cloths were woven. The Greek looms used to hang down vertically and the pattern of data on a histogram looks as if it is a set of stripes being woven on a vertical blanket.

In a later chapter we will explore the possibility that the pattern of bars on the histogram of *Figure 1.21* may be describable by a mathematical function. However, at this stage of our exploration of "Chaos and Complexity", we will regard the bar pattern of our histogram as a pattern of events discovered experimentally by simulating our problem using a stochastic mathematical model.

When we simulated the movement of our drunk from the lamp post using random numbers, we were beginning to use a powerful technique developed by mathemati-

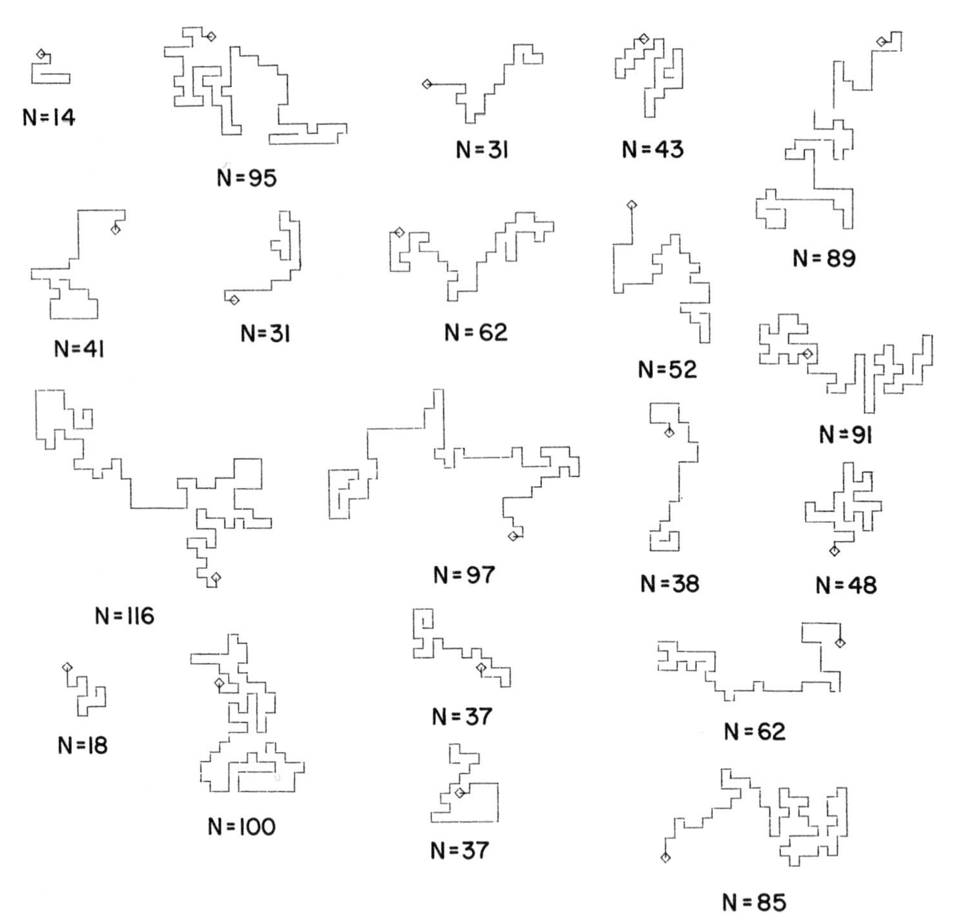

Figure 1.20. A wandering pixel undergoing a self-avoiding random walk on a square lattice often reaches a "dead end" position relatively quickly, as shown by the above record of self-avoiding random walks.

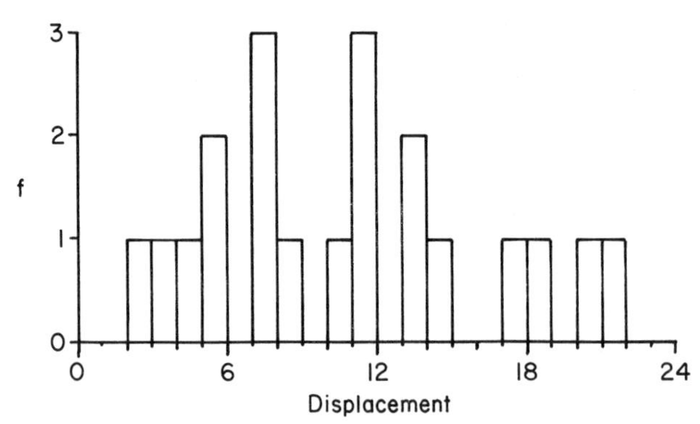

Figure 1.21. The distances travelled by pixels undergoing a self-avoiding random walk in two-dimensional space can be conveniently summarized in a histogram.

cians for studying stochastic processes known as *random walk theory* [8]. To explain the basic vocabulary and concepts used in random walk modelling of a difficult problem let us imagine that we were unaware of Gamow's proof of the average dispersal of a set of staggering drunks. Let us assume that we had been asked to discover experimentally the average distance moved by a drunk in our problem. In the real world this could be a study of how toxic gases moved away from a bomb exploded in a city. In this case, we would be interested in knowing how the molecules of the poisonous gas jiggled their way backwards and forwards under the molecular bombardment of the ordinary molecules of the air. Our interest in the average distance of dispersal would be to discover the level of the poisonous gas at different distances from the point of explosion of the original bomb at various times. Using known speeds of movement of gases and patterns of molecular bombardment, we could simulate the dispersal of the toxic gas molecules. Mathematical procedures for modeling a complex problem which cannot be solved theoretically are known as *Monte Carlo routines.*

The name for this technique comes from the fact that at the beginning of World War II, scientists were faced with the problem of deciding if a nuclear reaction was possible. They knew that many neutrons moving through uranium would undergo capture at random with subsequent emission of further neutrons. They could not predict from theory whether the net result would be the breeding of a neutron flux which would trigger an atomic explosion, or if the neutron flux would weaken with the consequence that the uranium intended as super explosive would sit on the ground like a lump of wet bread. Scientists studying this problem used the first large computer ever built to model the random walk of the neutrons through a lump of uranium, modeling mathematically the random encounters between neutrons and uranium atoms. The project was top secret and the scientist working on the project invented a code name for the project. Monte Carlo is the capital of the tiny country of Monaco in Europe, famous for its gambling casinos. Since the scientists trying to solve the problems of the atomic bomb were taking a big gamble, they called their project Monte Carlo. Since that first large project using a computer and random walk theory to arrive at a probable solution to a difficult and complex physical problem, these types of mathematical experiments have been called Monte Carlo methods. McCracken, in a review of the early Monte Carlo methods, states

> *The Monte Carlo method in general is used to solve problems which depend in some important way upon probability – problems where physical experimentation is impracticable and the creation of an exact formula is impossible* [35].

In this book we will often use Monte Carlo routines to discover the surprising patterns of chaos and complexity.

Section 1.9. Empirical Facts and General Truths

In his book in which he outlines the development of fractal geometry, Mandelbrot tells us that his thinking was stimulated by the work of a scientist who was generally unknown to the public – Lewis Fry *Richardson* (1881 – 1953) [36]. Richardson studied at Cambridge, England and obtained his Bachelor of Science in physics, mathematics, chemistry, biology and zoology. He began his scientific career at the Meteorological Office. At first his career evolved in the traditional manner, and he became the director of an institution conducting weather research. However, he was a Quaker and refused to take part in the First World War. Furthermore, even after the war he refused to continue his work on weather forecasting because he discovered that the people who were interested in using poison gas as a weapon of war wanted to use his data to study the effect of wind upon clouds of poisonous gas, and so he resigned from the meteorological service. In later life he was independently wealthy and able to continue his research in many fields. However, because of his isolation from the mainstream of science and his views on pacifism, he received little recognition during his lifetime. Amongst the papers he left unpublished at his death was a short study of the problems of measuring the coastlines of various parts of the world. In this study he discovered experimentally that there was no answer to a simple question; "How long is the coastline of an island?". In his original studies Richardson studied the coastline of Australia, South Africa, the German land frontier, the west coast of Britain and the land frontier of Portugal.

Essentially, the technique that Richardson considered as a basic procedure for characterizing the length of a coastline can be appreciated by considering attempts to measure the length of the coastline of Great Britain shown in *Figure 1.22*. One can imagine that a giant given the task of estimating the length of the coastline of Great Britain would stride around the coast and measure the coastline in terms of the number of steps necessary to complete his circuit around the island. We could mimic the behavior of such a giant on a smaller scale map of an island or lake by using a pair of compasses with the separation of the points set to simulate the stride of the giant. The size of this stride we will assume is λ_1. We would then mark on the island the point at which the giant would set off on his journey. The subsequent strides would create a polygon around the perimeter as shown in the figure. The perimeter of this polygon becomes our estimate of the perimeter of the island at an inspection resolution λ_1. If we repeat our exploration of the island with a smaller step size λ_2, we generate a different polygon, the boundary of which is the revised estimate of the length of the coastline at the new inspection resolution of λ_2.

If we were measuring a very large island, our giant's strides would be measured in hundreds of kilometers and the perimeter estimate would be thousands of kilometers. If we were measuring a very small island, using a dwarf to step around the island instead of a giant, then the strides would be several centimeters and the perimeter

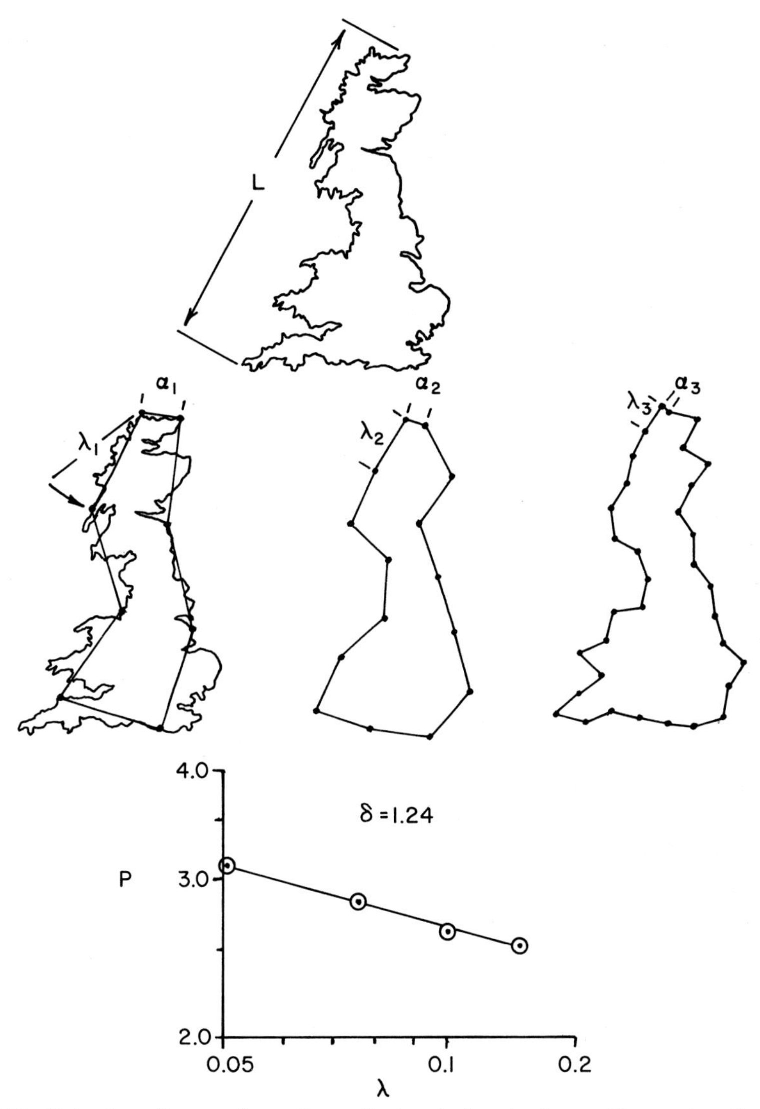

Figure 1.22. Richardson discovered experimentally that the length of a coastline, such as that of Great Britain, depends on the technique used to measure the coastline. λ = measuring step. P = perimeter

a small number of meters. If we then tried to compare the structure of the small island to that of the big island, we would have to use two different graph papers of very different scales. Mathematicians overcome this problem when they are interested in comparing things of very different sizes by using a mathematical trick known as *normalization.*

The word normalization comes from a root word meaning "a standard or something made according to rule". We say that someone's behavior conforms to the norm, meaning their behavior is standard for a specific society. When a variable in science is normalized it is presented in a standard manner. If magnitudes for islands are variably reported in kilometers, miles, centimeters etc. this can be confusing. The mathematical trick of normalization converts numbers to a dimensionless form and makes them all fractions of a given reference magnitude. When we normalize measurements on an island, or some shape, the standard reference we use in this book is the maximum length of the object. All measurements are expressed as a fraction of that length.

The quantity used to convert measurements into a dimensionless form is called the *normalization factor*. For our island profile of Figure 1.22 the normalization factor is the maximum projected length, L, shown in the diagram. Using this reference value our strides around the islands and the resultant perimeter estimates are expressed as a fraction or multiple of this maximum projected length. When we carry out this normalization, we can plot the explorations of the coastline of Great Britain on the same graph as the exploration of the coastline of the lake on which Laurentian University (Ontario, Canada) is built, Lake Ramsey, which is about 7 kilometers long. Richardson discovered experimentally that when he plotted the data for the coastlines and political boundaries as he explored in more detail the data gave better estimates of the length of the coastlines and boundaries and, when plotted on log–log scales, generated a straight line data plot heading for infinity. In Figure 1.22 data for the study of the coastline of Great Britain is plotted on this type of graph. Richardson regarded this data line tending to infinity, generated by his experimental investigations, as a novelty and did not anticipate that there would be any significance in the slope of the line on the graph paper.

An experimental result regarded as an experimental discovery which, at the specific stage of the development of science, does not have a theoretical explanation, is known as an *empirical fact*. If later developments in the theory of science enable the scientist to predict the pattern of events previously known as empirical fact, then the status of the pattern changes from empirical truth to scientific law or relationship. For example, historically, the result that a randomly diffusing object on average travels a distance given by $d_a = \lambda\sqrt{N}$ could have been discovered experimentally before it was proven theoretically. It would have started off as an empirical relationship before it became a scientific relationship or law predicting diffusion. The history and interrelationship of the words experiment and empirical are illustrated in *Figure 1.23.*

When discussing the question "how long is a coastline?", Mandelbrot was able to show that the slope of the line is very characteristic of the ruggedness of the boundary being explored. Using modern terminology, Mandelbrot pointed out that implicitly everybody had anticipated that there was an answer to the question "how long is the coastline?", with the added assumption that any uncertainty in the

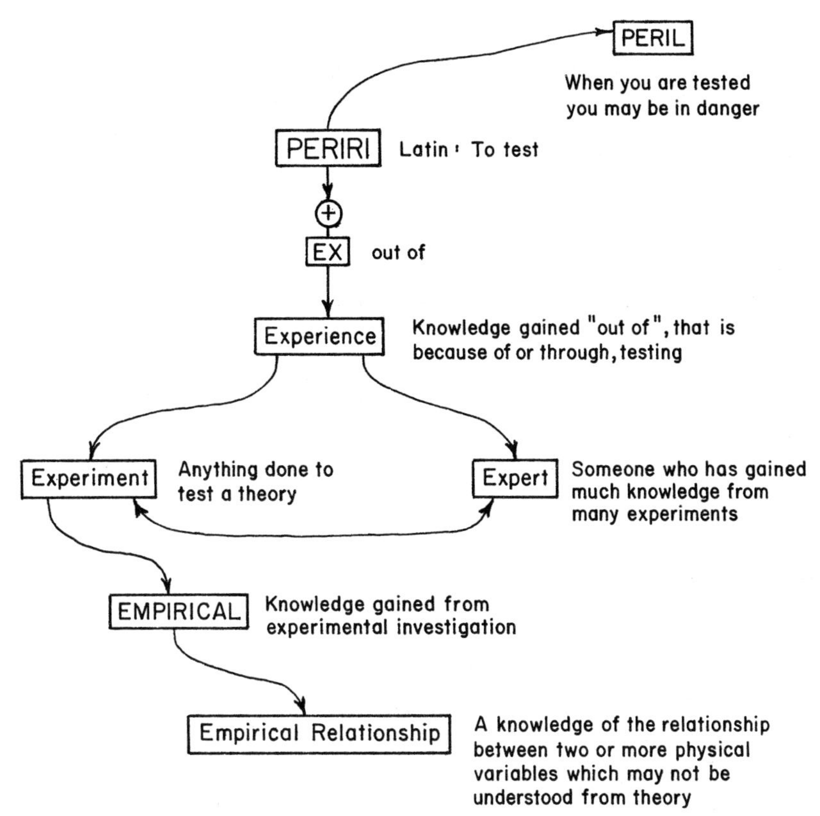

Figure 1.23. Empirical facts, arrived at by experiment, are the basis of experience.

coastline estimate was experimental error. What Richardson had demonstrated experimentally was that the magnitude of a coastline depends on how we measure it. Mandelbrot generalized this experimental discovery to the statement that one can never answer the question "how long is a coastline?" because all coastlines are in fact infinite in extent if one has the patience to take into account the contour of every grain of sand. All that one can do in a real situation is say that "if I estimate the coastline in a particular manner, then the answer I derive from my measurements is an operational definition of coastline magnitude".

Mandelbrot then went on to point out that what is really important in a study of the magnitude of a coastline is the rate at which our estimate of the magnitude of the coastline increases as we scrutinize the boundary with higher and higher resolution. When looking at the data of Figure 1.22, the slope of the data line, which measures the rate of increase in perimeter as we use smaller and smaller steps to explore the coastline, is an important parameter for describing the structure of the coastline.

It is rather interesting to compare the work of Planck on radiation theory and Richardson on coastlines. Both scientists thought they had discovered a single empirical truth and did not fully appreciate, initially, the importance of their own work. Planck was fortunate enough to live to see his work as the foundation stone of a new branch of physics – quantum physics. Lewis Fry Richardson, unfortunately, did not live to see his work blossom into fractal geometry. However, we remember the pioneering work of Richardson by the fact that a graph of the perimeter estimates against the resolution of inspection of a measurement presented in normalized fashion on log–log graph paper is known as a *Richardson Plot*. For a rugged boundary such as that of Great Britain, it will be shown later in this book that the slope of the data line can be interpreted as a fractal dimension quantifying the ruggedness of the boundary by the relationship

$$\delta = 1 + |m|$$

Where m is the slope of the data line, δ is the fractal dimension and $|m|$ is a mathematical symbol which indicates that one uses the positive value of the quantity inside the pair of lines. The surprising pattern of data points that Richardson plotted was an important step in the evolution of the general theory and truths of the subject which we have come to know as fractal geometry.

Section 1.10. Escape From Time Space into Phase Space

The title of this section seems to hint at the exciting possibility that we are going to discuss how human beings may be able to escape from the confines of space and travel into a new dimension of reality called "phase space". I am sorry to have to disappoint the reader, however, because what we are about to discover in this section is not a new form of space travel, but a novel technique for presenting well known facts. Some commentators on western civilization suggest that our civilization is dominated by the concept of time. Certainly we have made great advances in science by recording the fluctuating behavior of objects as a time series. Consider, for example, the behavior of a simple pendulum oscillating back and forth. The behavior of such a pendulum is described in terms of the *frequency of oscillation* and the *amplitude of the oscillation*. The meaning of these terms are illustrated in *Figure 1.24*(a). Many students begin their exploration of experimental physics by measuring the frequency of a simple pendulum, and by studying the behavior of clocks. If a pendulum is set swinging in a vacuum, the variations in the position of the bob of the pendulum will be a perfect sine wave of the type shown in Figure 1.24(b) (i). However, if the pendulum is swinging in the air, friction between the pendulum and the air gradually reduces the oscillation of the pendulum until it has zero velocity and displacement

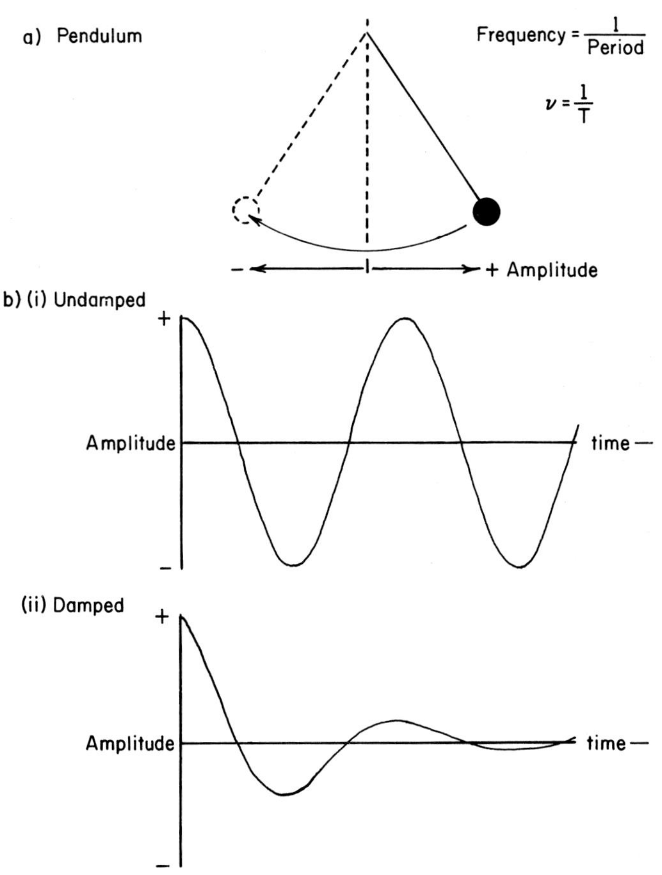

Figure 1.24. The swinging behavior of a simple pendulum is usually summarized as an amplitude–time graph. a) A simple pendulum and related terminology. b) Graphs of the position of the pendulum bob versus time for an undamped (i), and a damped (ii), pendulum.

and is said to be at rest. In standard textbooks used by physics students the behavior of a pendulum both in a vacuum and in air is recorded as a time series of the magnitude of the amplitude or velocity of swing. In Figure 1.24(b) (ii) the time–amplitude graph for a pendulum swinging in air is shown. This type of oscillation is known as damped vibration. A physics textbook describes a *damping of a vibration* or oscillation as a dying away of the amplitude of the oscillation as energy is withdrawn from the system.

To help understand what the mathematician means by *phase space*, we will summarize the behavior of the pendulum when swinging freely in a vacuum and when damped by air friction in a phase space diagram. The word phase is another word which has wandered into science from everyday speech, and which in science has a special meaning. A dictionary gives a definition of phase as: the appearance of the

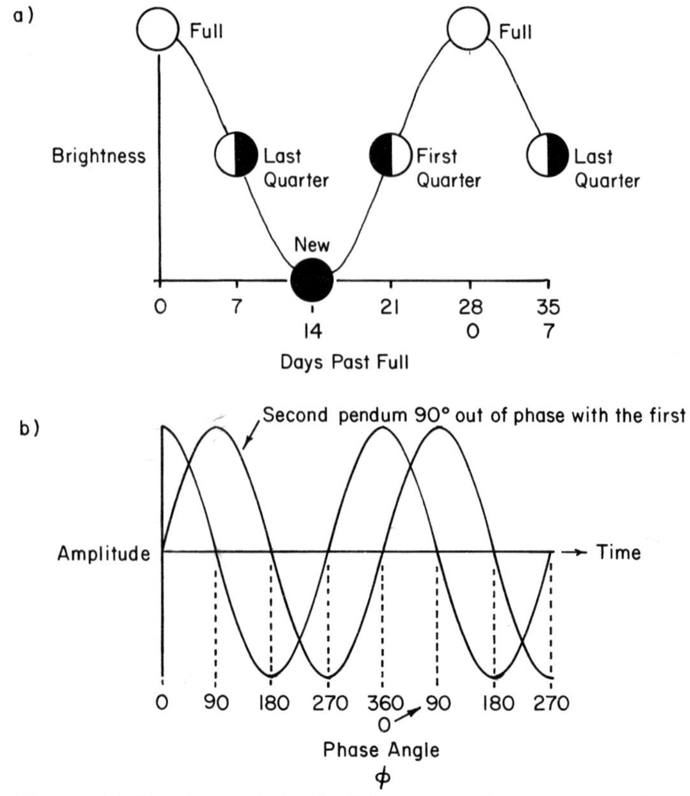

Figure 1.25. The word "phase" entered the English language from astronomy. It was first used to describe the amount of the moon that was shining in a lunar cycle. a) Phases of the moon. b) The phase angle, usually written ϕ, is a measure of the stage of oscillation at a specific time in relation to the starting point of the pendulums cycle.

moon or a planet at a given time according to the amount of illuminated surface exhibited; any transitory state or stage in a regularly recurring cycle of changes. It is derived from the Greek word phasis, meaning "to shine".

Thus the original meaning of the word developed amongst star gazers who were interested in the regular changes in the appearance of the moon. When scientists began to plot graphs they started to develop graphs of brightness of the moon at different stages. They developed graphs such as those shown in *Figure 1.25*(a). Such graphs would naturally have the ordinate labelled "phases of the moon", meaning the shining brightness of the moon. We have chosen to label our y-axis simply "Brightness". When scientists then went on to study the variations in the amplitude of a swinging pendulum they started to plot graphs of the type that we have already presented in Figure 1.24(b). The graphs of the amplitude of the pendulum as a time series looks very similar to the phases of the moon. To identify a particular stage of oscillation of the pendulum, it was natural to start using the vocabulary already developed in astronomy. So physicists started to describe a state of a pendulum at any particular instant in time as the *phase of the oscillation.*

When scientists began to write down the equation describing the curves such as those shown in Figure 1.24(b), it was found to include an angle in the mathematical formula which was related to the point in the cycle of the pendulum where it was set going. This angle became known as the *phase angle of the oscillation* or vibration. Again, in the language of mathematics, it turned out that one complete oscillation of the pendulum involved changes in the phase angle that vary from 0 to 360°; therefore the phase angle could also be marked on the time axis in the way shown in Figure 1.25. When two pendulums are set in motion at slightly different times, it is said that the two pendulums are swinging "out of phase". Pendulum 2 of Figure 1.25(b) has maximum velocity and zero displacement at time equals zero. Pendulum 1 has zero velocity and maximum displacement at time $T = 0$. For two pendulums swinging in the time pattern of Figure 1.25(b), the pendulums are said to be 90° out of phase with each other.

We have developed the terminology for describing the waves of Figure 1.25(b) by discussing the behavior of pendulums. Exactly the same vocabulary is used to describe wave motion in general. In *Figure 1.26* two plots of the behavior of the same pendulum are shown. The plot of amplitude changes with time is out of phase with the plot of velocity with time by 90°. To record the variation in behavior of the

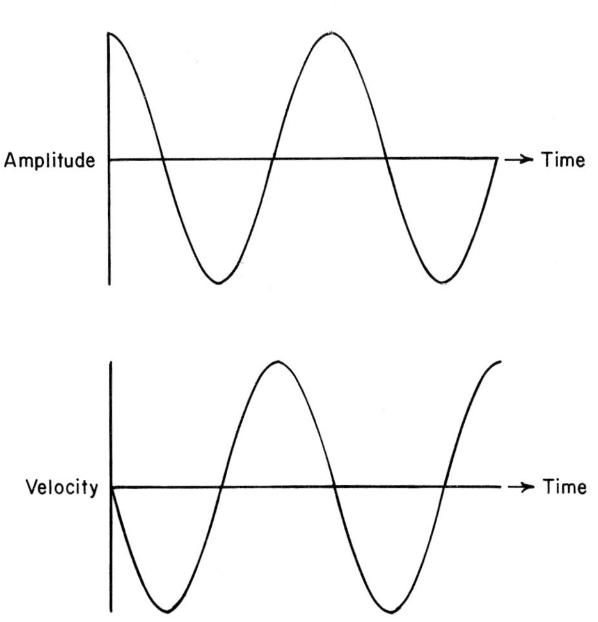

Figure 1.26. The time series representing the amplitude and velocity of a pendulum are out of phase with each other by 90°.

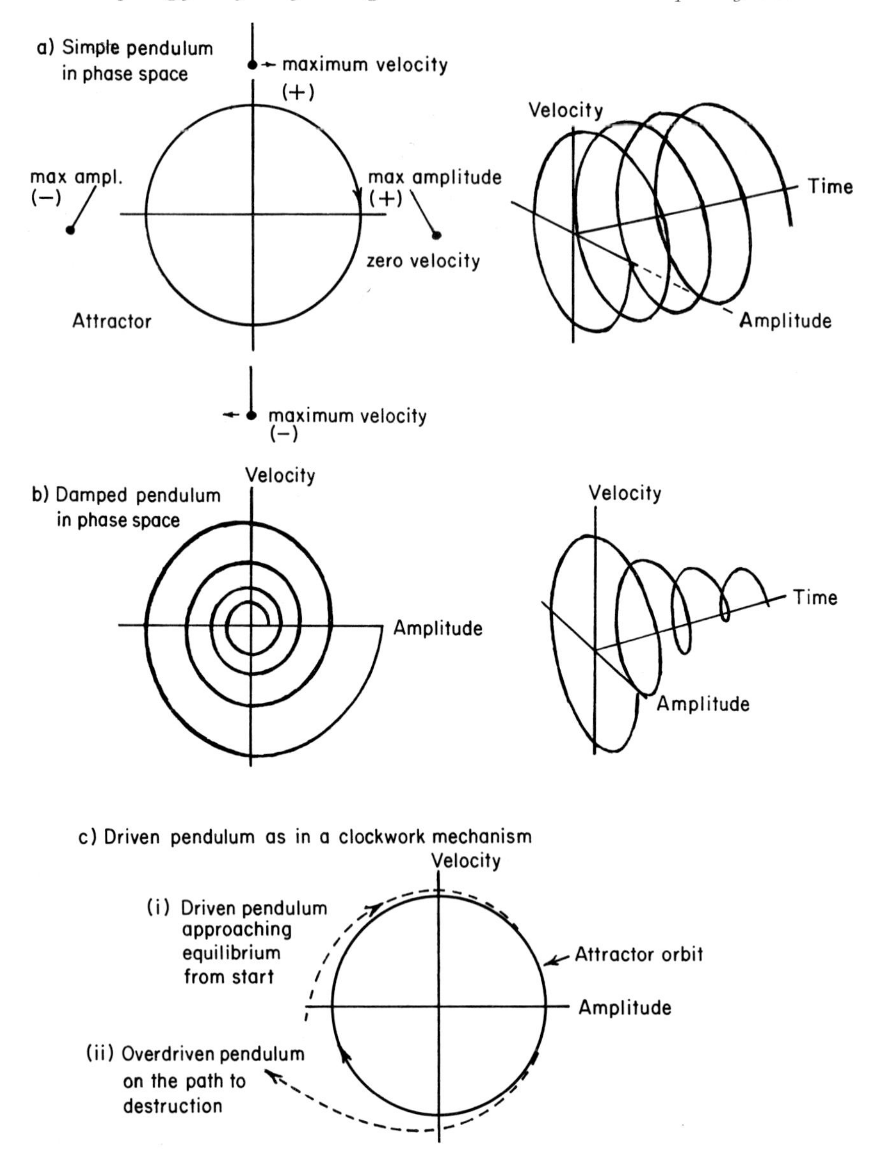

Figure 1.27. The behavior of a simple pendulum can be summarized in an attractive manner using phase space diagrams. a) Phase space diagram of an undamped pendulum. b) Phase space diagram for a damped pendulum. c) Phase space diagram for a driven pendulum.

pendulum during different stages of its swing, we can use amplitude and velocity to describe the motion instead of time. In *Figure 1.27*(a) the movement of the pendulum is recorded by plotting the velocity and the amplitude of the motion on two axes as illustrated. It is this type of graph which is known as a *phase space graph*. One does not know anything about the frequency of the oscillation, one only knows the pattern of repetition. That is, one knows the various phases of the motion but not the speed of change. As can be seen from the phase space graph of Figure 1.27(a), the track of an undamped pendulum in phase space is a circle. The radius of this circle is the amplitude of the oscillation. A pendulum oscillating with a larger amplitude of motion is a larger circle in phase space. Note that if we were given two circles in phase space which differed in amplitude, we would not know the frequency of either pendulum. To plot the information on the frequency of the pendulum as well as the amplitude, we would have to add a third axis for time. The total behavior of the pendulum in velocity, position and time space would be a spiral spring coiling out to infinity as shown in Figure 1.27(a). To create the simple phase space velocity position graph this spring is collapsed into a circle in the two-dimensional space of the phase diagram.

If we plot the oscillations of a damped pendulum in phase space, the resultant diagram is a spiral moving in from the circle representing the path of the undamped pendulum to the centre of the phase space diagram as shown in Figure 1.27(b).

Mathematicians started to develop phase space representation of oscillators long before the beginning of deterministic chaos studies. The student encountering this new way of displaying data will often ask the question "Why bother with a new way of presenting the vibration of the pendulum when everybody already understands the time series presentation?" The answer is that for some purposes phase space presentation is a more efficient summary of the data. For example, if one is interested in the overall decline of the amplitude of a damped oscillator, then the time series graph of this decaying amplitude could be a very long graph reaching out over many pages. The phase space record of the declining energy of the pendulum, however, is all captured in a convenient one-page graph. If we had used a time axis with our phase space diagram, our diagram would have been a spiral declining out in the time axis as illustrated in the sketch of Figure 1.27(b). It is useful to remember for future discussion that the visible pattern of oscillation in the phase space is a collapse of all possible oscillations in time onto the phase space map. In an article on the use of phase space to summarize the behavior of complex oscillators, Hale and LaSalle make the following comment.

> *What makes the phase space representation useful is that, without solving the equations that govern the motion, one can often generate the shape of the solution curves and thereby learn qualitative facts about the motion* [1].

Another type of pendulum that people are very familiar with is the one attached to a clock such as the familiar grandfather clock. To set such a clock in motion, one pulls the pendulum to one side and releases it. If the initial swing is large enough, the clock settles down to a steady amplitude swing even if the initial amplitude of the displacement used to start the clock is greater than the amplitude of the clock when ticking steadily. Even though the pendulum of such a clock is swinging in air, the regular push received from the weights driving the clock ensure that the pendulum swings with the same amplitude over a long period of time. For such an oscillator we can describe the motion of the pendulum in phase space by means of the line shown in Figure 1.27(c) (i).

Another type of behavior that could be observed in such a swinging pendulum, is one which would occur if a child unfamiliar with the mechanism of the clock were to keep pushing the pendulum until it swings with larger and larger amplitude, and finally hits the side of the wall of the container and disintegrates. The behavior of such a driven pendulum can also be shown quite easily in phase space as seen in Figure 1.27(c) (ii). When mathematicians started to study the patterns of behavior in phase space, they started to use a rather picturesque vocabulary which, although meaningful to them, has probably caused confusion for generations of students. When they looked at the declining spiral of the damped pendulum as illustrated in Figure 1.27(b) the mathematician started to imagine that the track of the swinging pendulum was being attracted to the central point of the spiral. They started to call the point of zero displacement an *attractor in phase space*.

To those uninitiated into the mysteries of advanced mathematics, the word attractor implies a force of attraction. It is essential to realize that when mathematicians talk about an attractor in phase space there is no physical attractive force involved. It is simply that the points of the graph seem to be drawn into a given point, which is described as an attractor. The apparent inability of the spiraling track to avoid this point does not involve any forces other than the attractive images of the mathematician. In the same way the circle for the stable oscillation of the driven pendulum of our grandfather clock is known to the phase space mathematicians as an *attractor orbit*. The idea is that it is a probable path to which the driven oscillator is "attracted" (again, no forces of attraction are involved).

Possible orbits of complex oscillation in a three-dimensionally structured graph using velocity, position and time would be an irregular spiral staggering out into time space which, when collapsed into phase space, gives us an attractor orbit with a finite thickness, although many possible orbits (often infinity) are confined within the limits of the width of the possible orbits in phase space. When mathematicians discovered such intricate infinitely variable pathways within a given region of phase space, they called such regions *strange attractors*! Now that mathematicians are more familiar with such complex orbits and odd points to which some systems tend, they are not really strange anymore, but we are stuck with the name. Generations of students will have to learn that strange attractors in phase space are not weird beings from outer space which are oozing with attractive allure for vulnerable human beings, but picturesque mathematical concepts.

Exercises

Exercise 1.1.

Select a region in linespace and split it into a set of pixels. Decide what is going to be your operational level for zero or infinity. Iterate the address of the pixels and then decide on a coloring scheme depending on how fast a pixel escapes to its goal of infinity or zero. If this is done as a class project for many students, each pair of students can take a set of pixels and the final rainbow can be assembled in the classroom.

Exercise 1.2. Student Experiment on Coin Flipping in a Freda-Fred Gambling Game

The students could play this game with actual coins or could simulate it on a simple computer. If the class did it with coin tossing, they could combine their data to produce a data summary of the type shown in Figure 1.13 (See Section 1.6). Because the coin flipping can go on for a long time before a zero state is reached, they could cut the experiment short by agreeing that once 100 flips had proceeded without a zero state they would call that "greater than 100" and begin the game again. This would have the advantage of limiting the graph plotting to 2-cycle logarithmic graph paper if the frequency of occurrence of a given number of coin flips between zero-sum states were to be expressed as a percentage of frequency. To help students play this game, a special simplified random number table containing only 1 and 2 is given in Figure 1.12.

Students should appreciate from this exercise that to state "on average nobody wins in such a game" is a sweeping statement that may require a very long series of events in a real time sequence to conform. This exercise will also give them experience in plotting data on log–log graph paper in a cumulative distribution form.

Exercise 1.3. Plotting the Growth of the Perimeter of a Koch Island on Log–Log Graph Paper

There are many types of *Koch islands* named after the mathematician Helge *von Koch*. These interesting mathematical curves have finite area and infinite perimeter when the construction algorithm used to create the curves are carried out an infinite number of times.

Mathematicians use the term *algorithm* to describe their mathematical recipes for carrying out calculations or building mathematical models. A mathematical dic-

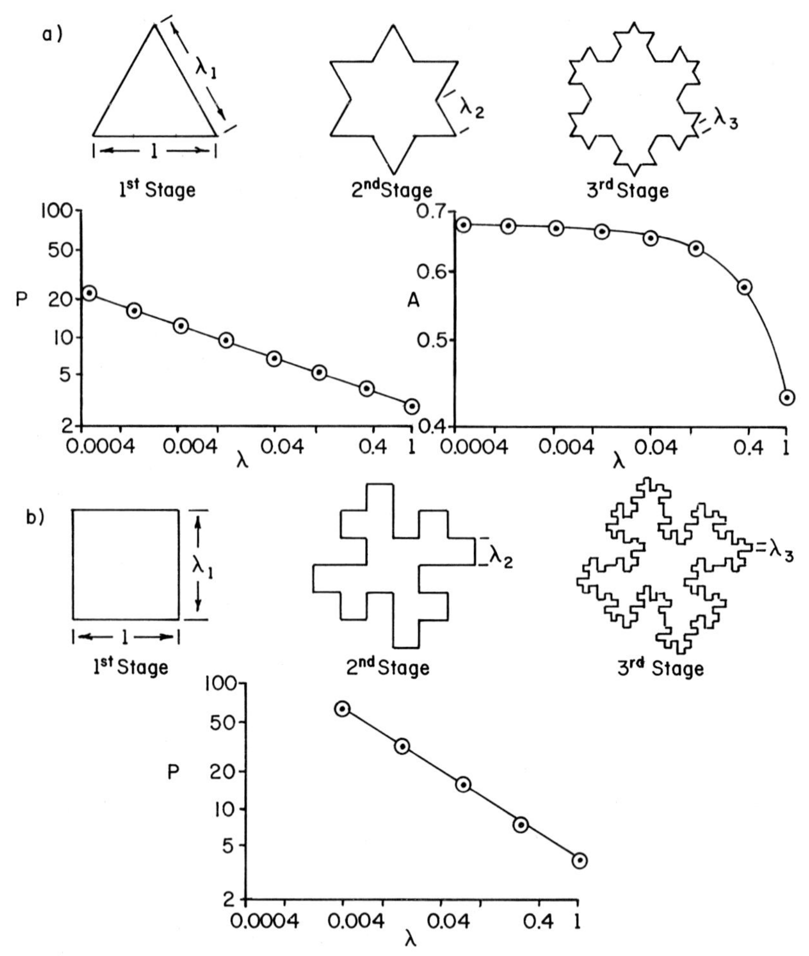

Figure E 1.3.1. Exploring the structure of Koch Islands. P = Normalized perimeter with respect to length of first stage. A = Normalized area with respect to length of first stage. λ = Length of new construction element as indicated in the diagram. a) The triadic island. b) The quadric island.

tionary defines an algorithm as "in modern mathematics, any method or procedure for computation".

The term is derived from a corruption of the name of a famous Arab mathematician who solved many problems in mathematics and invented algebra. His name was *Al-Khwarizmi*. He lived in Baghdad in the 9th century. Many people confuse algorithm with logarithms because of the similarity of the two words. The two concepts are totally unrelated.

The first three steps in the construction algorithm of what are known as the Koch triadic and Koch quadric islands are shown in *Figure E 1.3.1*. For the triadic island, each iteration consists of dividing each side of length λ into three sections, and placing a triangle with sides of length $\lambda/3$ upon the middle section. The quadric island is constructed in a similar manner.

It is useful to calculate the perimeter at each stage of the construction algorithm and to plot a graph of the perimeter of the island against the length of the component side of the islands. The data are summarized in the figure, and it can be seen that they generate a straight line. The teacher can use this exercise both to introduce the concept of perimeters trending to infinity when the construction algorithm is conducted n times, and to show that each time the construction is carried out the perimeter in the case of the Koch triadic Island increases by a factor of 4 over 3. At this stage the student should note that the slope of the data lines of the triadic island is given by $\log(4)/\log(3)$ and for the quadric island $\log(8)/\log(4)$, and that these different slopes characterize the main difference between the two islands. In the case of the quadric island, the perimeter doubles while the area stays the same. The teacher should point out that although both the triadic and quadric islands have perimeters that go to infinity, that of the quadric island goes to infinity faster than that of the triadic island.

Koch island curves were very important in the development of fractal geometry. They are obviously models for the physical processes of crystallization and the structure of items such as snowflakes. When the mathematician discovers a curve which can be used to model a physical system, he describes it as a paradigm. This term comes from a Greek word paradeigma, which means "to exhibit side by side". A mathematical model can be put alongside a physical system, and the resemblance of the two is so strong that one can work with the mathematical model just as well as with the physical system. The mathematician would state that the construction of the Koch island is a paradigm for the progress of crystallization in chemistry. Note that the Russians refer to a Koch island as a Van der Waerden curve. This difference in terminology is typical of the situation that arose when western and Russian scientists interpreted the development of science from their different perspectives [30].

Exercise 1.4. Iterate the Operation "Take the Square Root of ..."

Before the availability of pocket calculators, a calculation of the square root of a number used to be a difficult and time consuming routine. It is not surprising, therefore, that students of a generation ago were unacquainted with the surprising consequence of the repetition of the operation "take the square root of a number". The following experiment can be carried out with any small calculator. Let us start with two numbers, 8 and 0.8. If we iterate the operation "take the square root of a number" on these two numbers, the results are as given in *Figure E 1.4.1* (a).

It can be seen that for both the whole number and fractional number, the iteration of the operation "take the square root" results in a convergence of the calculated

a)

n Iteration	$x_{n+1} = \sqrt{x_n}$ $x_1 = 8.00$	$x_{n+1} = \sqrt{x_n}$ $x_1 = 0.80$
1	8.0000000	0.8000000
2	2.8284271	0.8944271
3	1.6817928	0.9457416
4	1.2968395	0.9724924
5	1.1387886	0.9861503
6	1.0671404	0.9930510
7	1.0330248	0.9965194
8	1.0163783	0.9982582
9	1.0081558	0.9991287
10	1.0040696	0.9995642
11	1.0020327	0.9997821
12	1.0010158	0.9998910

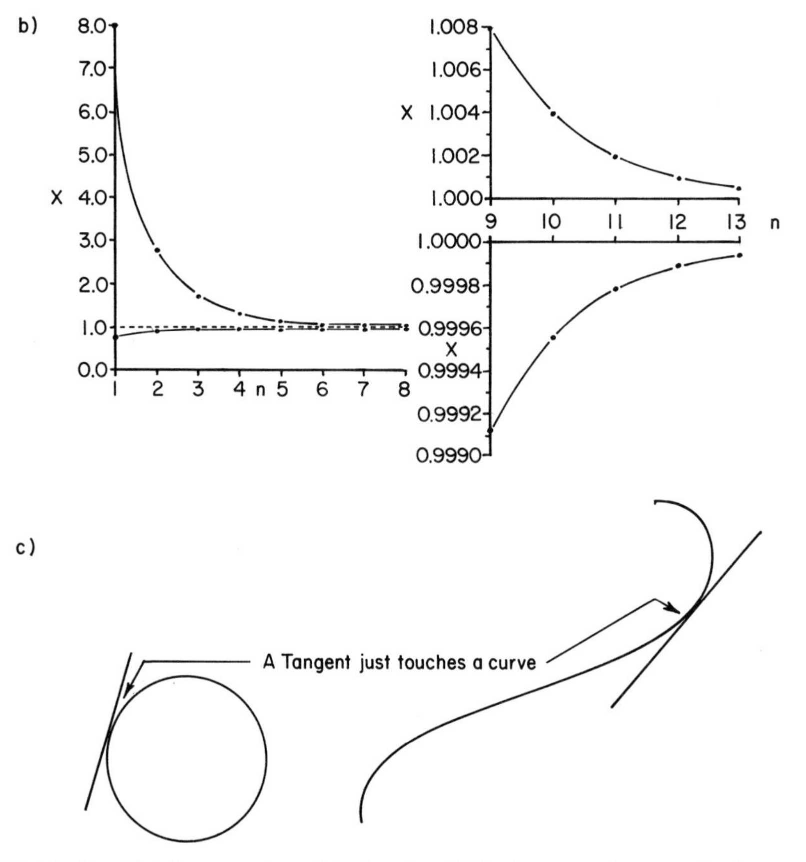

Figure E 1.4.1. Graphical demonstration of the iteration "Take the square the root of ..." converging onto the number one. a) Table showing the progress of the iteration of the operation on 8.0 and 0.8. b) Graphical representation of the calculations of (a) showing them converging to a value of 1. c) The appearance of a tangent to a curve.

value to the number 1. The convergence we achieve depends on two things: 1. Our patience in pressing the button marked square root. 2. The actual calculator we use, since different calculators use slightly different logic to achieve their calculations. The exact number of decimal places and the number of iterations required to reach the final result will vary slightly from calculator to calculator. The convergence of the values for 8 and 0.8 are illustrated graphically in Figure E 1.4.1 (b). It can be seen that after approximately 8 iterations both series have reached the value 1 within the accuracy with which we can plot the data on the first graph shown. If we use a much larger calculator, we will find again that the iteration never exactly reaches 1, with the difference between the iterated value and 1 disappearing into the distant decimal places.

Curves which come down to and almost reach a theoretical value are said by the mathematicians to be *asymptotic* to that value. This word comes from three Greek root words: a, meaning "not"; syn, "together"; and pipto, "to fall". An *asymptote* does not fall onto another line, but cruises alongside and towards a theoretical line but never reaches it. Another way of regarding an asymptote is that it is a line which is tangent to a theoretical value curve at infinity. The word *tangent* comes from a Latin word tangere, meaning "to touch". In everyday life *tangible* things can be touched and grasped. *Intangible* ideas cannot be taken hold of physically, but can only be imagined. In Euclidean geometry a tangent to a circle or a curve touches that curve at one point, as illustrated in Figure E 1.4.1 (c).

Note that the word *perimeter* comes from two Greek root words peri, meaning "around" and metron, "a measure". It is defined as the total length of the boundary of a figure. The term perimeter is used to describe general geometric figures, whereas the term *circumference* is used to describe the perimeter of a circle. This latter word comes from the Latin word circulus, meaning "a small ring" and ferre, meaning "to carry". A circumference carries the boundary line of a circle. The word peri has given us several words in the English language such as *periscope*, which is a device to enable us to look all around us. It is used especially by sailors in submarines when they are close to the surface of the sea.

The students can carry out the iteration of the square root operation with many different starting numbers, and they can study the way in which the calculated values converge onto the asymptote of 1. They can also be encouraged to use higher resolution arithmetic scales to plot more and more decimal places in their calculations. In Figure E 1.4.1 (b) we show an increased resolution plot for the last 5 values of the calculated data given above. In a large class students with different calculators can be invited to study the same convergence to detect slight differences between different calculators. At this stage of development, the fact that the iterated square root of a number converges to 1 can be regarded as an empirical discovery made possible by the speed with which the operation can be carried out with a hand-held calculator. In a class discussion the teacher could discuss the difficulty of carrying out such a calculation in the days before the availability of the hand-held electronic

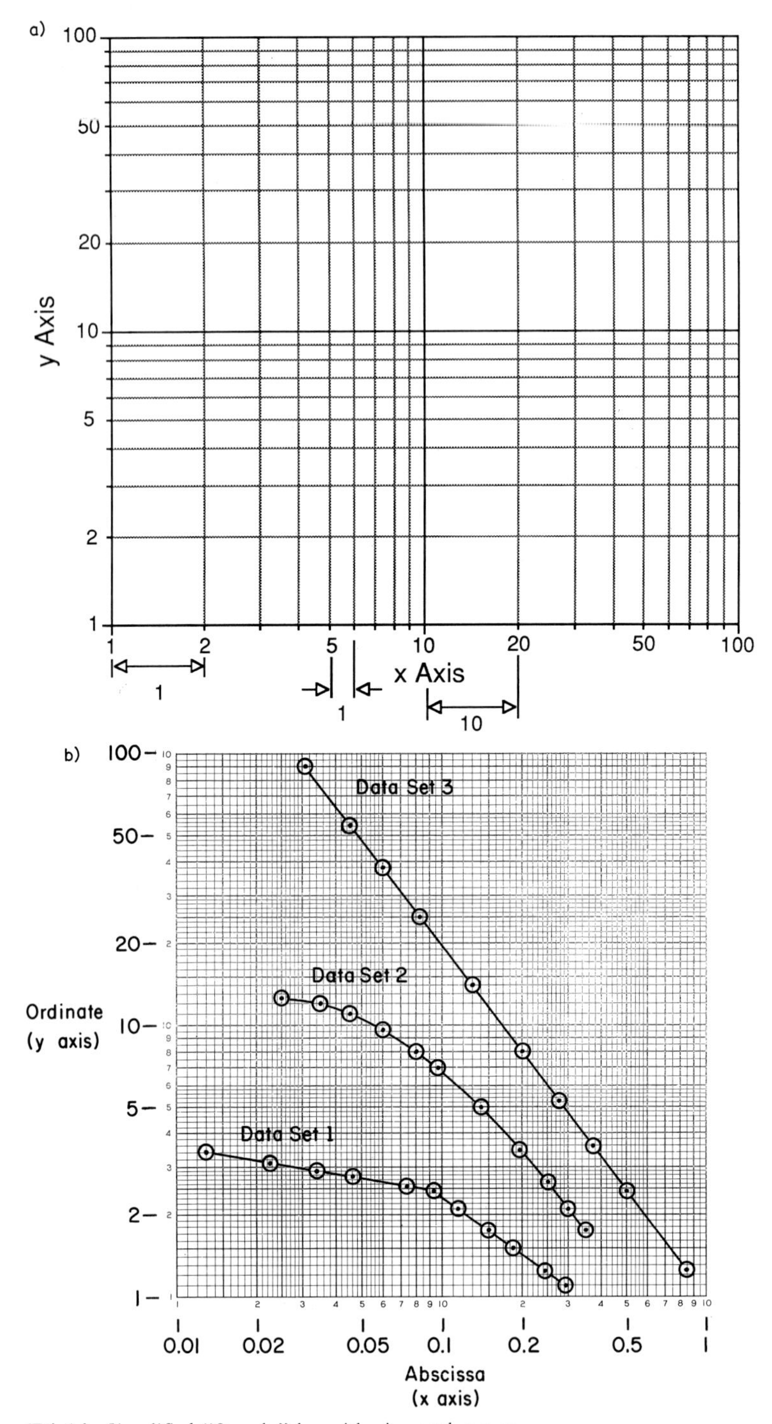

Figure E 1.5.1. Simplified "2-cycle" logarithmic graph paper.

Table E 1.5.1. Sample data for three lines to be plotted on two cycle logarithmic graph paper.

Data Set 1		Data Set 2		Data Set 3	
0.295	1.10	0.350	1.75	0.840	1.25
0.245	1.25	0.300	2.10	0.500	2.45
0.185	1.50	0.250	2.65	0.370	3.60
0.150	1.75	0.195	3.50	0.275	5.30
0.115	2.10	0.140	5.00	0.200	8.00
0.0920	2.45	0.0960	7.00	0.130	14.0
0.0740	2.55	0.0800	8.00	0.0820	25.0
0.0460	2.75	0.0600	9.60	0.0600	38.0
0.0340	2.90	0.0450	11.0	0.0450	55.0
0.0225	3.10	0.0350	12.0	0.0310	90.0
0.0130	3.40	0.0250	12.5		

calculator, and how modern computers are being used to discover mathematical patterns and relationships empirically.

Exercise 1.5. Plotting Points on Graph Paper with Logarithmic Scales

Students often find their first encounter with logarithmic scales difficult, because of the way in which the magnitude of a marked interval changes as one moves up the scale. On the 2-cycle logarithmic paper shown in *Figure E 1.5.1* the space between 1 and 2 on the first logarithmic scale is larger than the space between 9 and 10. When one moves above 10 into the second log cycle, units which represented 1 between 1 and 2 now represent 10 between 10 and 20, and so on. To help the students familiarize themselves with plotting data on logarithmic graph paper, the sets of data of Table E 1.5.1 should be plotted on 2-cycle logarithic graph paper.

References

[1] J. K. Hale and J. P. LaSalle, "Analyzing Non-Linearity", Chapter 6 in *Modern Science and Technology*, edited by R. Colborne, Van Nostrand, Princeton N. J., 1965, pp. 635–641.

[2] Biographical details for the scientists mentioned in this book are taken from the *Biographical Encyclopedia of Science and Technology*, by I. Asimov, Doubleday, Garden City, New York, 1972.

[3] L. Barnett, *The Universe and Doctor Einstein*, Harper and Row, New York, 1948.

[4] B. H. Kaye, *Fractalicious Structures and Probable Events – Determinism Lost and Regained*, in the Proceedings of the Conference of Mathematics Educators of Canada, Simon Fraser University, May 1990, Vancouver, edited by Martin Quigly, Newfoundland University, St. Johns, Newfoundland.

[5] J. Gleick, *Chaos, Making a New Science*, Viking Penguin, New York, 1987.

[6] F. C. Moon, *Chaotic Vibrations. An Introduction for Applied Scientists and Engineers*, 2nd ed., Wiley, New York, 1992.
[7] H. G. Schuster, *Deterministic Chaos, an Introduction*, 2nd edition, VCH, Weinheim, Germany, 1989.
[8] M. Schroeder, *Fractals, Chaos and Power Laws, Minutes from an Infinite Paradise*, W. H. Freeman, New York, 1991.
[9] In 1983 B. B. Mandelbrot published an updated and expanded version of *Fractals: Form, Chance and Dimension* under the title *The Fractal Geometry Of Nature* W. H. Freeman, San Francisco, 1983. This is considered by Mandelbrot to be the definitive book on the subject (personal communication).
[10] B. H. Kaye, *A Random Walk Through Fractal Dimensions*, VCH, Weinheim, Germany, 1989.
[11] For an introductory overview of Diffusion Limited Aggregation see B. H. Kaye, *A Random Walk Through Fractal Dimensions*, VCH, Weinheim, Germany, 1989.
[12] H. O. Peitgen and P. H. Richter, *The Beauty of Fractals – Images of Complex Dynamical Systems*, Springer, Heidelberg, 1986.
[13] Postcards showing fractal systems in color are obtainable from Art Matrix, P.O. Box 880, Ithaca, New York, NY 14851, USA.
[14] D. Carey,"Fractals Expand the Frontiers of Computer Graphics", *Canadian Data Systems*, March 1989, pp. 32–36.
[15] Media Magic, P.O. Box 507, Nicasio, California 94946.
[16] J. Briggs, F. D. Peat, *Turbulent Mirror*, Harper and Row, New York, 1989.
[17] I. Peterson, *The Mathematical Tourist*, W. H. Freeman, New York, 1988.
[18] B. H. Kaye, *A Random Walk Through Fractal Dimensions*, VCH, Weinheim, Germany, 1989.
[19] K. Kasahara, *Origami Made Easy*, Japan Publications, Tokyo, 1973.
[20] M. Mendes-France,"Les Courbes Chaotiques", *Images de la Physique*, suppl. 51, 1983, 5–8.
[21] Origamic fractals are described in K. Dekking, M. Mendes-France, A. Van der Boorten, "Folds 1", *The Mathematical Intelligencer* , Vol. 4, no. 3, 1983, pp. 130–138; "Folds 2, Symmetry Disturbed", *The Mathematical Intelligencer*, Vol. 4, no. 4, 1983, pp. 173–181; "Folds 3, More Morphisms", *The Mathematical Intelligencer*, Vol. 4, no. 4, 1983, pp. 190–195; Springer, New York.
[22] See discussion of bell curve modification of student grades given in [18].
[23] See article on G. Hardy in E. T. Bell, *Men Of Mathematics*, Vol. 1, Penguin Books, Harmondsworth, England, 1965.
[24] I. Stewart, *Concepts of Modern Mathematics*, Penguin Books, Harmondsworth, England, 1975.
[25] B. Evans, *Dictionary of Mythology*, paperback edition, Dell, New York, 1970.
[26] W. Brown, "New-Wave Mathematics", *New Scientist*, August 3, 1991.
[27] P. Beckmann, *A History of* π, Golem Press, Boulder, Colorado, 1970.
[28] D. G. Fink, *Computers and the Human Mind*, Anchor Book Co., New York, 1966.
[29] See "The Chudnovsky Algorithm For Pi", *Science Digest*, January, 1990, pp. 91–92.
[30] N. Ya. Vilenken, *Stories About Sets*, Academic Press, New York, 1965.
[31] For a pictorial tour of the real universe, from the smallest atom to the outer limits of the universe, see *Powers of Ten, About the Relative size of Things in the Universe*, P. Morrison, Phyllis Morrison and the office of Charles and Ray Eames, Scientific American Books, W. H. Freeman, San Francisco, 1982.
[32] M. J. Moroney, *Facts from Figures*, 2nd edition, Pelican Books, Harmondsworth, England, 1953.
[33] W. Feller, *An Introduction to Probability Theory and its Applications*, Vol. 1, Chap. 3, Wiley, New York, 1950.
[34] G. Gamow, *One, Two, Three ... Infinity*, Viking Press, New York, 1947.
[35] D. D. McCracken, "The Monte Carlo Method", *Scientific American*, May, 1955, pp. 162–165, reprinted in *Computers and Computations*, W. H. Freeman, San Francisco 1971.
[36] Mandelbrot (see [9]) outlines in his books the career and life history of Louis Fry Richardson.
[37] *Modern Science and Technology*, edited by R. Colborne, D. Van Nostrand, New York, 1965, pp. 633–641.

Chapter 2

Mandelbrot Sets
Julia Lace
and
Fatou Dusts

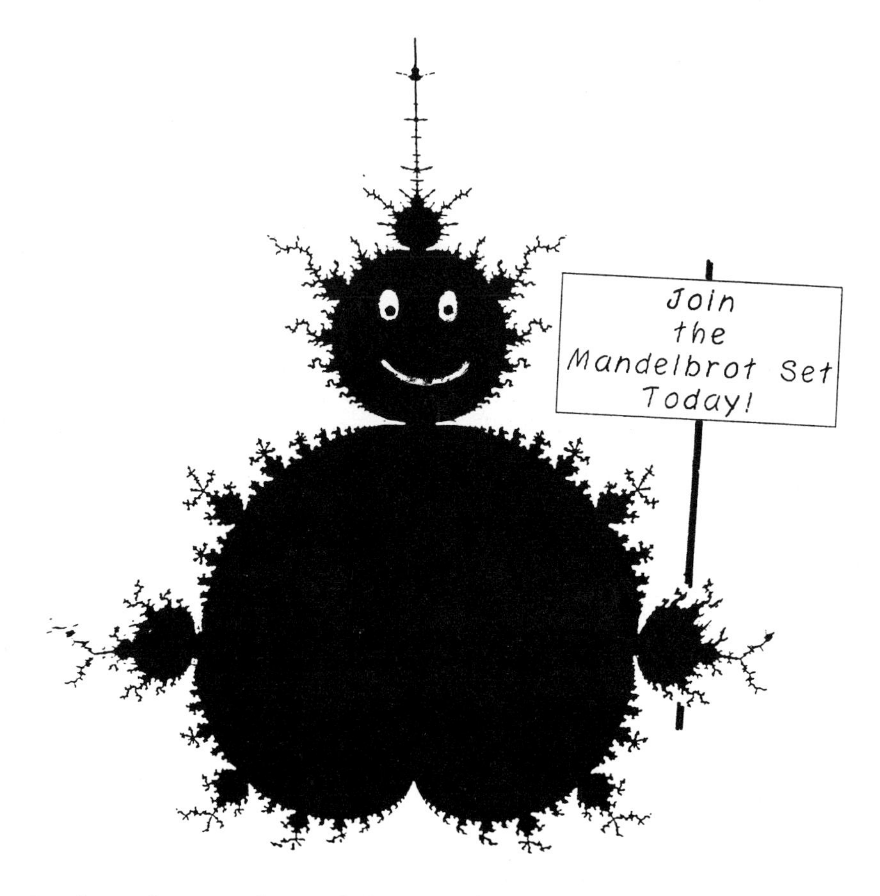

Explore the wonders of the Mandelbrot set and experience fractalicious joy!

(It's out of this world!)

Chapter 2 **Mandelbrot Sets, Julia Lace and Fatou Dusts** 79

Section 2.1 A Three-fold Path to Geometric Understanding 81
Section 2.2 The Mandelbrot Set – A Prison for Restless Pixels 85
Section 2.3 Julia Lace and Fatou Dusts 103
References 105

Chapter 2

Mandelbrot Sets, Julia Lace and Fatou Dusts

Section 2.1. A Three-Fold Path to Geometric Understanding

The title of this section refers to the fact that in the course of my professional life I have had to learn to describe geometric systems three different ways. My first encounter with geometry in high school was with traditional Euclidean geometry. It seemed at that time to be a dry subject full of obtuse triangles and parallel lines which never met each other, even at infinity. In my early geometry classes I learned how to prove ad nauseum many theorems, which, after they had been given a satisfactory proof, were signed off with the letters Q.E.D. Our teachers assured us that this was a Latin phrase, "quod erat demonstratum", usually shortened to Q.E.D., which means "that which was to be proved". However, one of my fellow students, who somehow managed to avoid Latin studies en route to his degree, used to insist that the letters Q.E.D. stood for "quite easily done" (he also thought that RSVP on a wedding invitation meant "rapidly send valuable present").

When I went to university, I had to relearn geometry again in a format where all geometric shapes had to be expressed in equations linking points in Cartesian coordinate space (when taught this way the subject is sometimes known as *coordinate geometry* or *analytical geometry*). I discovered that straight lines in Cartesian coordinate space were specified by equations such as

$$y = mx + b$$

I learned to interpret the various parts of such an equation so that I knew that m was the slope of the line, and that the intercepts of the lines on the x and y axes were given by the relationships indicated in *Figure 2.1*(a). I also discovered that the equation of a circle was

$$R^2 = x^2 + y^2$$

and that the circle could be regarded as the path of the vector R rotating with uniform angular velocity as shown in Figure 2.1(b).

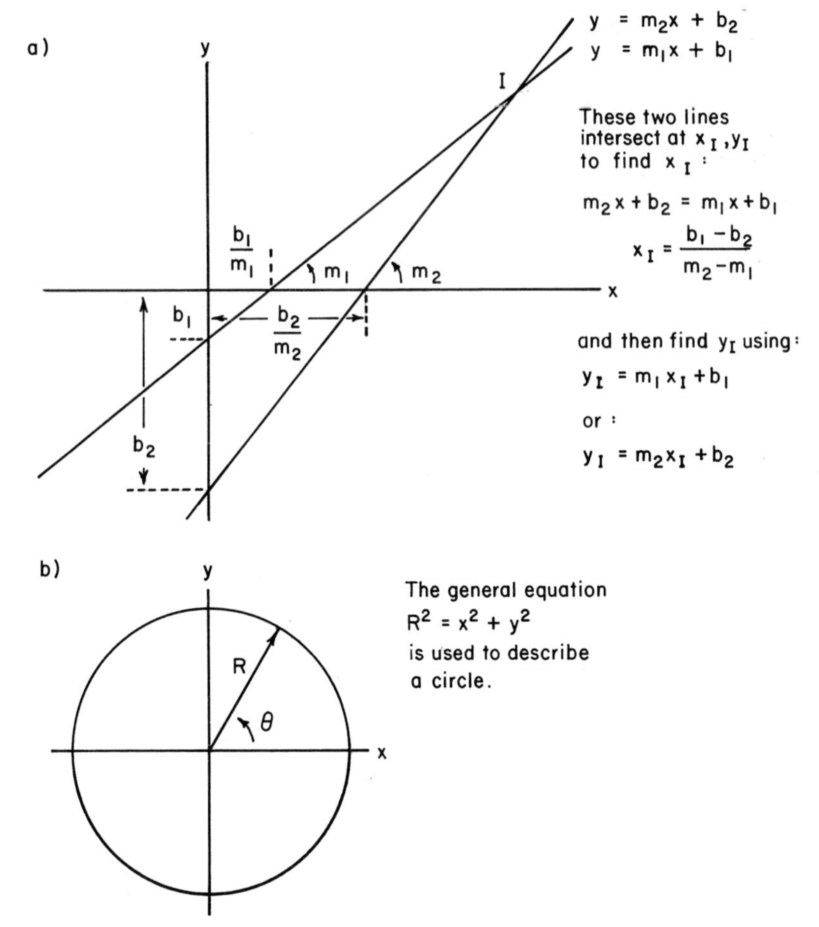

Figure 2.1. In analytical geometry, circles and lines become equations linking points in space. a) Equations for two lines in Cartesian coordinates. b) Equation for a circle.

After struggling through a university course on analytical geometry I thought that I was finished with learning the different forms of geometry that I needed to know. Then I read Mandelbrot's book. In it I discovered that mathematicians had yet another way of describing familiar circles and lines. In this third way of describing geometric shapes they used set theory. Set theory became very popular in the 1970s as high schools fell in love with what was called "new mathematics". Unfortunately, as I attempted to come to grips with the formulae in Mandelbrot's book on fractal geometry, I discovered that I knew nothing about set theory. In desperation I turned to an introductory book on the concepts of modern mathematics by Ian Stewart [1]. The chapter in that book on the "language of sets" proved to be a survival kit which enabled me to find my way through the jungle of formulae in Mandelbrot's book.

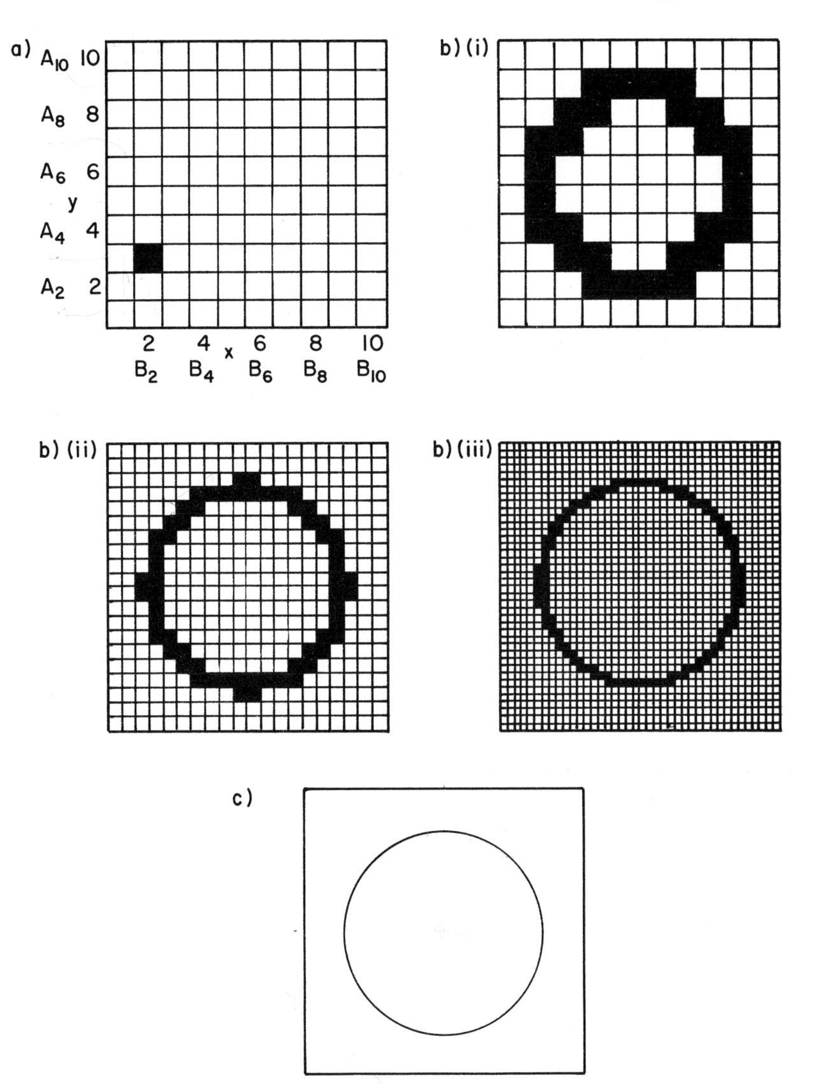

Figure 2.2. In geometry described from a set theory perspective, geometric shapes become subsets of pixels scattered in quantized space. a) Cartesian space defined as sets of rows and columns. b) The appearance of a circle in quantized space as the tile, or pixel, becomes smaller with respect to the circle. c) The ideal circle of Cartesian space and Euclidean geometry consists of an infinite number of tiny pixels which are a subset of the infinite pixels forming the two-dimensional space.

The chapter in Stewart's book ends with the statement

All of Euclidean geometry can be thought of as the study of the subsets of R^2

To understand what Stewart meant by this statement, let us look at the digitized circle drawn on the pixel mosaic of *Figure 2.2*. We can describe this mosaic as having rows of pixels A_1, A_2, etc., where the pixels in row 1 belong to the set $A_1 = (x, y)$ where $y = 1$ and x goes from 1 to 10.

Row $A_2 = (x, y)$ where $y = 2$ and x goes from 1 to 10 etc. We can also regard the mosaic as made up of columns where B_1 is the set of pixels with the addresses (x, y) where $x = 1$ and y goes from 1 to 10. We can now define a given pixel by regarding it as an intersection of the appropriate row and column pixel sets. The point shown in black at the bottom left hand corner is defined by the intersection of the set A_3 and B_2. Mathematicians write the fact that two sets intersect with a symbol, so that they would write that a pixel in the mosaic of Figure 2.2(a) is given by the intersection of two sets in the following way:

$$A_3 \cap B_2$$

We can also define two larger sets A and B which are "all pixels in rows" and "all pixels in columns". The mathematician then says that A_1 through A_{10} are subsets wholly contained within the set A. B_1, and B_2 are subsets of the set B and are wholly contained within that set.

To define the plane covered by all the pixels in A and B of our mosaic, mathematicians define what they call the *Cartesian product of two sets*. This name comes from the fact that this product defines Cartesian space. The mathematicians would write the mosaic of Figure 2.2 with the formula

$$A \times B$$

All mathematicians have agreed that *in set theory* this *multiplication sign* means a *very special process*. Students coming into a subject such as set theory bring with them preconceived ideas of what a symbol such as the multiplication sign means. Many difficulties occur in a new subject because they try to interpret the formula using the concepts of ordinary mathematics, when they need to realize that mathematicians have taken a familiar symbol and given it a new name in a new subject. The problem is that there are only so many conveniently written symbols to go around. Therefore creators of new subjects have to invest new meaning in old symbols. Whilst this turns out to be convenient for the mathematician, it creates havoc in the minds of students who inadvertently wander into a new field unaware of the tricks that the mathematicians have played with their old familiar symbols.

The Cartesian product is defined as the set of all possible *ordered pairs* formed by the act of selecting pairs from two different sets in a special way. If one carried out the operation to create the Cartesian product of the set A_{10} with the set B_1 through B_{10}, it would in fact define the row of pixels with the addresses (A_{10}, B_1), (A_{10}, B_2), (A_{10}, B_3) etc. In the same way, if you said "Create the Cartesian product of all A and B pixels", then the Cartesian product would in fact constitute the pixel addresses of every tile in the mosaic. Dr. Stewart tells us that if we let R stand for all real numbers, then the two-dimensional space is defined by the Cartesian product of $R \times R$. He also tells us that this is usually written in the simpler form R^2. To describe an object such as a circle in geometry from a set theory perspective, we indicate that the set forming the geometric figure is a subset of R^2. In our digitized circle of Figure 2.2(b) (i) the circle is defined by the collection of pixels drawn in black, the addresses of which are the intersections of the A and B sets of appropriate subscript. For the 20 × 20 two-dimensional space of Figure 2.2(b) (ii) our circle turns out to be a subset containing 60 members. Using such relatively crude quantum space, our circle does not look particularly circular. In fact in quantum space a true perfect circle of the type postulated in Euclidean geometry cannot exist. It is always slightly "saw toothed". However, using a smaller pixel the difference between the Euclidean circle with vanishingly small points and the approximate circle with a zigzagy border becomes vanishingly small. This is illustrated by the series of sketches in Figure 2.2. We see that as the constituent pixel becomes smaller, the number of pixels in the subset forming the circle increases towards infinity. In fact Euclidean geometry can be defined as the appearance of quantum geometry shapes when the pixels defining the quantum space become infinitely small. In Euclidean geometry, the subset of pixels forming the circle contains an infinite number of pixels chosen from a larger infinite set, and as Stewart pointed out in his book all Euclidean geometric figures are subsets of R^2. In the real world we are always using real pixels, even though they are very tiny, to depict space and the ideal Euclidean figures are only imagined from trends observed with smaller and smaller pixels.

When we start talking about infinite subsets of infinite sets we are into the subject known as *Cantorian set theory*. An interesting paradox that arises in geometric set theory with vanishingly small pixels is that in fact the number of pixels constituting the boundary becomes a smaller and smaller subset of the space in which it is drawn. One reaches the conclusion that although the boundary defining the circle contains an infinite number of pixels, the chances of choosing a pixel at random from the two-dimensional space almost never results in the selection of a pixel from the boundary, because the subset of pixels defining the boundary is such a small fraction of all possible pixels [2].

Now that we have learned how to describe geometric shapes in two-dimensional space from the perspective of set theory, we are able to discover what is meant by the Mandelbrot set. We will discover that Fatou dusts are in fact a set of trapped pixels sprinkled in space to form delicate patterns [3, 4].

Section 2.2. The Mandelbrot Set – A Prison for Restless Pixels

Part of our title for this section is taken from an article in *Scientific American* by Dewdney [5]. It draws attention to the fact that the core of the Mandelbrot set is a simple black and white affair showing the locations of pixels which cannot escape

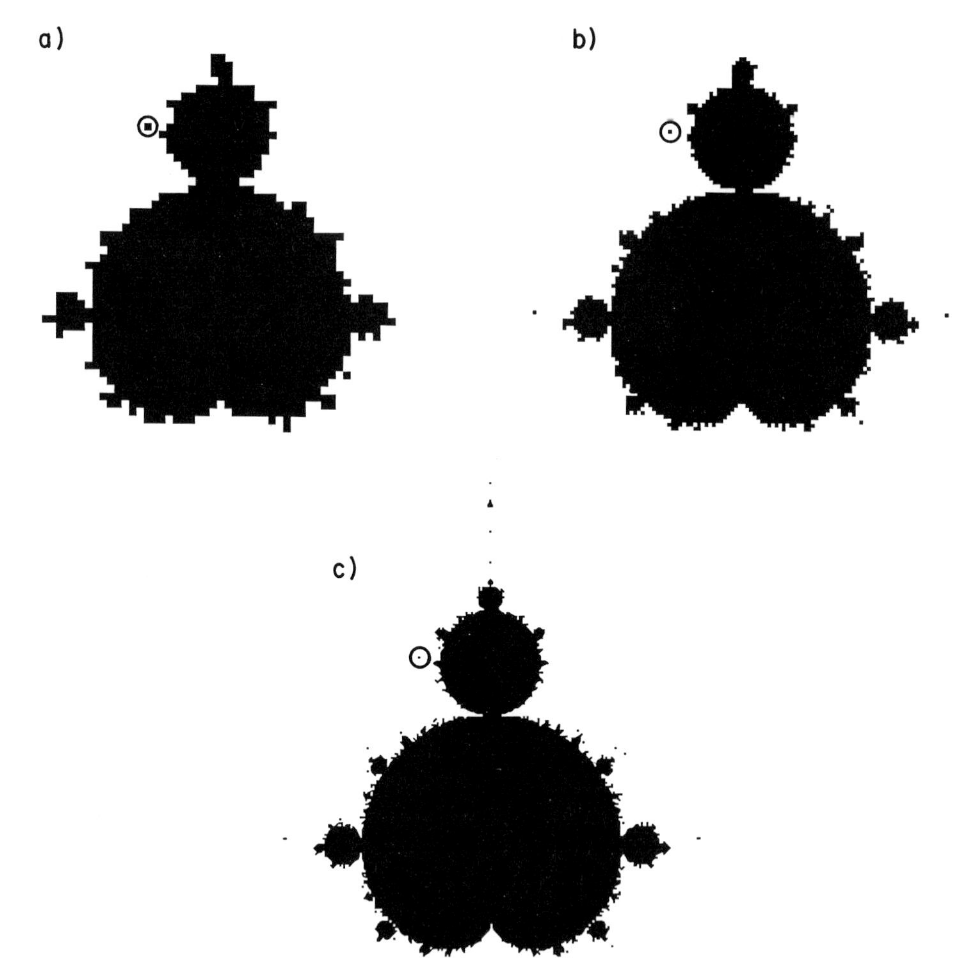

Figure 2.3. The visual boundary of the Mandelbrot set becomes increasingly complex as the pixel size used to construct it becomes smaller. The pixel size of each depiction of the Mandelbrot set can be seen from the size of the single pixel ringed in each figure. (When shown in this format the Mandelbrot set has been variously described as "The Spiderman", "The Pixel Alien From Outer Space" and "The Gingerbread Man"). a) Mandelbrot set at pixel size 4. b) Mandelbrot set at pixel size 2. c) Mandelbrot set at pixel size 1.

to outer space when they are subjected to a particular iteration process. This pattern of trapped pixels was discovered experimentally by Mandelbrot carrying out studies on large digital computers [6]. The structure of the Mandelbrot set in relatively crude quantized space, with the subsequent increase in complexity of the quantized boundary of the Mandelbrot set as the pixel size used to explore the structure of the set is reduced, is illustrated by the patterns shown in *Figure 2.3* (a, b, c). The glorious technicolor environment of the outer spaces of the Mandelbrot set, such as that shown in the color version of the Mandelbrot set shown on the cover of this book, are artistic creations of computer specialists who wanted to summarize graphically the speed at which pixels outside the Mandelbrot set accelerate into outer space by the mathematical operation carried out on their addresses.

To understand the operation to which the pixel addresses of the two-dimensional space were subjected by Mandelbrot en route to his discovery of the Mandelbrot set, we have to learn a modified way of describing the addresses in Cartesian space, involving what are known as complex numbers.

As the theory of mathematics developed, mathematicians found that a very strange quantity started to enter into their calculations. For example, they found that if one studied an equation such as

$$x^2 + 1 = 0$$

that the solutions of this equation involved a quantity $\sqrt{-1}$, whatever that was! Mathematicians began to write $\sqrt{-1}$ using the symbol i.

The Italian mathematician Girolamo *Cardano* in 1645 started to describe numbers involving $\sqrt{-1}$ as *imaginary numbers.* The term is unfortunate because to the ordinary person an imaginary number is something that is not real and does not exist. What the mathematician calls imaginary numbers are very real objects. By the early 1800s, the German mathematician Gauss showed that one could assign a physical meaning to $i = \sqrt{-1}$ if one looked at how one could represent numbers on a two-dimensional plane.

We have already discussed (see Section 1.4) the use of Cartesian coordinates to describe the address of a pixel. Another technique for describing the address of a pixel, originally suggested by Gauss, was to describe a point in two-dimensional space by the formula

$$P_1 = (a + ib)$$

In this address "a" is essentially the x coordinate of the point and "b" is the y coordinate when one agrees to interpret the i of the formula as a direction for the measurement of b at right angles to the direction of a. The relationship between the two ways of writing the address of a number are illustrated in *Figure 2.4.* A discussion of the reasons for choosing this form of pixel address is beyond the scope of this book, and the non expert reader is urged to accept the fact that the format is useful for many purposes. Mathematicians have agreed to use the symbol c for the address in this format and to call the relationship a *complex number*. Thus:

$$c = a + ib$$

We note from Figure 2.4 that the distance of the address from the origin is $\sqrt{a^2 + b^2}$. This distance is usually written in the symbolic form $|c|$. It is called the *modulus* of c. The word modulus comes from a Latin word meaning "a measure of". The modulus of c is a measure of its magnitude.

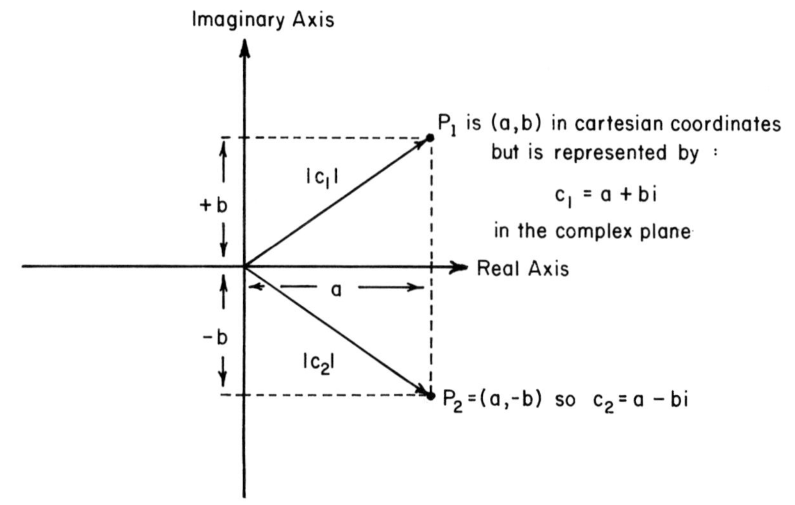

Figure 2.4. Using complex number notation the address of a point involves the symbol $\sqrt{-1}$ written as i.

The term *complex* as used by the mathematician is somewhat misleading in that this type of number is no more complex than any other number, but for historic reasons we are probably stuck with the name forever. The two parts of a complex number are also known as the *real* and *imaginary* parts. As Dewdney has pointed out, we could just as well call them Humpty-Dumpty numbers with a being the Humpty part and ib being the Dumpty [7].

Mathematicians have developed rules for what they call multiplication, addition and subtraction of complex numbers. In *Table 2.1* the operations which have to be carried out to add, subtract and multiply complex numbers are shown. It can be seen that the critical idea involved in the multiplication of complex numbers is that first we multiply each part of a number by each part of the other number. Then we make the substitution that $i^2 = -1$. The procedure for calculating the product of two complex numbers may appear odd and *arbitrary*, (arbitrary means "without a logical basis"), however, for now the readers are advised that they should be content with the fact that mathematicians know that the recipes given in Table 2.1 work and that they yield meaningful results. After the calculations have been carried out a few times the strangeness of the new, different multiplication process will wither away.

Now that we have discovered what is meant by a complex number, let us discover what happens to a point representing a complex number in a two-dimensional plane when we carry out an iteration of the form represented by the formula

$$z_{k+1} = z_k^2 + c$$

Table 2.1. The rules for carrying out mathematical operations on complex numbers differ from the rules encountered in "ordinary" arithmetic.

Complex numbers consist of *Real* and *Imaginary* parts and are usually written in the form:

$$\boxed{c = a + bi}$$

where: a is the Real Part of the complex number

b is the Imaginary Part and

i represents $\sqrt{-1}$

In two-dimensional space a is plotted on the x axis and b is plotted on the y axis.

Rules for mathematical operations on complex numbers

Addition

$$[\text{Real} + \text{Real}] + [\text{Imaginary} + \text{Imaginary}]\, i$$

Thus if: $c_1 = a_1 + b_1 i$ and $c_2 = a_2 + b_2 i$

then: $c_1 + c_2 = [a_1 + a_2] + [b_1 + b_2]\, i$

$$\begin{aligned} \text{e.g.: } [3 + (-2)i] + [4 + 1i] &= [3 + 4] + [(-2) + 1]\, i \\ &= 7 + (-1)\, i \\ &= 7 - 1i \end{aligned}$$

Subtraction

$$[\text{Real} - \text{Real}] + [\text{Imaginary} - \text{Imaginary}]\, i$$

Again if: $c_1 = a_1 + b_1 i$ and $c_2 = a_2 + b_2 i$

then: $c_1 - c_2 = [a_1 - a_2] + [b_1 - b_2]\, i$

$$\begin{aligned} \text{e.g.: } [3 + (-2)i] - [4 + 1i] &= [3 - 4] + [(-2) - 1]\, i \\ &= (-1) + (-3)\, i \\ &= -1 - 3i \end{aligned}$$

Multiplication

(1) Multiply both parts of c_1 by the Real part of c_2
(2) Multiply both parts of c_1 by the Imaginary part of c_2
(3) Transform any i^2 to -1 and collect them

Thus if: $c_1 = a_1 + b_1 i$ and $c_2 = a_2 + b_2 i$

$$\begin{aligned} \text{then: } c_1 \times c_2 &= [a_1 \times a_2] + [b_1 i \times a_2] + [a_1 \times b_2 i] + [b_1 i \times b_2 i] \\ &= [a_1 \times a_2] + [b_1 \times a_2]\, i + [a_1 \times b_2]\, i + [b_1 \times b_2]\, i^2 \\ &= [a_1 \times a_2] + [(b_1 \times a_2) + (a_1 \times b_2)]\, i + [b_1 \times b_2]\, (-1) \\ &= [(a_1 \times a_2) - (b_1 \times b_2)] + [(b_1 \times a_2) + (a_1 \times b_2)]\, i \end{aligned}$$

$$\begin{aligned} \text{e.g.: } [3 + (-2)i] \times [4 + 1i] &= [3 \times 4] + [(-2)i \times 4] + [3 \times 1i] + [(-2)i \times 1i] \\ &= 12 + (-8)\, i + 3i + (-2)\, i^2 \\ &= 12 + (-5)\, i + (-2)\, (-1) \\ &= 14 + (-5)\, i \end{aligned}$$

Table 2.1. continued

Magnitude (Modulus)

The Magnitude or Modulus of a complex number is the distance to the point from the origin in the complex plane and is written as $|c|$ and calculated using:

$$\sqrt{\text{Real}^2 + \text{Imaginary}^2}$$

If: $c_1 = a_1 + b_1 i$

then: $|c| = \sqrt{a_1^2 + b_1^2}$

e.g.: $|3 + (-2)i| = \sqrt{3^2 + (-2)^2}$
$= \sqrt{9 + 4}$
$= \sqrt{13}$

This formula is also sometimes written

$$z_{k+1} \leftarrow z_k^2 + c\,[n] \quad \text{or until} \quad z_{k+1} \geq m$$

These formulae summarize the following instruction set

1. Take 2 complex numbers z_k and c.
2. Square z_k and add c to it to create a new number z_{k+1} .
3. Using z_{k+1}, calculate the next value z_{k+2} in the same way.
4. Continue the process repeating n times, or until the calculated value exceeds the operational limit, m.

Thus, in the second iteration we take z_2 as calculated in the first iteration, square it, and add c to make the number z_3.

The student probably immediately asks the question "Why carry out this particular iteration?" Apart from anything else, the answer is that it leads to some surprising movements of the pixel address and that such an iteration leads to the fascinating beauty of the Mandelbrot set. A more technical answer is that when we carry out this iterative operation we are pursuing studies in the "theory of iteration of rational maps of the complex plane" whatever that means! This type of operation in an apparently abstruse area of mathematics received some extensive studies by the French mathematicians G. Julia and P. Fatou (1878–1929) in the early years of the 20th century (*abstruse* is defined in the dictionary as "hidden, remote from apprehension, difficult to understand"). G. *Julia* was one of Mandelbrot's teachers at the Polytechnique in Paris. After we have explored the Mandelbrot set, we will find that the same iteration carried out in a slightly different way generates some other interesting patterns called Julia sets and Fatou dusts.

In *Figure 2.5* a simplified version of the Mandelbrot set is shown. This simplified structure is sufficient and useful for our first attempt to characterize the behavior of pixels as we subject them to the mathematical operation which produces the Mandelbrot set. We first note that for all pixel addresses used to generate the Mandelbrot

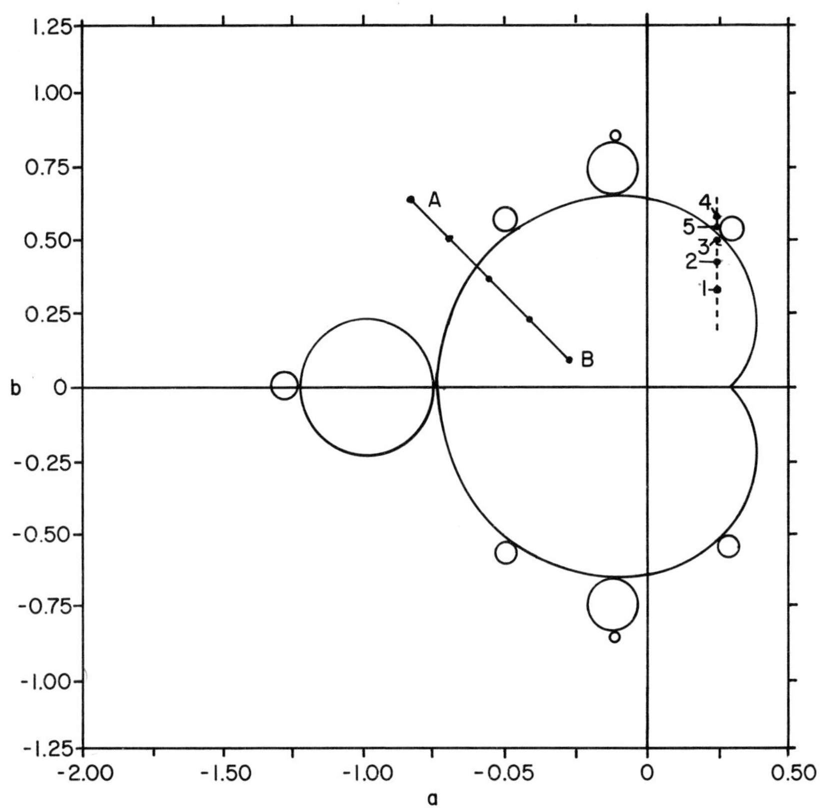

Figure 2.5. A simplified version of the Mandelbrot set in the complex coordinate plane is useful for locating pixels to be examined more closely as they are iterated to produce the Mandelbrot set. The points numbered on the diagram are: 1: $0.24 + i0.34$; 2: $0.24 + i0.44$; 3: $0.24 + i0.50$; 4: $0.24 + i0.59$; 5: $0.24 + i0.54965$.

set, z_1 equals zero. As we will discuss later in this chapter, the Julia sets and Fatou sets are generated by operating on z when c is fixed. The first pixel we will study in our exploration of the Mandelbrot set is marked by the number 1 inside the Mandelbrot set. It has the address $0.24 + i0.34$.

In the operation $z_2 \leftarrow z_1 + c$
We know that, since $z_1 = 0$ by definition that since $c = 0.24 + i0.34$ then:

$$z_2 = 0.24 + i0.34 \quad \text{1st iteration}$$
$$z_3 = (0.24 + i0.34)^2 + c$$

To calculate z_3 we first calculate:

$$z_2^2 = (0.24 + i0.34)\,(0.24 + i0.34)$$
$$= 0.0576 + i0.0816 + i0.0816 + i^2 0.1156$$
$$= -\,0.058 + i0.1632$$

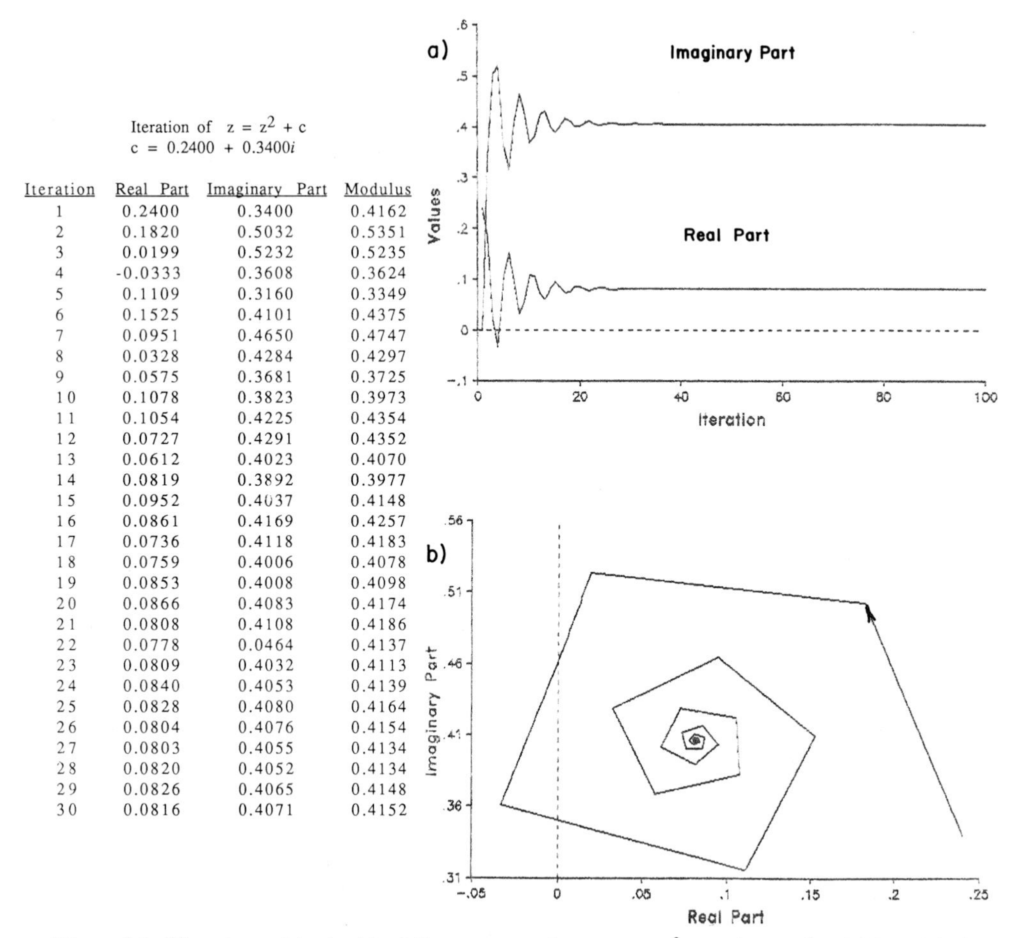

Iteration of $z = z^2 + c$
$c = 0.2400 + 0.3400i$

Iteration	Real Part	Imaginary Part	Modulus
1	0.2400	0.3400	0.4162
2	0.1820	0.5032	0.5351
3	0.0199	0.5232	0.5235
4	-0.0333	0.3608	0.3624
5	0.1109	0.3160	0.3349
6	0.1525	0.4101	0.4375
7	0.0951	0.4650	0.4747
8	0.0328	0.4284	0.4297
9	0.0575	0.3681	0.3725
10	0.1078	0.3823	0.3973
11	0.1054	0.4225	0.4354
12	0.0727	0.4291	0.4352
13	0.0612	0.4023	0.4070
14	0.0819	0.3892	0.3977
15	0.0952	0.4037	0.4148
16	0.0861	0.4169	0.4257
17	0.0736	0.4118	0.4183
18	0.0759	0.4006	0.4078
19	0.0853	0.4008	0.4098
20	0.0866	0.4083	0.4174
21	0.0808	0.4108	0.4186
22	0.0778	0.0464	0.4137
23	0.0809	0.4032	0.4113
24	0.0840	0.4053	0.4139
25	0.0828	0.4080	0.4164
26	0.0804	0.4076	0.4154
27	0.0803	0.4055	0.4134
28	0.0820	0.4052	0.4134
29	0.0826	0.4065	0.4148
30	0.0816	0.4071	0.4152

Figure 2.6. When iterated in the Mandelbrot set equation $z_{k+1} = z_k^2 + c$, the number $0.24 + i0.34$ is attracted to the value $0.0816 + i0.4171$ and cannot escape from the Mandelbrot set (note: the arrows drawn on the lines between successive iterated values and the lines themselves have no physical significance but merely portray the progress of the iteration to its next value). a) Real and imaginary parts plotted versus the number of iterations. b) Plot of imaginary versus the real part of the number as it is iterated.

$$\text{Therefore } z_3 = (z_2)^2 + c$$
$$= 0.1820 + i0.5032 \quad \text{2nd iteration}$$

Repeating the next iteration in the same way gives us

$$z_4 = 0.0199 + i0.5235$$

In the table of *Figure 2.6* the progressively changing values of $0.24 + i0.34$ as it is iterated 30 times are shown. The real and imaginary part of the complex number and the modulus of the number is also given in this table.

We can make two informative graphs out of this set of data. In Figure 2.6(a) the information is plotted as the change in the two parts of the number against the number of iterations. From one point of view, we could regard the iterations as events in time. Then our graph would be a record of the changing magnitudes of the real and imaginary parts of the iterated complex number as a time sequence. It can be seen from Figure 2.6(a) that the complex point $0.24 + i0.34$ appears to oscillate as it is iterated before settling down to a steady value of $0.0816 + i0.4071$. Another way of showing the movement of the complex point as it is iterated is given in Figure 2.6(b). It can be seen that when depicted in this way the pixel address appears to spiral into a constant value. Personally, I was very surprised the first time I saw the series of calculations and the graphic display of the value of the iterated complex number spiralling into a constant value as summarized in Figure 2.6. The mathematician says that the iterated complex number $0.24 + i0.34$ is attracted to the point $0.0816 + i0.4071$. I do not know of any reason why this particular number should be attracted to the specific attractor point; I regard it as a fascinating experimental discovery. The reader is reminded that the attraction between the two points does not occur because of a real force. The points only seem to be "mathematically" attracted to each other. The important thing to realize is that the iterated pixel does not escape into outer space, but remains trapped within the space limits determined by the Mandelbrot set. Another way of describing the movement of the iterated point into a constant value is to call the point to which the iterated values move the *drainage basin* of the mathematical process. The mathematician watching his point move round and round and into the strange attractor visualizes an element of water gurgling around and disappearing into the drainage basin formed by the mathematical attractor.

Eager to see what happened as we looked at other points in the plane, we (Garry Clark and I) now chose the complex number $0.24 + i0.44$. The two different types of graphs summarizing the behavior of this iterated point are shown in *Figure 2.7*. It can be seen that although the iterated point still moved to a strange attractor, it took longer to spiral in on its drainage basin than the number $0.24 + i0.34$.

We now continue our exploration of the Mandelbrot set by continuing to move along the dotted line representing the real number 0.24 with various imaginary parts, as drawn in Figure 2.5, by moving to the complex number $0.24 + i0.50$. The pattern created as we iterate this number is summarized in *Figure 2.8*. It shows that this value is attracted to a fixed point and probably lies within the Mandelbrot set. However, even after one hundred iterations we are still not too sure of the value to which the iterated number is converging. This lack of confidence in our ability to predict the movement of the iterated pixel arises from the fact that we are getting close to the interface of the Mandelbrot set with the pixels outside the set. The number $0.24 + i0.50$ is close to the interface between the pixels which are trapped in the Mandelbrot sets and those that can escape to infinity. This becomes very apparent when we try to study the behavior of the pixel at $0.24 + i0.59$. Because of the

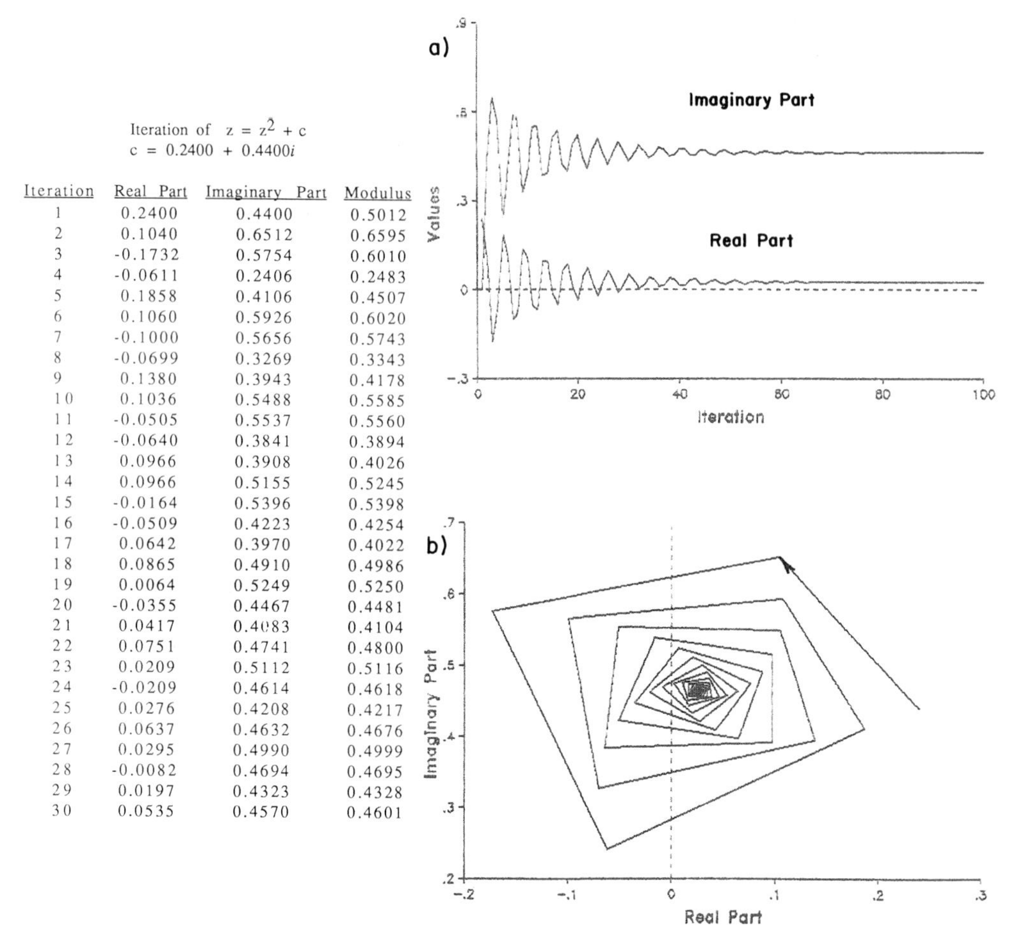

Iteration of $z = z^2 + c$
$c = 0.2400 + 0.4400i$

Iteration	Real Part	Imaginary Part	Modulus
1	0.2400	0.4400	0.5012
2	0.1040	0.6512	0.6595
3	-0.1732	0.5754	0.6010
4	-0.0611	0.2406	0.2483
5	0.1858	0.4106	0.4507
6	0.1060	0.5926	0.6020
7	-0.1000	0.5656	0.5743
8	-0.0699	0.3269	0.3343
9	0.1380	0.3943	0.4178
10	0.1036	0.5488	0.5585
11	-0.0505	0.5537	0.5560
12	-0.0640	0.3841	0.3894
13	0.0966	0.3908	0.4026
14	0.0966	0.5155	0.5245
15	-0.0164	0.5396	0.5398
16	-0.0509	0.4223	0.4254
17	0.0642	0.3970	0.4022
18	0.0865	0.4910	0.4986
19	0.0064	0.5249	0.5250
20	-0.0355	0.4467	0.4481
21	0.0417	0.4083	0.4104
22	0.0751	0.4741	0.4800
23	0.0209	0.5112	0.5116
24	-0.0209	0.4614	0.4618
25	0.0276	0.4208	0.4217
26	0.0637	0.4632	0.4676
27	0.0295	0.4990	0.4999
28	-0.0082	0.4694	0.4695
29	0.0197	0.4323	0.4328
30	0.0535	0.4570	0.4601

Figure 2.7. The number $0.24 + i0.44$ is attracted to $0.535 + i0.4570$ by the Mandelbrot iteration operation. Its movement to its attractor is slower than for the number of Figure 2.6 because in that case the number is deeper within the body of the Mandelbrot set.

extremely rapidly escaping track of this pixel, we have to use a different graph to show its behavior after only 21 iterations. After 21 iterations, we have to use a number which needs 40 zeros to express its value, or an exponential of 39 for the number 10. This is the notation used in *Figure 2.9*.

To show the problems of studying iterations of complex numbers in the vicinity of the interface between the trapped and escaping pixels, the data for 100 iterations of the number $0.24 + i0.54965$ are shown in *Figure 2.10*. Note that if we had stopped our iterations at 40 we could not have expected that the pixel was beginning to escape, and although continuing the iterations up to 100 shows some oscillation in the value, we are left with the fact that we really do not know what the pixel is going

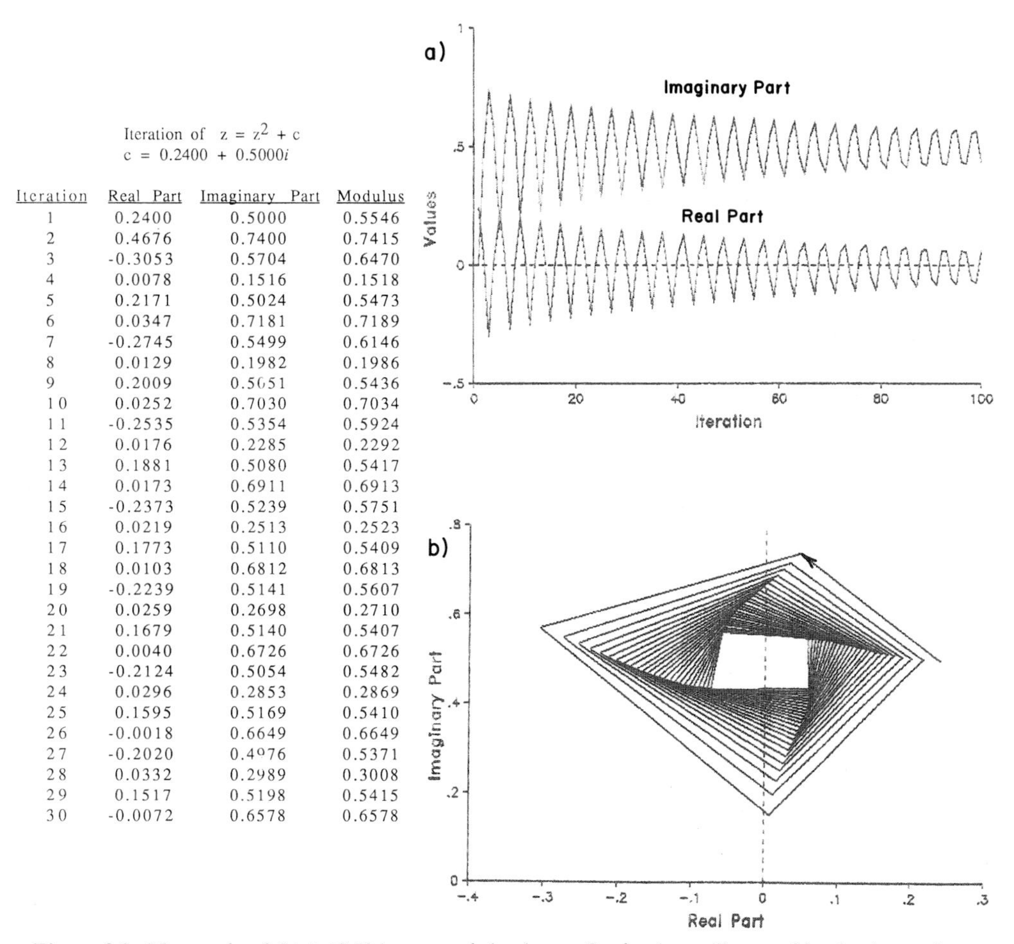

Iteration of $z = z^2 + c$
$c = 0.2400 + 0.5000i$

Iteration	Real Part	Imaginary Part	Modulus
1	0.2400	0.5000	0.5546
2	0.4676	0.7400	0.7415
3	-0.3053	0.5704	0.6470
4	0.0078	0.1516	0.1518
5	0.2171	0.5024	0.5473
6	0.0347	0.7181	0.7189
7	-0.2745	0.5499	0.6146
8	0.0129	0.1982	0.1986
9	0.2009	0.5651	0.5436
10	0.0252	0.7030	0.7034
11	-0.2535	0.5354	0.5924
12	0.0176	0.2285	0.2292
13	0.1881	0.5080	0.5417
14	0.0173	0.6911	0.6913
15	-0.2373	0.5239	0.5751
16	0.0219	0.2513	0.2523
17	0.1773	0.5110	0.5409
18	0.0103	0.6812	0.6813
19	-0.2239	0.5141	0.5607
20	0.0259	0.2698	0.2710
21	0.1679	0.5140	0.5407
22	0.0040	0.6726	0.6726
23	-0.2124	0.5054	0.5482
24	0.0296	0.2853	0.2869
25	0.1595	0.5169	0.5410
26	-0.0018	0.6649	0.6649
27	-0.2020	0.4976	0.5371
28	0.0332	0.2989	0.3008
29	0.1517	0.5198	0.5415
30	-0.0072	0.6578	0.6578

Figure 2.8. The number $0.24 + i0.50$ is attracted slowly to a fixed point as illustrated by the time series and spiral of attraction shown above.

to do without carrying out many more iterations. The data of Figure 2.10 begins to demonstrate the difficulty of locating the limits to the structure of the Mandelbrot set and the fact that we need to be able to carry out massive computations to be sure of the behavior of pixels located near the interface of the trapped pixels with those that can escape.

The basic technique for studying the interface of the Mandelbrot set with its pixel environment can be understood from the data summarized in *Figure 2.11*. We start our large scale exploration on the behavior of pixels near the interface of the Mandelbrot set by considering a 100 pixel group with the addresses the real part ranging from -0.5 to $+0.4$ and with complex values ranging from $i0.25$ to $i1.15$.

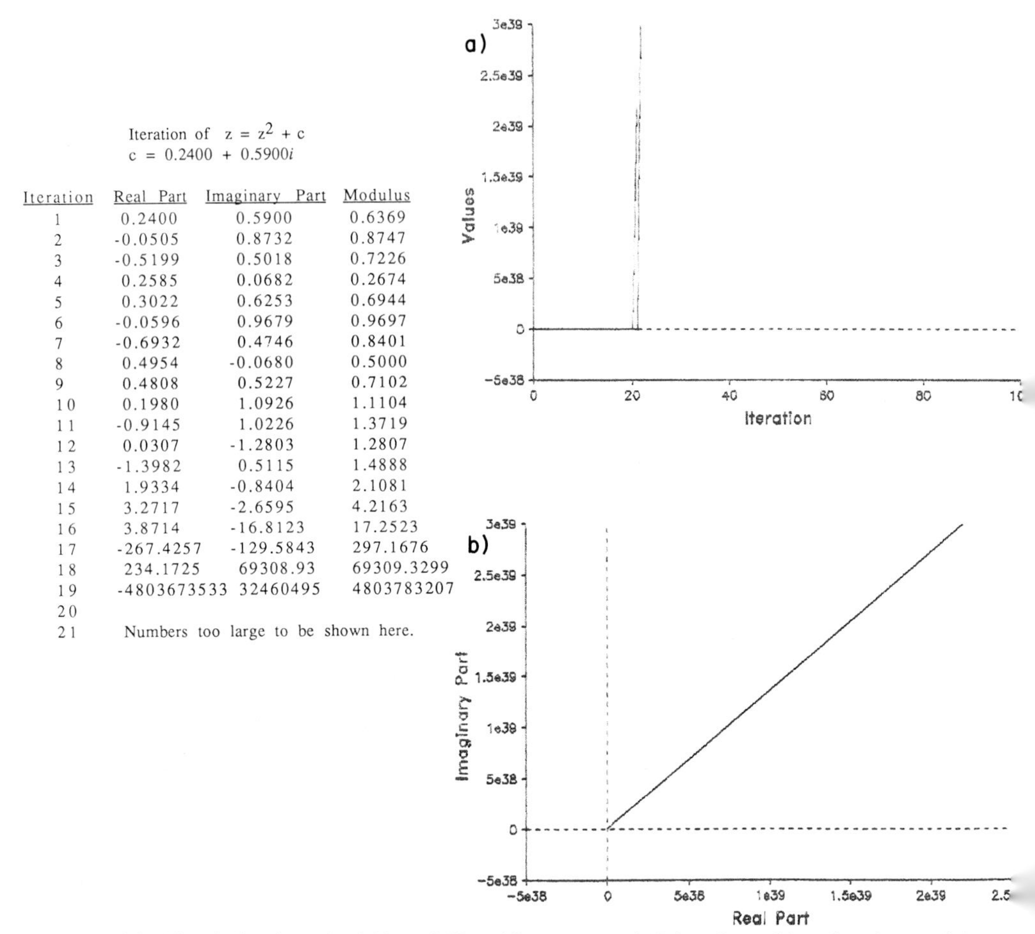

Iteration of $z = z^2 + c$
$c = 0.2400 + 0.5900i$

Iteration	Real Part	Imaginary Part	Modulus
1	0.2400	0.5900	0.6369
2	-0.0505	0.8732	0.8747
3	-0.5199	0.5018	0.7226
4	0.2585	0.0682	0.2674
5	0.3022	0.6253	0.6944
6	-0.0596	0.9679	0.9697
7	-0.6932	0.4746	0.8401
8	0.4954	-0.0680	0.5000
9	0.4808	0.5227	0.7102
10	0.1980	1.0926	1.1104
11	-0.9145	1.0226	1.3719
12	0.0307	-1.2803	1.2807
13	-1.3982	0.5115	1.4888
14	1.9334	-0.8404	2.1081
15	3.2717	-2.6595	4.2163
16	3.8714	-16.8123	17.2523
17	-267.4257	-129.5843	297.1676
18	234.1725	69308.93	69309.3299
19	-4803673533	32460495	4803783207
20			
21	Numbers too large to be shown here.		

Figure 2.9. The pixel at the point $0.24 + i0.59$ rapidly escapes to infinity when subjected to the Mandelbrot set iteration.

Our previous explorations were within the third column of pixels from the right in the top right hand side of Figure 2.11 (a). The pixel address of the top left hand corner of this block of pixels would be written as $-0.05 + i1.15$. This number was subjected to sufficient iterations to show that it escaped quickly to infinity. Open squares indicate pixels which fled to infinity in less than 20 iterations when iterated with the Mandelbrot operation. The iteration studies were carried out successively along each row of the pixel block. The first one that required more than 20 iterations to escape to infinity was located at the intersection of the third row down with the sixth column across. This diagonally hatched pixel in Figure 2.11 (a) required more than 20 but less than 30 iterations to disclose its behavior, whereas those that

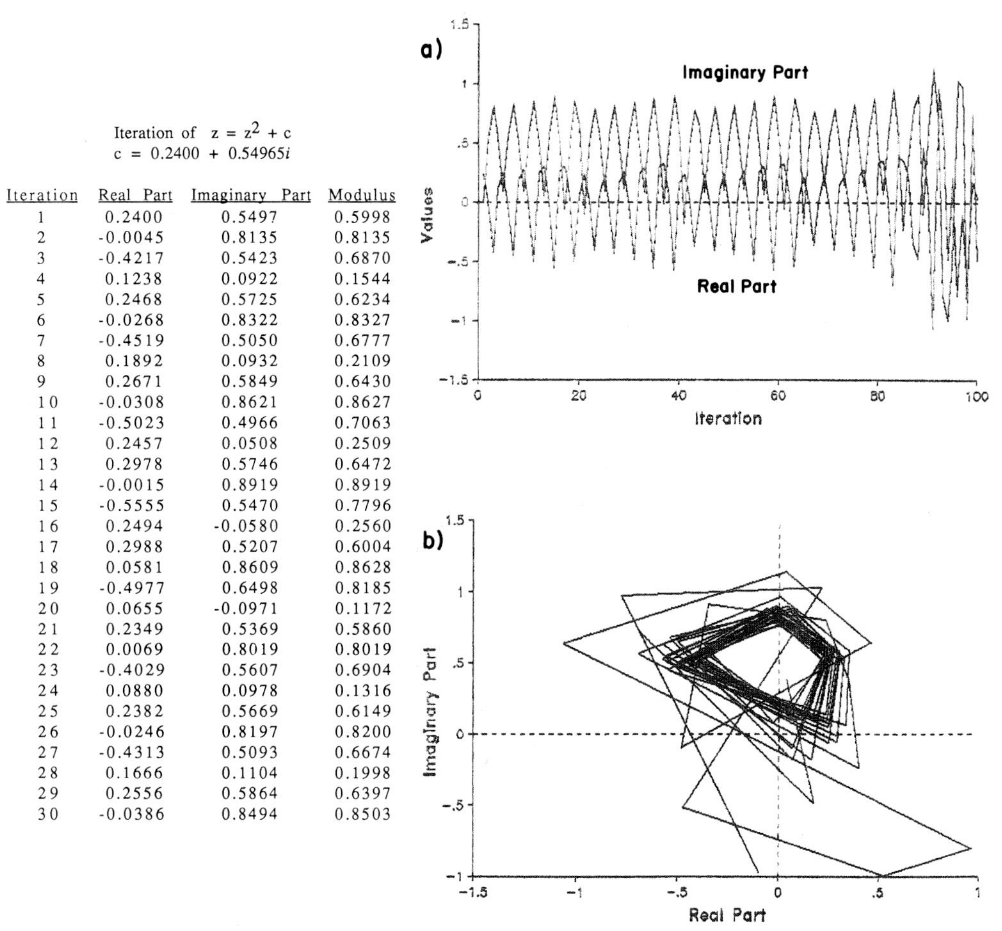

Iteration of $z = z^2 + c$
$c = 0.2400 + 0.54965i$

Iteration	Real Part	Imaginary Part	Modulus
1	0.2400	0.5497	0.5998
2	-0.0045	0.8135	0.8135
3	-0.4217	0.5423	0.6870
4	0.1238	0.0922	0.1544
5	0.2468	0.5725	0.6234
6	-0.0268	0.8322	0.8327
7	-0.4519	0.5050	0.6777
8	0.1892	0.0932	0.2109
9	0.2671	0.5849	0.6430
10	-0.0308	0.8621	0.8627
11	-0.5023	0.4966	0.7063
12	0.2457	0.0508	0.2509
13	0.2978	0.5746	0.6472
14	-0.0015	0.8919	0.8919
15	-0.5555	0.5470	0.7796
16	0.2494	-0.0580	0.2560
17	0.2988	0.5207	0.6004
18	0.0581	0.8609	0.8628
19	-0.4977	0.6498	0.8185
20	0.0655	-0.0971	0.1172
21	0.2349	0.5369	0.5860
22	0.0069	0.8019	0.8019
23	-0.4029	0.5607	0.6904
24	0.0880	0.0978	0.1316
25	0.2382	0.5669	0.6149
26	-0.0246	0.8197	0.8200
27	-0.4313	0.5093	0.6674
28	0.1666	0.1104	0.1998
29	0.2556	0.5864	0.6397
30	-0.0386	0.8494	0.8503

Figure 2.10. A more detailed examination of the area by applying more decimal places results in the data for $0.24 + i0.54965$.

remained trapped in pixel space, shown in black, all remained trapped within the confines of the Mandelbrot set after more than 35 iterations.

To increase our resolution of exploration of the pixels of the Mandelbrot set, we now move to the pixel block of Figure 2.11 (b). This pixel block is in the general region of the pixel of Figure 2.11 (a) which showed by its behavior that it represented a location near to the interface of the Mandelbrot set. Using a much smaller pixel, the behavior of a hundred pixels within this block were studied and now the behavior of the pixels became very unpredictable, with some of them lying within the Mandelbrot set and others escaping to infinity at various rates as shown by the various shadings of the pixel squares. To demonstrate that the behavior becomes

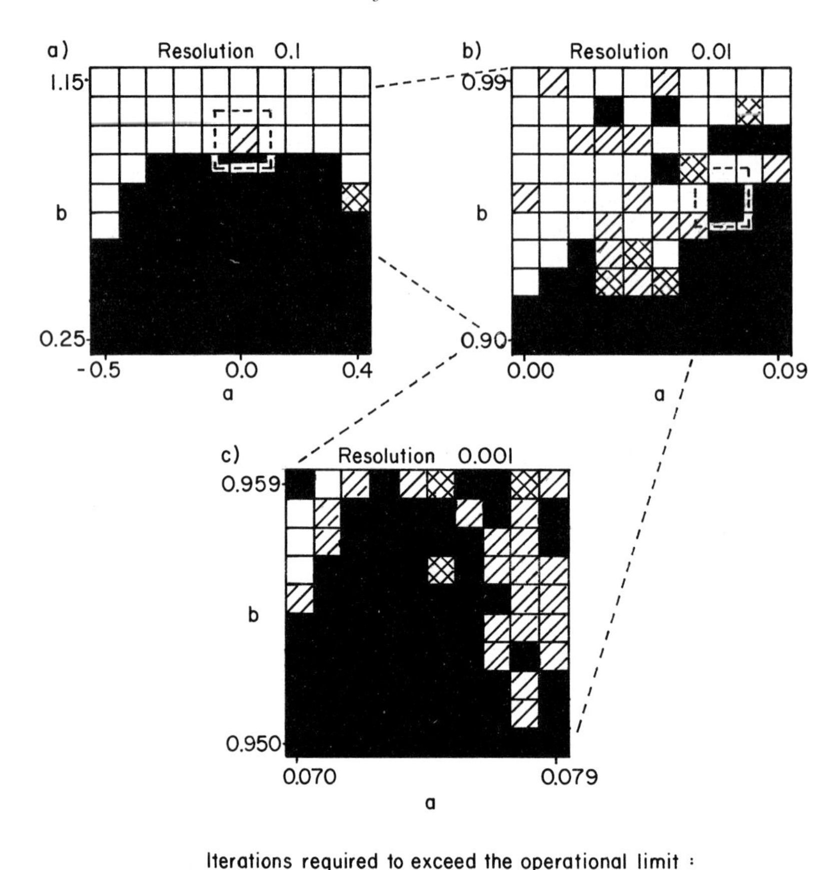

Figure 2.11. Determining the behavior of a pixel close to the interface of the Mandelbrot set with its pixel environment when iterated with the Mandelbrot set operation is a difficult problem. a) Region shown at a resolution of 0.1. b) Region shown at a resolution of 0.01. c) Region shown at a resolution of 0.001.

even more exotic at higher resolution, we consider the behavior of the pixel block of Figure 2.11 (c), which is at one order of magnitude higher resolution than that of Figure 2.11 (b). Again the behavior of the pixels appears to be very unpredictable, showing that the interface between the trapped pixels and those that can escape is very complex.

An alternative way of summarizing the behavior of pixels at the interface between the Mandelbrot set and its environment is shown in *Figure 2.12*. In this diagram the modulus of the complex number (its magnitude from the 0 point of the axis of the coordinate system) is shown after 40 or fewer iterations. It can be seen that the value of the modulus of z is a good indication of whether the pixel will escape or not.

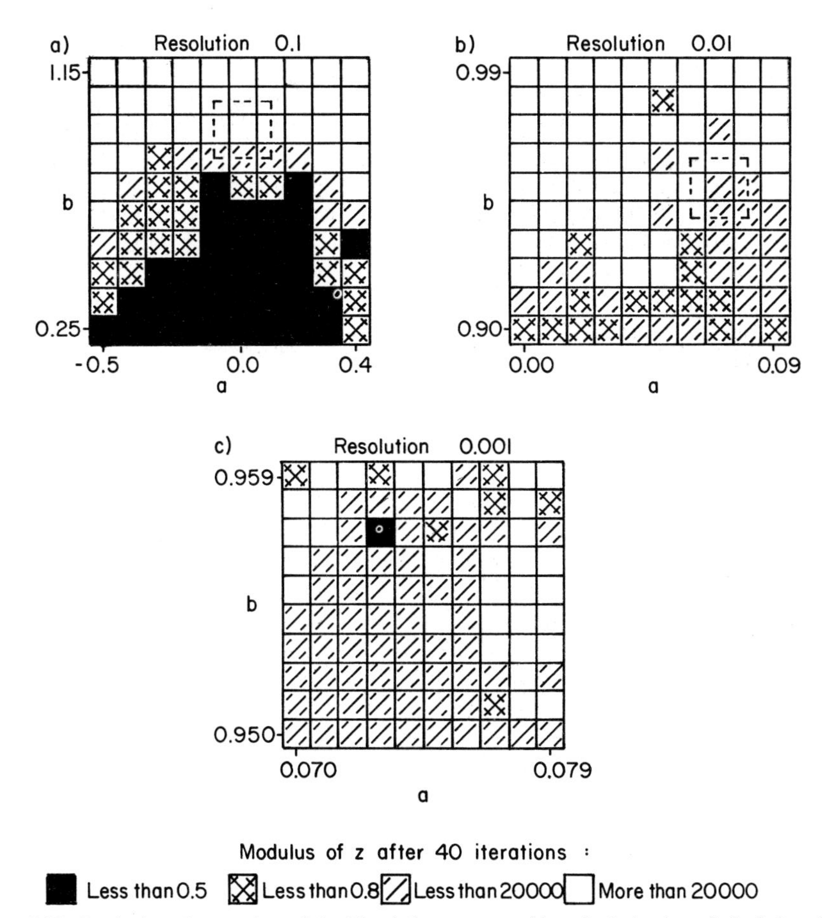

Figure 2.12. In the interface region of the Mandelbrot set, tracking the behavior of pixels is a difficult task. In this figure the modulus of the complex number after 40 iterations dictates the shading of the pixel.

Mathematicians have shown that if the value of the modulus ever reaches 2, then the pixel is sure to be able to escape. For pixels within the interface at high resolution, iterations of more than a thousand often fail to reveal whether the value is approaching 2 or not. Commercially available software can now be used with small computers to generate the basic structure of the set in the way that we have shown in the set of digitized outlines of Figure 2.3 [5, 7, 8]. When we look at the overall structure of the set as revealed by the first picture [Figure 2.3(a)] we see that there is one pixel which is apparently free floating. This pixel has been ringed to draw attention to the pixel size used in this diagram, and also to demonstrate that low resolution exploration of the set appears to create islands of trapped pixels well away from the basic Mandelbrot set.

When we move to the generation of the digitized image of the Mandelbrot set of Figure 2.3(b) using a smaller pixel size, we now see several disconnected pixels floating around the main set. Mandelbrot tells us that this was one of the surprising features that he observed when the very first Mandelbrot set was generated during experiments carried out at Harvard University in 1980. Mandelbrot has written an essay recalling the excitement and surprise that he felt when he first saw what is now known as the Mandelbrot set evolving on his computer screen. Writing in 1986 he tells us

> *I find it hard to believe that only six years have passed since I first saw and described the structure of the beautiful set – to which I am honored and delighted that my name should be attached* [2].

In general he points out that the beauty of many fractal systems such as the Mandelbrot set

> *is the more extraordinary for its having been wholly unexpected: they were meant to be mathematical diagrams drawn to make a scholarly point, and one might have expected them to be dull and dry ... It seems that nobody is indifferent to fractals, in fact many view their first encounter with fractal geometry as a totally new experience from the viewpoints of aesthetics as well as science.*

He goes on to say

> *Before my memory falters, let me retell the very beginnings of the beautiful set, the Mandelbrot set, featured in this book. After a few iteration steps on a rough grid, we saw that this set (which came to be called the Mandelbrot set) includes a very crude outline of the two discs. We also saw on the real line to the right and left of the discs the crude outlines of blobs which I called "atoms"* [2].

At first Mandelbrot did not know if these blobs, which are the free floating pixels of our pictures of Figure 2.3(a, b, c), were connected to the main discs. However, Mandelbrot tells us that he arranged for more advanced computations to be made at IBM, and that these experiments showed that the whole set was connected (very early graphics of the Mandelbrot set are reproduced in Peitgen and Richter's book [3]). Mandelbrot tells us

> *In 1980 I could not prove that the whole set was connected but I did not have the nerve to tell the scientific community that experiments had shown the connectedness of the sets.*

Not only was Mandelbrot unsure of the connectedness of the sets, but he did not anticipate the full complexity of the structure of the interface between the Mandelbrot set and the pixel space surrounding it.

To explore the fine details of the Mandelbrot set requires much bigger computers than will be available to the average student reader. However, the complexity of the

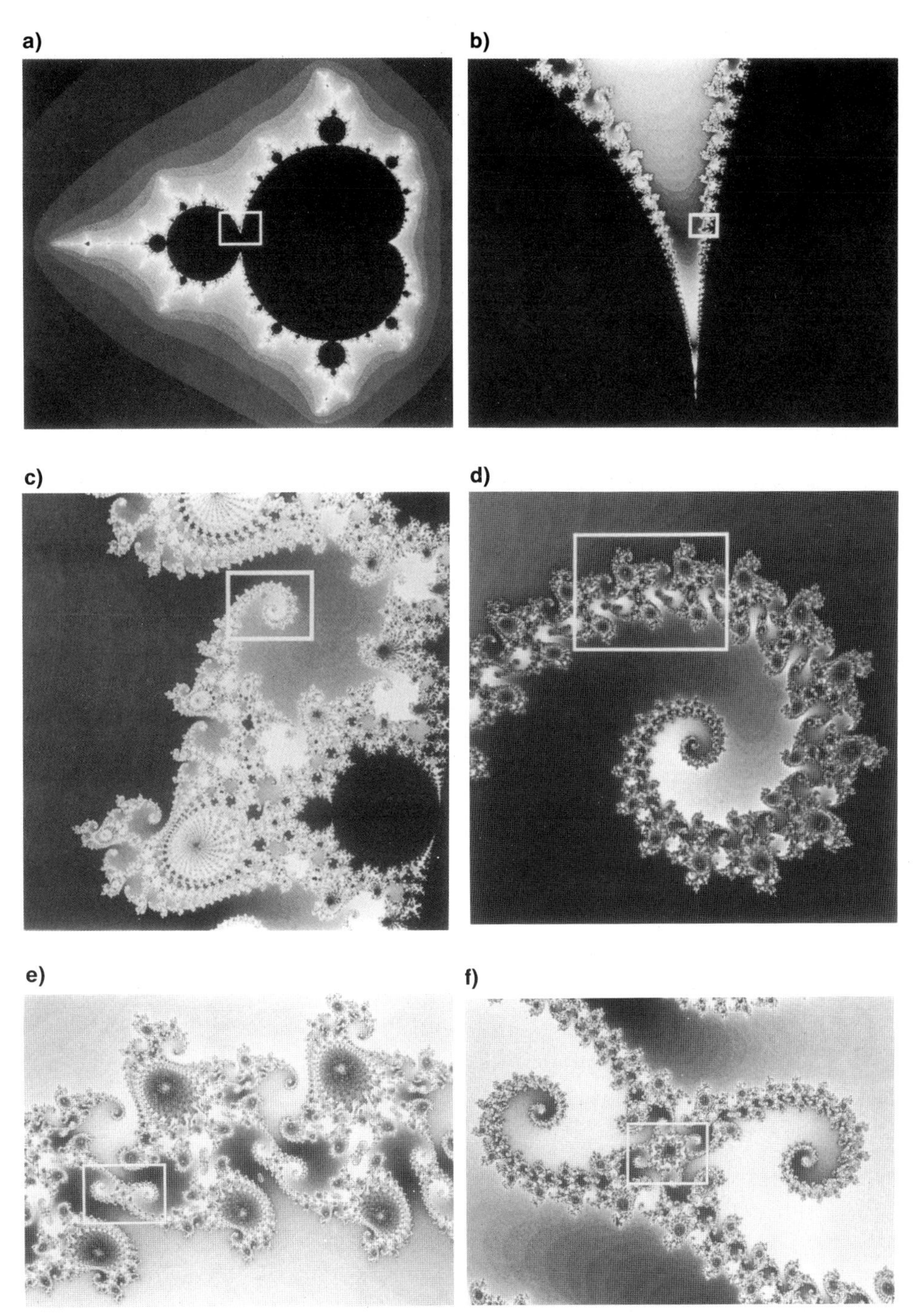

Figure 2.13. Higher and higher resolution inspection of the behavior of the pixels located at the interface of the Mandelbrot set reveal endless complexity and delicate interwoven patterns of behavior [11]. Peitgen, H.-O./Richter, P. H.: The Beauty of Fractals, 1986 © Springer-Verlag, Berlin-Heidelberg

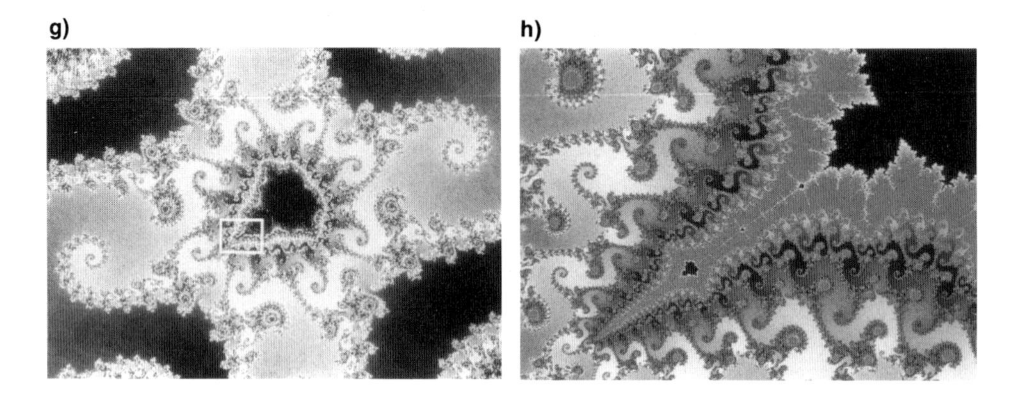

interface is illustrated by the diagrams reproduced from Peitgen in *Figure 2.13* [3]. Beautiful colour pictures of the same highly complex interface are available in many books. See for example those in the books on chaos by Gliek [10]. Several video tapes are available showing the evolution of the intricate structure of the interface of the Mandelbrot set with its environment as the resolution of inspection is increased. In these videotapes colour is used to code the volatility of the pixel environment, that is the speed with which the pixels escape. (Note the word *volatile* comes from the Latin word Volare meaning "to fly", a volatile liquid is one that flies into the vapour stage very quickly and volatile pixels are the pixels that fly away from the Mandelbrot set at high speed.)

One of the surprises awaiting computer specialists as they explored the Mandelbrot set in greater and greater detail was the discovery that the fringes of the Mandelbrot set were decorated by miniature Mandelbrot sets. Thus, in Figure 2.13(g) a miniature Mandelbrot set found at a very high resolution is shown.

The spirals and whirls of the Mandelbrot set seem to be endless, as shown by the fascinating patterns discovered as you explore at higher and higher resolutions, demonstrated by the various parts of Figure 2.13 in what is commonly known as a zoom.

Our brief exploration of the structure of the Mandelbrot set shows that it is not useful to talk about the boundary of the Mandelbrot set. Even at the highest resolution that mathematicians have been able to use in their computers the structure of the interface between fleeing pixels and trapped pixels is infinitely complex. In modern mathematical language we say that the boundary of the Mandelbrot set is itself a fractal structure of infinite complexity. Our discussion has been limited to two-dimensional space whereas the real Mandelbrot set can exist in three-dimensional space with our picture of Figure 2.3 being sections through the three-dimensional structure of the Mandelbrot set [12].

One of the exciting aspects of exploring the fractal structure of the interface of the Mandelbrot set is the fact that the experimenter can choose his own region in coordinate space and watch unique patterns develop on his computer as he chases the pixels around his experimental data space. The reader is warned however that discovering such patterns can lead to addictive behaviour so that excessive time is spent lingering over the computer.

Section 2.3. Julia Lace and Fatou Dusts

So far in our study of complex pixel patterns, we have focused on addresses which specify that z_1 should be equal to 0 in the iterative program and then we have varied c in our study of what happens to pixels in two-dimensional space. In this section we will explore briefly what happens when we fix a value for c in the iterative operation and vary the value of z_1.

When we take the basic equation and fix the value of c and vary the value of z, we again obtain a pattern enclosing pixels which cannot escape to infinity. In

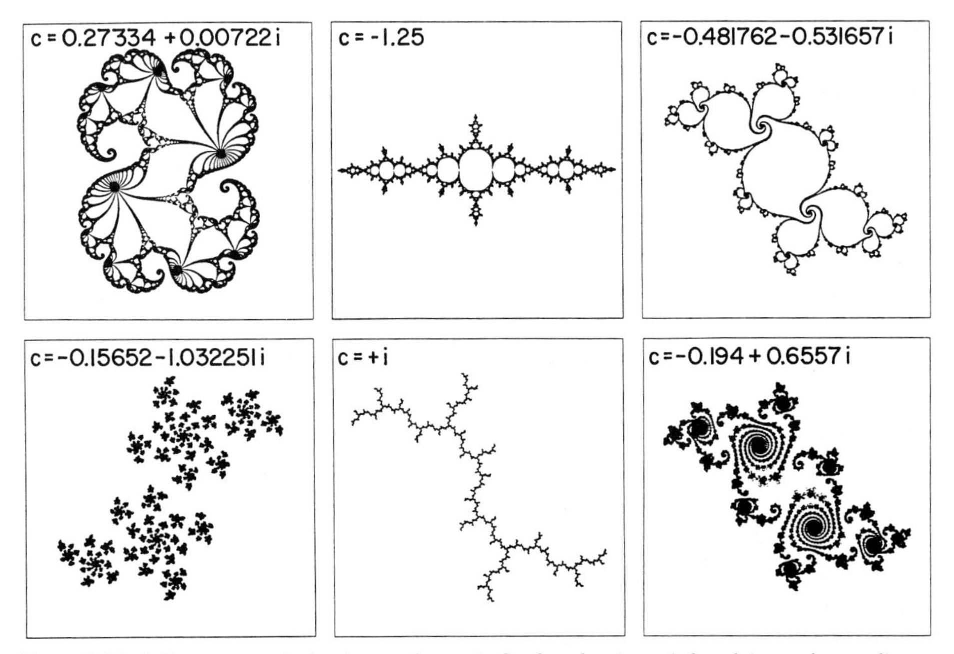

Figure 2.14. Julia sets are pixel prisons when c is fixed and z is varied and iterated according to $z_{k+1} = (z_k)^2 + c$ n times or ad nauseum. The black parts of the Julia sets are points that cannot escape. If c is in the Mandelbrot set, the Julia set is a connected lace (Figures a, b, c, and e). If c is not in the Mandelbrot set the Julia set is a Fatou dust (d and f). The Fatou dusts are self-similar and have fractal boundaries.

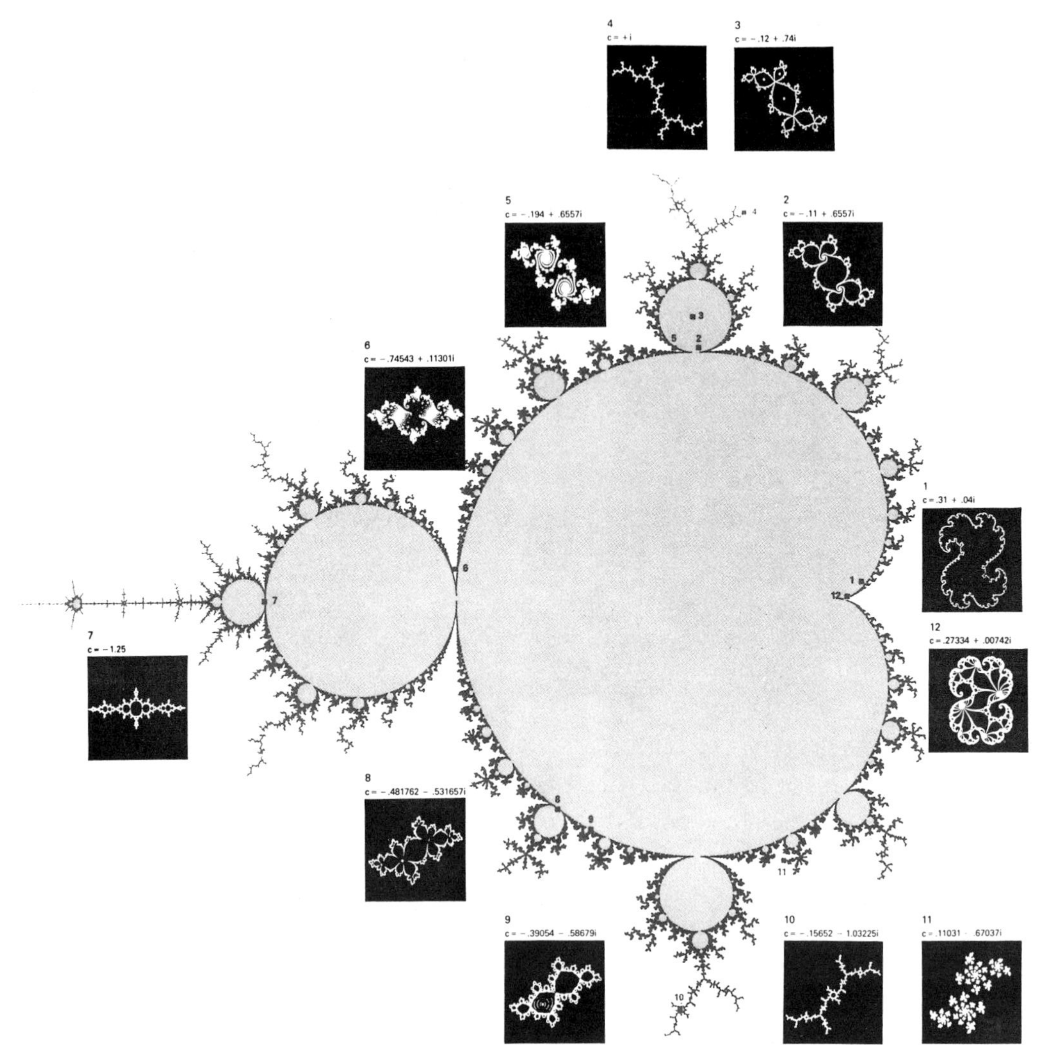

Figure 2.15. The type of Julia lace or Fatou dust generated by iterative calculation of the equation summarized in Figure 2.13 depends on the location of the chosen value of c in the expression $z + c$. Peitgen, H.-O./Richter, P. H.: The Beauty of Fractals, 1986 © Springer-Verlag, Berlin-Heidelberg

Figure 2.14 the patterns of such anchored pixels are shown for the six values of c given by the frame of each pattern. It can be seen that for some values of c the pattern is an interconnected set of lines which look like lace. Other patterns are sprinkled around the data space like the footprints of a dizzy wolf. All of the patterns of Figure 2.14 are technically described as Julia sets, but one can usefully differentiate between the connected patterns, which we will call Julia lace, and the scattered patterns, which are known to mathematicians as Fatou dusts. When describing the many varied patterns of the Julia sets, Ivars Peterson makes the following comment:

By varying c an incredible variety of Julia sets can be generated. Some look like fat clouds, others are like twisted bushes, many look like sparks floating in the air after a flare has gone off. With a little imagination the person flipping through a book of these Julia sets may glimpse the shape of a rabbit, the jagged form of a dragon and the curls of a sea horse [12].

Mathematicians have now established that the Mandelbrot set consists of all the values of c that have connected Julia sets. The way in which the Julia set varies with the roaming of the value of c with respect to the Mandelbrot set is illustrated by the information summarized in *Figure 2.15*. The discussion of how we can generate our own Julia set is beyond the scope of this book, and the interested reader is referred to the information in References [8], [9], and [12] which will get them started on an exploration of their own Julia sets. A specific type of Julia set is discussed briefly in Chapter 14.

References

[1] I. Stewart, *Concepts of Modern Mathematics*, Penguin, Harmondsworth, England, 1975.
[2] See discussion in [9] of Chapter 1.
[3] H. O. Peitgen, P. H. Richter, *The Beauty of Fractals – Images of Complex Dynamical Systems*, Springer, Berlin, 1986.
[4] J. Briggs and F. D. Peat, *Turbulent Mirror*, Harper and Row, New York, 1989.
[5] A. K. Dewdney, "A Tour of the Mandelbrot Set Aboard the Mandelbus", *Scientific American*, February 1989, pp. 108–111.
[6] See description of Mandelbrot's explorations of the set which bears his name as set out in [3].
[7] A. K. Dewdney, "Beauty and Profundity – the Mandelbrot set and the Flock of its Cousins Called Julia", *Scientific American*, November 1987, pp. 140–144.
[8] R. L. Devaney, *Chaos, Fractals and Dynamics, Computer Experiments in Mathematics*, Wesley, New York, 1990.
[9] Media Magic, P.O. Box 507, Nicasio, California 94946.
[10] J. Gleick, *Chaos, Making A New Science*, Viking Penguin Inc., New York, 1987.
[11] Many three-dimensional versions of the Mandelbrot set are to be found in [3].
[12] I. Peterson, *The Mathematical Tourist*, W. H. Freeman, New York, 1988, pp. 158–163.

Chapter 3

The Normal Distribution of Drunks

85039	35246	27026	13873	35350	94513	38339
41583	27563	73009	52091	86401	64081	10484
86454	57429	72880	60952	61152	02839	30079
18132	29785	26865	42058	97353	43889	41507
08991	95359	03425	39800	53545	16848	26169
70241	10522	30170	[illegible]	28314	84072	86847
41446	43982	[illegible]	[illegible]	88922	76378	76329
46822	21407	[illegible]	[illegible]	95903	71550	87652
92492	06469	[illegible]	92453	68124	98263	75567
88502	80485	[illegible]	93014	94452	18065	42134
27189	95527	49801	[illegible]	[illegible]	74343	51864
77989	88732	[illegible]	[illegible]	[illegible]	81044	77751
10506	33585	[illegible]	93165	[illegible]	06743	65506
00091	74367	[illegible]	04084	[illegible]	82641	32061
72757	59927	[illegible]	87841	[illegible]	96364	60506
16516	[illegible]	[illegible]	[illegible]	[illegible]	68394	32666
03482	[illegible]	23288	39594	[illegible]	81274	99982
91988	[illegible]	34581	79674	85790	22454	87109
02004	[illegible]	13700	89012	70678	[illegible]	89488
[illegible]	43501	65527	82167	15386	[illegible]	[illegible]
63437	81051	56704	59752	67516	93433	82926
82941	47088	75897	62126	14258	15839	82679
04187	79955	30747	81231	26399	40379	16182
64467	48089	89590	15504	16287	02513	03315
34712	90626	62663	68216	39411	28219	98681

Random, random is the trend
Cause with cause together blend
Stir and mix with gay abandon
See appear the bell shaped phantom!

Chapter 3 The Normal Distribution of Drunks 107

Section 3.1 Can Tumbling Coins Tell Us Anything About Staggering Drunks? 109
Section 3.2 Seductive (Often Meaningless) Patterns in Random Time Sequences 112
Section 3.3 Technical Names for the Bell Curve 114
Section 3.4 Mathematical Description of the Structure of the Gaussian Probability Curve 115
Section 3.5 Stretching Data Mathematically with Gaussian Probability Graph Paper 120
Section 3.6 When are we Likely to Encounter Gaussian Probability Distribution in Stochastic Systems? 124
Section 3.7 Gaussianly Scattered Drunks 126
Section 3.8 Gaussian Variations in the Structure of Powder Mixtures 127
Section 3.9 Using The Gaussian Probability Curve to Characterize Uncertainty in Experimental Data 133
Section 3.10 The Power of Self-cancelling Errors to Improve Experimentally Based Estimates of a Physical Quantity 137
Section 3.11 Using Dot Counting to Estimate Irregular Areas 140
Section 3.12 Finding the Percentage of a Given Material in an Ore 145
Section 3.13 Aperture Size Distribution in a Woven Wire Sieve 147
Section 3.14 Characterizing The Weight Variation in a Population of Candy Covered Chocolate Buttons 150
Section 3.15 Buffon's Needle – A Surprising Pattern of Events 155
Section 3.16 Exploiting the Efficiency of Antithetic Variates to Improve the Efficiency of the Estimates of π by Buffon's Needle Experiments 158
Section 3.17 How Much Work in Needed to Improve One's Confidence in an Average Value? 163
Exercises 167
References 175

Chapter 3

The Normal Distribution of a Set of Staggering Drunks?

Section 3.1. Can Tumbling Coins Tell Us Anything About Staggering Drunks?

The picture created in the mind of the reader by the title of this chapter is probably that of an array of drunks distributed along the sidewalks of the main drinking areas of a large city. This is not the intended meaning of the title. In this chapter, we will set out to see if there is any pattern to be discovered in the resting positions of the 20 drunks of *Figure 3.1* when the drunks have taken 49 steps of the same size. We will discover that the apparently chaotic scattering of the drunks about the lamp post conforms to a surprising pattern which occurs in many stochastically fluctuating systems in the world around us.

If we look at the set of drunks in their resting positions after their staggering dispersal as summarized in Figure 3.1 (a), we can define a new variable to describe their success or failure to reach the most probable distance from the lamp post. This new variable y is equal to the distance of the drunk from the expected dispersal radius as shown in Figure 3.1 (b). We now ask whether there is any mathematical equation which describes the values of y for the dispersed drunks. To help us in our search for a possible pattern of meaning in the observed positions of the scattered drunks, it is useful to develop some ideas which are drawn from the subject of probability. Tossing a coin to see whether it falls on the ground with its head or its tail showing is an age-old technique for making decisions when logic and desire fail to help us decide which choice to make in a given situation. When we flip a coin to let its final position guide us in our decisions, we are exploiting the fact that it is virtually impossible to predict the outcome of the flipped coin.

From a strict theoretical point of view, the outcome of the coin tossing should be predictable, since the movement of the coin must obey the laws of mechanics and motion. In theory, if we knew all of the various factors governing the coin's behavior, from the time it leaves the finger to the time when it comes to rest on a surface, one should be able to predict if its final resting position would show a head or a tail. In reality, however, our ability to predict the outcome of flipping the coin

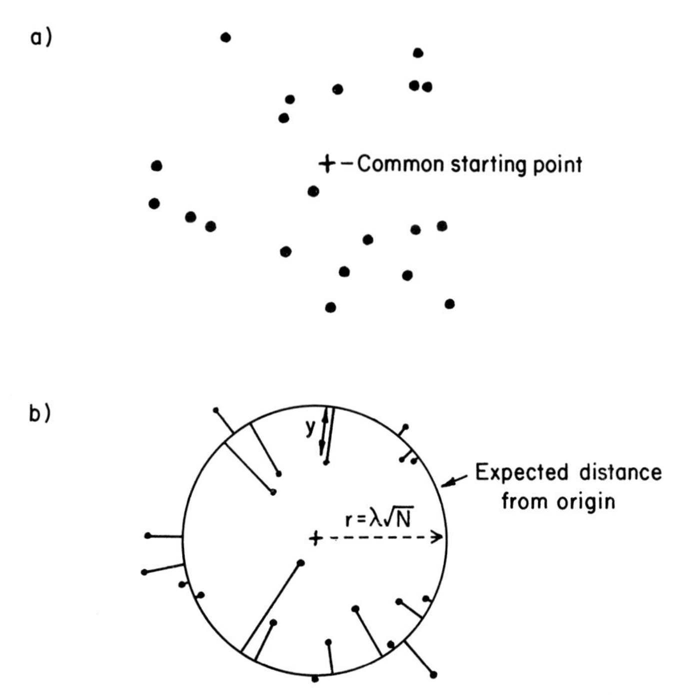

Figure 3.1. Looking for patterns in the dispersal behavior of a set of drunks can lead to some surprising stochastic relationships. a) Locations of 20 drunks after each drunk has taken 49 random steps of equal size. b) Location of the drunks with respect to their expected distance of $r = \lambda\sqrt{n}$ where λ = step size and n the number of steps.

is nil, because our ability to predict the coins movement is paralyzed by the fact that the final outcome is very sensitively linked to very small variations in the initial conditions, and all we can do in practice is study experimentally what happens when a coin is flipped.

Making bets on the possible outcome of randomly varying events such as the flipping of a coin is a widely practiced form of gambling. The need to recognize or know of probable patterns of events in gambling situations led to the study of what is now known as *probability theory* [1]. Today we can recognize that probability theory was the first step taken by mathematicians to develop pattern recognition in the behavior of deterministic chaotic systems. As Parratt has pointed out in his book on probability,

> *Around 1650 gambling was very popular in fashionable circles of French society. Games of dice, cards, coin tossing, roulette, etc., began to be highly developed. As personal honor and increasing amounts of money were involved, the need was felt for some formulae with which gambling chances could be calculated. Some of the influential gamblers sought the help of the leading mathematicians of the time such as Pascal,*

> *Fermat and later Dalembert and de Moivre. From this strange amalgam of gamblers and mathematicians we had the development of a scientific study of the laws of chance* [2].

We have become so familiar with the term "*laws of chance*" that we forget how contradictory this term is to the uninitiated. In an essay on chance by *Poincaré*, it is pointed out that

> *Chance is only the measure of our ignorance*

and that

> *Fortuitous phenomena are by definition those whose generative laws we do not know* [3].

Perhaps the difficulty of interpreting statements such as "the laws of chance" could be avoided if we did not use the word "law". What is described as the law of chance in a study of stochastic systems is really a probable pattern of events that occurs in an apparently random fluctuating system.

What kind of pattern, if any, occurs if we extend our coin flipping to look at what happens when we tumble 25 coins from a container and look at the pattern of heads and tails which are manifest on the surface receiving the coins? Most of the factors that determine a head or tails position of rest for a single flipped coin are present 25 times in the tumbling cascade of coins that we flip out of a container. Added to all of these causes is the interaction of these causes and the effect of coin to coin collisions as they tumble onto the surface. If we could put ourselves back into the position of the gambling aristocrats of the 1600s, without any of the subsequently developed knowledge of the patterns of probability which are described in our statistical textbooks, could we, at that time, have guessed that any pattern would emerge from the successive throwing of 25 coins? If we could find a probable pattern of events in the tumbling of the coins we might be able to decide what odds to accept on the possibility that there would be 17 heads manifest in the set of 25 tumbled coins.

To enable us to study not only the pattern of events when tossing 25 coins, but also the actual state of any one coin in the set of coins, an experiment was carried out with 24 pennies and 1 dime. When the coins were tipped onto the table the state of the dime was easily recognizable amongst the overall pattern of heads and tails manifest by the coins (a *dime* is a 10 cent silver coin used in North American coinage, it is nearly the same size as a cent).

In *Table 3.1* the pattern of events observed when the 25 coins were tumbled on a surface 20 times are recorded. In *Figure 3.2* the histogram of these observed populations is given. Superimposed on this histogram is the famous (or infamous, depending on your point of view) bell curve with which teachers are said to whip their students into uniform grade performance. As I have discussed elsewhere, teachers

Table 3.1. Experimental data summarizing the outcome of 20 throws of 25 coins.

Experiment	Number of Heads	Condition of Dime
1	11	Tails
2	14	Heads
3	10	Tails
4	17	Heads
5	12	Tails
6	12	Tails
7	10	Tails
8	13	Tails
9	10	Heads
10	13	Heads
11	14	Tails
12	7	Heads
13	15	Tails
14	8	Heads
15	13	Tails
16	13	Tails
17	14	Tails
18	9	Tails
19	12	Heads
20	14	Heads

who use bell curves to suppress the number of A's in a class should appropriately be called ding-dongs! (see the discussion given in [4] pages 154 to 156). A reader can be forgiven for being suspicious at a rush to fit a bell curve to the sparse data of the histogram of Figure 3.2. The writer must confess that the main reason for claiming the fit of such a curve to sparse data is that he knew the answer already. The reason we set out to fit a bell curve to such a small set of data is that this way we can keep things simple, so that the reader can more easily follow the discussion without being drowned in data. The skeptical reader is invited to perform the experiment 1,000 times to discover for himself that the fitting of the bell curve to the data of Figure 3.2 was an exercise in anticipatory creativity!

Section 3.2. Seductive (Often Meaningless) Patterns in Random Time Sequences

Before moving on to a discussion of the possibility that the bell curve can be used to discuss the distribution of the population of heads observed in the tumbled coins, it is interesting to note the sequence of heads and tails recorded for the dime. This

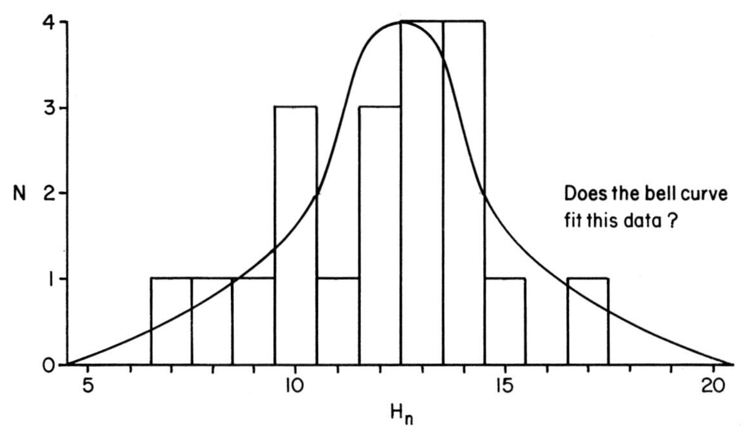

Figure 3.2. The pattern of the recorded population of heads in a set of 25 coins tumbled at random onto a table 20 times may be describable by the "bell curve". H_n: the number of heads appearing in a set of 25 coins.
N: the number of times out of 20 that H_n occurred.

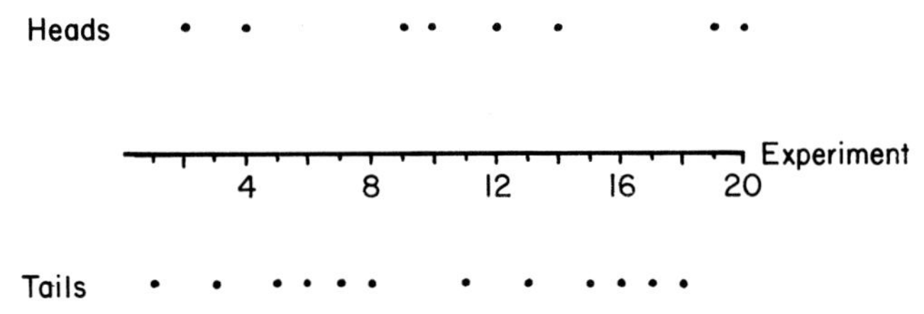

Figure 3.3. Stochastic patterns which "appear" in short run data can sometimes seduce investigators into generating grand theories based on sparse data.

sequence is set out as a time series in *Figure 3.3*. The student who carried out the experiment with the tumbling coins was fascinated by the fact that the pattern of heads and tails appeared to repeat itself with a strange rhythm. The pattern beginning at ll is an exact replica of the pattern in the first 10 events. However, when I asked him why he didn't carry out the experiment for another 10 throws to see if the pattern repeated again, he admitted with a smile that he knew that the chances of a repeat of the pattern were extremely rare and that he didn't really believe that there was any significance in the pattern. In spite of this admission, he still insisted that the repeated pattern was interesting. Unfortunately, even mature scientists, specialists in their own fields but unfamiliar with stochastic events, can get very excited about repeating stochastic patterns such as Figure 3.3. They can build wonderful theories upon similar shaky stochastic foundations. When one is observing the pattern of heads and tails in a set of coins, it is quite easy to repeat the experiment again to show that the pattern had occurred by random chance. However, if one is looking at the pattern of cold and warm winters in successive years, one cannot repeat the experiment easily. It is all too tempting to see meaning in a weather cycle when chance has been scrambling the variations. In the early 1960s people who had observed the weather since the 1940s were predicting the possibility of an ice age because they saw a cooling trend in the data. By the 1980s the predictions had shifted to a discovery of what was seen as a global warming in the weather data that promised to frizzle the agricultural endeavors of the western and to flood the cities of the eastern United States. The scientific community must now await another 20 years of fluctuations to see if we are to fry or freeze, but as we wait there is no shortage of people willing to predict disaster before the next set of data is available (see the discussion of long-range weather patterns in Chapter 15).

Stochastic time sequences of the type shown in Figure 3.3 are as seductive and as dangerous as the *Sirens* of Greek mythology. In Greek mythology a Siren was half woman and half bird. They lived on an island near the twin nautical hazards of *Scylla* and *Charybdis* [5]. Scylla was a six-headed sea monster. Each head had three rows of teeth and the lower limbs of the monster were snakes and barking dogs. Charybdis was a whirlpool. As the Greek mariners approached these two hazards, placed on the opposite sides of a narrow channel, the Sirens sang a melodious song that lured men to their deaths by making them oblivious of the dangers on either side of them. Physical scientists must exercise great discipline when faced with interesting stochastic patterns, to avoid being seduced by the data into wrecking their careers as effectively as if they had encountered Scylla or Charybdis! Above all stochastic time sequences one should perhaps print the warning

"BEWARE THE STOCHASTIC SIRENS"

It requires every skill in statistical analysis of data to separate the illusory stochastic time sequence sirens from the real rhythms of the universe.

Section 3.3. Technical Names for the Bell Curve

The *bell curve*, which we will find turns up as a surprising pattern in many stochastic situations, has several names. The term "bell curve" is obviously derived from the shape of the curve. Technically it is known variously as the *Gaussian probability curve* or a *normal probability curve* [1, 2]. The term normal probability distribution comes from the fact that so many stochastic systems, such as the observed population of heads in 25 coins tumbled many times, can be described by this mathematical curve. Because of its frequency of occurrence, many scientists came to expect such a probability function to be the normal type of probability that occurs in nature. The term Gaussian probability curve comes from the fact that a formula for describing the structure of the curve was derived mathematically by the German mathematician Karl Friedrich *Gauss* (1777–1855). Gauss, the son of a bricklayer, exhibited such great talent as a child that the Duke of the area of Germany in which he lived paid for his university education. In 1807, at the age of 30, Gauss became Professor of Mathematics and Director of the Astronomy Observatory at the University of Göttingen. He was one of the pioneer developers of the theory of electromagnetism, and helped to develop telegraphy. The unit of magnetic flux density known as *the Gauss* is named after him. Although the honor of discovery is usually given to Gauss, the curve bearing his name was first described mathematically by the English mathematician *de Moivre* [6]. Abraham de Moivre (1667–1754) was born in Champagne and educated in Brussels. He lived most of his adult life in London as a refugee. He is sometimes referred to as an Englishman of French extraction. It is said that

> *He was a poor man – a lover of books and greatly interested in the theories of probabilities*

He was probably a Hugenot (Protestant refugee from religious persecution in France). He was elected to the Royal Society of Great Britain in 1697, and was later chosen to be a referee in the controversy between Newton and Leibnitz as to who had invented differential calculus (see the discussion of this subject in Chapter 14). De Moivre published his description of the bell curve in 1733, after studying the patterns of chance events in gambling games. He probably discovered it experimentally as a surprising pattern emerging from the deterministic chaotic events at the gambling table.

In these notes we shall describe the bell curve probability distribution as the Gaussian probability curve. This name is preferred to avoid any implication that all interacting stochastic systems generate variables that can be described by this type of curve, and to emphasize that other probability distributions are not abnormal. Although historically de Moivre should be credited with the discovery of this curve, the use of Gauss's name to describe this particular function is so wide-spread that it would not be useful to insist on a new terminology honoring de Moivre.

Section 3.4. Mathematical Description of the Structure of the Gaussian Probability Curve

The nonmathematically inclined reader can skip this section; the mathematics of the curve are given here for the sake of complete presentation and to familiarize the student reader with certain mathematical terms and data processing procedures.

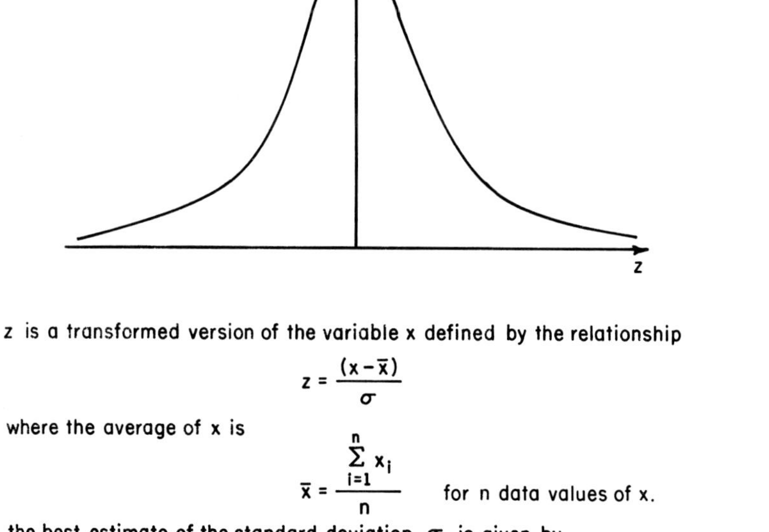

y
z
z is a transformed version of the variable x defined by the relationship
$$z = \frac{(x - \bar{x})}{\sigma}$$
where the average of x is
$$\bar{x} = \frac{\sum_{i=1}^{n} x_i}{n}$$ for n data values of x.
the best estimate of the standard deviation, σ, is given by
$$s_n^2 = \frac{\sum_{i=1}^{n} (x - \bar{x})^2}{n}$$ (σ^2 is known as the variance)
$$y = \frac{1}{\sqrt{2\pi}} e^{\frac{-z^2}{2}}$$
where
$$e = \frac{1}{0!} + \frac{1}{1!} + \frac{1}{2!} + \frac{1}{3!} \cdots \doteq 2.7183 \text{ and } n! = n(n-1)(n-2)\cdots(2)(1)$$
($\doteq$ means "is approximately equal to")

Figure 3.4. The mathematics of the equation which describes the structure of the Gaussian probability function can be quite intimidating to those new to statistical relationships.

In *Figure 3.4* the mathematical equation which generates the shape of the Gaussian probability curve is shown. We first note that in presenting the data from an experiment, such as the observed populations of heads in a set of coins, it is useful to present the variables in a standard form which makes use of the concepts of an average value and a standard deviation, a quantity used to describe the range of variation in the observed data. Some of the symbols used in the equations of Figure 3.4 will be familiar to many readers, but since this book is intended for use in self study, and because one cannot be sure that everyone shares the same mathematical background, the meanings of the various symbols will be discussed in detail.

We first note that the average of the number of heads which are manifest in a throw of the coins is variously represented by symbols $\bar{x}$ and μ. To calculate the average from the observed values, we first carry out the operation which is written symbolically:

$$\bar{x} = \sum_{i=1}^{i=n} x_i$$

In this formula the symbol $\sum$ is the capital form of the Greek letter S, known as *sigma* (the lower case form of this letter is σ). The symbol $\sum$ instructs the mathematician to carry out the operation of adding all the items following the symbol, that is, the values x_i. On the large letter $\sum$ the subscripts $i = 1$ at the bottom and $i = n$ at the top tell the mathematician to add a series of values of x_i with i varying from 1 to n. To obtain the average, this sum is divided by the number of pieces of information included in the sum. In the general case this is n. In our coin tumbling experiment, our average population of heads in 25 coins thrown onto the table 20 times would be calculated from $i = 1$ to $i = 20$ and the sum would be divided by 20.

Using the data of Table 3.1 we obtain an average value of $\bar{x} = 12.6$. As a consequence of averaging, we note that we have less than a whole head in the value of the average. Familiarity with averages makes us sometimes forget how strange it is to say that on average we can expect 12 whole heads plus 0.6 of a head. To the uninitiated a 0.6 coin is a strange result requiring fractured coins!

To describe the range of variations in a variable such as the number of heads appearing in a set of data, mathematicians use a quantity known as the *standard deviation of the data*. The formula for estimating the standard deviation of a set of data, usually denoted by σ is given in Figure 3.4.

Let us now look at the various parts of the formula which tells us how to calculate the standard deviation. First of all, we notice that the quantity

$$(x_i - \bar{x})^2$$

is the difference of a given value of x_i from the average value of x multiplied by itself (i.e., squared). In other words it is "the difference from the mean" squared. To calculate the standard deviation, these squared differences are added and then divided by the number of pieces of information, n. This sum of squared differences is obviously larger when some of the values are very different from the mean, and smaller when the spread of the values cluster tightly around the average value. Mathematicians call the square root of these squared differences the standard deviation of the data set. The different values of x_i used to calculate the average and standard deviation constitute the *data set* (note that we have to use a number based on the squared values of the differences from the mean to study the range of values present in a set of data, because adding up the differences of x_i from the mean will give an expected value of 0). In Figure 3.4 the steps in the calculation of the standard deviation for a data set such as that of Table 3.1 is shown.

If one were to repeat several times the experiment of throwing the coins 20 times, the value of the standard deviation calculated for each set of data would vary from one set of data to another. Because of this fact, our particular set of 20 values of H_a, the average number of heads, gives us only an estimate of the true or exact value of the standard deviation of all possible sets of head population in randomly tumbled sets of 25 coins. The theoretical value of the square of the standard deviations for an infinitely large set of experiments is given a special name and symbol by the mathematicians. This name is the "*variance*" of the data and is represented by the symbol σ^2. In the language of the mathematician the value of s, the standard deviation of the data set of Table 3.1, is described as an estimate of the square root of the variance of the population of all possible experiments of which the data set in Table 3.1 is a sample set. When carrying out calculations using the theoretical formula described in Figure 3.4, the value of σ is usually not known, and we have to use s in our calculation of the transformed version of the variable x to calculate the values of z. When looking at information such as the data set of Table 3.1 and the transformed data sets, experimentalists often refer to the *original data,* i.e., the recorded value of x, the number of heads visible in a throw of the coins, as the *raw data*. The values of z calculated from the values of x are then referred to as the *transformed data*.

Mathematicians who have studied statistical data for many years have found out that they can improve their estimate of σ from the value of s by using what is known as *Bessel's correction*. It is named after Frederich Wilhelm *Bessel*, a German astronomer (1784–1846) famous for the development of *Bessel functions* used in advanced physics studies. Bessel's correction of the estimated variance allows for the fact that the value of s is calculated from only n items of information (in our case 20 sets of data). The numerical value of Bessel's correction is

$$\sigma^2 = s_c^2 \left(\frac{n}{n-1} \right)$$

where σ is the best estimate of the theoretical value of σ calculated from s_c, the calculated value of s based on n items of data. We can see that if we applied Bessel's correction in our experiment, where $n = 20$, the Bessel correction is 1.03. If we had an experimental set involving 100 throws of the coins, the Bessel correction factor would drop to 1.005. The actual formula for the bell curve using our transformed variables is

$$y = \frac{1}{\sqrt{2\pi}} e^{-z^2/2}.$$

This includes two famous irrational numbers, π and another irrational number which is represented by the symbol e. Like our friend π, the irrational number e has the habit of turning up in unexpected places. John *Napier* (1550–1617), a Scottish mathematician who invented logarithms for use in multiplication and division, used this number e as the basis of his logarithms. Another British scientist, Henry Briggs (1561–1630), suggested the change from the base e to the base 10 for everyday logarithms. Strictly speaking, the logarithmic tables placed at the back of textbooks should be called *Briggsian Logarithms*. Logarithms to the base e are described as *natural* or *Napierian* logarithms, although many students can't see anything natural about them!

Two different symbols are used when writing down the logarithm of a number to indicate whether the scientist is using a logarithm based on the number 10 or on the number e. These various conventions are as follows:

$$\log_{10} N, \quad \mathrm{Log}\, N, \quad \ln N$$

where ln indicates logarithm to the base e where

$$\mathrm{e} = \tfrac{1}{0!} + \tfrac{1}{1!} + \tfrac{1}{2!} + \tfrac{1}{3!} + \cdots = 2.7183$$

Just why the number e turns up in so many odd places is one of the mysterious and surprising aspects of natural science. Many scientists have wondered why this number appears to be one of the basic nuts and bolts of the physical universe. No satisfying answer has ever been discovered.

The reader exploring the mysteries of mathematics for the first time may find it strange to see exclamation marks put down by the side of each number in this series. The sign which is familiar as the exclamation mark in literature has a special meaning in mathematics. It symbolizes what is known as the *factorial operation*. When a mathematician sees the sign "!" after a number, he or she knows that this is an instruction to carry out the mathematical operation of multiplying the number with the exclamation mark by a series of terms which are created by subtracting one from the listed number, and then from that new number, and so on until the final act of subtracting one creates the number one. Thus 4! means "calculate factorial 4" which is

$$4 \times 3 \times 2 \times 1,$$

again

$$6! = 6 \times 5 \times 4 \times 3 \times 2 \times 1.$$

We note that mathematicians have all agreed that the term 0! and the term 1! shall mean 1. Using what we know about the factorial operation, we can write the series for calculating e given in Figure 3.4 in the form

$$\mathrm{e} = 1 + 1 + 0.5 + 0.16667 + 0.04167 + 0.00833 \ldots$$

For the purposes of this book, we will take the value of e to be 2.7183. The fact that the bell curve structure can be derived from well-known mathematical items such as π and e should not distract us from the real justification for using the bell curve to describe natural phenomenon, such as the population of heads in a throw of 25 coins: It can be shown experimentally that this type of curve describes the type of pattern which emerges from many deterministic chaotic systems. If we did not have the theoretical formula for describing the structure of the bell curve, we would still have the experimental fact that this structure exists. From experiments, we would know it to be useful for describing the patterns that emerge when studying the behavior of some deterministic chaotic systems. On the other hand, if we had been able to derive the bell curve from theory, it would have been a useless theoretical exercise in the juggling of mathematical constants if we could not discover any pattern of events which fitted the bell curve. It is interesting to note that the mathematician E.T. Whittaker made the statement

> *Everybody believes in [the Gaussian probability distribution], the experimentalists because they believe that it can be proved by mathematics, and mathematicians because they believe it has been established by observations* [2].

In this quote I have altered the original term used by Whittaker. He used the term the "exponential law of errors" where I have written "Gaussian probability distribution". This is because the older style of terminology referring to the law of errors has been superseded in modern technical English by the term Gaussian probability distribution. See the discussion of experimental error in Section 3.10.

Section 3.5. Stretching Data Mathematically with Gaussian Probability Graph Paper

If students had to work directly with the mathematical formula for the bell curve to test that the probability distribution of observed experimental data could be described by the Gaussian probability function, the use of Gaussian probability statistics would probably be left to the professional statisticians. Fortunately, however, beginning experimentalists can learn, in a relatively painless manner, how to test and manipulate experimental data with respect to Gaussian probability distribution functions by using a special graph paper. This graph paper is known variously as *Gaussian probability paper*, *normal probability paper* and *arithmetic probability paper*. As a first step towards understanding the structure of Gaussian probability paper, it should be noted that the type of data display shown in Figure 3.2 is called a *distributed histogram* of the data. This is because the data has been shown spread out along the abscissa of the graph. An alternative type of histogram used by the scientist is the *cumulative data histogram*. When dealing with data such as that in Table 3.1, one can work with either the *cumulative undersize histogram* or the *cumulative oversize histogram*. The term *undersize* in this context means observed populations, in our case the sets of coins with a given number of heads, equal to or less than a stated value. The *oversize* cumulative histogram records the frequency of populations of heads greater than or equal to a stated value.

Table 3.2. It is often convenient to convert experimental data into what are known as cumulative distributions. The data from Table 3.1 is converted into what are known as the cumulative oversize distribution and the cumulative undersize distribution.

H_n	Frequency	H_0	H_0 [%]	H_u	H_u [%]
7	1	20	100	1	5
8	1	19	95	2	10
9	1	18	90	3	15
10	3	17	85	6	30
11	1	14	70	7	35
12	3	13	65	10	50
13	4	10	50	14	70
14	4	6	30	18	90
15	1	2	10	19	95
16	0	2	10	19	95
17	1	1	5	20	100

H_n represents the number of heads appearing when twenty-five coins are tossed at once. H_0 represents the oversize distribution of the number of heads appearing, that is the cumulative number of experiments showing the same or more heads than the stated number H_n. This may also be written as: $H_0 \geq H_n$. H_u represents the undersize distribution of the number of heads appearing, or the cumulative number of experiments showing the same or less heads than the stated number H_n. Similar to the above this may be written as: $H_u \leq H_n$.

To create the cumulative forms of the histogram, one adds up the occurrence of given populations in the appropriate manner. If we were to work with the undersized cumulative histogram for the data of Table 3.1, we would be interested in the number of experiments which had populations of heads equal to or less than a certain value. On the other hand, if we were working with the oversized information, we would calculate the number of populations that had a number of heads greater than or equal to the stated value. For illustrative purposes both the oversize and undersize cumulative populations for the original data of Table 3.1 are calculated and displayed in *Table 3.2*. In *Figure 3.5* the two cumulative histograms are shown.

If one marks the midpoints of the bars on the cumulative histograms, it can be noted that the bell curve of Figure 3.2 becomes an S curve in Figure 3.5. If we now imagine that one were to draw the cumulative histogram on a sheet of rubber which was thicker in the middle and thinner at the two ends, then one could stretch out the curves of Figure 3.5 so that they became straight lines. In fact mathematicians have used the equations of the Gaussian probability curve to calculate how to stretch out the probability axis (the frequency-of-occurrence axis) so that the S curve becomes a straight line. The data from Table 3.2 are plotted on a sheet of this Gaussian probability paper as shown in *Figure 3.6*. The reader should note very carefully how the significance of the divisions between numbers on the abscissa are changed drastically as one moves from left to right. The difference between 1 and

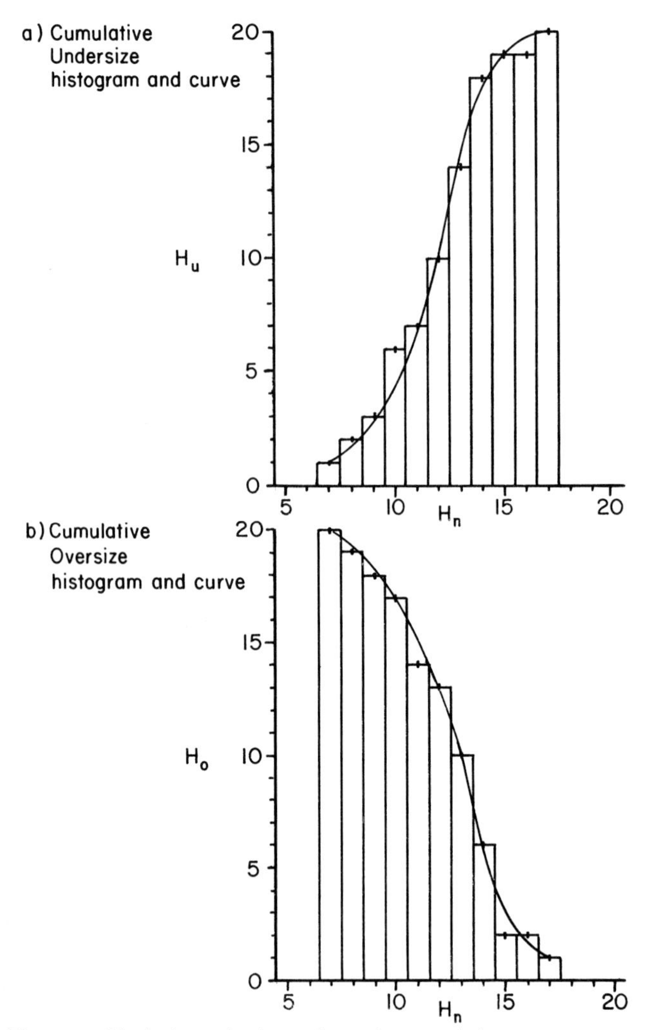

Figure 3.5. The tops of both the undersize and oversize cumulative occurrence curves for the tumbled 25 coins trace out "S" curves. a) Cumulative undersize histogram curve. b) Cumulative oversize population histogram curve. H_n : the number of heads showing in a set of 25 coins. H_u : the number of trials out of 20 showing the same or fewer heads. H_0 : the number of trials out of 20 showing the same or more heads.

2% occurrence is the same size physically as between 40 and 50% occurrence in the middle of the axis. Again, by the time one gets out to the far right of the abscissa, the difference between 98 and 99% is the same as the difference between 40 and 50%.

Technologists are often in a hurry. Because words like probability are rather long when writing information down by hand, in many engineering textbooks the name

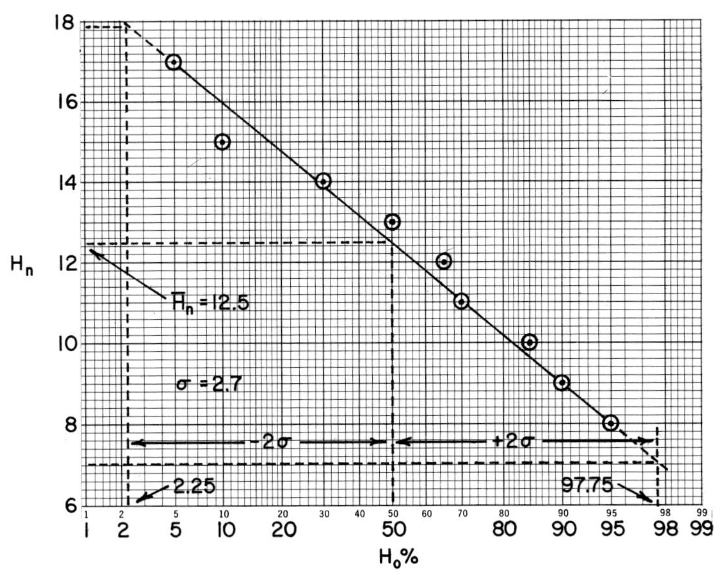

Figure 3.6. When experimental data can be described by the Gaussian probability function the cumulative frequency distribution of the data generates a straight line on Gaussian probability paper.

of the graph scale for the probability, as in Figure 3.6, is shortened to the "*probit*" scale.

To plot the data for the cumulative-greater-than population of the observed number of heads in the tumbled set of 25 coins, we use the oversized data of Table 3.2. The first piece of information that we have is that populations of 17 or more heads were 5% of the observed data. This point is plotted by moving up the 5% probability line until one crosses the 17 head line as shown in Figure 3.6. The next item of information from Table 3.2 is that 10% of the observed population had 15 or more heads leading to the data point located in Figure 3.6, as illustrated. Plotting all of the data from Table 3.2 we come down to the final point that we are able to plot: 95% of the observed coin populations had 8 or more heads (note that we cannot plot 0 or 100 percent on the probabilty scale).

It can be seen from Figure 3.6 that a dataline can be drawn through the points. This fact strongly suggests (but does not prove) that the variations in the observed number of heads in the thrown coins is a Gaussianly-distributed fluctuating population (this is a term often adopted by scientists discussing systems which can be described by the Gaussian probability distribution function). That is, the variations can be described by the bell curve.

It should be noted that one can quickly estimate the standard deviation of a set of data using this type of graph paper. It can be shown that

The range $\bar{x} \pm 1\sigma$ includes 68.3% of all possible values.
The range $\bar{x} \pm 2\sigma$ includes 95.5% of all possible values.
The range $\bar{x} + 3\sigma$ includes 99.7% of all possible values.

To use this fact, we first mark off 95.5% of possible values by making a mark at 2.25 on the probability scale and erecting a line to cut the dataline, in this case at 17.9 heads. We then move to 97.75 on the probit scale and construct the vertical line that cuts the dataline at 7.1 heads. We take the difference of these two dataline cuts and divide by 4 to obtain the value of ± 2.7 for the estimate of the standard deviation.

We note that this graphical technique of averaging the data gives us a most probable value for the population of heads in the thrown coins of 12.6. We know that the theoretical value is 12.5, so that we have come within 1% of the most probable value with this data plotted on the Gaussian probability graph paper (this good agreement is probably partly fortuitous but certainly encouraging!).

When using the estimate of the standard deviation derived from the graphical treatment of the data we do not use the Bessel correction, since the scatter of the data has been partially allowed for by drawing the best line through the scattered datapoints.

To obtain the estimate of the standard deviations, we had to extend the datalines beyond the last points. This procedure is known as *extrapolation* of the experimentally constructed dataline. For the purposes of this calculation, extrapolation is a relatively safe procedure but it should normally be carried out with great caution. We will discuss the problem of scatter in the tails of Gaussian distribution data later in this chapter.

The slope of the dataline on Gaussian probability paper corresponds to the range of variation in the data summarized on the graph. If the range of variability decreases, the slope of the line decreases. In the extreme case, when there is no variability, the dataline would consist of a line parallel to the probit axis.

Early on in our discussion of the tossed coins, we posed the question "What odds should a gambler accept that the coins will manifest 17 or more heads?". From the graph of Figure 3.6, it can be seen that a population of heads equal to or greater than 17 in the 25 coins occurs with a frequency of 1 in 20. Therefore a gambler should obtain odds of better than 20:1 to have a positive chance of winning in a sequence of throws of the 25 coins.

Section 3.6. When Are We Likely to Encounter Gaussian Probability Distribution in Stochastic Systems?

Now that we have discovered that some stochastically fluctuating systems manifest patterns of events that can be described by the Gaussian probability distribution,

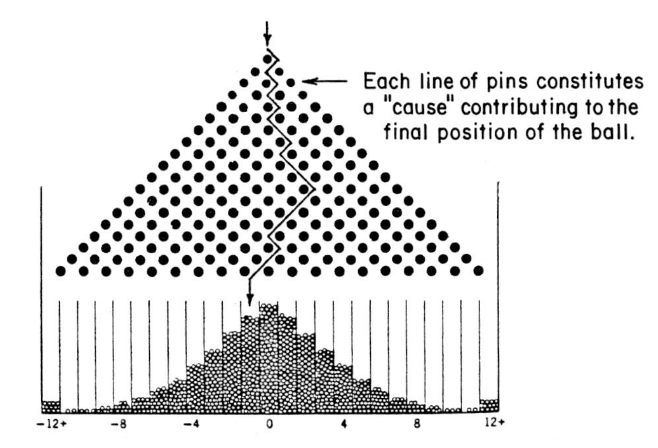

Figure 3.7. The "bell curve" distribution can be anticipated when many small causes interact randomly to produce a final effect (in this case the rest location of a ball falling through an array of deflector pins) [2].

have we any way of anticipating which type of deterministic chaotic system will be described by the Gaussian probability curve? From a study of various systems which generate patterns describable by the Gaussian probability distribution, it is probably safe to say that one can anticipate a set of events describable by the Gaussian probability curve when the pattern of events is generated by the random interaction of many small causes of comparable magnitude acting independently of each other to produce the observed results (see the cartoon at the beginning of this chapter). Outcomes of chaotic interaction which require the favorable combination of causes to produce certain events manifest other types of probability distribution functions (see Chapter 4). The physical system illustrated in *Figure 3.7* can help students new to stochastic systems to grasp the fact that multiple interactions of simple causes of approximately the same strength can generate a Gaussian probability distribution. Consider what happens when a ball is allowed to bounce its way down through a set of pins. When the ball hits a given pin it has a 50% chance of going left or right. The path taken by a ball down through the pins is a combination of the causes at each line of pins causing it to go left or right at that pin. The position that a ball reaches at the bottom of the array of pins is therefore dependent on many interacting causes of approximately the same strength. Each line of pins can be considered to be a cause acting on the ball at each stage of the pin table. Repeating this experiment many times always results in a bell curve distribution of balls at the bottom of the array of pins, as shown in the diagram. The bell curve distribution of the balls is a surprising pattern of organization appearing in a chaotic system of randomly interacting causes of the same order of magnitude (for a more extensive discussion of this type of system, see p.152 of [4]). The fact that the randomly interacting multiplicity of small causes produces events which vary in such a way that they are Gaussianly distributed about a most probable value is known to the mathematician as the *central limits theorem* [2].

The scientist who pioneered the use of the Gaussian probability distribution in applied science was Sir Francis *Galton*. Galton (1822–1911) was a first cousin of Darwin, who established the theories of evolution. Galton compiled statistics on the properties of identical twins, the frequency of yawns, lifespan, the sterility of heiresses (a major mechanism in the accumulation of wealth amongst the aristocracy) and the inheritance of physical and mental characteristics. Commenting on the fact that so many different systems could be described by the Gaussian distribution function, Sir Francis Galton once said

> *I know of scarcely anything so apt to impress the imagination as the wonderful form of cosmic order expressed by the law of frequency of error [his name for the Gaussian probability distribution]. The law would have been personified by the Greeks and defied if they had known of it. It ranged with serenity and incomplete self-efacement amidst the wildest confusion. The larger the mob and the greater the apparent anarchy, the more perfect is its sway; it is the supreme law of unreason. Whenever a large sample of chaotic elements are taken in hand and marshalled in the order of their magnitude, an unsuspected and most beautiful form of regularity proves to have been latent all along* [7].

Section 3.7. Gaussianly Scattered Drunks

Finally, after many detours to acquire the relevant theory and mathematical skills, we can come back to the question posed at the beginning of this chapter: Is there any pattern to the scatter of the drunks about their expected dispersion distance? The position reached by any staggering drunk is the consequence of 49 steps, each taken at random to produce a probable value of distance from a point. We can regard any individual step as a contributory cause to the final position of the drunk. Therefore, the final position of the drunk can be seen as having been caused by 49, equally strong, causes interacting at random. Therefore, from what we have discovered about the Gaussian probability distribution, we can anticipate that the scatter diagram of the drunks is such that their distance from the expected dispersal distance, the variable y, should be describable by the Gaussian probability distribution. In *Figure 3.8* the values of y are plotted on Gaussian probability graph paper. It can be seen that the values are describable by the Gaussian distribution. If we used the other term for the Gaussian distribution, we could say that the data of Figure 3.1 show the normal distribution of drunks!

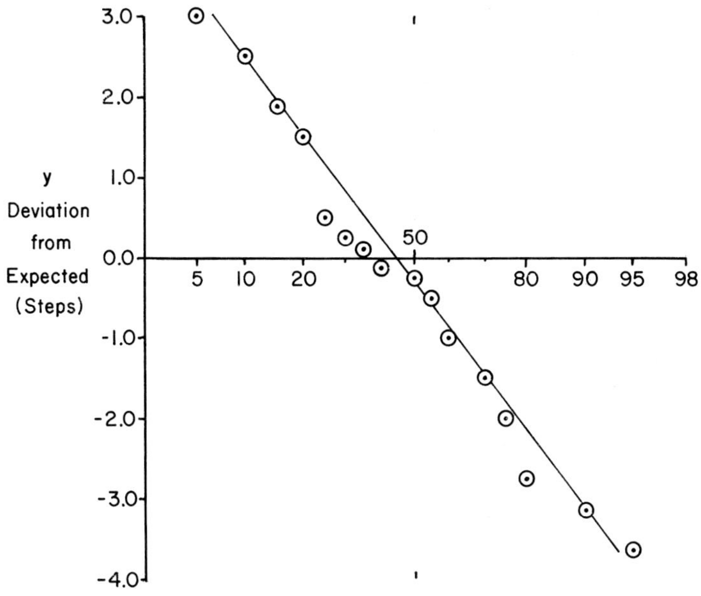

Figure 3.8. The distribution of dispersing drunks can be summarized on Gaussian probability graph paper.

Section 3.8. Gaussian Variations in the Structure of Powder Mixtures

An important industrial problem involves the making of mixtures of two or more powders. For example, in the pharmaceutical industry a powder of aspirin is often mixed with another harmless, tasteless powder, such as starch, to create a tablet of manageable size. Systems as diverse as dry rocket fuel, cake mixes, powdered soup and garden fertilizer are composed of different powdered ingredients that must be mixed together. Face powder is made from a mixture of powders. Metal alloys are sometimes made from powder mixtures. An important problem to be tackled when carrying out a powder mixing process is to see whether fluctuations observed in the concentration of one powder in a sample of the mixture occur by random chance, or because the mixer used to prepare a mixture is not functioning adequately. Consider, for example, the problem of mixing a powder (A) which will constitute 10% of the final mixture into another powder (B). If the two are placed in a blender, then from time to time a sample of the mixture would be taken out and examined to see how much of powder A was contained in the sample. It can be shown that, for theoretical reasons, one should always try to have the components of a powder mixture of the same physical size. To simulate how the structure of a sample taken from a mixture varies by chance, even when the best possible random mixture is being created by the blender, we will simulate the structure of various samples taken from the mixture. We will assume that if we were to take a small piece of the random number table, we could simulate a 10% mixture of a particular ingredient (A) by turning every 4 in the random number table into a black square representing the component of interest. We would then leave the other squares white, representing the rest of the mixture. For illustrative purposes we will assume that the sample of powder taken out of the mixer can be represented as a cube. We will express the size of the sample taken in terms of d, the size of grains represented by the black pixels. Let us assume that the smallest sample that we would take would be a cube of side length $\gamma = 3d$ where d is the size of simulated grains of powder, that is the pixel size. We could then assume that we could chop the cube into three equally thick slices of thickness d. We could then simulate the structure of the three slices of the sample volume by taking three adjacent sets of nine digits from the random number table. The structure of the $3d$ sample volume could be simulated by the array of digits shown in *Figure 3.9*(a). The expected concentration of the digit 4 in each of the slices of the simulated volume is the same: 10%. The measured content of a series of simulated sample cubes represents the variation by random chance that would occur when sampling a mixture using a size of $3d$ where d is the size of the grains being dispersed in the mixture. Four of the possible variations simulated in this way are presented in Figure 3.9(a). The reason that the actual numbers of the digit 4 vary in the table is due to the interaction of the random procedures being used to generate the digits' equal probability. This corresponds to the fact that in making the powder mixture the position of any one grain is supposed to be randomized without regard to the position of the other grains. The word "supposed" is used in this context because one of the real problems in dealing with actual powders is that very often the grains of the powder have a high attraction to each other. Unless the mixer incorporates mechanisms to separate the strongly attracted grains of the powder to be dispersed, the mixture ends up less than well dispersed according to the criteria that each grain of the powder is completely randomized without reference to the position of the other grains. From our knowledge of the factors causing the variations in content of the simulated sample volume, we would anticipate that the variations in content of the ingredient of interest should be describable by the Gaussian probability function. In Figure 3.9(b) the fluctuations in the content of active ingredient in 100 samples of size $3d$ are shown. It can be seen that the fluctuations are Gaussianly distributed, and that in this situation even 100 pieces of information are not sufficient to pin down the percentage value expected of the sampled volumes, because the average value for the 100 samples is not the expected value of 10%. Most people shown the graph of Figure 3.9 are surprised at how varied the sample of mixture content can be. If one were to take a sample of powder from a real mixer with the sample size being equal to $3d$, then even a 25% ingredient

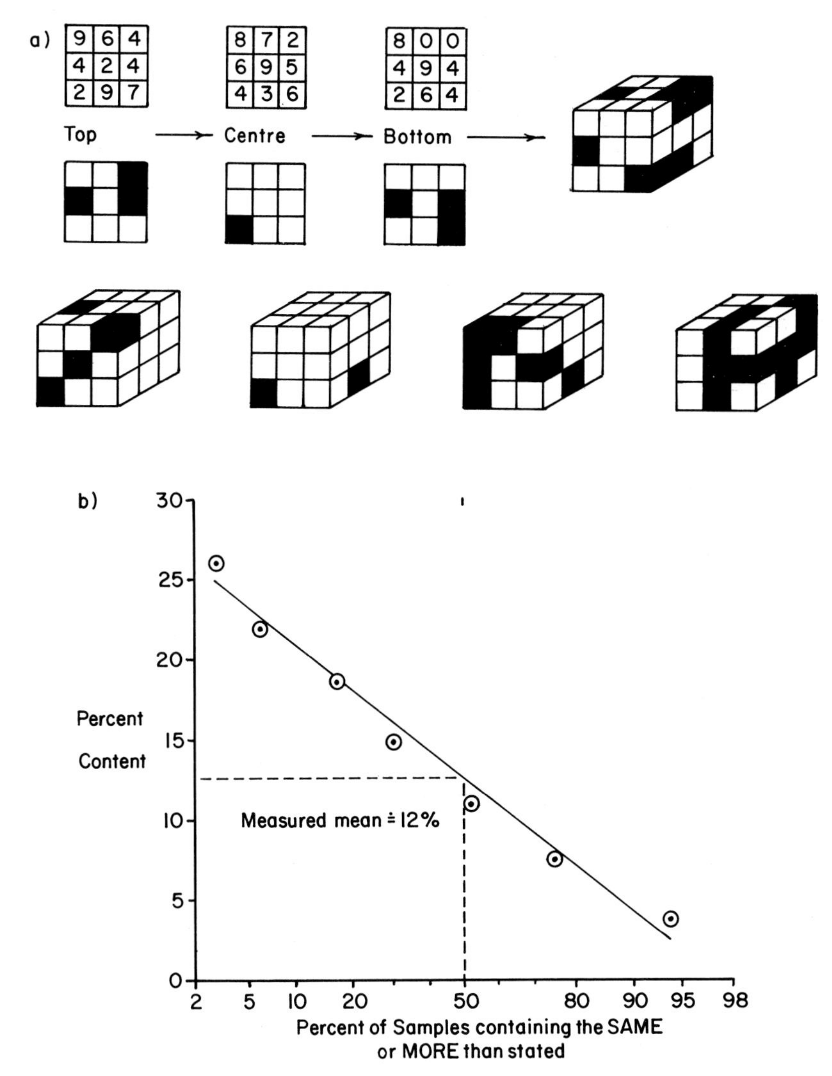

Figure 3.9. The legal variations in random mix of two powder constituents are distributed according to the Gaussian probability function. a) Simulated sample cubes of size $3d$. b) Legal variations in mixture richness for a 10% mixture in a sample of size $3d$ where d is the size of the component grains.

content, when one is anticipating a 10% ingredient level, has a one in fifty random chance of occurrance. Therefore such a sample would not be an indication that the mixer was not working. The physical significance of the variations of Figure 3.9(b) depends upon the situation encountered, but when working with powder mixtures one would not normally like to have to make tablets as small as $3d$ from the mixture, because of this wide variation in real content.

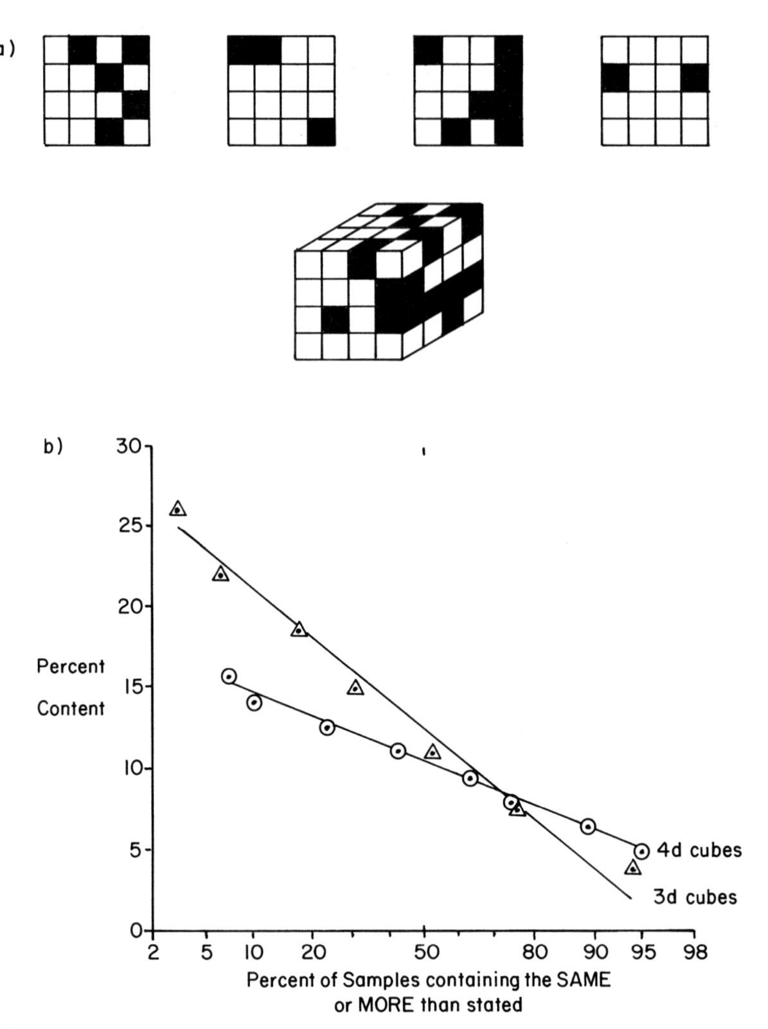

Figure 3.10. As the sample size increases, the recorded variations of the mixture richness decreases. a) Simulated sample of volume $(4d)^3$. b) Simulated variation in 100 samples of a 10% mixture for two different sample sizes.

To find out how big a tablet one would have to make to have the variations by random chance fall to within a given range, one can continue the simulation experiment by simulating the concentration fluctuations in samples of larger and larger volume. In *Figure 3.10* the variations in a sampled volume of size $4d$ are compared to those of a sample size of $3d$. The appearance of the four slices of such a sample of simulated powder mixture is shown in Figure 3.10(a). It can be seen that again the variation is Gaussianly distributed, but that the standard deviation for the sampled volumes has dropped dramatically as we increase the sampled cube from $3d$

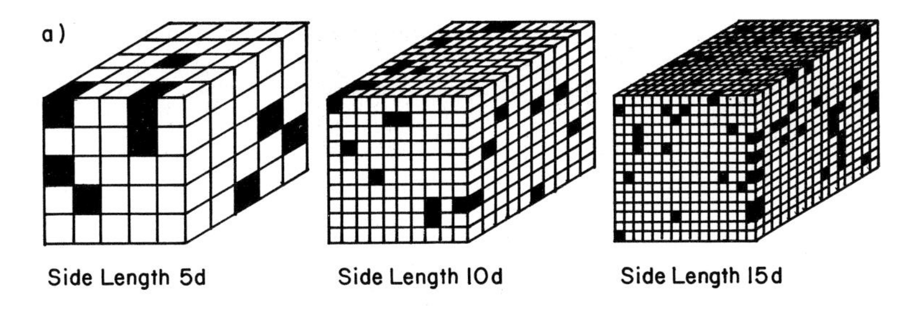

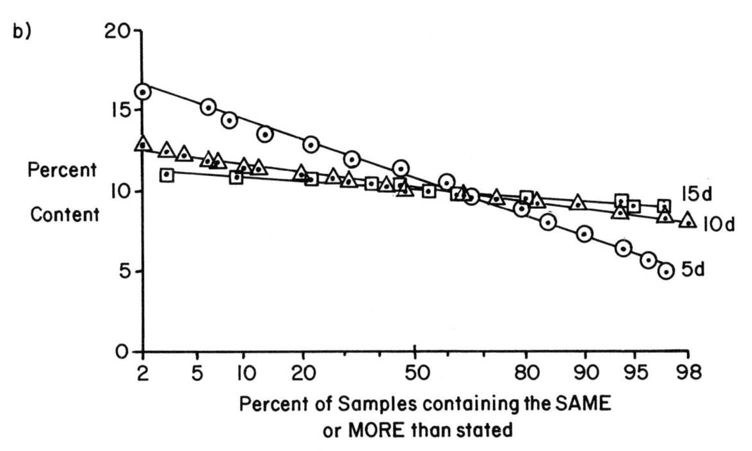

Figure 3.11. Permitted variations in measured mixture richness decline rapidly as the size of the inspected sample increases. a) The appearance of 5, 10, 15*d* cubes. b) Comparison of the distributrions for 5, 10, and 15*d* cubes.

to 4*d*. In *Figure 3.11* the appearance of the simulated sampled volumes of sizes 5*d*, 10*d*, and 15*d* are shown. The fluctuation even at 15*d* is such that in 1% of the samples one could obtain a value as low as 8.4% active content (black cubes) and a value as high as 10.5% content.

When evaluating the performance of a powder blender, one can evaluate the richness of a sample taken from a mixture and compare it with the possible variations about an expected value. For example, if one were generating a mixture of two powders in which one ingredient had an expected percentage of 10%, and if a sample of size 15*d* from the mixer with a content of 12% were obtained, this would indicate that the mixing process was probably incomplete. If many subsequent mixture samples continued at this high level, then the design of the blender would have to be re-evaluated and perhaps altered (a colleague, N. I. Robb, has developed a program for IBM compatible computers which generates a graph of legal variations at any mixture richness and inspection sample size, and also generates and displays the appearance of the mixture as in Figure 3.11 [8]).

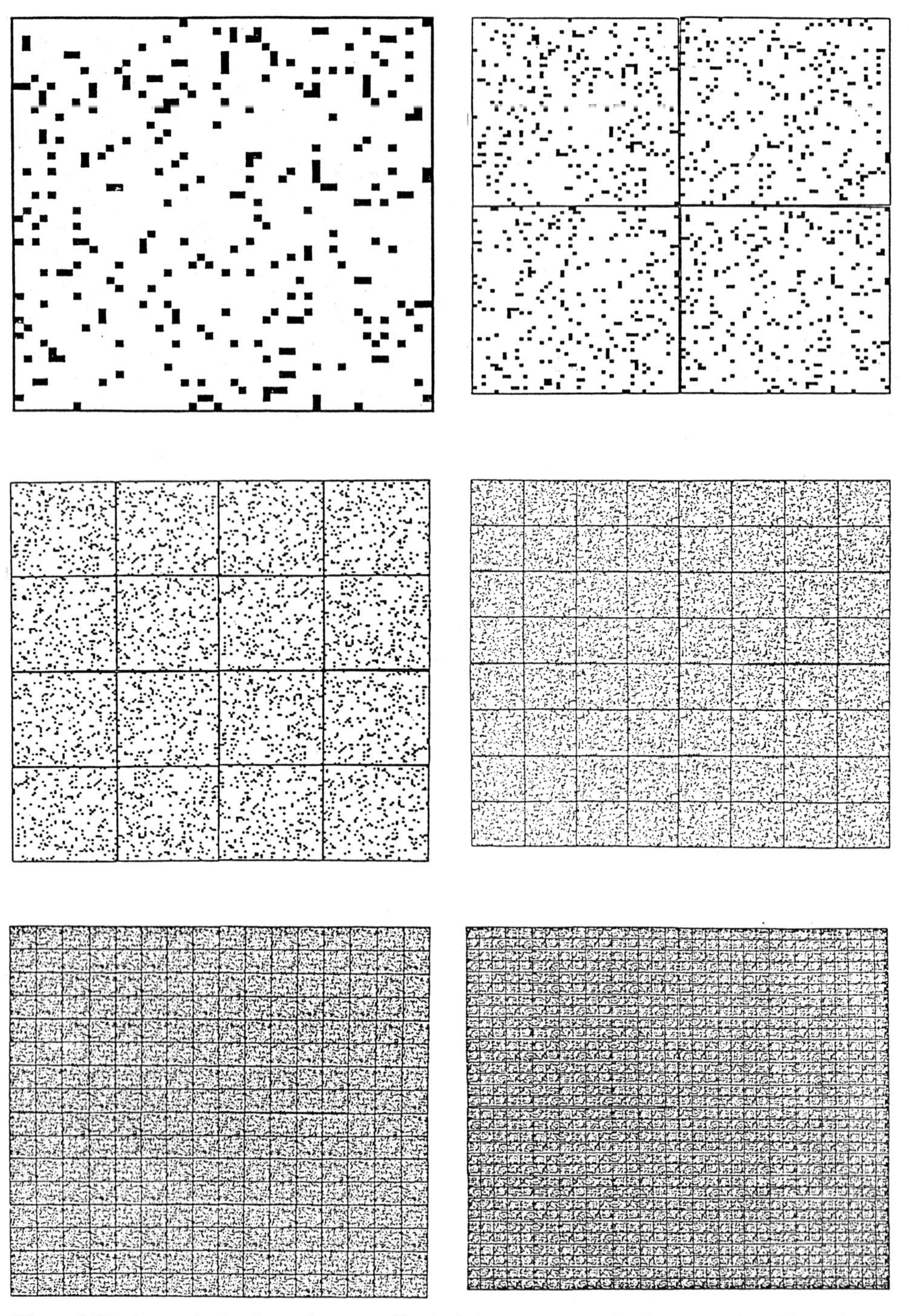

Figure 3.12. A quantized universe becomes effectively homogeneous when large samples of the universe are inspected.

When studying the properties of powder mixtures, one is actually looking at a quantized universe in which the quantum of material is relatively large compared to the container in which it is being mixed, as compared to the quantum of energy in the physical universe around us. Looking at the structure of a sample of the mixture that is only $3d$ in edge dimension, is similar to looking at the distribution of energy in a quantized universe. The quantized structure of the mixture has to be taken into account at such high resolution inspection. As one looks at larger and larger samples of a mixture, one is covering the transition from the behavior of a quantized universe to the behavior in one in which the system can be treated as being homogeneous and continuously variable. Although this discussion of mixture richness appears to be a relatively simple system dealing with an industrial problem, the study prepares the student for tackling a quantum universe at high resolution and helps the student to understand how a quantized universe becomes effectively a continuous variable system when one is dealing with a system which is large compared to the quantum size. When teaching students the relationship between quantum physics and classical physics, I have found the picture in *Figure 3.12* useful. In this figure the appearance of a quantized universe at various levels of inspection are illustrated. By the time the composite universe contains just over 1,000 fields of view, the universe has become almost a uniformly grey background.

When discussing the structure of a system such as that of Figure 3.12, the system is sometimes described as *pseudohomogeneous*. This means that the system is essentially homogeneous (of the same texture throughout) for the purpose of a given operation or perspective. The words *heterogeneous* and *homogeneous*, which mean the exact opposite of each other, can cause confusion to many students. To help clarify the proper usage of both words, several words related to the original Greek root words homo and hetero are shown in *Figure 3.13*.

Section 3.9. Using the Gaussian Probability Curve to Characterize Uncertainty in Experimental Data

One of the first areas of applied science where a student is likely to encounter the bell curve is in the discussion of uncertainty in a physical measurement. The reader should note that the term used here is "uncertainty" and not "error". When discussing the confidence that one can place in a measurement of a physical quantity, many people use the terms error, accuracy, precision, and uncertainty in a sloppy interchangeable manner. The term *error* implies that a mistake has been made in the measurement procedure. Very often when evaluating data from an experiment, a person is really discussing uncertainty when they are talking about error. For example, if we were looking at tumbled coins on the table and we counted the

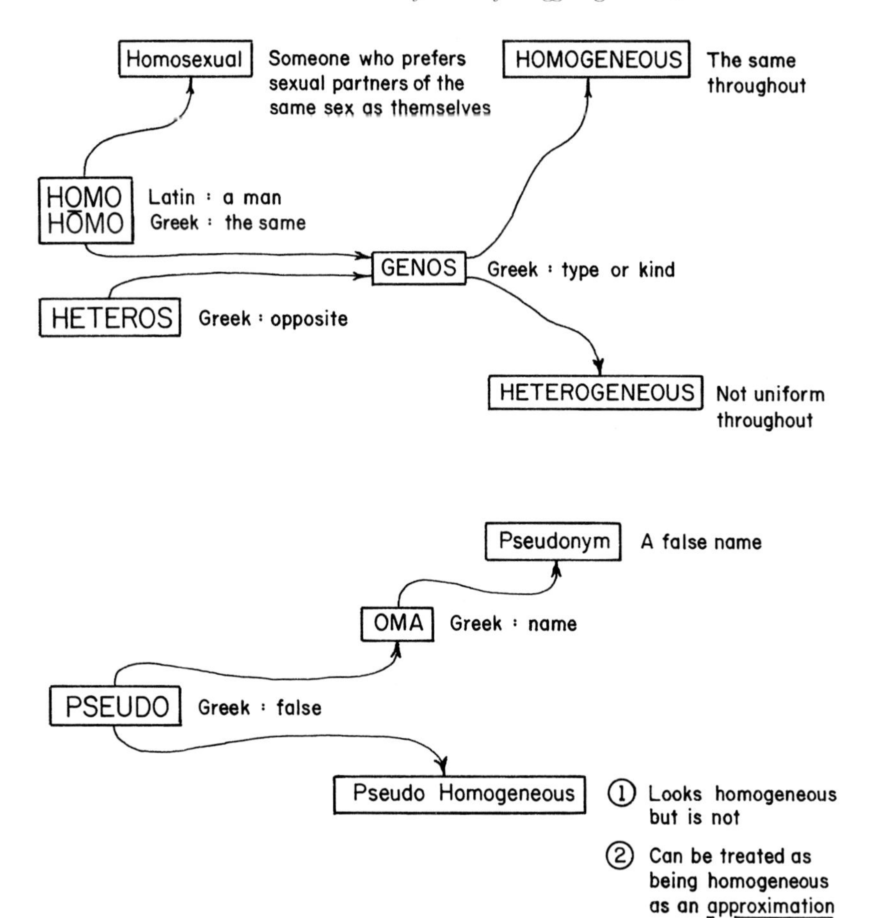

Figure 3.13. Homo and hetero are a pair of Greek prefixes which cause confusion to students; a confusion confirmed by the fact that homo in a modern word can also come from a Latin root word meaning "man".

number of heads and recorded the number as 16 when it was in fact 15, then that type of mistake is an error (the Latin root word from which error is derived is the word errare which means "to stray away from"). If one cannot move the coins around to carry out the count, it is quite easy to make errors in counting the number of heads manifest in a tumbled set of 25 coins. If one were to carry out the experiment by tumbling the coins and to count the number of heads without touching them, the observers would probably all agree in the first few minutes of carrying out repeat experiments. It can be shown, however, that within half an hour the observers could not be relied on to count the number of heads shown in the 25 coins because of fatigue to the eye and brain system and the way in which the eye scans the field of view. A very important problem comparable to counting the heads amongst the set

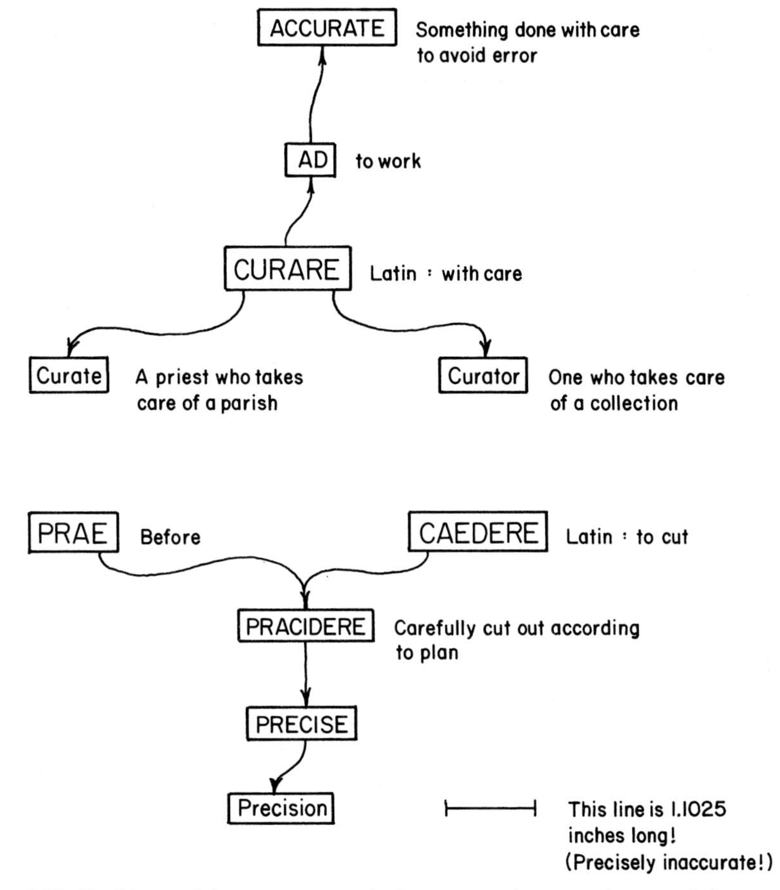

Figure 3.14. Precision and Accuracy are words that are not always used accurately!

of coins without touching them arises when people are asked to count the number of dust fine particles deposited on a microscope slide that has been put out to collect dust settling from the atmosphere. It has been shown that if humans are asked to carry out this task on a repetitive basis, they start to make mistakes. Counting the number of fine particles present in a field of view using human operators for long periods of time (more than 30 minutes) is not a reliable method of measuring dust levels (for a full discussion of this problem see Reference [9] and [10]).

The origin of the words *accuracy* and *precision* are summarized in *Figure 3.14*. One can see that when one makes an accurate measurement, it is one that has been carried out with great care to avoid error. The accuracy of a measurement is concerned with the correspondence of the measurement made with the reality that one is trying to quantify. When trying to measure the height of a person, one must have a well calibrated ruler or tape and then one can make the measurement without systematic error by making sure that the person is standing firmly on the ground, without shoes, and that the measuring rod is brought down to the top of the person and touches the head and not the tip of the hairs standing up on the top of the head. Furthermore, one must make sure that the position of a rod brought onto the persons head, with respect to the wall containing the marks from which the height is to be determined, must be level and not slanting upwards or downwards. When all these factors have been carefully arranged, then one can hope that the measured height will bear some direct relation to the actual height of the person; that is, it will be an accurate estimate of the person's height. The word *estimate* means to evaluate the worth or magnitude of something. A person who is used to measuring the height of other people can often make a good guess at the accurate height of a person. Such a guess by a skilled person is known as a "*guesstimate*".

If the person is about 1.80 meters in height (approximately six foot), then a statement that the person has a height of 1.82 meters is an accurate statement. The way this result is stated will convey to scientists reading the result that the person carrying out the measurement claims an accuracy of approximately one in two-hundred parts; that is, half a percent. In scientific circles, the number of digits used to report a measurement is always linked to the *claimed accuracy* of the result. Unfortunately, in many instances today the student of science tends to report all the decimal points displayed on a calculator without linking the decimal places which should be quoted and the accuracy of the result. For example, it may be that one student called upon to measure the height of the person had to use a yard stick and then convert the yards to meters. When he carried out this calculation on his calculator, he may have reported the results that the person was 1.822864 meters high. If the student writes this down as the measurement that he has made, then a scientist interpreting his data would assume that the student is claiming an accuracy of one part in two million. This level of accuracy is highly improbable, because the meter rule or yard stick is unlikely to be accurate to more than one part in one thousand. The claimed accuracy for a measurement is often described as the precision of the measurement. The precision, however, refers to the exactness of the description and is unrelated to the accuracy. For example, even if laser beams and all the modern arts of science were used to measure the height of a person, if one reported the height of the same person we have already been discussing as being 1.723, we would be claiming a precision of one part in two thousand in a measurement that we would know to be inaccurate to within two parts in twenty. The accuracy is only 10%, whereas the precision of the measurement may well be one part in two thousand. Progress in experimental science often involves the struggle to avoid inaccuracies and to increase the precision of all measurements made.

Let us suppose now that all the students in a class were going to measure the height of one person. The way in which each individual student would carry out the measurement is bound to lead to small differences in the reported height of the student. A particular student may not place the rod exactly at the midpoint of the student's head, the ruler may not be exactly level, the person being measured may slouch growing tired with time, etc. As a consequence, the measured height could range in value from, say, 1.818 to 1.826, with the individual students claiming that their measurements were precise to within one millimeter. In this situation, the differences between the various heights reported by the class of students are arising from the interaction of many small causes. Therefore, one would anticipate that the variations in measurement, which represents the uncertainty in the height to be assigned to the person being studied, would follow a Gaussian distribution. Note, again, that the term uncertainty is being used here because one has attempted to reduce error and to make the measurements as accurate and as precise as possible. The residual variation after error has been eliminated is not properly called error. It is more usefully described as *uncertainty*. Historically, it was recognized experimentally that residual uncertainty in a series of measurements, such as the height of a person evaluated by many people, could be described by the Gaussian distribution function. In fact some experimentalists still use the term "the law of errors" to describe what we are describing in this book as the Gaussian probability function. It is unfortunate that most students meet the Gaussian distribution function for the first time in the analysis of uncertainty, since they often associate statistical reasoning with experimental error, that is the confidence level in their data, when in fact statistical reasoning and the behavior of stochastic systems should be studied seperately. Error analysis in uncertainty quantification is a subbranch of stochastic reasoning. It cannot be stated too firmly that not all variations in uncertainties when studying stochastic systems are due to error. When we study the characterization of the fractal structure of coastlines in Chapter 8, we will find it very necessary to distinguish non-error variability from error in our estimates of the perimeter of a coastline. When error has been eliminated, and when one has achieved all possible accuracy and precision, it is common practice to report the residual uncertainty in a set of measurements by stating the mean and standard deviation of the set.

Section 3.10. The Power of Self-cancelling Errors to Improve Experimentally Based Estimates of a Physical Quantity

It is not always appreciated by experimental scientists that if the uncertainties involved in a set of measurements can be described by the Gaussian probability distribution function, then the addition of a series of measurements, each of which is subject to uncertainty varying according to a Gaussian distribution function, can result in a quantity emerging from the addition process which may be of higher

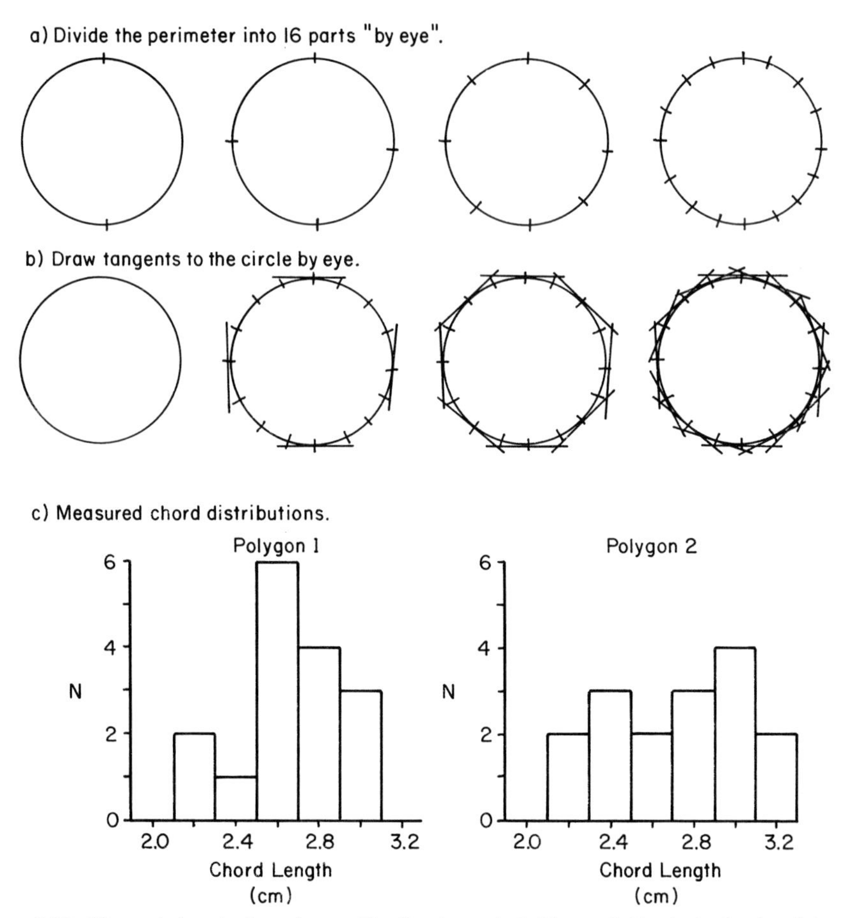

Figure 3.15. The variations in the polygon sides fitted to a circle "by eye" (that is, by free hand sketching rather than by accurate geometric construction) are Gaussianly distributed (that is, are describable by the Gaussian probability distribution function). a) Dividing the circle into 16 parts by eye. b) Draw tangents to the 16 points on the circle. c) Results for chords generated from two polygons drawn by the above method.

accuracy and precision than the individual measurements. This is because when uncertainties are distributed according to a Gaussian distribution function, positive and negative errors of the same magnitude are equally probable and tend to cancel out when added together. This process is described as improved accuracy from *compensating errors* or, as it is sometimes described, *self-cancelling errors*. Traditionally, scientists are taught to eliminate errors. To adopt strategies in which they can live with uncertainties because of self-cancelling aspects of the experimental procedure is alien to the traditional training guidelines drilled into the developing scientist. However, it has been pointed out that very often scientists do achieve amazingly

Table 3.3. Variations in the side length of a polygon drawn freehand (by eye) around a circle.

Polygon 1 [cm]	Polygon 2 [cm]
2.9	2.7
2.9	3.1
2.6	2.8
2.5	2.3
2.3	2.8
2.6	2.5
2.7	3.0
3.0	2.3
2.2	3.0
2.8	2.7
2.7	2.4
3.0	3.2
3.1	2.8
2.9	2.4
2.6	3.1
2.7	3.2

accurate values for physical quantities because, without their knowing it, their experimental procedure incorporates self-cancelling uncertainties. A simple geometric experiment which illustrates both the generation of variations describable by the Gaussian distribution by a multiplicity of random interactions of small causes, and at the same time demonstrates how a quantity being measured can be more accurate than the contributory measurements because of self-cancelling errors, is illustrated in *Figure 3.15*. The aim of the experiment is to assess the perimeter of an equisided polygon drawn on the perimeter of a circle. If asked to draw a sixteen sided polygon on a circle, most engineering students would painstakingly divide the circle into equal portions using compasses etc. They would then very carefully construct the tangents of the sixteen points to form an equisided polygon. One can show, however, that one can usually do as well as the careful engineering student by sketching in the polygon and exploiting the Gaussian distribution of variations in the polygon's sides. First, one would judge the divisions of the circle "by eye" in the way suggested by the sketches of Figure 3.15(a). One would then proceed to draw in the tangents, judged "by eye" as shown in Figure 3.15(b) leading to the final sketched polygon of Figure 3.15(b) (note one could probably draw the tangents with greater accuracy than that shown in Figure 3.15(b), but then the "by eye" construction would have looked too regular!) One then measures the length of the sides of the polygon to estimate the overall perimeter of the polygon. In Figure 3.15(c), the measured lengths of the sides of the polygon constructed on a circle of diameter 9.5 cm in two separate experiments, as measured to the nearest millimeter are shown. In Figure 3.15(c) the two sets of measurements shown in *Table 3.3* for two polygons

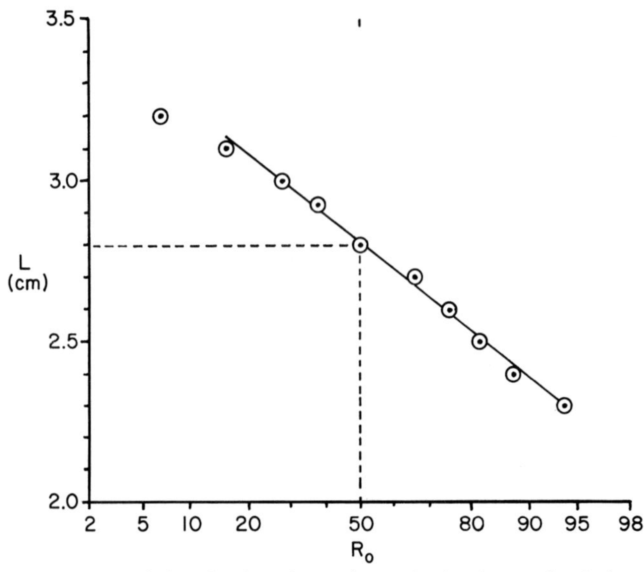

Figure 3.16. The variations in side length of a polygon drawn freehand around a circle can be described by the Gaussian probability function. L: Length of the polygon side. R_0: Percent of sides having the same or longer length.

sketched round the same sized circle are summarized. These two sets of data yield estimates of the total perimeter that differ by only 2%. Using graphical techniques one cannot normally work to better than 1% accuracy and precision. The truth of this statement can be checked by asking a class of students to construct a square inside a circle using rulers and compasses as carefully as possible (a relatively simple task) and comparing the results obtained by the various members of the class.

In *Figure 3.16* the 2 sets of data of Table 3.3 have been amalgamated to generate an undersized distribution and the data plotted on Gaussian probability paper. It can be seen that the most probable value of the chord size is estimated at 2.8 cm, leading to an estimate of the polygon perimeter which compares well with the value determined by straight addition of the various side lengths of the polygon.

Section 3.11. Using Dot Counting to Estimate Irregular Areas

In experimental science the technologist is often faced with the task of measuring the area enclosed by a meandering boundary. The word *meander*, which means "to wander aimlessly", has an interesting history. It is the name of a river in Turkey in a region known to the Romans as Phrygia. This river was famous for its many twists and turns as it crossed the plain to the sea. Ovid, a Roman poet, described the path of the river in these terms

> *The limpid Meander sports in Phrygian fields. It flows backwards and forwards in its varying course and meeting itself behold its waters that are to follow until it fatigues its wandering current now pointing to its source and now to the open sea.*

As we will discover in Chapter 8, fractal geometry helps us to relate the meandering of a river to the ruggedness of the terrain through which it flows. We will also find that the fractal dimension of the path of a river is a useful way of describing the way that the river has moulded the landscape around it. In *Figure 3.17* the meandering boundary of Lake Ramsey is shown (Laurentian University is built on the shores of Lake Ramsey). How can we estimate the area of the lake? When asked to measure the area of the lake, many students would lay a grid of squares over the profile of the lake and count squares. Counting the squares that lie completely within the profile is easy enough, but what about those that straddle the boundary? One way of estimating the area of the lake lying partially within the boundary squares would be to move painstakingly around the profile, trying to estimate visually how the partial square coverage adds up to form a number of complete squares. This scrutiny of the partial coverage of some of the boundary intersecting squares is not necessary, because there is a more efficient measurement procedure which exploits the fact that

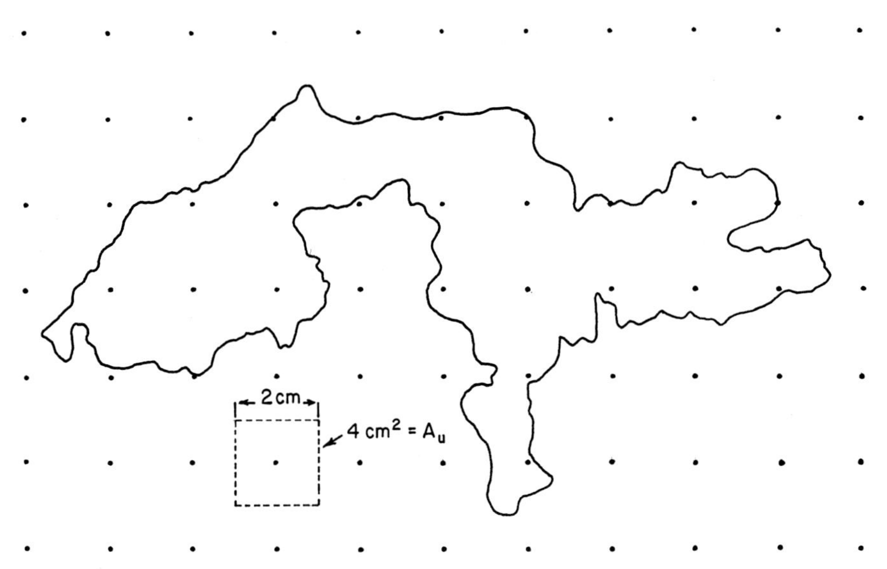

Figure 3.17. Dot count arrangement for Ramsey Lake. The dots were originally spaced at 2 cm, therefore the unit area for estimate $A_u = 4$ sq. cm. Area estimate is given by $A_{RL} = (N_I + N_B/2)\, A_u$ where N_I is the number of dots within the lake profile and N_B is the number of dots which lie on the boundary.

on the average half of the boundary squares will be less than half covered, and half of them more than half covered. Because of this fact, an efficient and unbiased way of estimating the area of the lake is to add half of the boundary squares to the number of the squares lying completely within the lake boundaries. Thus:

$$A_{RL} = \left(N_I + \frac{N_B}{2}\right) A_u$$

A_{RL} = area of the lake
A_u = area of the unit measurement square
N_I = number of squares inside the profile
N_B = number of squares crossed by the boundary

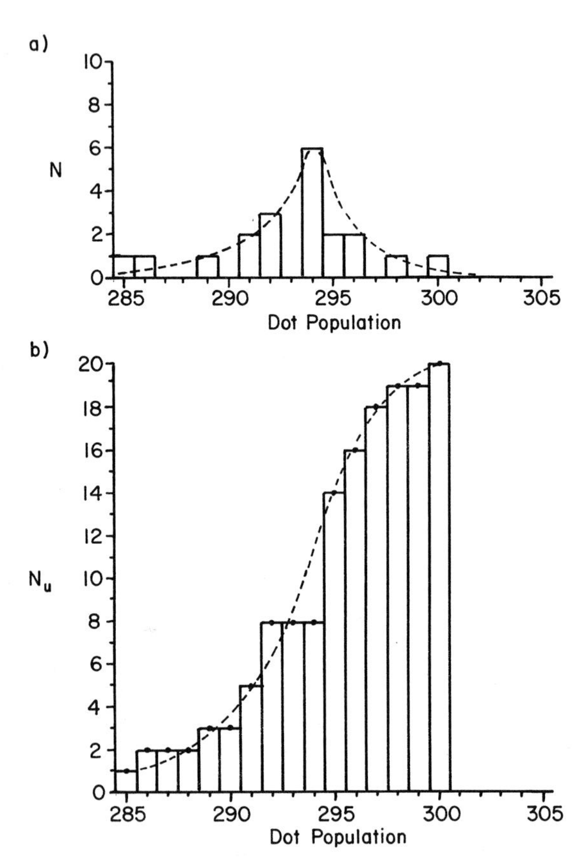

Figure 3.18. The variation of the number of dots falling within an irregular profile varies with random positions of the grid of dots over the profile. a) Distribution histogram for the dot count variation recorded when placing an array of dots over the profile. N = number of dots. b) Cumulative histogram of the variations in dot count data used to estimate the area of the profile of Lake Ramsey. N_u = number of trials with dot populations equal to or less than the stated values.

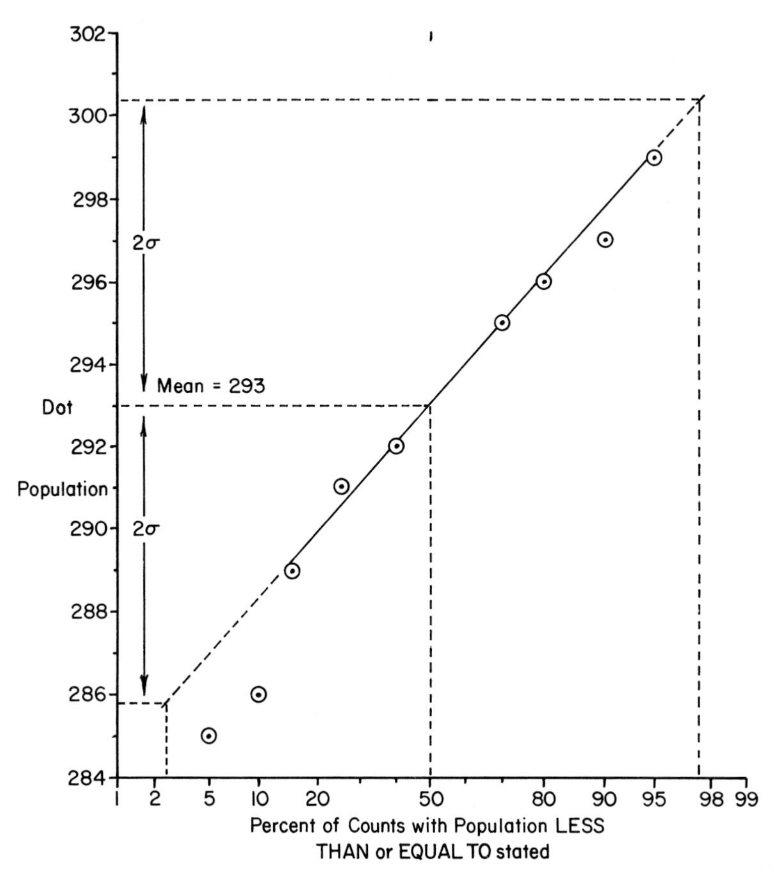

Figure 3.19. The cumulative undersize distribution of the number of dots in an estimate of the area of Lake Ramsey can be described by the Gaussian distribution function.

An alternate efficient experimental strategy for measuring the area of a profile such as that of Lake Ramsey is to represent each search square by a point at its center, as shown in Figure 3.17. Estimating the area now involves counting the number of dots falling within the profile. With this experimental strategy only a few dots fall on the boundary, and again an efficient and unbiased estimate of the area takes half of the dots in doubt. When estimating the area this way, one does not have to make the boundary as thin as possible and the dots as vanishingly small as possible in an attempt to minimize uncertainty.

I have taught students the "dot counting" strategies for measuring areas for many years, and I know that many students are reluctant to abandon partial square estimating because they feel that by putting more effort into the estimating process they must be improving the accuracy and precision of their estimate. Dot counting seems too simple. Once they are persuaded that dot counting can work, their next

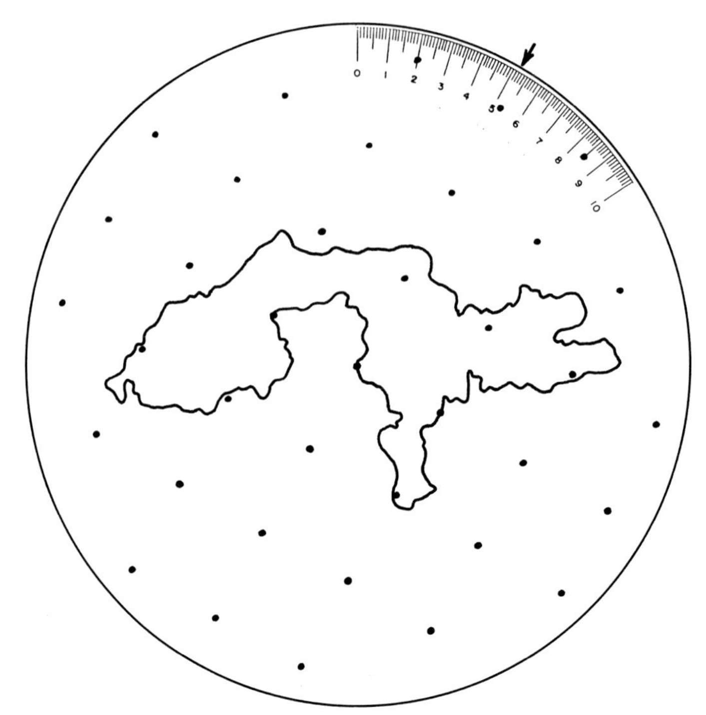

Figure 3.20. Regular positioning of a circle of dots on a randomly meandering profile yields the same variations in dot population as positioning the grid at random.

temptation to "improve accuracy" is to use so many dots in their estimate that the profile to be evaluated looks as if it has a bad case of measles. One can achieve high accuracy and precision in dot counting of irregular areas with a relatively coarse grid of dots by exploiting the fact that variations in the dot population of an irregular profile as the position of the grid on the profile is varied can be described by the Gaussian probability distribution. Therefore, by taking a series of measurements using various positions in the grid, one can plot the variations as a cumulative distribution function on Gaussian probability graph paper. One can then take the average value of the dot population from the graph. In *Figure 3.18* a series of area measurements made with a measurement grid of 0.5 cm placed over the lake profile 20 times are summarized as a histogram. It can be seen that the variations in dot population can be described by the Gaussian distribution as shown by the data of *Figure 3.19*. Note that the variations in the dot population for successive positions of the grid do not represent error in the measurement of the area, but uncertainty. Using the Gaussian probability plot of the varying dot populations does not estimate error, but reduces uncertainty by exploiting the way in which the uncertainty fluctuates.

In an earlier section of this chapter, it was indicated that one can anticipate Gaussian fluctuations when observations are being generated by the random interaction of many small causes. In the case of our dot population falling within the boundaries of the profile, the many small causes generating the variation in the population of the dots falling within the profile are the random meanderings of the profile across the space occupied by the grid of dots.

When asked to generate a set of randomly varying dot populations using a profile such as that of Lake Ramsey and a regular grid of dots, most students will try to randomize the position of the grid of dots by some such strategy as closing ones eyes when placing the dots over the profile. However, if the meanderings of the boundary profile are essentially random, then one can generate a random set of varying populations by systematic rotation of a set of dots. Again, students are often suspicious of this fact and sometimes they need to try the experiment for themselves before believing in the efficiency of the strategy. To compare random placing with systematic rotation, a class of students can be split into two halves. Each group can be given the lake profile or a similar profile with a fractal boundary. One group should try placing the grid of dots "at random" on the profile, whereas the others can use a system such as that of *Figure 3.20* to rotate the dots systematically. The two halves of the class can then compare data. Good irregular profiles which can be used in their studies are the outlines of Ireland, Great Britain and the Norway-Sweden peninsula.

Section 3.12. Finding the Percentage of a Given Material in an Ore

In the mining industry one frequently needs to estimate the *richness of an ore* (an ore is a piece of rock which contains valuable minerals). That is, one must assess the amount of valuable mineral in an ore specimen. The richness can sometimes be estimated chemically, but more often it is done physically. To do this, the geologist or the mining engineer takes a piece of the ore and cuts it with a diamond saw. The exposed section through the ore looks something like the system shown in *Figure 3.21*. The system of Figure 3.21 is actually an artificial system created by locating the profiles of 5 carbon black pigment fine particles in the circle. These profiles will be studied extensively when we come to the fractal geometry section of this book. It is useful to begin to study their properties by using them to illustrate the technique used by the mineralogist to assess the richness of the rock. For instance, we can imagine that the profiles of Figure 3.21 are sections through nickel sulfate material present in the ore body. They could also be holes in a piece of coke used in the metallurgical industry; the size and number of holes in the coke (almost pure carbon

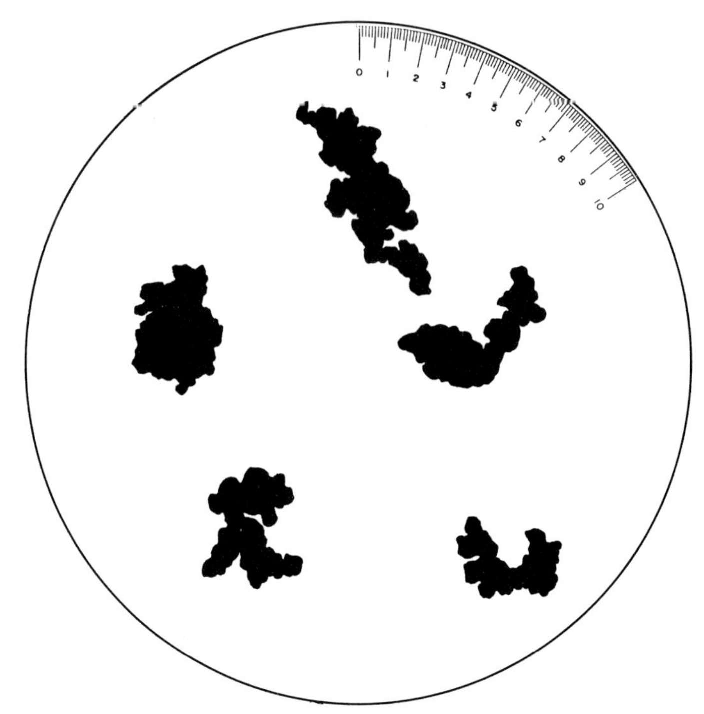

Figure 3.21. A synthetic section through a piece of ore illustrates the task facing the mineralogist as he seeks to characterize the richness of the ore (in this case the fractional area of the field of view occupied by the black profiles).

made by heating coal) is an important piece of information when looking at the use of the coke in the manufacture of iron from iron ore. The system could also be holes in a filter or pigment in a rubber tire (carbon black is used to reinforce the rubber).

To assess the portion of the area of the circle which is occupied by the black areas, we can use dots like those of Figure 3.20. If we are only interested in the fractional area occupied by the carbon black profiles, we can count the number of dots falling within the various profiles plus half of those in doubt. Then the area occupied by these profiles is the number of dots within the profile divided by the total number of dots. In the metallurgical industry and in the mining industry analyzing the richness of a piece of ore is called *modal analysis*. The technique of using dot counting to estimate areas is often referred to as *Chayes' dot-counting technique* after the scientist who originally developed a series of ideas involved in this technique for studying the structure of rocks [11]. A better estimate of the area coverage can be obtained by rotating the dot map into several positions and exploiting the fact that the variation in the dot population will be Gaussian. Because of this fact, the average of the area covered taken from the Gaussian distribution of the various estimates constitutes an unbiased and efficient estimate of the areas of the profiles. A set of data illustrating this point is summarized in *Figure 3.22*.

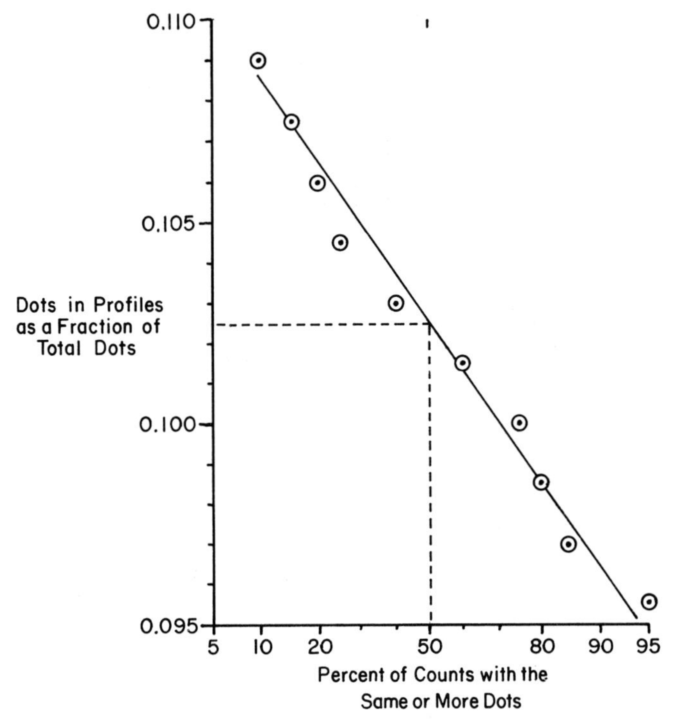

Figure 3.22. The variation in dot population within the profile areas of Fig. 3.21 is Gaussianly distributed.

As one carries out this experiment, one can also obtain the area of the individual profiles by counting the dots within each profile and plotting a graph of the various estimates. One can find comparable pictures to those of Figure 3.21 in many mineralogical textbooks on which measurements can be made. One can also use a map of a set of islands to carry out the measurement. A typical set would be the map of the Scilly Isles located off the tip of Cornwall in Great Britain, shown in *Figure 3.23.* The same dot counting technique is used to estimate a percentage of different trees in an aerial photograph of a forest or of mixed meadow and tree-covered land. It can also be used to estimate the size of the craters on the moon's surface.

Section 3.13. Aperture Size Distribution in a Woven-Wire Sieve

An important process in industrial technology uses sieves to separate powders into different sized fractions. Many of these sieves have surfaces made of woven-wire

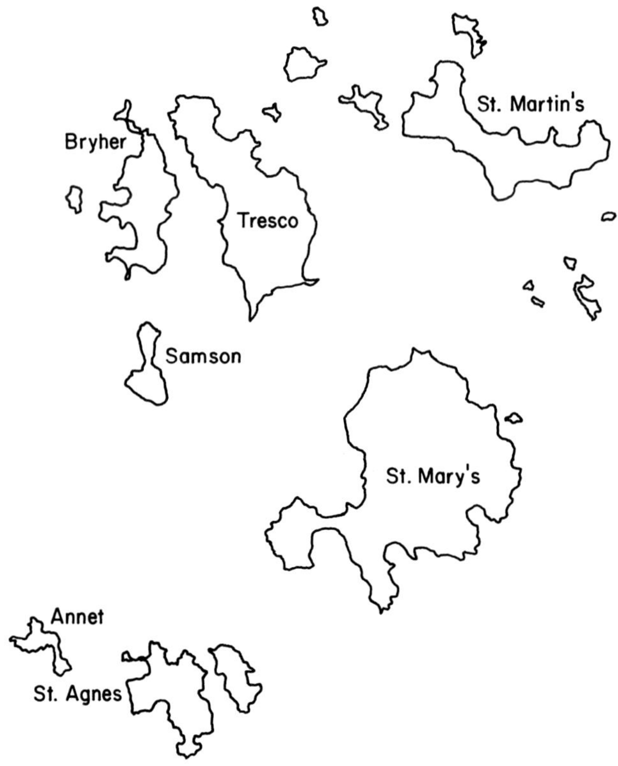

Figure 3.23. One can use the dot counting technique to measure the percentage of an area of the sea occupied by a group of islands such as the above map of the Isles of Scilly which are to be found off of the tip of Cornwall, England.

cloth. The manufacturer of this cloth exercises tight controls over the weaving process and the dimensions of the wires used to weave the cloth, in an attempt to make all the holes in the sieving surface exactly the same size. This is a difficult task and any real woven-wire cloth has a range of apertures present in the cloth. In *Figure 3.24*(c) a magnified image of a typical piece of woven-wire cloth is shown and the method of weaving the wire is shown in Figure 3.24(a). From Figure 3.24(c) it can be seen that the apertures are not exact squares, but look like trapezoids. The word *aperture*, meaning air opening or hole in an object, comes from the Latin word aperire, "to open". The word trapezoid is defined in the dictionary as any four sided figure that is not a parallelogram. Since the word trapezoid comes from the Greek word trapezion, meaning "a small table", this word suggests that many of the earlier mathematicians worked at irregularly shaped tables. The small departures from identical structure in the many apertures of a real sieve are caused by the random interaction of small changes in manufacturing technology, such as the tension in the loom used to weave the cloth, slight variations in the diameter of the wires fed to

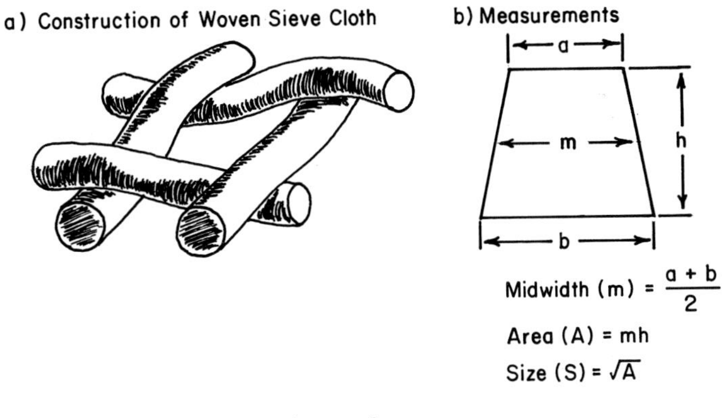

Figure 3.24. Measuring the projected area of a wire mesh sieve aperture distribution is an important industrial problem. a) Detail of the weave of a sieve cloth. b) Measuring the dimensions of an aperture. c) Photo of a magnified woven-wire sieve surface.

the loom, small shifts in the wires as the weave is formed, etc. Therefore one would anticipate that the size distribution of the apertures of the sieve will be describable by the Gaussian distribution function. Extensive experimental investigations have confirmed this fact [12]. When attempting to describe the size of an aperture in a sieve, one can make direct measurements on the images of the sieve mesh by determining the mid-width dimension of the hole as illustrated in Figure 3.24(b). Modern computerized image analysis allows one to characterize the aperture of the sieve by measuring the area and then taking the size of the aperture to be the square root of the area. In *Figure 3.25* the size distribution of the mesh of Figure 3.24 as measured by this technique is shown.

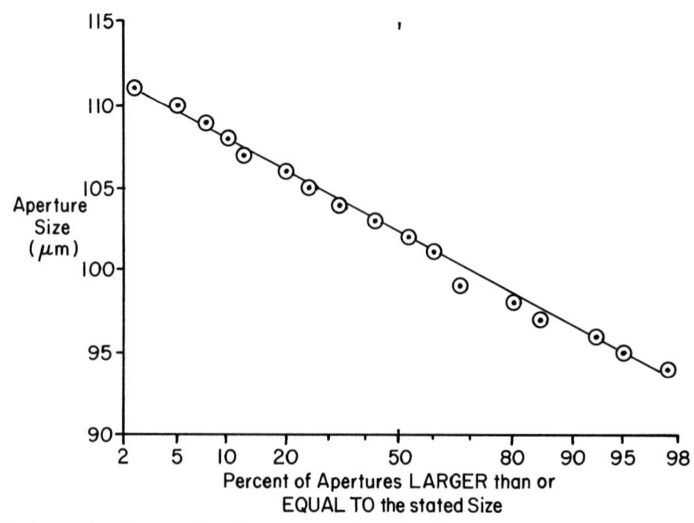

Figure 3.25. It can be shown that the aperture size distribution of a sieve is a Gaussian probability function.

Section 3.14. Characterizing the Weight Variation in a Population of Candy-Covered Chocolate Buttons

Several different makes of pill-sized chocolate buttons, covered with candy, are available commercially. These chocolate buttons are usually brightly colored and there are several different colors in a box. In this study a 60 gram box of "*Smarties*®" chocolate buttons manufactured by Rowntrees of York, England, were used, but other brands of candy could be used to generate similar data. The buttons studied in this experiment were circular and approximately 1 cm in diameter and 0.62 cm thick. The box contained 62 buttons. Apart from the color variation, all the buttons looked alike. It is obvious that the manufacturers try to make all the buttons the same size, but how successful are they?

Before beginning an investigation into a problem such as "What are the weight variations in a set of chocolate buttons?" a scientist usually tries to develop a theory which will predict the variations in the weight of the buttons. Many of the words we use in experimental science come from terms that were used by Greek scholars more than 2000 years ago. Up until 100 years ago most scientists took Greek and Latin as part of their studies, and the meanings of the technical terms derived from Greek were self-evident to these scientists. Today many of the words are unintelligible to the average student, and it is worth pausing for a few moments to explain the technical terms used by a scientist when developing his or her theories. The word

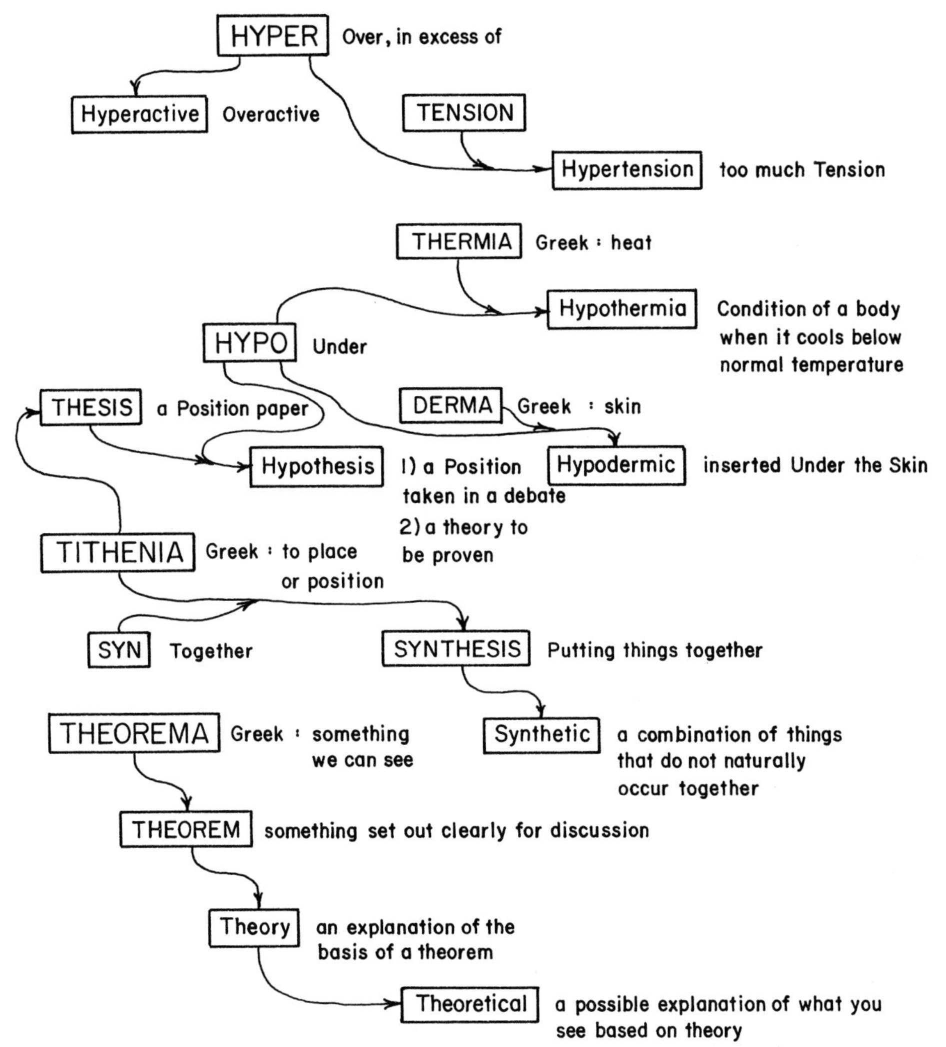

Figure 3.26. The Greek root words hypo and hyper have given the English language many technical terms which, because of the similar spelling of the two prefixes, often give students problems when trying to remember these words.

theory comes from the Greek word meaning "something which can be seen clearly". The idea was that in a debate the first speaker would set out clearly for all listeners the position that was being taken. As a consequence, the word *theorem* came to mean something which is set out clearly for discussion. A theory was an explanation of the basis of a theorem. The term *theoretical* is used to describe a possible explanation of "what we see" before those explanations are tested by experiment. The Greeks used to describe a new idea put forward for discussion as a *hypothesis*. This comes from two Greek words, meaning "something placed under a position". It is rather interesting to note that in modern English we talk about a "*position paper*" as being "something set forward for discussion". In political discussions in North America, we also talk about a "plank" of a party program. Obviously we don't mean that the planks of a program are the literal supports for a political stage; they are the ideas which form the basis of a position adopted by the party. In modern English we use almost exactly the same ideas as the Greeks, and the hypothesis is just like a main plank in a position adopted by a group of scholars. The two Greek root words hypo and hyper often confuse students, and a few words related to hypothesis are illustrated in the wordweb in *Figure 3.26*.

Presumably the small variations in the weight of the individual candy buttons comes from the interaction of many small causes acting at random in the manufacturing process. Therefore, it would seem to be a reasonable hypothesis that the weight distribution of the individual buttons would be describable by a Gaussian

Table 3.4. The weight distribution of 62 buttons in a Smarties® box.

Mass m [g]	Frequency f	Cumulative Σf	Percent [%]
0.80	0	62	100
0.81	0	62	100
0.82	1	62	100
0.83	1	61	98.4
0.84	2	60	96.8
0.85	5	58	93.5
0.86	3	53	85.5
0.87	6	50	80.6
0.88	4	44	71.0
0.89	5	40	64.5
090	1	35	56.5
0.91	4	34	54.8
0.92	4	30	48.4
0.93	4	26	41.9
0.94	2	22	35.5
0.95	2	20	32.3
0.96	2	18	29.0
0.97	1	16	25.8
0.98	0	15	24.2
0.99	7	15	24.2
1.00	3	8	12.9
1.01	3	5	8.1
1.02	1	2	3.2
1.03	0	1	1.6
1.04	0	1	1.6
1.05	1	1	1.6

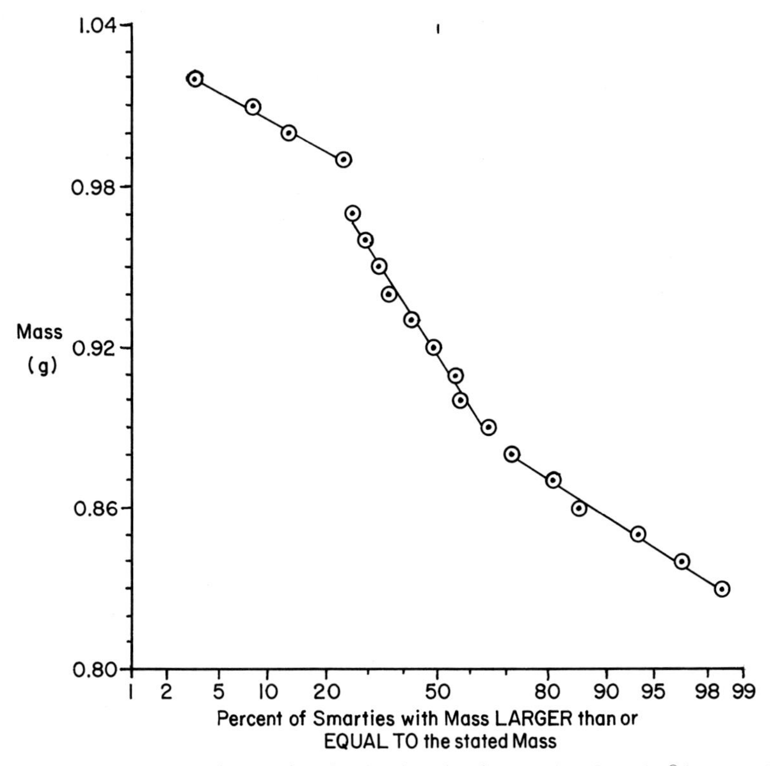

Figure 3.27. When the size distribution data for the chocolate buttons in a Smarties® box are plotted on Gaussian graph paper, the bimodal structure of the graph indicates the possibility that more than one machine is being used to produce the buttons.

probability function. To test this hypothesis, all of the buttons in one box of candy were weighed. The cumulative weight distribution of the buttons is summarized in *Table 3.4*. This data was plotted on Gaussian probability paper to yield the display shown in *Figure 3.27*. It can be seen that the dataline was not a simple relationship, and that apparently three straight line portions could be fitted to the data as shown in the figure. A person experienced in the use of Gaussian probability paper immediately recognizes the datalines of Figure 3.27 as being due to what is known as a *bimodal distribution*. A *mode* in a set of data is a point in the data set where information clusters. In a simple Gaussian probability distribution the centre of the bell curve is a mode about which the data clusters. When we have a bimodal distribution we have a camel with two humps instead of a bell curve! In *Figure 3.28* the histogram of the weight distribution of the buttons clearly shows this bimodal distribution. When this type of data is plotted as cumulative distribution, there is a break in the first part of the data curve as the effect of the second distribution function distorts the original data curve. Note that the third dataline appears to be a continuation of the first

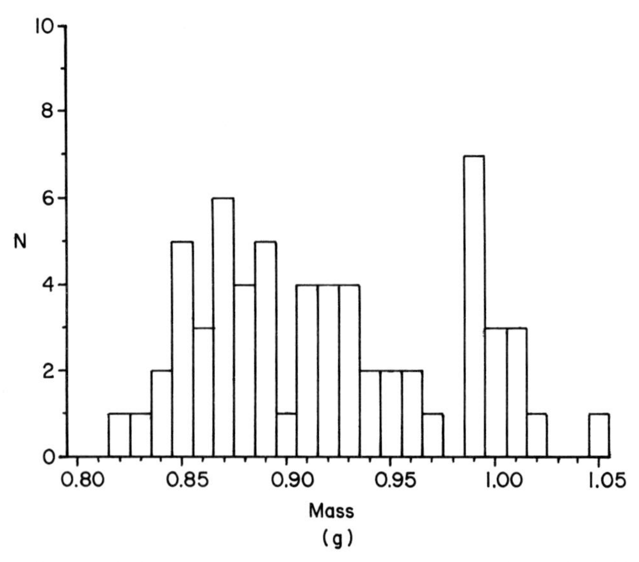

Figure 3.28. The distributed histogram of the weight of the Smarties® clearly indicates the bimodal distribution of the weight of the candy buttons. *N*: Number of buttons of the stated weight.

dataline, with the middle portion being due to the higher than expected number of buttons with weight in the region 0.97 to 1.03 grams.

When one has a bimodal distribution such as that of Figure 3.27, one must modify one's hypothesis to account for the observed data. In this case the most probable modified hypothesis which will explain the data is that more than one machine is being used to produce the buttons. Each machine probably produces buttons with a Gaussian distribution, but when the output from more than one machine is combined to generate a product, then the weight distribution is bimodal. When one looks at the data of the table it would seem reasonable to guess that the manufacturer is really trying to make buttons which have a weight of approximately 1 gram and that a newer machine was indeed producing many buttons with a tight range of from 0.97 to 1.02 in weight, and that an older machine, probably with worn parts or less tightly controlled production, was responsible for the lower weight buttons of a wide size range. Again, however, this is only a hypothesis, and to test such a hypothesis one would have to go into the candy factory and measure the output from each machine producing the candy buttons.

There are many similar systems to the Smarties® box and one can investigate the possibility of either single or bimodal Gaussian distributions by studying the variations in weight of the members of a population. If sensitive weighing equipment is not available to the student, a set of dog biscuits of nominally the same size can be used to study the weight variations in a population of objects. If one switches one's attentions to a natural product, such as the varying weight of peanuts, there is sometimes a larger weight variation but often the variations in weight will still be describable by the Gaussian distribution function.

In a class discussion of the variation in weight of the candy buttons, a student asked whether medical pills had the same weight variation, and if so what that would mean for the patient taking a dose of medication by swallowing a pill. The student asking the question was challenged to investigate his own question by buying a population of pills (that is a bottle of pills) of any medication available over the counter and carrying out a study of the weight variations from pill to pill (for a discussion of the significance of the variation of the numbers of each color of Smartie® from box to box of the chocolate buttons, see Section 5.3).

It should be noted that the students carrying out the study of the weight variations in Smarties® reported that the experiment helped them to appreciate the magnitude of a 1 gram weight when studying scientific weights and measures.

Most students expressed surprise at the actual weight variation present in the buttons. Another interesting experiment for the students to carry out is to use a pair of callipers to measure the width of the buttons to see how well the manufacturer was able to maintain the dimensions of the manufactured object.

Section 3.15. Buffon's Needle – A Surprising Pattern of Events

In 1977 a special conference was held in Paris to celebrate the 200th anniversary of the publication of a scientific article by the French scientist *Buffon* (1707–1788). Georges Buffon initially studied law and then became the keeper of the royal gardens for the Kings of France. He is best known for his work on nature which was published in forty-four volumes. In 1733 he wrote a short memoir entitled "Resolution of Problems with Regard to Games of Chance". Apparently as a young man Buffon had been introduced to the mathematical problems of gambling by a Genevan mathematician called Gabriel Cramer. Cramer was a friend of the Bernoulli family, a very famous mathematical family, several members of which had become involved with the study of the probable patterns in gambling games. At that time a great deal of discussion centered around what is known as the St. Petersburg paradox. Apparently the problem first arose amongst the aristocratic gamblers of St. Petersburg in Russia. This paradox said that if a game of chance went on forever the winner's gains would be infinite. We are told that Buffon reacted to this problem not with theoretical criticism but from the practical point that a game cannot go on forever. In his essay on "Games of Chance" Buffon discussed the prediction of the problem of the pattern of events generated when needles fall on the floor. He did not publish these essays until much later in a book published in 1777, hence the date of the scientific meeting celebrating the famous problem [13].

The problem studied in what is now known as *Buffon's needle* is the frequency of intersections that occur when one tosses short straight objects, such as needles, onto a set of parallel lines, such as those formed by wooden strips used in hardwood flooring [14, 15]. One wonders how Buffon was led to study such an odd scientific problem. I am told that compulsive gamblers will gamble on the outcome of anything that attracts their attention. They have been known to make wagers on the outcome on races between cockroaches, or which rivulet moving down a window pane in the rain will reach the bottom of the window first. Perhaps some bored nobleman watched a needle fall from the hand of a seamstress onto the floor and made a wager with his friend as to how often a falling needle would cross a line between floorboards. We shall never know the stimulus to this specific study, but whatever it was, this study of a rather odd problem of needles falling onto straight lines has become the basis of a modern branch of mathematics which is known as *geometric probability* [15]. I first met the problem of Buffon's needle in a book by George Gamow where he discusses the problem in terms of matchsticks falling onto the stripes of the American flag [14]. I can remember being very surprised by the fact that if the matchsticks were the same length as the width of the stripes on the flag, then after throwing the matchsticks onto a straight flag the number of intersections of matchsticks with the lines of the flag was given by the formula.

$$\frac{2}{\pi} = \frac{\text{Number of intersections}}{\text{Number of matchsticks}}$$

We now have a method for evaluating π from the rearranged formula

$$\pi = \frac{\text{Number of matches}}{\text{Number of intersections}} \times 2$$

This result can be generalized for needles of length l, some fraction of the distance L between planks in the floor, so that after n throws of the needles giving a total of N intersections:

$$\pi = \frac{2nl}{N}$$

and if we were to use a needle length greater than the distance between the planks the general relationship is

$$\pi = \frac{2n(lL)}{N}$$

The fact that mathematicians initially found the consequences of this Buffon needle experiment surprising is indicated by the comments made by Augustus DeMorgan when he reviewed the problem in his book "The Budget of Paradoxes" published in 1872 [16].

It should be noted that DeMorgan used the word paradox in a special sense to mean any curious tale about science and scientists that he had come across in his extensive reading, any piece of gossip, choice examples of lunacy and assorted riddles and puns (see comments by James R. Newman on Augustus DeMorgan in Volume 4 of the *World of Mathematics* [17]).

In his review of Buffon's needle, DeMorgan tells us that

> *In 1855 Mr. Ambrose Smith of Aberdeen made 3,204 trials with a rod 3/5ths of the distance between the planks. There were 1203 clear intersections and 11 contacts on which it was difficult to decide. From this data the value of π is 3.1553. If all the 11 doubtful contacts had been taken as intersections, the result would have been $\pi = 3.1412$, which is exceedingly near to the known value of π.*

DeMorgan also tells us that

> *A pupil of mine made 600 trials with a rod of the length between the seams and arrived at a value of $\pi = 3.137$.*

DeMorgan comments on this method of measuring of π as follows

> *This method will hardly be believed until it has been repeated so often that there never could have been any doubt about it.*

The Buffon's needle approach to the measurement of π has been reviewed in a book on geometric probability by Kendal and Moran [15]. They present a table of the various attempts to achieve an evaluation of π by the throwing of needles onto parallel lines. Kendal and Moran discussed the apparent high accuracy of the value of π achieved by these experiments and state that

> *The results of these experiments are mainly due to an adroit use of optional stopping, that is, the termination of the experiment when the estimated value of π was close to the known value!*

Kendal and Moran point out that if the true value of π was unknown before the experiment, so that optional stopping could not be used, a better method of estimating π is to cut out a large circle of wood and use a tape measure.

The power of the adroit usage of optional stopping combined with the knowledge of π to begin with is illustrated by the experiment of Bridgeman who, by judicious choice of needle length, was able to obtain an accurate value of π after two throws! (See discussion in [15]).

Obviously the Buffon's needle problem is no longer of interest to working scientists as a method of measuring π, but it is of great interest to us because it introduces us to the study of the pattern of events when the position of objects are varying and random in two or more dimensional space. The same type of logic embodied in the study of Buffon's needle has become the basis of important techniques for characterizing fineparticles when they are inspected by means of a television camera. The line scan of a television camera generates the sets of parallel lines and one can evaluate the perimeter of an irregular profile by counting the number of intersections of the TV scan lines with the profile. This technique will be discussed in greater detail in the next section.

Section 3.16. Exploiting the Efficiency of Antithetic Variates to Improve the Efficiency of the Estimates of π by Buffon's Needle Experiments

An interesting variation of the original Buffon's needle technique can be used to lay the basis for an understanding of the design of efficient techniques for characterizing the structure of irregular profiles by means of intersection frequencies when a profile is placed upon a grid of lines (an important technology both in air pollution control and in medical science). Let us consider what would happen if we were to scatter twenty-four needles of length equal to the distance between a set of parallel lines by sprinkling them at random on the lines. For the data we will present later in this section, sprinkling at random involved closing one's eyes and dropping short

Table 3.5. Intersection frequency for 24 needles thrown onto a set of parallel lines.

Intersects N	Occurrence f	Cumulative Σf	Percent [%]
10	0	100	100
11	4	100	100
12	3	96	96
13	5	93	93
14	13	88	88
15	13	75	75
16	14	62	62
17	24	48	48
18	10	24	24
19	7	14	14
20	7	7	7
21	2	3	3
22	0	1	1
23	1	1	1
24	0	0	0
25	0	0	0

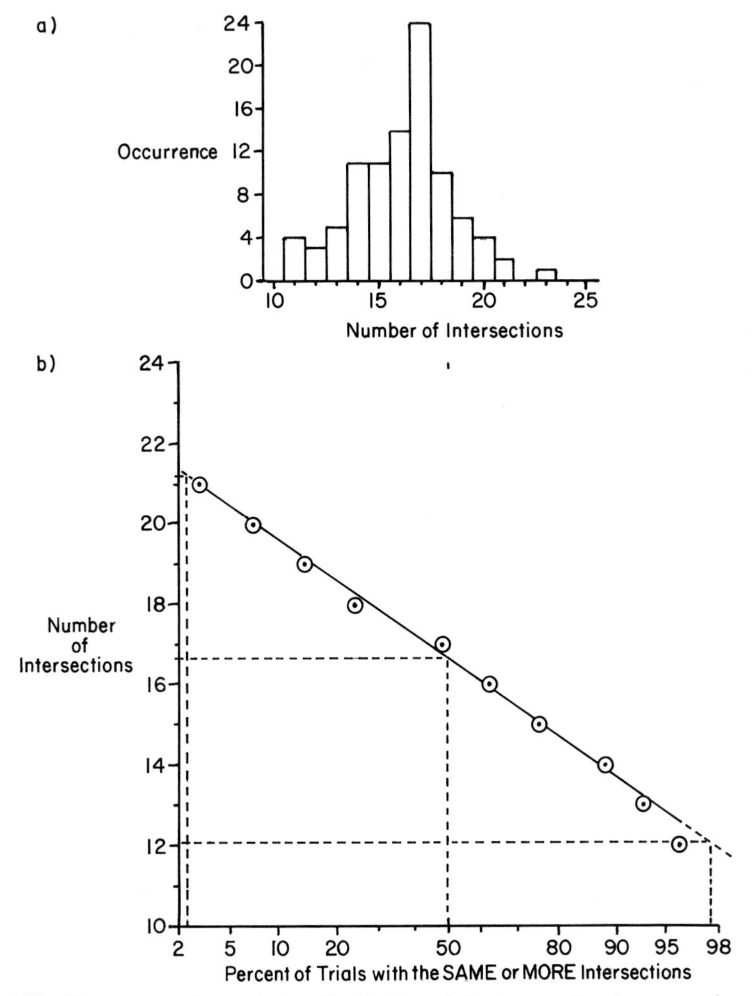

Figure 3.29. Histogram and cumulative distribution of the intersection frequency data of Table 3.5 showing that the variations are Gaussianly distributed.

plastic sticks of the appropriate length, at random, from a height of five or six inches onto a set of straight lines. The number of times that the lines were intersected was counted, then the sticks were gathered up and the experiment is repeated. The data generated in 100 experiments of this kind are summarized in *Table 3.5*. The variation in the number of intersections in each throw of twenty-four needles is due to many small causes acting at random. Therefore, from what we know of the Gaussian distribution, it is reasonable to anticipate that the variation in intersections from set to set of the randomly tossed twenty-four needles will be Gaussianly distributed. The histogram of the values of Table 3.5 and the cumulative version of the data plotted on Gaussian probability paper are shown in *Figure 3.29*. It can be seen that the data is indeed Gaussianly distributed, but that the average value is 8% different from the theoretically predicted value. This is not surprising, since 2,400 pieces of information is still a relatively small amount of data when one is looking for a stochastically varying quantity. The standard deviation of the data is plus or minus 13.8%.

Hammersley and Morton have discussed how one can use a technique called *antithetic variate characterization* for measurements of this kind. The word *antithetic* means "placed at right angles" to another object. The basic aim of an antithetic variate strategy is to make two measurements together in such a way that if one measurement is likely to be high, the other is most probably low. To exploit antithetic variate measurement techniques in Buffon's needle experiment, we use two needles joined together in the middle instead of single needles. In essence, the probability of cutting a line by means of an antithetic linkage of two needles is increased because if one needle lies in such a direction that it does not intersect the parallel lines, the other needle has a very high probability of intersecting the lines.

To show the power of the antithetic strategy in the Buffon needle type of experiments, the twenty-four plastic sticks used in the experiment to generate the data of Table 3.5 were made into a set of equiarmed crosses by joining them at the centre. Then the twelve crosses made in this way were thrown at random onto the set of lines 50 times to generate the data of *Table 3.6*. Again, the histogram for the data of

Table 3.6. The scatter of intersection data decreases when pairs of needles are made into crosses and thrown onto the parallel lines.

Intersects N	Occurrence f	Cumulative Σf	Percent [%]
10	0	50	100
11	1	50	100
12	1	49	98
13	3	48	96
14	6	45	90
15	8	39	78
16	8	31	62
17	11	23	46
18	4	12	24
19	3	8	16
20	4	5	10
21	1	1	2
22	0	0	0
23	0	0	0
24	0	0	0
25	0	0	0

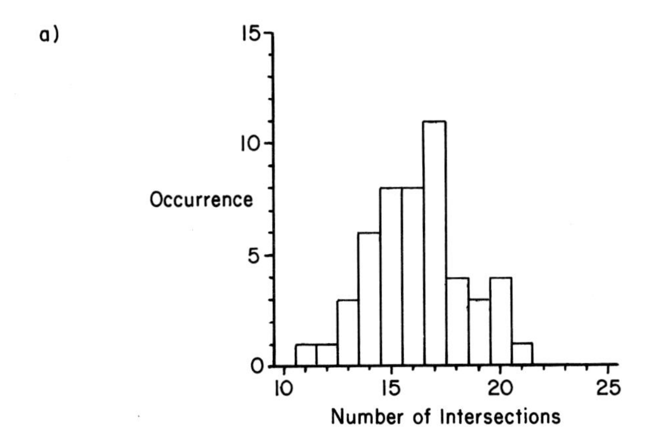

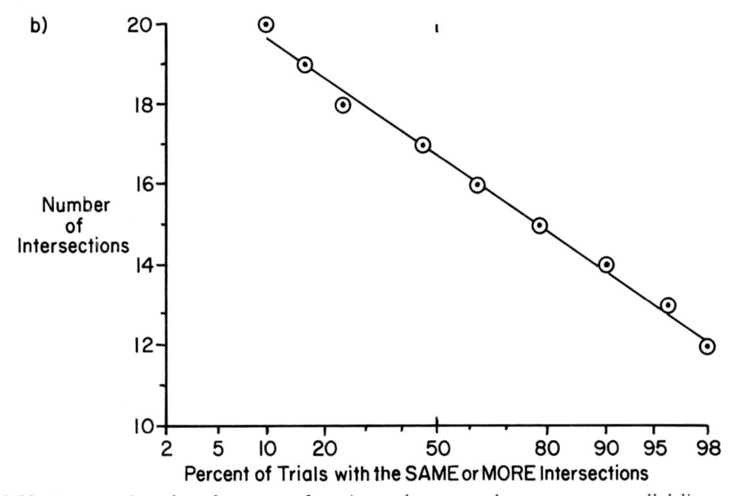

Figure 3.30. Intersection data for a set of equiarmed crosses thrown onto parallel lines.

the 12 crosses and the plot of the data on Gaussian probability paper is shown in *Figure 3.30*. It can be seen that the base of the bell curve for the crosses as compared to that of the single needles has narrowed because the data has a smaller range of variation. Again the data fits a Gaussian probability distribution line. The advantages of the antithetic technique can be increased further by now using three needles joined together to make a six armed "spider" with a 60° angle between adjacent arms. Eight spiders are thrown onto the parallel lines 33 times to generate a comparable set of data.

A little thought will show that instead of creating crosses to improve the estimate of π by Buffon's needle, one could achieve the same effect by adding an extra set of

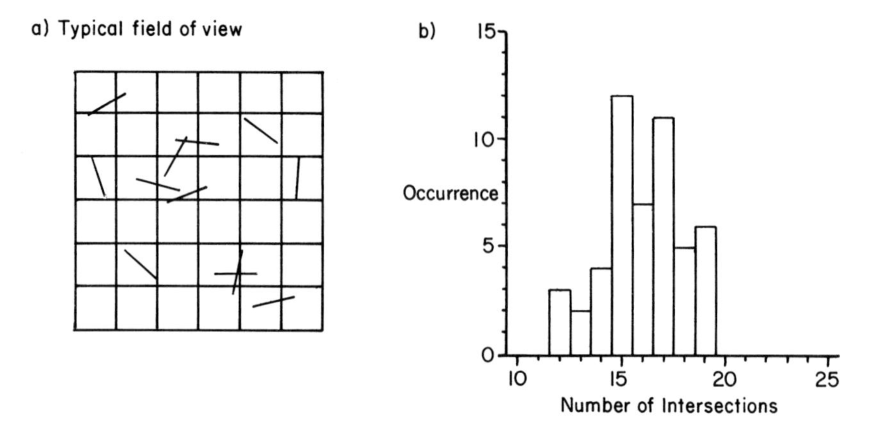

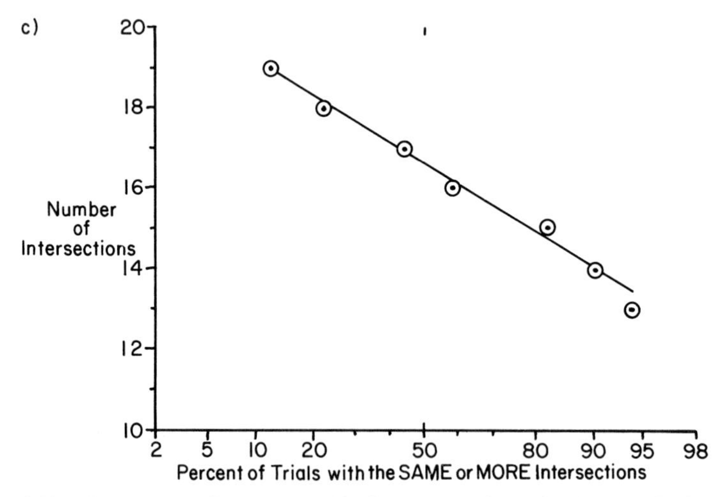

Figure 3.31. Throwing needles onto a grid of squares exploits the theoretical advantage of antithetic variates. a) A typical field of view. b) Histogram of the number of intersections. c) Gaussian distribution of the intersection frequency.

lines to the original simulated planks at right angles to the original planks. When a single needle is now thrown onto the grid, if the needle misses a parallel line in the horizontal direction, it is very probable that it will hit a vertical line. In *Figure 3.31* (a) a set of 12 needles thrown onto a square grid is shown, and the set of data generated by 50 experiments of this kind are shown in Figure 3.31 (b) and summarized in *Table 3.7*. The Gaussian distribution for this set of data is shown in Figure 3.31 (c), and it can be seen that it is essentially the same as the data generated by the antithetic experiment using the crosses on a single set of parallel lines.

Table 3.7. Intersection frequency data for needles thrown onto a grid of vertical and horizontal lines as shown in Figure 3.31.

Intersects N	Occurrence f	Cumulative Σf	Percent [%]
10	0	50	100
11	0	50	100
12	3	50	100
13	2	47	94
14	4	45	90
15	12	41	82
16	7	29	58
17	11	22	44
18	5	11	22
19	6	6	12
20	0	0	0
21	0	0	0
22	0	0	0
23	0	0	0
24	0	0	0
25	0	0	0

Section 3.17. How Much Work is Needed to Improve One's Confidence in an Average Value?

In Section 1.9 the reader was warned that it is all too easy to say "on the average" something will tend towards a given value. In this section, we will explore experimentally how much work is needed to improve the confidence we can place in an average value calculated from a set of data. Consider the problem of calculating the average value of a set of digits taken from a random number table, such as the following sequence:

9 5 4 2 9 0 5 0 2 3 4 2 4 4 5

Since each digit from 0 to 9 is equally probable, then on the average the sum of a set of random digits should tend to

$$(0 + 1 + 2 + 3 + 4 + 5 + 6 + 7 + 8 + 9)/10 = 4.50$$

If we add the digits of the first short sequence given above we have an average value of 3.87. Is this value for the average of the 15 digits close enough to the expected value 4.50 for us to assume that the sequence of $9 \rightarrow 5$ is typical of the value we would obtain by taking any 15 digits at random from the table, or should we suspect that the random number table is biased towards the lower digits? (The reader might ask an acquaintance if the sequence given from 9 → 5 above is biased. Because of the absence of the digits 6, 7 and 8, many people unacquainted with the properties of random systems will voice a suspicion that the sequence is biased). If we take the next sequence in the random number table it is

8 1 4 7 9 0 6 5 8 2 5 7 8 3 2

The average value of this sequence is 5. Note the "big increase" of the occurrence of 6, 7 and 8 compared to the first sequence. Are 5 and 3.8 permissible variations in the average of 15 random digits when the anticipated value of the set is 4.5? How many digits should be added to be sure that the value is 4.5 and not 4.6 for a sequence being studied? The statisticians have solved this problem theoretically, but we can discover the answer for ourselves by carrying out the following set of experiments. How would one expect the average of a set of numbers to vary if we considered many sets of digits containing the same number of digits? If we consider the 15 numbers of our set of digits, then from our point of view each digit is a contributory cause to the average, and the average is caused by the random iteration of 15 causes. Therefore we expect the variation in a set of average values based on 15 digits to be describable by the Gaussian probability distribution. Therefore we can create a set of averages by adding up the digits in a set and then the distribution of averages could be plotted on Gaussian probability graph paper. To illustrate this basic concept, let us consider the sample case of the variation in the average value of 5 digits chosen from a random number table. Using the sequence already presented, our first six values in the experiment would be

95429 → 5.8
05023 → 2.0
42445 → 3.8
81479 → 2.9
06582 → 4.2
57832 → 5.0

In *Figure 3.32*(a) the histogram of one hundred 5 digit averages is shown. It can be seen that the data look as if they are describable by a bell curve. This is confirmed by a plot of the data as a cumulative equal-to-or-less-than distribution on Gaussian probability paper in Figure 3.32(b). It can be calculated from this graph that the 100 averages and 5 digits is 4.51 ± 1.29, where 1.29 is the calculated standard deviation of the data. No one is surprised by such wide variation in the average when considering only 5 digits, but how quickly does the standard deviation fall as the number of digits increases? In *Figure 3.33*(a) the distributions of the variations in 100 sets of averages based on increasing numbers of random digits are shown.

In Figure 3.33(b) the standard deviations measured from the data distributions are plotted against the number of digits included in the average. It can be seen that a

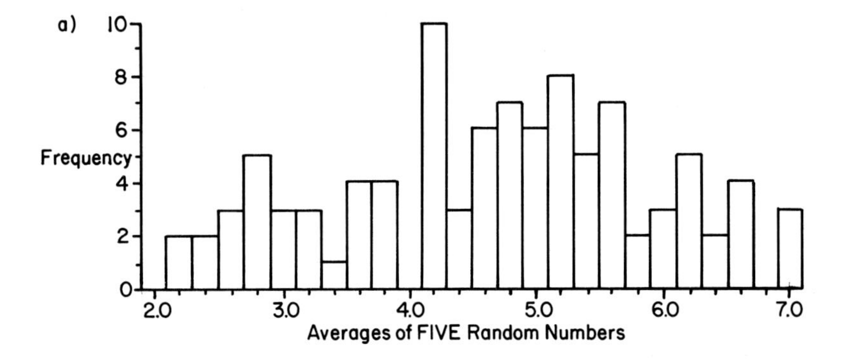

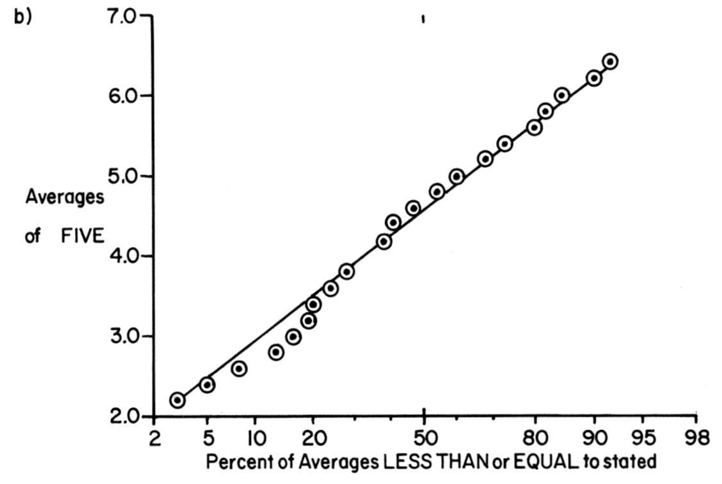

Figure 3.32. The averages of many sets of five random numbers produce values discibable by the Gaussian distribution function. a) Histogram for a set of 100 averages of five digits. b) The data of (a) plotted on Gaussian probability graph paper.

straight line relationship is obtained on log–log graph paper. The slope of this line equals − 2. Expressed as a mathematical relationship this result is summarized in the formula

$$n = K\sigma^{-2}$$

In the language of the mathematician this means that the size of the standard deviation is inversly related to the square root of the number of items included in the average. This means that to halve the standard deviation one must square the number of items of information. Looking at some of the experiments that we have carried out so far, if we wish to halve the standard deviation of the data based on 20 items of information, we must take 400 items of information.

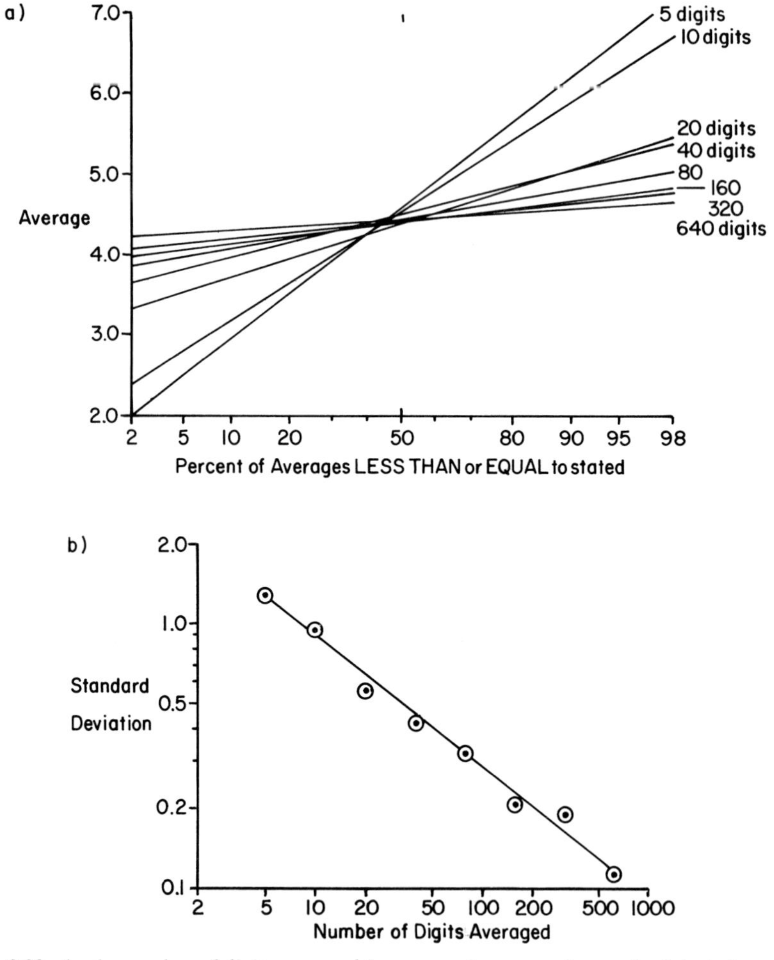

Figure 3.33. As the number of digits averaged in a group increases, the standard deviation of the data decreases. a) Data for 100 sets of averages of 5, 10, 20, 40, 80, 160, 320 and 640 random digits plotted on Gaussian probability paper. b) Standard deviation of the different sets of data of (a) versus the number of random digits averaged plotted on log–log scales.

The reader should note that even after 640 digits, the standard deviation of the average value is 0.11 in 4.5, which is still 2.5 % of the expected answer. Most students are surprised that even after 640 digits there is still a large amount of uncertainty in the expected average. This demonstrates that when mathematicians state that the average value of a sequence of random digits tends to 4.5, they have in mind a much longer sequence of digits than most students would be prepared to tackle. Mathematicians have a phrase for a very long sequence of digits which must be added to arrive at an exact value such as 4.5: an *ergodic sequence*. This comes from a Greek word meaning "tiresome" or "boring". To generate the dataline of averages based on 640 digits, the student carrying out the experiment had to add up 64,000 random number digits, a process which he certainly described as tiresome and boring!

The relationship that we have discovered between the standard deviation and the number of items of information helps us to understand why it is relatively easy to obtain values of the order of magnitude of an average and standard deviation describing the behavior of a stochastic system, but to reduce the uncertainty requires very heavy investment of time and effort in experimental study.

Exercises

Exercise 3.1.

Construct 20 random walks for a "drunkard" consisting of 25 steps each. Compare the average displacement from the point of the migrating drunk to the expected value (five steps). Find the distances of the individuals from the average displacement (quantity y). Plot the cumulative distribution of y equal to or less than a given value on Gaussian probability paper.

Exercise 3.2.

Take a set of coins such as pennies. Throw them 50 times, counting the number of heads at each throw. Plot the cumulative undersize population of heads in the various throws of the set of coins on Gaussian probability graph paper. Estimate the mean and standard deviations from your graph.

Exercise 3.3.

Students can simulate the structure of many different powder mixtures using random number tables by appropriate transformation rules. They can then study the variations encountered from one simulated sample to another at various sample sizes. Variations can be summarized on Gaussian probability graph paper.

Exercise 3.4.

Measure the volume of a pebble by hanging it by a string in a graduated cylinder half full of water. The volume change shown on the scale of the cylinder will indicate the volume of the pebble. Ask a class of students to each carry out the experiment. Tabulate the data and calculate the mean and standard deviation of the volume estimates. Plot the cumulative undersize data on Gaussian probability paper and calculate the mean volume and standard deviation of the estimates from the graph.

Exercise 3.5.

If a class project were to be arranged to carry out the experiment to study the variation in chord lengths when constructing a polygon by eye, it would be found that the polygon perimeter estimates made by the individual students would themselves be Gaussianly distributed. The class data could be amalgamated to create a bell curve from the individual estimates of the perimeter.

Exercise 3.6.

If a class were to carry out the experiment of constructing "by eye" the polygon on a circle, half of the class could do various inscribed polygons, whilst the other half could construct "exscribed" polygons. The data from the polygon perimeter versus the number of sides in the polygon could then be used to draw the graph of *Figure E 3.6.1.(b)*, which can then be used to deduce the best value of π obtained by looking at the convergence of the perimeter estimates of multi-sided polygons. The more the number of sides to the polygon, the better the polygon approximates the circle (see discussion of Archmedes' method for measuring π in *Computers and the Human Mind*, by Fink [18]).

Exercise 3.7. Use of Gaussian Probability Paper to Study "Student Marks"

In the lists below are the final marks of two sets of students. The first set is for 46 students who took a compulsory first year course in physics. The other set of marks are for an elective course in physics dealing with environmental science.

Student marks in compulsory course: 46, 26, 57, 69, 44, 59, 83, 26, 52, 50, 49, 72, 97, 46, 59, 44, 69, 49, 56, 63, 0, 84, 61, 56, 59, 61, 56, 59, 61, 60, 59, 55, 53, 56, 93, 54, 34, 60, 37, 39, 88, 75, 49, 20, 67, 42, 40, 13, 70, 78,

Student marks in elective: 80, 68, 54, 77, 67, 57, 83, 51, 51, 73, 0, 62, 40, 72, 57, 77, 74, 30, 64, 68, 72, 62, 68, 62, 74, 52, 66, 68, 63, 50, 50, 75, 64, 82, 85, 81, 50, 70, 44, 68, 61, 68, 74, 63, 70, 77, 68, 70, 72, 76, 58, 61, 54, 88, 74, 63, 58, 65, 17, 75, 74, 74, 81, 50, 71, 65, 76, 74, 73, 70, 63.

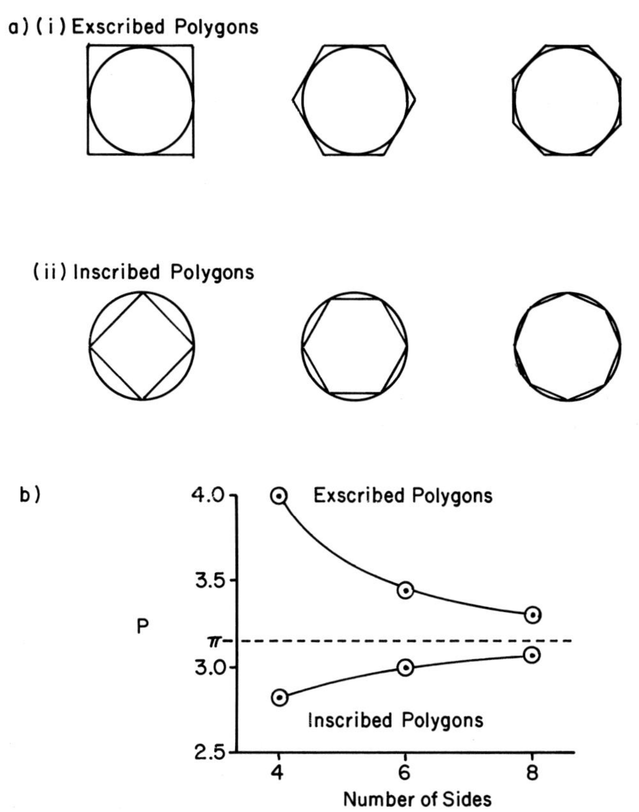

Figure E 3.6.1. One can estimate π by studying the convergence of estimates of P/D for a series of decreasing side length (increasing number of sides) polygons drawn on a circle as illustrated above. The perimeter of the polygon, P, divided by D, the diameter of the circle, becomes the estimate of π at inspection resolution λ, where λ is the side of the polygon of side number n.

Plot the cumulative less-than-or-equal-to percentage frequency distribution function for both sets of marks on Gaussian probability paper. Discuss the appearance of the curves in the light of the following information: On average 34% of first-year students in Ontario do not return for a second year (this information was valid during the 1970s and may have changed since). Using the less-than-or-equal-to probability distribution one cannot plot the highest mark. For the case of the compulsory course, from your dataline, predict the probability of a student obtaining 97%. The student who obtained this mark was taking a second degree program. Comment on the probability of obtaining this mark with and without this background. Comment on the fact that in the compulsory course there is a higher than expected number of very low marks. In the elective course 73 students registered for the course but two withdrew. Comment on the average mark in the compulsory and the elective course, including a comment on the withdrawals in the elective.

Exercise 3.8. Gaussian Distribution of Lake-Melting Data

Laurentian University is situated on Lake Ramsey. During the winter months the surface water freezes to a depth of several feet. In the following table the dates on which the lake was declared ice-free are given:

Year	Day	Year	Day
1980	May 2	1961	May 7
1979	May 3	1960	May 3
1978	not available	1959	April 29
1977	April 23	1958	April 17
1976	April 23	1957	April 24
1975	May 7	1956	May 10
1974	May 2	1955	April 21
1973	April 21	1954	April 26
1972	May 11	1953	April 20
1971	May 6	1952	April 28
1970	April 29	1951	April 26
1969	April 28	1950	May 9
1968	April 18	1949	April 24
1967	April 30	1948	April 25
1966	April 24	1947	May 8
1965	May 5	1946	April 4
1964	April 29	1945	April 5
1963	April 24	1944	April 30
1962	April 27	1943	May 7

Let R = cumulative percentage of years for which the ice has vanished by a given day. Plot a graph of R against the date on Gaussian probability graph paper. Comment on the result (is it surprising to you? If so, why? If not, why not?). Comment on the probability of the melting dates of 1946 and 1945. In 1981 the ice melted on April 14th, what does that fact tell you about the weather in March 1981? Would you suspect that the climate is warming up or cooling down from the data of the table? Justify your answer.

In Sudbury the local newspaper, the Sudbury Star, runs a competition in which prizes are awarded to the person who guesses the date and time to the nearest minute at which the lake is declared ice-free by a local judge. The prize-winning time is often judged to within a minute or two between competitors. Comment on the usefulness of the graph you have drawn as an aid to winning the ice judging competition. Do you think skill is involved in winning the prize?

Exercise 3.9.

In Section 3.14 the weight distribution in a set of candy buttons was explored. It was shown that the weight distribution was a bimodal Gaussian probability distribution. If one takes this type of candy button set and measures the diameter and thickness of the buttons using a micrometer gauge (an instrument readily available in a physics laboratory), then the diameter and thickness distribution functions are

Table E 3.9.1. Data for the measured thickness and diameter of a set of Smartie® buttons.

Measurements of Smarties Diameter [cm]	Frequency f	Cumulative Σf	Percent [%]
1.40	1	64	100
1.41	0	63	98.4
1.42	0	63	98.4
1.43	0	63	98.4
1.44	2	63	98.4
1.45	5	61	95.3
1.46	6	56	87.5
1.47	5	50	78.1
1.48	13	45	70.3
1.49	7	32	50.0
1.50	11	25	39.1
1.51	5	14	21.9
1.52	4	9	14.1
1.53	2	5	7.8
1.54	2	3	4.7
1.55	1	1	1.6

Thickness [cm]	Frequency f	Cumulative Σf	Percent [%]
0.55	1	64	100
0.56	3	63	98.4
0.57	4	60	93.8
0.58	9	56	87.5
0.59	8	47	73.4
0.60	9	39	60.9
0.61	14	30	46.9
0.62	8	16	25.0
0.63	5	8	12.5
0.64	0	3	4.7
0.65	1	3	4.7
0.66	2	2	3.1

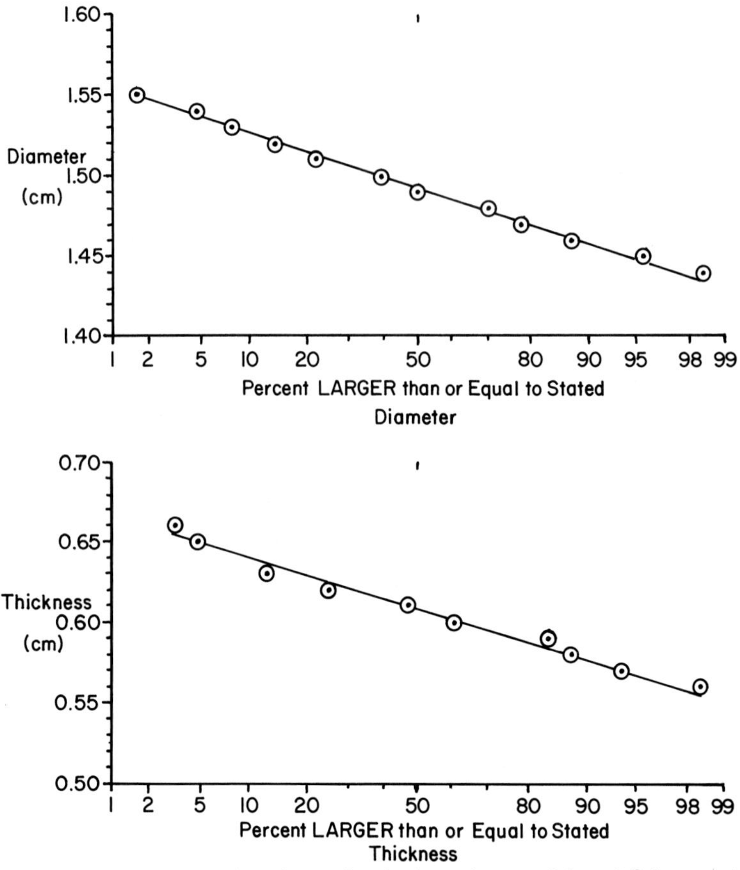

Figure E 3.9.1. The diameter and thickness distribution of a set of Smartie® buttons is a Gaussian probability function.

again describable by the Gaussian probability distribution function as illustrated by the data sets of *Table E 3.9.1* and *Figure E 3.9.1*. In a class project students should buy a set of candies such as M & Ms®, and the different distribution functions representing weight and the dimensional distributions can be measured and plotted on Gaussian probability paper.

Exercise 3.10. The Normal Number of Words in a Line of a Printed Page

It can be shown that the number of words in a line of print on a page of a book can be desribed by a Gaussian distribution. When selecting a passage to test this idea, one must select a print layout in which the line has a sufficient number of words (the short columns of many newspapers are not suitable for this study). In *Figure E 3.10.1* the number of words per line on fifty lines of text beginning on p. 182 of the first

a)

Number of Words per Line
for
50 Lines of Text
starting on
Page 182
of
A Randomwalk Through Fractal Dimensions

Number N	Frequency f	Cumulative Σf	Percent %
10	1	50	100
11	5	49	98
12	4	44	88
13	7	40	80
14	10	33	66
15	10	23	46
16	5	13	26
17	5	8	16
18	2	3	6
19	1	1	2
20	0	0	0

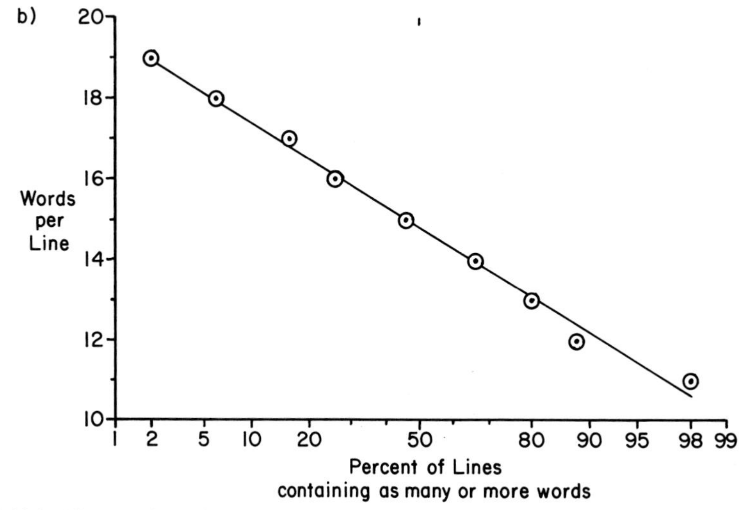

Figure E 3.10.1. The number of words per line on the printed page of a book follow a Gaussian probability function.

edition of *A Random Walk Through Fractal Dimensions* is shown [4]. From the graph of Figure E 3.10.1 it can be seen that the word frequency per line is describable by the Gaussian distribution. The reader can repeat this experiment with various printed pages. One should exclude from this type of study the partial lines at the end of a paragraph.

Table E 3.10.1. Number of words per line for 50 lines of text starting on page 182 of A Random Walk Through Fractal Dimensions.

Number N	Frequency f	Cumulative Σf	Percent [%]
10	1	50	100
11	5	49	98
12	4	44	88
13	7	40	80
14	10	33	66
15	10	23	46
16	5	13	26
17	5	8	16
18	2	3	6
19	1	1	2
20	0	0	0

If one looks at the number of words in a sentence then the distribution is not usually Gaussian but is usually an example of another stochastic relationship described in Chapter 4 – the log-normal distribution.

Exercise 3.11.

An important industrial task in the preparation of powdered or crushed materials is to use a sieve to create fractions of powder which are of approximately the same size. A *sieve* is defined in a dictionary as a vessel with a meshed or perforated bottom used to seperate the fine particles of anything from the coarse particles. The dictionary defines a *mesh* as an opening between the threads of a net. The surface of industrial sieves are often made by weaving an open type cloth using wire thread. A typical piece of wire mesh sieving surface used in industry is shown in Figure 3.24. The holes in the middle of the mesh are also called the sieve apertures. It is obvious that if the sieve is to do its job properly, all the apertures of the sieving surface must be identical. A real sieve, even a new one, will obviously have a range of aperture sizes. The variations in the aperture dimensions of a real sieve will be caused by many small variations in the manufacturing technique. In this chapter we have discovered that one can reasonably anticipate that a system will be describable by a Gaussian distribution when the parameter being studied is being generated by the random interaction of many causes. The reader will therefore not be suprised to discover that the variations in sieve mesh apertures can be described by means of the Gaussian distribution function. The students should make their own set of measurements on the apertures of Figure 3.24 and construct their own graphs from their data.

References

[1] For a readable introduction to the theory of gambling games see M. J. Moroney, *Facts From Figures*, 2nd revised edition, Pelican, Harmondsworth, England, 1953.

[2] L. J. Parratt, *Probability and Experimental Errors in Science*, Dover, New York, 1971.

[3] See article on "Poincaré" in *World Of Mathematics*, edited by J. R. Newman, Simon and Schuster, New York, 1956.

[4] B. H. Kaye, *A Random Walk Through Fractal Dimensions*, VCH, Weinheim, Germany, 1989.

[5] B. Evans, *Dictionary Of Mythology*, paperback edition, Dell, New York, 1970.

[6] See article on "de Moivre" in [3].

[7] Quoted in B. J. West, M. Shlesinger, "The Noise In Natural Phenomena", *American Scientist*, 78, January–February 1990, 40–45.

[8] Copies of this program are available from Professor Robb, Physics Department, Laurentian University, Sudbury, Ontario.

[9] T. Allen, *Particle Size Measurement*, 4th edition, Chapman and Hall, London, 1990.

[10] B. H Kaye, *Direct Characterization of Fineparticles*, Chapter 9, Wiley, New York, 1981.

[11] F. Chayes, *Petrographic Modal Analysis*, Wiley, New York, 1956.

[12] See discussion of sieve aperture measurement in Chapter 4 of B. H. Kaye, *Direct Characterization of Fineparticles*, Wiley, New York, 1981.

[13] M. J. Kendal, P. A. P. Moran, *Geometrical Probability*, Hafner, New York, 1963.

[14] Buffon's needle is discussed in G. Gamow, *One, Two, Three ... Infinity*, Viking Press, New York, 1947.

[15] For a highly mathematical discussion of Buffon's needle problem see M. J. Kendall, P. A. P Moran, *Geometric Probability*, No. 10 of *Griffith's Statistical Monographs in Courses*, Charles Griffith, London, 1963.

[16] This material is reprinted in *World of Mathematics*, edited by J. R. Newman, Simon and Schuster, New York, 1956.

[17] See discussion in Volume 4 of *World of Mathematics*, edited by J. R. Newman, Simon and Schuster, New York, 1956.

[18] D. G. Fink, *Computers and the Human Mind*, Anchor-Book Co., New York, 1966.

Chapter 4

The Rare Events of Cooperative Chaos

Chapter 4 The Rare Events of Cooperative Chaos 177

Section 4.1 "Double or Quits" – A Desperate Gamble 179
Section 4.2 The Surprise of a (Half) Lifetime! 184
Section 4.3 Making Things Normal Again with Logarithms 189
Section 4.4 How Long is a Game of Snakes and Ladders? 193
Exercises 201
References 206

Chapter 4

The Rare Events of Cooperative Chaos

Section 4.1. "Double or Quits" – A Desperate Gamble

The word fallacy is defined in the dictionary as

an apparently genuine but really illogical argument; a delusion, something that deceives or misleads

The fallacy which seduces many gamblers into being wiped out at the gambling table is the hope that after a run of adverse events the gamblers "luck" must change. For example, if a gambler was playing a simple game of pitch and toss of a single coin, in which bets were being laid on an outcome of the tossed coin, then after losing six times in a row the gambler, convinced his "luck must change", is often tempted to say "let us go double or quits at the next toss of the coin". If he has lost $ 600 by the successive tosses of the coin by wagering $ 100 on each throw of the coin, he wagers that on the next coin he will either get everything back that he has wagered, or the other player will double his profits to $ 1,200. Somehow, the gambler deludes himself into believing that there just cannot be another tail in the sequence after six successive tails from random tossing (always assuming that the coin is not biased). But, unfortunately for the gambler, there is absolutely no difference in the chances of a tail on the seventh throw as there was on the first throw; the odds are still fifty-fifty that a head will be produced by the 7th toss of the coin. A mathematical dictionary gives the following definition of "odds": In betting, the ratio of the wager of one party to that of the other, as to lay or give odds of, say, 2 to 1 (if A bets $ 2.00 and predicts correctly the outcome of a gamble, B pays $ 1.00. If A is wrong, B keeps the $ 2.00). The word "odd" has an odd history. It comes from an old English word meaning a triangle or a point. An *odd number* was one which left a spare digit when divided by two. By extension, an *odd person* was originally a player in a game without a partner. Then an odd person came to be one who was so strange he could never find a partner. Just how the word odds came to mean the number ratio expressing the chances of winning in a gambling game is a mystery. In fact, the number of times that one can get a long run of apparently rare events, such as seven tails of a flipped coin in a row, is not as uncommon as one would think – unless one has studied the patterns of chaotic events [1]. To produce a long run of tails in the tossing of a coin, it is almost as if the random causes producing the outcome of the flipped coin suspend their competition for a short time and act cooperatively to produce a rare event. From one point of view, a run of seven tails is a pattern produced by the "blind cooperative" contributions from the chaotic conditions producing the coin tosses (note that this is only an *image* which helps the newcomer to the study of deterministic chaos understand how the pattern of events are generated. In reality the random causes are still acting at random, but we can say that the pattern is being produced by what we choose to call *cooperative interaction* to produce relatively rare events). In this chapter we will discover strange patterns of stochastic events which, if one did not know the concept to be false, one would feel were being generated by the fiendish cooperation of the devils of random chance (see Feller's comments on coin tossing sequences reviewed in [2]).

The problem of long runs of improbable sequences of heads or tails in gambling was first studied by Buffon, of needle tossing fame (See Section 3.15). Buffons original work was reviewed in an essay by Augustus De Morgan [1].

A table given by De Morgan in his discussion of the problem, containing theoretical and experimental data generated by Buffon and three other mathematicians, is summarized in *Table 4.1*. Before looking at the pattern of events discovered by Buffon in coin-tossing experiments, it is useful to explore the meaning of words used by gamblers when making bets on the outcome of a game.

The word *gambler* comes from an old English word gamenian, meaning "one who likes to play games". In modern English a gambler is one who likes to play games to win money. To win money, a gambler guesses the outcome of a game involving chance events, and deposits with another person an amount of money known as his *wager* (originally a *wage* was a pledge to pay someone a set amount of money when a given amount of work was performed). The wager is lost by the gambler if his guess is wrong, but if he is correct it is returned to him along with his winnings. In the tossing of a simple unbiased coin, the probability of the outcome of a head or a tail is the same. Gamblers describe such a situation as the odds of the two alternatives, heads or tails, being even. On the other hand, if one is making a wager on the outcome of the throw of a die, then the chance of a 6 being thrown is 1 in 6. A gambler would be delighted to play with someone who wagers a dollar that a throw of the die will produce a 6 at winning *odds* of 4 to 1 (usually written 4 : 1). This description of the odds when playing the game means that each time the throw of the die is not 6, the wager of one dollar is lost; each time the throw is a 6, the player receives $ 4.00. In many games of chance the player accepting the wagers, and paying out the odds if the gambler is correct, is known as the *banker*. Playing the game "throw a six for a win" with odds of 4 to 1 will soon convince the reader that the banker will on average be the big winner (see discussion of gambling games in [3]). If the odds of a "6 to win" is set at 6 to 1, there will be no winner on average. If the gambler tries to find a banker to give him odds of 7 : 1 on the chance of throwing a 6, the would-be gambler is not likely to find a banker daft enough to play with him.

Table 4.1. Summary of data on the outcome of a sequence of coin tossing experiments reviewed by De Morgan [1].

Sequence	Theory [*]	Buffon	Charlie	Fred
TH	512	494	507	480
T^2H	256	232	248	267
T^3H	128	137	99	126
T^4H	64	56	71	67
T^5H	32	29	38	33
T^6H	16	25	17	19
T^7H	8	8	9	10
T^8H	4	6	5	3
T^9H	2	0	3	4
$T^{10}H$	1	0	1	0
$T^{11}H$	0	0	0	0
$T^{12}H$	0	0	0	0
$T^{13}H$	0	0	1	0
$T^{14}H$	0	0	0	0
$T^{15}H$	0	0	1	0
Total Data	1023	987	1000	1009

[*] Note that only whole number values are quoted. In the above table the superscipt on the "T" represents the number of tails in a row. For example: T^5H represents the sequence TTTTTH.

Bankers usually know a lot more about gambling odds of a given game than the average player.

From De Morgan's theoretical data we are told that the chances of the next throw also being a tail after 6 tails in a row is 8 chances in one thousand, or odds of 128 : 1. Buffon's original experimental data gave odds slightly more favorable than this (8 in 987 = 123 : 1). If you were present where a coin had tossed heads 6 times, and you could be sure that the coin was unbiased, then if you took odds of a 126 : 1 that the next toss will give a head you have a very small chance of winning.

If one looks at the columns of Table 4.1, although the events are generated by a chaotic system generated by tossing coins, there is a remarkable uniformity between the columns, and one asks the question "Is there any mathematical pattern that we can discover in the recorded events?"

If we look at the data of column 2, the theoretically probable set of events, we can see that the frequency of occurrence of a sequence of tails decreases drastically as the specified number of tails increases, with each number lower in the table being half the number above it. In fact, the probabilities of increasing numbers of tails in a given sequence follows a pattern which is known in science as a *logarithmic decay*. This name comes from the fact that when we have a decay in the magnitude of a set of numbers of the type shown by the data in Table 4.1, it can be shown that if we plot the logarithm of the probability of the given event against the number of tails in a

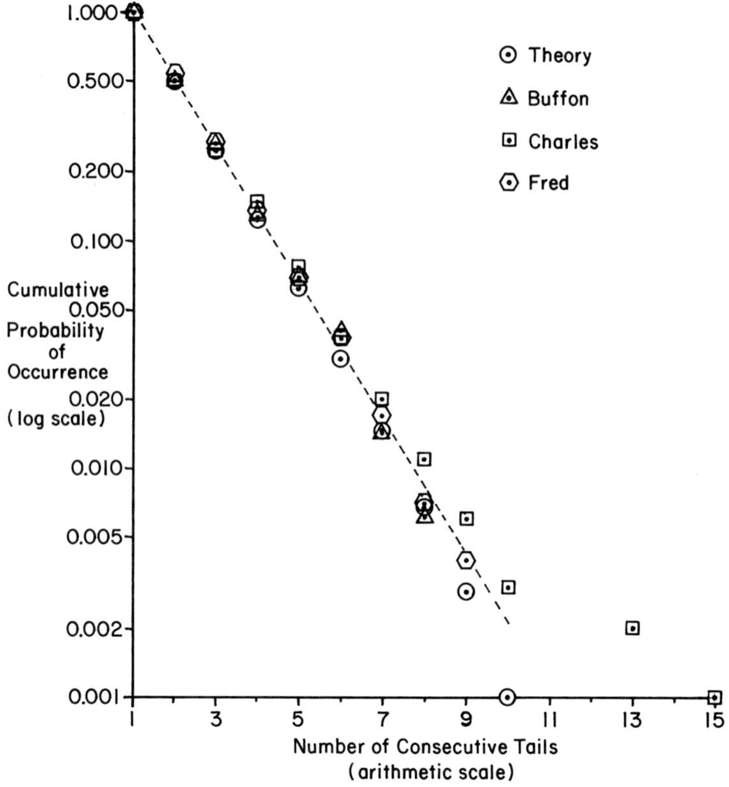

Figure 4.1. The probability of occurrence of a specified sequence of "tails" in a coin tossing experiment (data from Table 4.1 plotted in cumulative format on log-linear scale graph paper) generates a pattern of events known as a logarithmic decay.

sequence on the abscissa, we obtain a straight line as shown in *Figure 4.1*. This type of decline in a variable such as the probability of a run of tails in a sequence of tossed coins is also known as *exponential decay* from the fact that the basic equation describing the curve can be written in the form

$$P_N = k\,\mathrm{e}^{-N} \text{ which transforms to } \ln P_N = k' - N$$

where P_N is the probability of N tails in a sequence as a consequence of tossing the coin N times, k is a constant and e is the irrational number that we encountered in Chapter 3. As we discussed in Chapter 1, when we have a number such as e raised to a power such as n, this is described as exponential notation. Hence the alternative term exponential decay, since in the equation we have a regular exponent (for a discussion of the basic concepts of logarithmic notation see Chapter 1).

A famous story illustrates how most people are unprepared for the very rapid changes in a variable which is subject to either exponential growth or decay. The story is told that a peasant invented the game of chess to please the emperor of China. The Emperor so enjoyed the game that he offered to give the peasant his weight in gold as a reward for inventing chess. We are told that the crafty peasant declined this gift of gold on the grounds that, as a poor peasant, he would not know how to deal with so much gold, but as a human being he needed rice to live. He therefore requested that the Emperor of China give him a grain of rice for the first square of the chess board, two grains of rice for the second, four grains of rice for the third and so on to the end of the board. It can be shown that this means that after N squares the peasant would have N_{T} grains of rice given by $N_{\mathrm{T}} = 2^{N}$. The chess board has 64 squares. Not knowing how vast an amount of rice was involved in this doubling process at each square, the emperor readily agreed to the rice reward. If one completes the calculation with the help of a calculator, one finds that there was not enough rice in all of China to pay the reward.

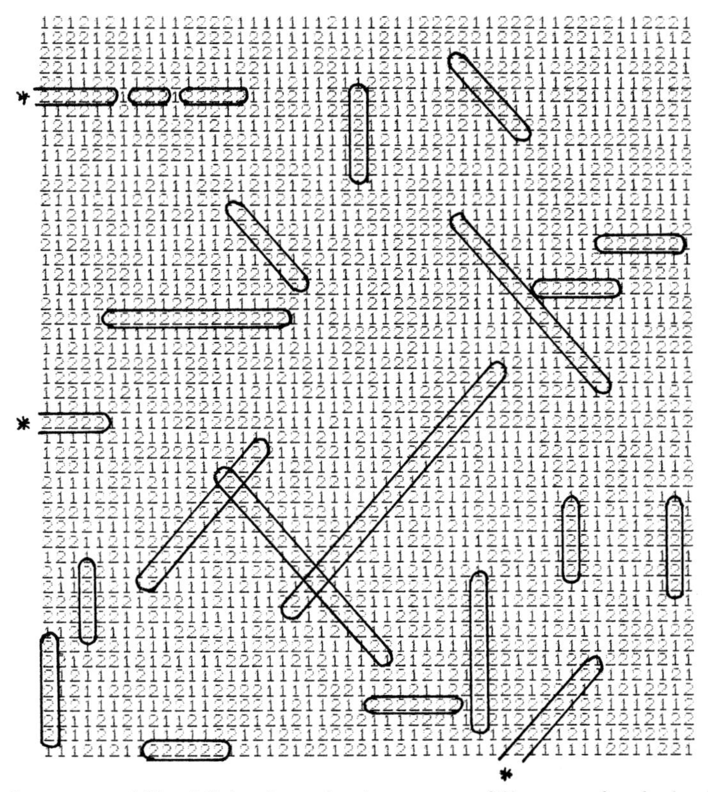

Figure 4.2. Long runs of "heads" (as shown by the patterns of 2's assumed to be heads) are not as uncommon as some people think. Those marked with * should not be used in any study of "long runs" because they start or end on the boundary of the table.

The special random number table consisting of only 1 and 2, which was introduced in Figure 1.12 of Chapter 1, can be used as the basis for a project to discover the frequency of occurrence of long sequences of heads or tails in a coin-flipping experiment. This table is shown in *Figure 4.2*, and some of the sequences of heads (assuming that the two represents a head) have been circled. The sequences can occur in either vertical, horizontal or diagonal sequences. The reader can block out any sequences which occur in the same way. Caution must be exercised with patterns marked with an asterisk, since they represent sequences at the end of the table. They should not be used when computing the frequency of occurrence of a run of a set number of heads, since it is not known if an extension of the table would make them members of even longer running sequences. A completely independent set of data can be obtained by blocking runs of 1 and then of 2. The two sets of data for the 1's and the 2's can be combined to estimate the frequency of occurrence of long runs of heads or tails in a gambling game. The reader with some computer skills could generate the same data by causing all the 1's in a computer version of the table to vanish from the table. Likewise, a class project could place a transparency sheet over the random number table and one part of the class could look for sequences of 1's, whilst the other part of the class looked for runs of 2's.

Section 4.2. The Surprise of a (Half) Lifetime!

There is no doubt that one of the most surprising aspects of the structure of the universe awaiting the scientists of the 19th century, as they worked to isolate the various atoms making up the universe, was the discovery of the spontaneous disintegration of some atoms of some elements [4]. This spontaneous disintegration of an atom is now called *radioactivity*. The first radioactive substance to be discovered by the scientific community was a compound of the element uranium. Its strange properties were first investigated by the French physicist Antoine Henri *Becquerel* (1852–1908). We are told that whilst investigating the *fluorescence* of certain crystals (their ability to give out light when they receive radiant energy from the sun or other source of light), he put some crystals on a photographic plate and wrapped them up in dark paper and placed them in a drawer. He intended to take the wrapped plate out of the drawer on a day when there was bright sunlight, to see if the bright light from the sun could go through the paper and activate the fluorescent crystals (the origin and meaning of words such as fluorescence, phosphorescence and radioactivity are illustrated in the wordweb of *Figure 4.3*). The fluorescent material that Becquerel was using was a compound of uranium known as potassium uranyl sulfate.

Becquerel could not carry out his intended experiment because of a period of dull weather. He then became impatient to see what was happening to his photographic

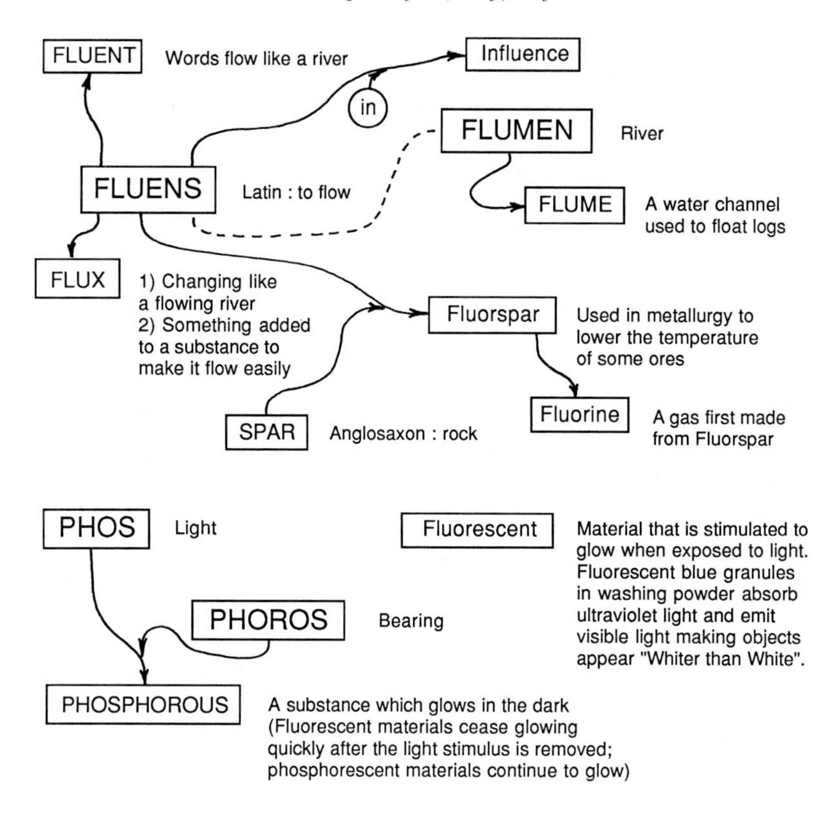

Radiation : Energy like that of the suns rays (Radius was the latin for the spokes of a wheel; shafts of light from the sun look like spokes of a wheel. Beams of light were originally called radii which with common usage became rays.)

Radioactive material can give out three types of "beams" which historical reasons are called α, β and γ rays
α rays are helium atoms stripped of thier electrons
β rays are short wavelength electromagnetic waves
γ rays are high speed electrons

Figure 4.3. The surprising discovery of radioactive materials made it necessary to make new words to describe the behavior of such materials.

film. Hoping that perhaps there had been some faint residual fluorescence in his material, he developed the photographic film without exposing the wrapped film to light. He was very surprised to find a very extensive activation of the film from some unknown radiation coming from his crystals. We now know that the radiant energy which activated his photographic film was coming from the energy generated by the spontaneous disintegration of some of the uranium atoms in his crystals. Radioactivity occurs in an atom when the nucleus of the atom contains so many components that it is somewhat unstable and has a tendency to spontaneously disintegrate.

However, for some strange reason not all the atoms of a radioactive substance disintegrate at the same time. By studying radioactive materials, scientists discovered experimentally that if one started off with a given number of atoms, then in any given time period a fraction of these atoms would disintegrate. They also discovered that in successive periods of time, the fraction of remaining atoms which disintegrated remained the same. If we had n atoms to start with in a period of time t, the number of atoms which would remain at the end of the period would be $n(1 - P)$ where P is the probability that an atom of the substance would disintegrate. In the second time period, the number of atoms which would survive would be $n(1 - P)^2$. This type of disintegration over a long period of time can be shown to be another form of logarithmic decay.

From a human perspective, the survival of any given atom appears to be dependent on a favorable combination of the chances of survival. If the atoms could think and have emotions, then at the end of each period of time they would wonder how they had managed to survive that period, and would wonder about their destiny in the next period. Using our picturesque concept of cooperative chance producing rare events, survival for a long period of time – when there is a fixed rate of disintegration – would appear to require cooperative combination of the chances of survival in successive time periods.

We can simulate the radioactive decay of a set of atoms by using a random number table. If we look at the top of a random number table we can consider 100 digits along the top of the columns to represent 100 atoms. We can simulate the disintegration of the atoms over a period of time in which there is a 10% chance (that is a 0.1 probability) of disintegration by moving down the column and deciding that a disintegration occurs whenever we meet the number 9. Stated in another way, at the end of a given simulated period, when the chance of disintegration is 1 in 10, on average 90% of the atoms will still be left untouched at the end of the period. Moving along our table of *Figure 4.4*(a), we discover that at the end of the first time period on a random chance basis we actually have 88 atoms left. For the sake of presenting the data from our simulated experiment, we will consider that one minute elapses before we move down the table to look at the next row of digits. When we move to the second row, we look for 9's underneath a non-9 position in the top row. This exercise results in the simulated set of survivors after 2 time periods shown in Figure 4.4(b). If we plot the number of atoms left after each disintegration period, we have the pattern of events shown in *Figure 4.5*(a). This data illustrates clearly the

Figure 4.4. The random radioactive disintegration of a set of atoms can be simulated using a random number table. In this table the survival of atoms over a period of time when the probability of disintegration per time interval is 0.10 (that is, 10% of the atoms will disintegrate) is simulated. a) The original 100 × 50 random number table used for this simulation. b) The appearance of the table after the simulation in which the occurrence of a 9 is assumed to represent the decay of the atom in that column.

a)

6295402943655879364871459799802840113928598011809778466833269658988727726554968282338285830152211248
3477192872875617071061725517336991124290945428591278722476030347729528639560125529840323701803282051
3815677341984133724311140659739957750568048742374406401873977289832441888617180355626778472992576106
8929747718917048441801356544433151582462460198266304754144893942094211580424318460977748010593173822
4592757240451075616657669086814141328347789198003297118067748774360156380371620135137545136802334380
4057007999277092229225318364160096841343114419755771923493135971395710059756498279253938763370784956
36258865551307808623998738008429711237565690961690578870773280963767080491687529050175343128259376606
0889032086671672840765670741445148266902954956566161208171723318792523698952828477956458770804458292
1886292089038700596124762940886516963298718376622505033242147423233769063045467670026419574011264451
1150937950468186686493752318302126504910762267378695983133913036951035173421172988647705413602623113
0443656894380594650776027999822346450154969900304519946554487504924541430071197531797053734589325665
4184469445275606018999491153268924743452891452864903580672060963584612626118672477411994155890186160
5044115736195429934618367981285035630649443138656956823498192028322272001149200724369005430541792251
6057716095124640358315893866734803428918851839245887633317619986531488499470606032083479440044906550
5661884121432625887824560685572756035890344637134595152101335269322480503364950206296764141835805313
9542995393699313145435457321143540846677988795974205639241757943892151576020359025465547807909873594
4473371660721005364880716316016869992500896092038865927555256727022907550976317734119942168486110457
7598238721090367040545720844861904408000527978705459005290003839580367341124186630170846588444103926
8959356259105420909516192822027872125394674406107206626426829152245955232630976426824829361681313878
0740739827305922402339305048740537835151478293730811306027323418904228585417257389071581666715929633
4559503178938800352813480469950382750746770076883645143565459461742752312091438081477369524493497435
6422462101833744974242754646988433218130583565815197612871752261384175964138333612408893184171390669
9980189888704296638086753442045926320599371937505769703562442069404571939918602745435354497297898928
8097762350238491747793234780065763028011462536168362498022571917906313079004673321413636274873492174
6100084473474882790881003633242393854787893103714580306192571869054653072531989489909764155724602621
8340308404009730621027249216673161815522506191210284114848845012968942932647234498794167918386132007
4778270170214440678481758747358733063034934708503558437888079214188677028640205252843216922137329408
0941079959130619996382403752617439012908446037353615702534575425647485233997169256806708014154894809
8171981141203786171741156374916227775353426420992959966412878700126462940002845709469804160589787970
1728286203077305403214819881509526343141178397708243641059208134602507253499337002064627194574062080
0165232499019864219992277308717797164075024617553025501501134460087433006837910722315421622551698434
7876697737701894920107477208528406128354840069246698294408491252512673700071450906862864390426849495
5350403566410867046339285785637554289512664361771052582899853786360788274124750271577135423107713353
3392822958180411319290017324550931083995897746998226510836221254600489893041555009482075046458892562
8121431882502461824892838750152388688938716074762324755715433166783624270856590532386307751968941437
8051588953176526469313376135254277809456604317906638399887965885618220510160497654062491099556348244
8111728474038250556485545409902714466674841051237118889878145223258244952060594726622013812874438374
9330129964147810047164049628158004894571498729108051774896940265463161546223111351160290156473007750
2187660343247674052368568335194385217202970878491388375377106161969820131319040645206063860639866652
3610124555246915531480656876949204210967802277843510416703459330787360534526117946645915213243074795
2965405268266215128654016927040693592320092449833499258037785467451499136593643484760764466402697534
6715693151115607693271539996765377322484611988320158923183199509211840876468533058143420232897230657
4681207692969348391955687402572999151477379968788739407411264446978889544623860958555697932725811296
7386486418904131972798760070649370924719043413651229372707811955495559630883584560162212284829605006
5935963881474673446945954509790224294325730728382660728776846926582198500872115132444344320377379360
2891842344185385084817113432121343766905091558364807089029480369971255139448659789052888903553542145
4527138945954206071454523412848193914482211210832835757374472986004546287422777310215097744342622961
3382986076094524070214006301045118106850981733857758797708787816791942151083744199254271219235894254
4499726649799017698168225014943886471460893196015432281978602302961409818870205063060452464631230378
8223786534271535154364074413500575300264591866711793531085143024434641298207212662419932332653721179

b)

62 5402 4365587 36487145 7 802840113 285 801180 778466883326 658 88727726554 68282338285830152211248
34 71 2 7287561 07106172 5 336 1124 0 54285 27872247603 347 2 52863 560 2552 840323701803282051
38 56 7 41 8413 72431114 6 73 7750 8 87423 406401873 7 28 3 44188 617 8035 62677847 2 2576106

[illegible]

TIME	SURVIVORS
1	88
2	79
3	73
4	66
5	64
6	54
7	48
8	46
9	41
10	39
11	37
12	32
13	26
14	25
15	21
16	18
17	18
18	18
19	18
20	16
21	14
22	12
23	10
24	9
25	9
26	8
27	8
28	8
29	8
30	8
31	6
32	6
33	6
34	5
35	5
36	5
37	5
38	4
39	4
40	4
41	3
42	3
43	2
44	2
45	2
46	2
47	2
48	1
49	1
50	1

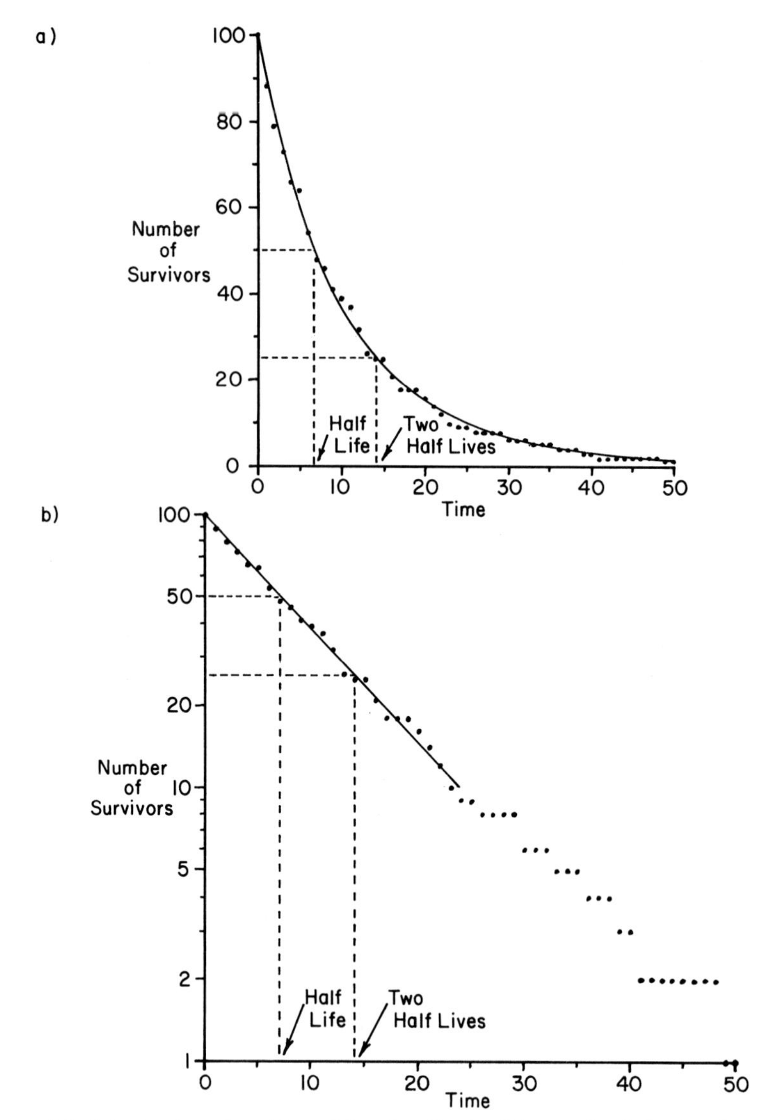

Figure 4.5. Scientists have found it useful to describe the disintegration behavior of a set of radioactive atoms in terms of the half life of the material. This is the time required for half of the original atoms to disintegrate. a) Simulated survival of 100 atoms subjected to radioactive decay. b) When plotted on a logarithmic scale against time on an arithmetic scale, a straight line can be drawn through the data.

drastically falling population resulting from the logarithmic decay of a radioactive material. The survival rate plotted on a logarithmic scale as shown in Figure 4.5(b) generates a straight line.

Scientists have found it useful to use a quantity known as the *half-life* of a radioactive system to characterize its rate of decay. By definition, the half-life of a radioactive material is the time required for the number of surviving atoms to equal half of the population of atoms at the beginning of the experiment. It can be seen from Figure 4.5(b) that for our simulated radio active decay, the half-life of the set of atoms is of the order of 7 minutes. The half-life of real radioactive materials varies enormously from a fraction of a second to millions of years. But whatever the half-life of the substance, radioactive disintegration patterns all follow the characteristic logarithmic decay illustrated in Figure 4.5. Perhaps when we have a better understanding of the dynamics of the tiny sub-units of an atom we may be able to relate the individual half-lives to the inner workings of the atom. Until then, the logarithmic decay of radioactive material will remain an empirical discovery (originally surprising!) that awaits a fundamental explanation in terms of cause and effect.

It should be noted that the datapoints of Figure 4.5(b) start to deviate from the decay line drawn through the survival data when the number of survivors drops to about 20. This is because once we are dealing with a small number of survivors, such as 20, it is not really a valid data processing technique to use statistical averaging. The pattern of survival in 20 survivors is more erratic than when considering 2,000,000 survivors. The readers can verify this for themselves by repeating the survival pattern of Figure 4.5 using different digits to generate 10 sets of survival data. One could show that, down to about 20 survivors, all of the curves give the same value for the slope of the line. Moreover, the scatter for the survival after 20 will average out to continue the dataline if 10 sets of data are amalgamated to make the survival pattern for the study of a population which initially contained 1,000 atoms. Again the straight line relationship will hold until the number of survivors is small (about 20). It should be noted for such a curve that the time required to halve the population remains the same as one goes from 1,000 to 500 and then from 500 to 250. This consistency of time required to halve the population illustrates the usefulness of the concept of the half-life of a radioactive material.

Section 4.3. Making Things Normal Again With Logarithms

To discover another fascinating pattern occuring in chaotic systems when large values of a variable can only occur when "chance cooperates", let us consider the pattern of events which occur as we set out from one digit in a random number table to see how far we can travel before we encounter another digit of the same magnitude. Since all digits are equally probable, on average one should go 9 steps before reaching another 9 (not 10, because the digit 9 is one of the possible digits). If we were to start off in the top left-hand corner of the table at the digit 9, for the random number table of *Figure 4.6*(a), then as we move 1 step below the 9, there is a chance

a) 95429050234244581479065825783219864096554609189898
54933303766021712916710344149300810407462232898099
49242618153687872093634791584781812726448593582053
41240220566687948908449091282075666703828660021025
020493822231089913677143607401606527802815378870695
27000125887267750662960475120961781227069483424773
91285543453403437310552916339998036177111595726772
44887269957423788921180379266490519132015426892950
72892367484254400765298260458216897055070011595513
17812437570702978410937620960633152811057969822892
95921104138621585039352462702613873353509451338339
40023282869294341583275637300952091864016408110484
80560817227287086454574297288060952611520283930079
47712678992190018132297852686542058973534388941507
80406138228195608991953590342539700535451684826169
56642537305671070241105223017047951283148407286847
74682443936916741446439824548523758889227637876329
45524464091292846822214070465303029959037155087652
57139421138563792492064693598992453681249826375567
00705177435868788502804852235293014944521806542134
05549701019294527189955274980105829597627434351864
22327554809590777989887329356775074324148104477751
86018858358836710506335853156893165638320674365506
70839189513348400091743677289204084597708264132061
67411886457778972757599273987687841515959636460506
28259118225305816516201967109863699630756839432666
04503790286340403482766772328839594661198127499982
63288931349137891988121573458179674857902245487109
63701544298699202004283511370089012706789268689488
96540711599347897919435016552782167153863352712041
16744202368950663437810515670459752675169343382926
75423864818414782941470887589762126142581583982679
48129767515349404187799553074781231263994037916182
81433532486364664467480898959015504162870251303315
81800279366047434712906266266368616394112821998681
07751337314428596913330779300652025194207728692896
26480310029389849074163542882409677575321489843899
09458049231790470053690338176654408038190400869508
91478940429094563299865726730355861787724664120170
36033561054164123328598884718572830719505956497746
22404607018974920791608980696275470239081835108554
94658816090544672224936835222417750481384162090644
39921054933454493414028310269506558134379127930904
36744751789113127882879818235962162723379892174546
16956159784614332638469172308769066999850173520241
33291762225208807482833089896040993842067631191982
63079726058950742852146811879578102306301617577998
47767522036266332254319661580599528340026923977643
76432631902994804588272771139309048552107157995191
04180187175857332422386158566603020279310166044950

Figure 4.6. The pattern of interdigit track lengths in a random number table is not a Gaussian "bell curve". a) Random number table used in the track length between identical digits experiment. b) Inter-digit track length is not Gaussianly distributed. c) The histogram of track lengths between digits looks like a slumping bell curve.

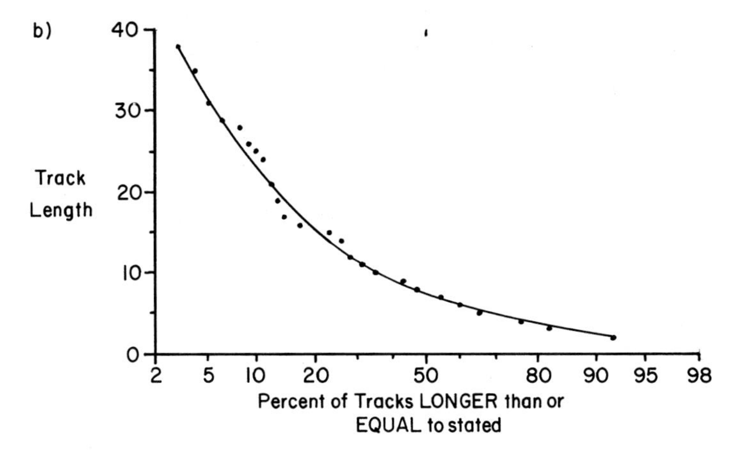

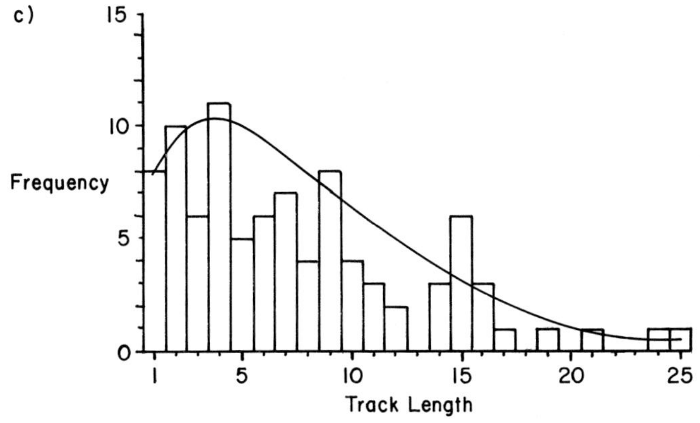

of 1 in 10 that we will discover another 9. In fact, we see from the table that the next digit is 5. As we move on to the next square, we have travelled 2 squares and we still have a chance of 1 in 10 of discovering another 9. This time the number is 4, and we move 1 more square looking for a 9. So far in our journey we have travelled 3 steps and each time our explorations could have been cut short by discovering a 9. The fact that we are still travelling after three steps has required "cooperation of chance events" such that we do not meet a 9. As we continue to travel down the column of the random number table, we find that the first 9 is encountered after 5 steps. If we continue the search for a 100 9's, the distribution of tracks between the 9's are as shown in *Table 4.2.*

Is there any mathematical pattern to be discovered in the frequency of the various path lengths? If we plot them on Gaussian probability paper, the resulting dataline is curved as shown in Figure 4.6(b). From this data display we assume that the interdigit track lengths are not Gaussianly distributed. The distributed histogram for the track lengths is shown in Figure 4.6(c). It can be seen that it looks like a bell

Table 4.2. Distances between the digit 9 in the random number table in Figure 4.6(a).

Track Length L	Occurrence f	Cumulative Total Σf	Track Length L	Occurrence f	Cumulative Total Σf
1	8	100	20	0	12
2	10	92	21	1	12
3	6	82	—	—	—
4	11	76	24	1	11
5	5	65	25	1	10
6	6	60	26	1	9
7	7	54	27	0	8
8	4	47	28	2	8
9	8	43	29	1	6
10	4	35	30	0	5
11	3	31	31	1	5
12	2	28	—	—	—
13	0	26	35	1	4
14	3	26	—	—	—
15	6	23	38	1	3
16	3	17	—	—	—
17	1	14	—	—	—
18	0	13	54	1	2
19	1	13	55	1	1

curve that has been knocked sideways. Statisticians described such a structure as a *skewed bell curve.*

Mathematicians have shown that very often a skewed bell curve can become a bell curve if one uses the logarithm of the magnitude of a variable instead of its original value when constructing a histogram. It is not recorded who first discovered that skewed bell curves can often be treated as Gaussian probability curves by using a logarithmic transformation. Perhaps a mathematician looking at the skewed bell curve realized that if the abscissa was squeezed to the left using a logarithmic scale then the skewed bell curve became a true bell curve.

In the language of mathematicians, distributions of events which can be described as being Gaussianly distributed when using the logarithm of a variable are described as being *log-normally distributed.* It can be shown that the track length data of Table 4.2 can be described by a log-normal distribution function. A mathematical demonstration and proof of this statement is beyond the scope of this book, but to demonstrate the validity and utility of the log-normal description of systems, we will make use of a special graph paper designed by mathematicians which enables us to anticipate straight-line data curves when the variables being studied are describable by a log-normal distribution function. This graph paper is known either as *log probability graph paper* or *log-normal graph paper*. The cumulative frequency of tracks equal to or greater than a stated size for the track data between 9's in the random number table

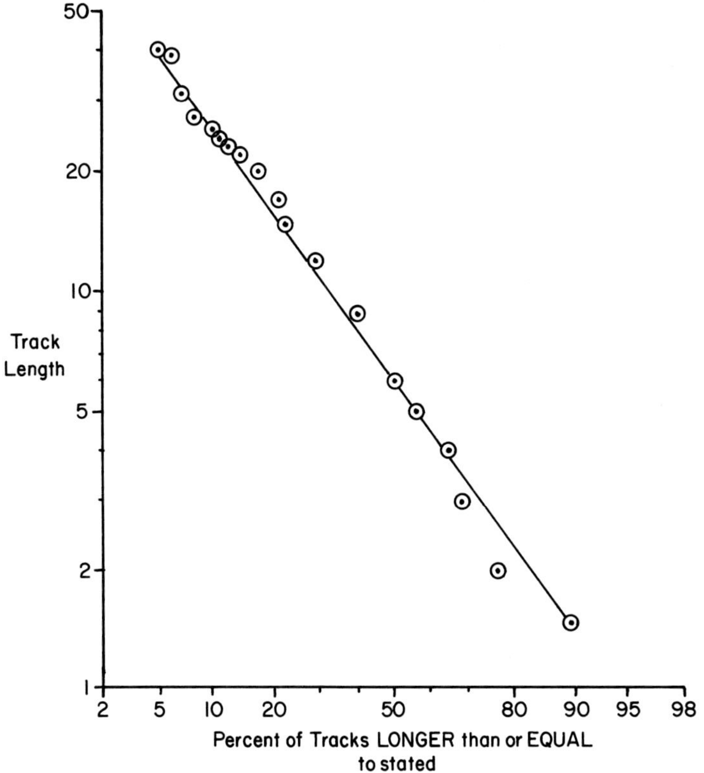

Figure 4.7. The frequency distribution of tracks between the digit 9 on a random number table can be described by the log-normal probability distribution function.

are plotted in *Figure 4.7* on log-probability graph paper, and it can be seen that the data points can be linked with a straight line. In fact, the size distribution of a variable which requires favorable (cooperative) interaction of random causes for large values of a parameter can often be described by the log-normal distribution function. In the next two sections we will explore such log-normal patterns generated by "cooperative" interaction of chaotic variables.

Section 4.4. How Long is a Game of Snakes and Ladders?

In *Figure 4.8* a well-known childrens game, snakes and ladders, is shown. The rules for playing the game vary slightly from one commercially available game set

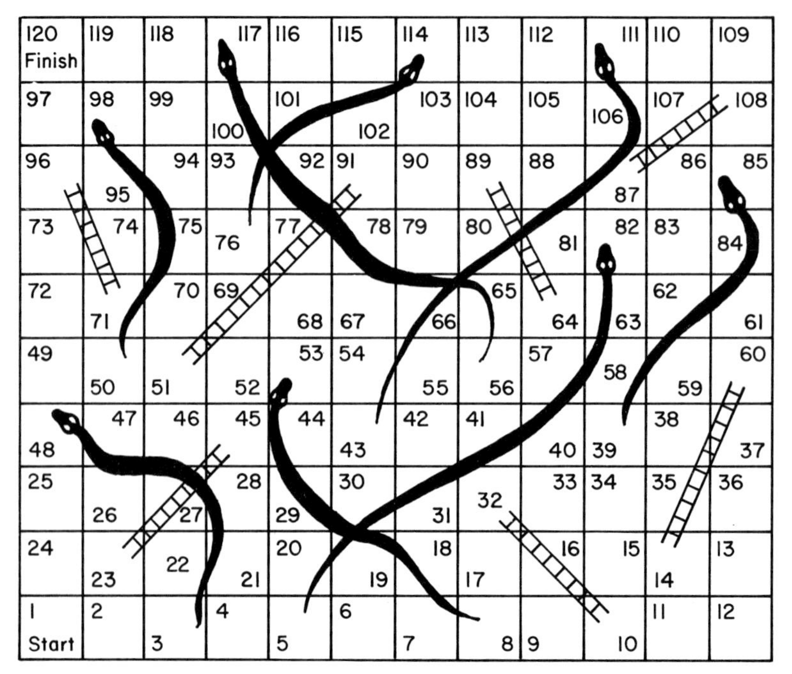

Figure 4.8. A game of snakes and ladders involves a Markovian chain of events, the length of which is long when stochastic events "cooperate" to prolong the game.

to another, but we will use the rules as set out for the game in an authoritative book on games entitled *The Board Game Book*, by R.C. Bell [5]. Bell tells us that the game of snakes and ladders was invented in India. It was originally called "moksha-patamu". Teachers of the Hindu religion used to use the game to teach children the religious truth that good and bad behavior exist side by side. According to the Hindu religion, the aim of life is to reach Nirvana (*Nirvana* is a state in which all desire and emotion are suppressed; it is not exactly analogous to the Christian heaven). According to the Hindu beliefs wickedness, symbolized by a snake, leads to reincarnation into a lower animal form and sets the individual back in the game of life.

> *In the original Indian game each ladder had a name, thus faith went from square 12, reliability from square 51, generosity square 57, etc. In the same way the snake heads indicated vices, with disobedience taking you down to square 41, vulgarity to 49, theft to 52, drunkenness to 62, etc.*

(Numbers given here do not correspond to those of Figure 4.8, but to the board in Bell's book). In the western world the moral lessons embedded in the original game of snakes and ladders have been lost. It is now simply a game of chance played with a die.

In attempts to simulate the processes of human thinking and learning, the developers of computers have put great effort into trying to program computers to be able to play games. In particular, the challenge of making a computer learn how to play chess has occupied many people for many man-hours. In general, the development of programs to simulate the progress of a game played by humans is of great help in developing an appreciation of computer logic. It also helps us to understand the problems involved in simulating the progress of complex systems in everyday life. A complete discussion of this branch of mathematics, which is known as *game theory*, is beyond the scope of this book. It is, however, useful to introduce game theory, and the general concept of simulating the progress of the game by computer, by attempting to discover a mathematical pattern in the outcome of a simple game such as snakes and ladders where no decision-making skills are involved in the progress of the game.

Consider the question, "How long is a game of snakes and ladders?" Immediately, we have to recognize that the question itself has no answer. As so often happens in game theory, the experimentalist must learn to refine the questions to be able to pose a question that has an objective answer. From the computer's point of view, a game does not last a number of seconds, but a number of moves. Therefore a more appropriate question is "How many moves does it take to complete a game of snakes and ladders?" Again, this question has no unique answer. A game of snakes and ladders can, in theory, last forever. In practice, one must study experimentally the range of answers to this question in order to be able to give probabilities that a game can last for a specified number of moves.

Before one can simulate how many moves are needed to complete a game, one must set out the rules for the game in a clear manner. In this example, we will follow the rules set out by Bell in his book on board games. These are as follows:

1. The game can be played by any number of people (for the system shown in Figure 4.8, up to four markers are allowed).
2. The aim of the game is to reach home, which for the board of Figure 4.8 is number 120.

Rules of Play

Rule 1

The players each throw a die once. The player with the highest score has the first turn. The players sit around the table in a clockwise order of play determined by the original throw of the die, the caster with the lowest score sitting to the right of the opening player (this rule is not relevant to our mathematical simulation since we are not concerned with who will be player 1 or 2 etc., but in a real game the first player to throw the die can have a real advantage over the other players).

Rule 2

Before players can start to move their markers along the squares they must each throw a 6. A player who has thrown a 6 then throws again, moving the marker the number of squares indicated by the second throw. Each player throws the die in turn, and moves the marker according to the number shown on the die. If such a movement results in the marker moving to the bottom of a ladder, the marker climbs to the top of the ladder. If at the end of a move the marker lands on a square where there is a snake's head, then the marker must slide down to the tail. If a player throws a 6, his or her marker is moved 6 places, taking either a ladder or snake if the count takes it to the bottom of the ladder or the head of a snake. The player then takes a second throw of the die and moves the appropriate number of steps. Each time a six is thrown, the same player has an extra throw of the die.

Rule 3

The first player to reach 120 wins the game and the others continue playing for places. An exact throw of the die is needed to win. If the number thrown is more than required, the marker is moved backwards from square 120 for each unwanted point.

We assume that during the progress of the game, markers can occupy the same site on the board without interference. Some players play the game such that if more than 1 marker is on the same square the first marker can block the second marker. Other people let the second marker knock the first marker down to 0 if it lands on an occupied position. These variants can be built into computer models, but it makes the game more complicated and will not be considered here.

Assuming that the markers do not interfere with each other, one can first of all generate the distribution function of the variation in the number of moves required to reach home by playing the game many times as if there were only a single player. The special random number table shown in *Figure 4.9* can be used to simulate the act of throwing a die with six faces. In *Table 4.3* the distribution of the number of moves required to complete 1000 single games as played on a computer are summarized.

Can we anticipate which, if any, mathematical function can be used to describe the range of the number of moves required to complete a single player game summarized in Table 4.3? The prolonging of the game (from the point of view of a player) requires an unfortunate combination of probabilities to delay arrival at home. In this case, the number of moves required to win the game can be prolonged by an undesired cooperation of the random causes. Therefore, the distribution of values in Table 4.3 will probably be describable by the same function that describes the steps between a specified random number in our journey through the random number table described in the previous section. In *Figure 4.10* the data of Table 4.3 are

41322	46655	44653	34365	12133	46344	55653	26626	15663	63266
33621	55324	12136	14441	51446	13626	52155	13133	25624	61233
41126	44622	23215	13125	51645	24334	64245	22333	43425	22123
33634	25415	63616	63354	11224	26554	31262	54223	22346	26165
15235	12665	64131	41525	24334	32421	44365	64514	65561	64646
33366	52132	61346	13252	46355	22222	12631	52414	43531	51164
62322	32522	42126	14265	42556	62442	31541	55231	55226	26556
23252	56513	13526	13664	14414	63566	41664	26341	55262	55625
24526	51546	56253	25451	63562	21633	31335	25543	46135	62663
36261	36245	45335	51146	42326	34322	52645	26113	31236	51321
43516	35221	15536	26151	25553	14514	35322	64354	33346	55341
31321	31533	31526	66253	35242	55312	24516	45246	13214	16164
13243	45426	24351	33251	33452	15631	11224	56355	23164	23154
51146	24151	24655	16263	25236	46223	12253	42536	12111	51431
36442	31351	64341	46422	65454	36312	14363	32634	31662	56636
26325	33434	34141	15345	31144	36113	65421	14143	43561	64624
54554	44614	36544	22624	32433	33563	55321	35412	12155	51554
41335	52561	53562	53265	55444	42145	62632	35455	16246	26342
15621	45446	66662	16654	53264	26143	16232	13565	65311	16621
56162	36335	31335	11426	22413	54621	54242	35561	26241	12641
41635	35531	14512	56545	42362	23321	65252	14124	61541	22466
44526	12544	42432	45442	41665	24123	34356	12456	36221	36353
23413	43544	12152	12453	61135	65141	16512	45115	64455	12336
63412	21366	11151	25625	23112	36622	51633	54426	42641	13522
41462	66261	63223	34332	15322	42532	63631	23512	23153	13264
44466	36141	65661	23323	12644	54663	15254	61513	66522	14426
41123	34325	12564	24212	45132	21542	52512	26463	53636	16212
22364	33463	42366	55111	53523	52362	63323	64363	34541	35635
14114	14522	12331	65652	66251	45412	13225	46112	24145	23513
26413	36314	55156	51135	62561	65512	43222	44363	35362	24452
44151	33251	13224	64166	54361	65124	54114	51166	54556	45515
23313	21536	41412	23254	56244	26135	45442	31566	36235	64265
31615	36613	52254	32545	21145	43663	52443	41126	41661	64516
54562	66612	26122	14436	52624	25535	24313	56353	61156	32511
51243	25251	25623	56441	25321	64532	44161	54261	42426	51252
16614	62444	53654	53515	55423	64446	64534	33313	62533	61314
14344	36342	42621	53345	32543	63243	41646	51165	52364	41216
52646	26415	25461	53415	31654	65622	32555	61355	23651	26436
63344	51351	26123	53453	14331	32156	52422	16563	34461	13153
56231	12253	55345	41655	55565	53312	66545	62521	12523	16353
13414	55314	55424	56415	14646	32451	63241	54541	51226	24241
42361	43333	15234	12142	34513	45411	31165	66225	23416	35144
22221	55641	16136	62563	54516	34524	65433	43125	32253	42124
66661	33236	23232	45111	34163	65635	31333	14261	23135	26115
21613	53152	25123	56663	22122	16661	36133	66426	53643	11611
36124	56523	11436	41512	34323	55342	26652	23133	41255	43426
16212	43646	32653	43224	16434	41532	16164	41561	53244	35663
41346	11414	11222	34312	63145	64463	45663	64353	62322	35322
52554	25613	46143	16111	34416	16452	64335	13434	52422	14223
44553	23455	56533	32232	13555	61356	12315	53343	34265	24552

Figure 4.9. A random number table with only six digits randomized in position is useful for simulating games usually played with a die having six faces.

plotted. It can be seen that there is a very good fit of the distribution of the data points to the log-normal distribution function.

It is interesting to note that the student who generated the data for Figure 4.10 during a student project at Laurentian University (Mahlon Bryanton) commented in his laboratory report

> *Looking at the graph it is amazing to see how well the data fits a straight line. Most students would be extremely surprised at such a distinct pattern emerging out of the chaos of the game.*

The reader should note that for a set of data which is describable by the log-normal distribution function, the average and the 50% value of the distribution are no longer the same quantity. In such a distribution the middle number for the set of data

Table 4.3. Record of the number of moves required to finish a game of snakes and ladders when 1,000 games were simulated on a computer.

Length of Game N	Occurrence f	Cumulative Σf	Percent [%]	Length of Game N	Occurrence f	Cumulative Σf	Percent [%]
15	8	8	0.8	160	14	896	89.6
20	43	55	5.5	165	3	899	89.8
25	34	89	8.9	170	13	912	91.2
30	64	153	15.3	175	9	921	92.1
35	47	200	20.0	180	5	926	92.6
40	71	271	27.1	185	3	929	92.9
45	65	336	33.6	190	9	938	93.8
50	55	391	39.1	195	10	948	94.8
55	49	440	44.0	200	2	950	95.0
60	44	484	48.4	205	3	953	95.3
65	42	526	52.6	210	3	956	95.6
70	33	559	55.9	215	4	960	96.0
75	29	588	58.8	220	1	961	96.1
80	36	624	62.4	225	3	964	96.4
85	29	653	65.3	230	4	968	96.8
90	32	685	68.5	235	4	972	97.2
95	29	714	71.4	240	2	974	97.4
100	22	736	73.6	245	3	977	97.7
105	21	757	75.7	250	3	980	98.0
110	21	778	77.8	255	2	982	98.2
115	12	790	79.0	260	0	982	98.2
120	19	809	80.9	265	0	982	98.2
125	7	816	81.6	270	2	984	98.4
130	14	830	83.0	275	2	986	98.6
135	13	843	84.3	280	0	986	98.6
140	8	851	85.1	285	3	989	98.9
145	12	863	86.3	290	1	990	99.0
150	9	872	87.2	295	0	990	99.0
155	10	882	88.2	300	0	990	99.0

is described as the *median value* of the data. The most frequent value (the peak of the skewed bell curve) is known as the *mode* of the set of data. *Figure 4.11* shows the skewed bell curve for the single player set of moves for playing snakes and ladders with the board of Figure 4.8 and the rules set out in this section. The mode, median and average values of the set of data are indicated in the diagram.

To simulate the results of games between two players, we select two games from the single player set of data at random. This can be done by giving a number between 0 and 999 to the "move sets" generated earlier. Then, one selects at random two numbers from 0 to 999, and the appropriate move sets become the two games of the opposing players. From the random number table of Figure 4.6(a) we choose 954

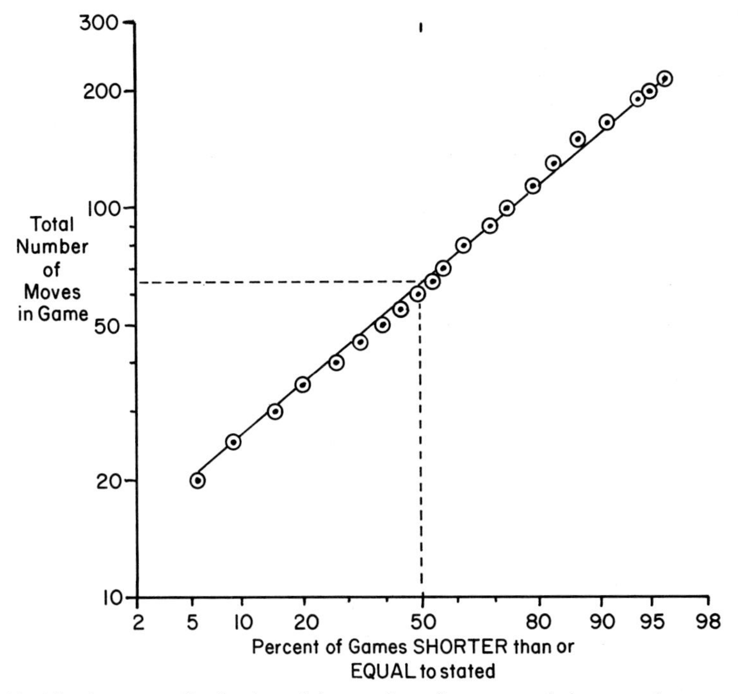

Figure 4.10. The frequency distribution of the number of moves needed to complete a game of snakes and ladders when a single player plays the game as a "pastime" can be described by the log-normal probability distribution function. Data based on 1,000 games. Median number of moves 58.

and 290, which selects 63 and 312 as the move sets for our two players. From this data we decide that, in this simulated match, the winning number of moves was 63 (in a real situation the losing player would not normally continue play, but we need to know the sequences of play for both players so that we can decide the winner). The data for the winning number of moves for 500 games simulated in this way are shown in *Figure 4.12*. Again, the number of moves needed to win are log-normally distributed. Note, however, that the median number of moves required to win when two players compete with each other is 43 moves. This compares to a median value of 65 moves needed to win when a single player plays the game.

To simulate the number of moves required to win when four players compete in a game, we extend the strategy used to simulate competition between two players. We now choose 4 random numbers between 0 and 999, and enter the data summarized in Table 4.3 to list the number of moves for the winner and each player completing the game in turn. The shortest number of moves amongst the 4 values chosen is the number of moves for the winner. In *Figure 4.13* the distribution of the winning number of moves for 4 players is shown (in such a situation the number of games needed to become the second and third player home may be important. The

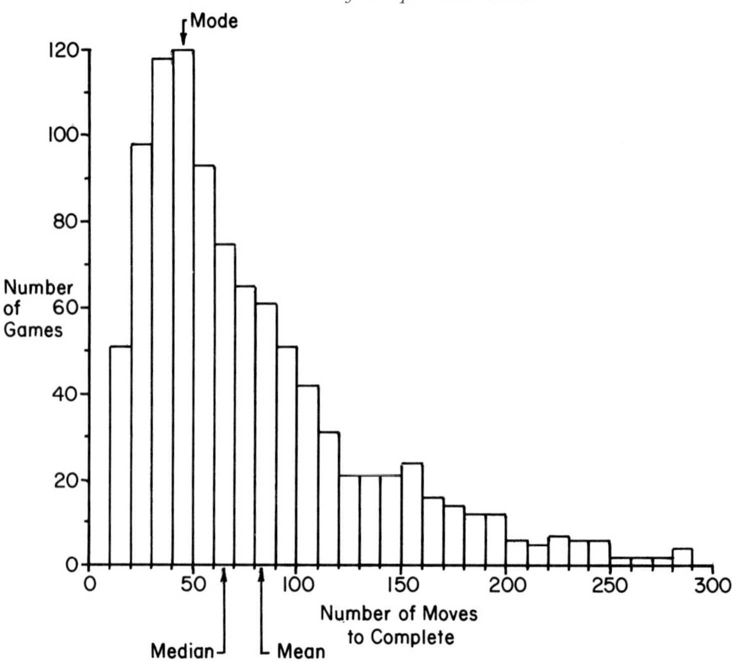

Figure 4.11. When quoting average statistics for a set of data describable by the log-normal distribution function, it is necessary to distinguish carefully between the mode, the mean and the median value of the distribution. The differences between the various parameters of the set of data for the game of snakes and ladders are indicated clearly in the above histogram.

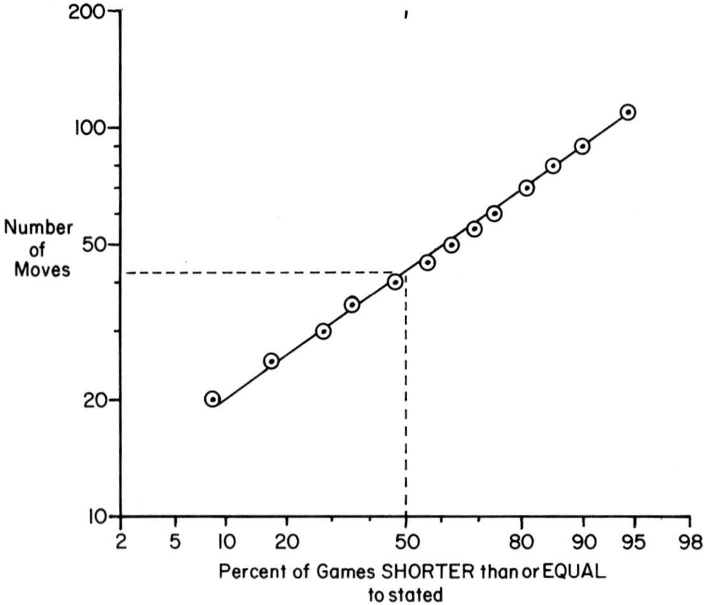

Figure 4.12. When two players compete in the game of snakes and ladders the distribution function of the winning number of moves is again log-normally distributed. Data based on 500 games. Median 43 moves.

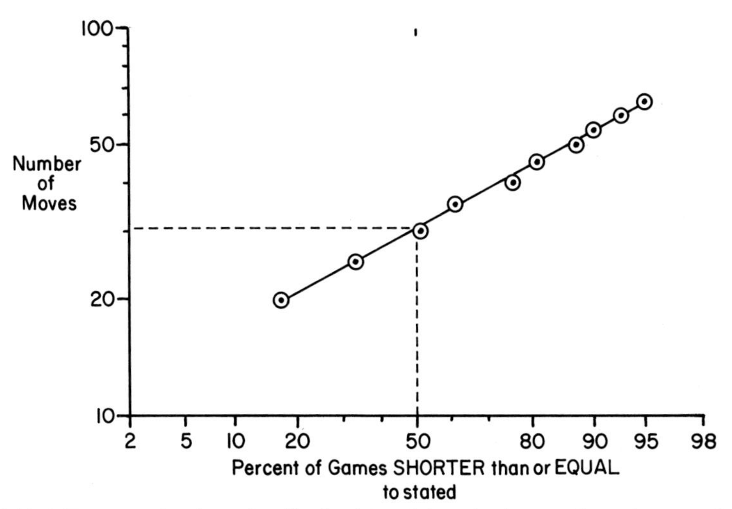

Figure 4.13. When one simulates the distribution of the winning number of moves for four players competing with each other, the log-normal distribution of the data has a median value of 31. Data based on 250 games.

readers can determine for themselves how the number of moves for the second home and the third home compare with the number of moves required to win).

We can see from Figure 4.13 that the winning number of moves when 4 players compete is again describable by the log-normal distribution function, and that the median value of the number of moves required to win has dropped to 31. It is left as an exercise for the reader to discover how the median number of moves needed to win decreases as the number of players decreases. Although the game usually only takes place with four players, the reader can look at the distribution of winning "move sets" when even more players compete.

Exercises

Exercise 4.1. Studying Envelope Decay

Many organizations try to economize on the cost of office stationary by using internal circulation envelopes which can be used many times to send messages back and forth within the office. The envelopes usually have an outer cover with a set of lines organized so that, in theory, the address of various people to whom the envelope travels are written in sequence down the envelope. Often these envelopes carry messages such as "Do not tape or staple". In the *New Scientist* of June 9th,

1990, Dr. William Sutherland of The School of Biological Sciences at the University of East Anglia studied the life expectancy of such envelopes. Each time he received an internal circulation envelope, he noted how many times the envelope had been used. He was able to show that the number of envelopes which had been used for a given number of times suffered a logarithmic decay in utilization rates. Students should be encouraged to discover if the companies where their parents work or if the school board of which their school is a part uses such internal office mail. They should collect information on the frequency with which envelopes have been used for a given number of times. They should assemble a table of the form shown below:

Table of Results

Table E 4.1.1

N_U = Number of times the envelope was used.
N_F = Number of envelopes used N_U times.

Students should investigate whether the envelopes do indeed suffer an exponential life-expectancy decay as they are used within the office system being studied. Students should be encouraged to look up the original story in *New Scientist* and compare their results with this. The students could then discuss why it is that this relatively simple way of economizing on envelopes is often not efficiently pursued in an office situation. In particular, they could discuss why secretaries often ignore the instructions and continue to staple the envelopes to death so that they often have to be withdrawn from circulation because of their delapidated condition long before they have been used for the number of times available if all of the address windows down the front of the envelope were used. The students could also discuss why more use is not made of recycled envelopes in the mail system. For example, they could discuss the use of robotic mail sorters, which in Canada results in the decoration of the front of the envelope with florescent lines so that any label that is to be used for recycling would have to be fairly large to cover up not only the stamp and the last address, but also the robotic strips glowing from the front of the envelope. This exercise would have the dual advantage of stimulating ecological discussions and generating an understanding of logarithmic decay of objects subjected to random erosion. The students might also discuss how one would approach the secretarial staff of a large organization to explain to them the evidence they have collected showing that the envelope reuse system was not working efficiently. They could draft a letter explaining in nontechnical language the meaning of the graph that they generated from the study of the circulation efficiency of envelopes. The teacher could use this exercise to generally explore the difficulties of explaining scientific conclusions to nontechnical groups.

The students can discuss how many envelopes could be saved if they were all used for all of the available address space on the envelope, as compared to the envelopes in circulation when only a fraction of them reach the ultimate death of being completely used. If one could find an organization using such envelopes, one could estimate the total number of envelopes in use by taking a survey of envelopes in actual daily use. From this the number of trees that could be saved by complete usage of multi-trip envelopes could be estimated. This could be a suitable science-fair project for students with access to an organization with re-circulating envelopes amenable to participation in this study.

Exercise 4.2. Robots Unlimited

"Robot assembly" is a game that can be played with two or more players using a pencil, paper and a die (the word *robot*, meaning a mechanical device programmed to carry out work procedures, comes from the Czech word robata, meaning "work"). The robot constructed in the game is shown in the diagram below. It has the following different parts: 1 body, 1 head, 2 eyes, 1 mouth, 2 arms, 2 legs, 2 antennae and 1 control box. To play the game with two players, each player throws a die in turn. The numbers which a player must throw to be permitted to draw a part of the robot are as follows:

The sequence of construction as the game proceeds is that a body must be available before one is allowed to use a 1 thrown by the die to start drawing the control panel and its two knobs. Again, the player must generate a head with a throw of 3 on the die before he is allowed to use a 5 to construct antennae and a 2 to draw the mouth and eyes. The first player to construct a complete robot shouts "ROBOT!!" and that is the end of the game. The score for one complete robot is 48. The score for an incomplete robot is arrived at by adding the value of the parts of the incomplete robot as illustrated for the three incomplete robots of *Figure E 4.2.1.*

The robots drawn during the game will not look exactly like our prototype robot. They will be more like the sketches given in the diagram. Part of the fun of the game when several players are playing a sequence of robot constructions is to award a prize for the funniest robot or the fiercest. To generate data which can be used to illustrate some of the data processing techniques developed in Chapters 3 and 4 of this book, there are many variations possible with this game. If two players are competing against each other, they can play several games with a die and record the distribution of scores generated by the losing player. They should be Gaussianly distributed. The winner and the loser could keep a record of the throws of the die, and they could show that on average each number appeared equally probably in the Markovian chain of events. The number of throws necessary to win should be log-normally distributed in a sequence of games.

Before getting down to such detailed analysis of individual moves, the game can be a useful procedure to introduce members of a class to each other at the beginning

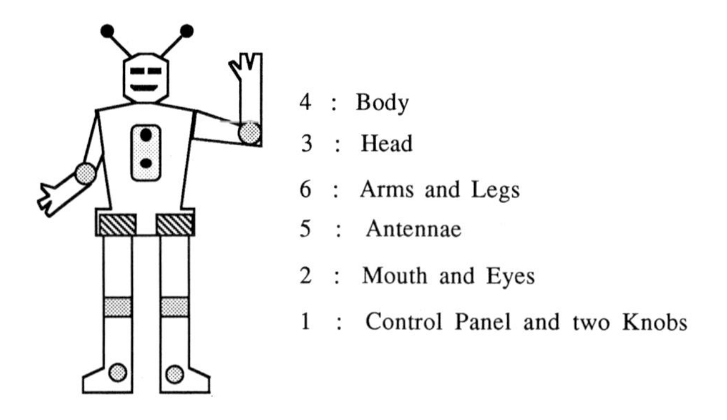

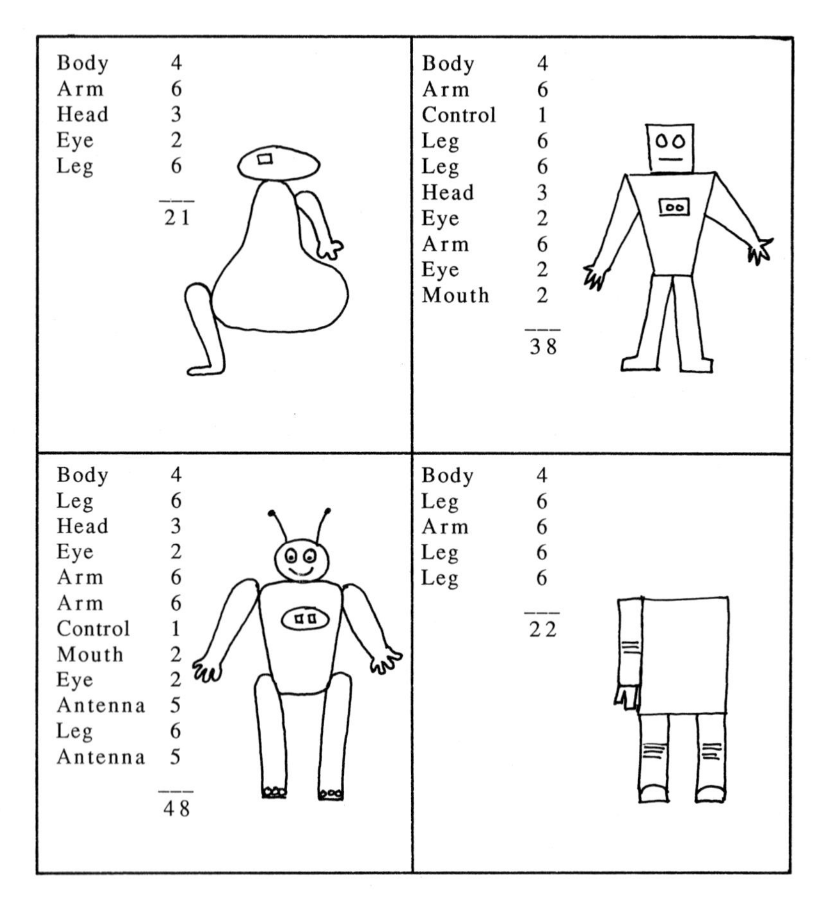

Figure E 4.2.1. A typical game card for robot assembly should contain some strange-looking unfinished robots. The game score for a partially assembled robot when someone else in the game completes a robot is illustrated above.

of a programme of lectures (a kind of stochastic mixer!). The players are provided with a card on which they are able to draw a series of robots as illustrated in Figure E 4.2.1. To describe the dynamics of the robot production game, assume that we have sixteen players grouped into sets of four, and that each are provided with a die. The players operate in pairs so that at location one for the first robot construction players one and three will cooperate, as will players two and four around the locations. When playing the robot game, a die is thrown and once the number appearing on the die has been noted it is passed to the next player, and so on around the group. The throw of the die for cooperative players contributes to their robot. As soon as the cry of ROBOT! arises from one location, the score for the particular robot of a player is tallied and put into a score box of the playing card (note that the word score for the point gained by a player and the word tally are both related to an old way of keeping records before writing was invented, and afterwards by illiterate workers. Workers used to count their cattle by cutting notches on a stick, and to score originally meant to make a scratch or mark. The Latin word for a cut is talea, and a tally was originally the number of notches cut onto the stick they were using as a record-keeping device. Gun fighters in the wild west used to cut notches on their gun to keep a record of how many people they had shot in gunfights).

To mix up the players for the next robot construction effort, the winners from each location move around the group of players clockwise and rearrange themselves at the new location so that the one winning player teams up with a losing player who has stayed at the location where they lost the competition. This is shown in the configuration map of *Figure E 4.2.2*. The game can be continued until everyone has met everyone else (notice that meeting everyone else is different from the probability of having played with everyone else). The students can choose a given number of players and select winners at each table by the flip of a coin and mix them up and keep a record of individual player movements, and decide how many times a game must be played for everybody to:

a) sit at the same table as everybody else and
b) play with everybody else.

A typical set of "progressive opportunities" is shown, and the students can suggest how to vary the rules if they have an odd number of players or if they have two more players than needed to make sets of four (hint: all the players are given a number and substitution of the extra players proceeds either according to a systematic substitution schedule or a random number schedule). At the end of a given number of games the students can use the data they have tallied to look at the distribution of scores, and can construct histograms and try to plot the data on log-normal and Gaussian graph paper. They can discuss the physical significance of the pattern of data. This type of game can also be used by the teacher to introduce the concept of simulating the game after one has played it physically by introducing students to random number tables and Markovian chains. The students can then

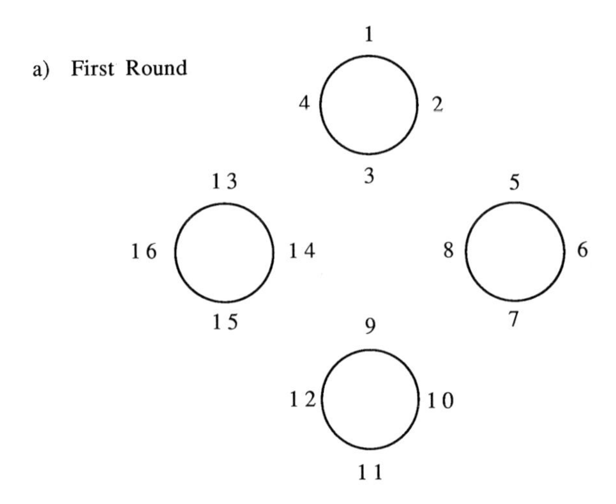

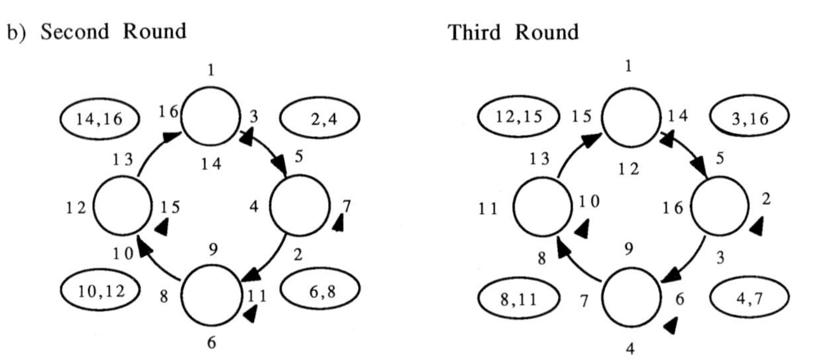

Figure E 4.2.2. Robot assembly can be used as a stochastic mixer in class studies. The numbers circled indicate the winning pairs from the previous round.

simulate the simple competitive game between two players on the computer and proceed to more and more complex simulations according to their interest in the project.

References

[1] See the article on De Morgan in *World Of Mathematics*, edited by J. R. Newman, Simon and Schuster, New York, 1956.
[2] B. H. Kaye, *A Random Walk Through Fractal Dimensions*, VCH, Weinheim, Germany, 1989.
[3] M. J. Moroney, *Facts from Figures*, 2nd revised edition, Pelican, Harmondsworth, England, 1953.
[4] See, for example, a review of the history of the discovery of atomic radioactivity in: D. Halliday and R. Resnick, *Fundamentals of Physics*, Second Edition, Wiley, New York, 1981.
[5] R. C. Bell, *The Boardgame Book*, Exeter Books, Simon and Schuster, New York, 1983.

Chapter 5

Prussian Horses
and
Fishy Statistics

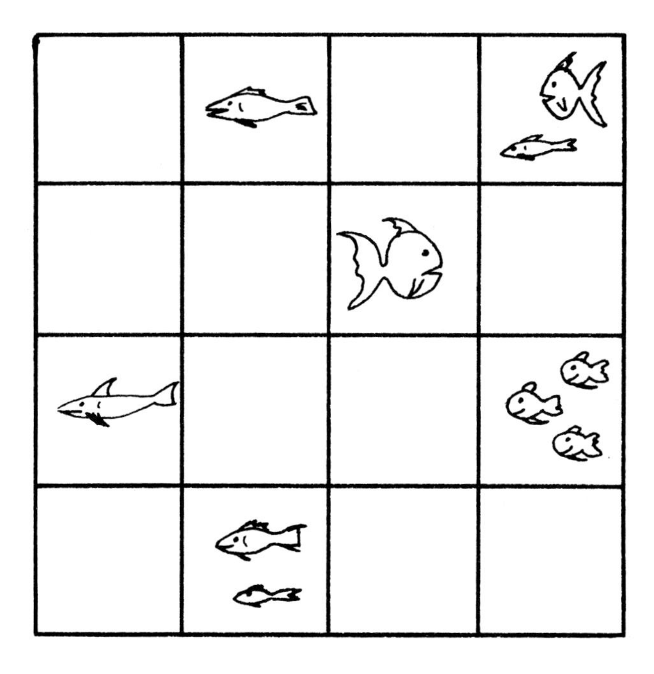

For sparse populations
Poisson fluctuations
Mark the count variations
In the zones of inspectation

(Poisson fluctuations describe the variations in population from one inspection zone to the next when populations are low)

Chapter 5 Prussian Horses and Fishy Statistics 207

Section 5.1 Death Rates from Horse Kicks in the Prussian Cavalry 209
Section 5.2 Using Poisson Graph Paper in the Assessment of Dust Deposition Density in Air Pollution Studies 213
Section 5.3 Stingy with the Yellow Buttons? 215
Section 5.4 Using Poisson Trackers to Monitor Chaos in a Powder Mixer 220
Section 5.5 Segregated and Tumbled Jelly Beans 223
Section 5.6 A Cautionary Tale of Tails 226
Exercises 227
References 229

Chapter 5

Prussian Horses and Fishy Statistics

Section 5.1. Death Rates from Horse Kicks in the Prussian Cavalry

In this section we will look at some statistics collected on the deaths from horse kicks in the Prussian cavalry. Via these statistics the reader will be introduced to one of the surprising patterns of chaos known to statisticians as the *Poisson distribution.* It is said that if there is anything worse than a bad joke it is one that has to be explained. Nevertheless, to those unfamiliar with French, it has to be pointed out that the word Poisson, which was the name of a famous French mathematician Simeon Denis *Poisson* (1781 – 1840), means fish. Hence the terrible pun in the title of this chapter! Poisson originally studied medicine but we are told that "when his first patient died under his ministrations he naturally took a dislike to the profession" [1].

When Simeon abandoned medicine, it is said that the Poisson family tried to make him become a lawyer on the theory that "he was fit for nothing better". Simeon Poisson finally studied mathematics. He entered École Polytechnique (Paris) in 1798, and became a professor in 1808. We are told that from a theoretical consideration of the wave theory of light he deduced that there should be a bright spot at the centre of the shadow of a small opaque object. This seemed to him to be so absurd that he rejected the wave theory of light. In fact, the bright spot can be found at the centre of a shadow, and Poisson could have been famous for establishing the wave nature of light if he had believed his own logic. His failure to follow his own logic serves to emphasize the fact that even great men are conditioned by the thought systems of their time, and that sometimes their conclusions can be wrong.

Statisticians have discovered that the Poisson distribution is useful for describing variations in the occurrence of events which are relatively rare and which on average occur with a frequency z in any set data. The Poisson distribution is based upon the irrational number e that we met in our study of the Gaussian distribution of a set of scattered drunks. Poisson discovered that if the expected average number of events in a particular situation was given by the number z, then populations of the magnitude 0, 1, 2, 3, etc. would arise with the probability given by the equation

$$1 = e^{-z} e^{2} = e^{-z}\left(1 + z + \frac{z^2}{2!} + \frac{z^3}{3!} + \cdots\right).$$

The expected values of the various populations calculated using the formula given above are actually the successive terms in the mathematical series given by

$$e^{-z}, \quad z e^{-z}, \quad \frac{z^2}{2!}, \quad \frac{z^3}{3!}, \ldots .$$

In his book *Facts from Figures*, M. J. Moroney states that the fact that the probabilities of the fluctuations of rare events can be described by the Poisson equation is "a very remarkable fact" [1]. I first came across the Poisson distribution when using a Geiger counter to study fluctuating observations in a low rate of radioactive disintegration. I found it a very surprising pattern in what I would have expected to be unstructured chaos.

Nearly every textbook on statistical techniques and data handling introduces the Poisson distribution to the student using a study of death from horse kicks in the Prussian army. The data was collected when armies still used large numbers of cavalry to fight wars [1]. Over a 10-year period, the Prussian army collected statistics on the number of soldiers that were kicked to death in 20 different army corps. The data constituted 200 sets of "corps-year" data. In *Table 5.1* this data is summarized. From the data it can be seen that the average number of deaths per "corps year" was 0.61. This value can be substituted into the Poisson equation to generate the probable occurrence of observed populations of 0, 1, 2, 3 and 4 deaths in a given year

Table 5.1. The recorded death rates of Prussian cavalry soldiers from horse kicks is often used in textbooks of statistics to illustrate the variations of observed populations of a variable which is fluctuating in a manner predictable from the Poisson distribution function [1].

Deaths per Year N	Frequency f
0	109
1	65
2	22
3	3
4	1

The average number of deaths per year is: $z = 0.61$. Therefore: $e^{-0.61} = 0.543$. The probability of each number of "Deaths per Year" is given by the successive terms of: $e^{-z}(1 + z + z^2/2! + z^3/3!! + \cdots)$, so we have: $0.543\,(1 + 0.61 + 0.61^2/2! + 0.61^3/3! + \cdots)$.

Deaths per Year	0	1	2	3	4
Calculated Probability	0.543	0.331	0.101	0.021	0.003
Expected Frequency	109	66.3	20.2	4.1	0.6
Observed Frequency	109	65	22	3	1
Cumulative Frequency	200	91	26	4	1
Observed Probability	1.000	0.455	0.130	0.020	0.005

in any given corps. The basic calculation of these probabilities and their values are summarized in the second part of Table 5.1. From these calculated probabilities, one can calculate the frequency with which deaths should occur within the 200 pieces of information. These calculated data are summarized in Table 5.1. When one compares the actual observed deaths with the predicted deaths from the Poisson equation, the agreement is amazing (it is rumored that at one lecture on statistics a student asked how the horses knew how often they had to "kick to kill" to be in agreement with Poisson fluctuations. It is said that the speaker's answer was that the horses were noted for an abundance of horse sense!).

The Poisson formula and the calculations of the probable fluctuations in events from a known expected population as set out above and in Table 5.1 will probably intimidate the average reader. If it was necessary to carry out the calculations of the probability of the various fluctuations of observed events which would be consistent with the Poisson distributions, the reader would probably decide to move on to the next chapter. However, the use of Poisson statistics has been greatly simplified by the availability of *Poisson probability graph paper* [1]. The structure of this graph paper, shown in *Figure 5.1*, is fairly complicated since it involves the crossing of three different sets of lines. The first set of lines, which come from the ordinate and are parallel to the abscissa, generates a Gaussian probability scale. The scale on the abscissa is a straightforward logarithmic scale. Crossing the 2 sets of parallel lines are

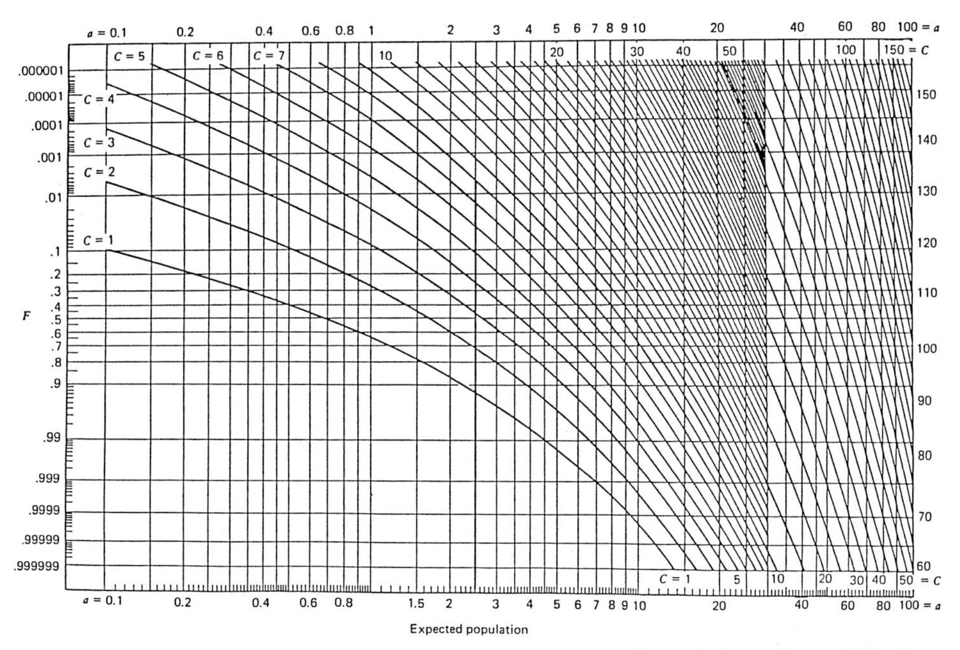

Figure 5.1. Poisson graph paper is useful for calculating expected populations of a variable from recorded fluctuations of the variable in a set of samples.

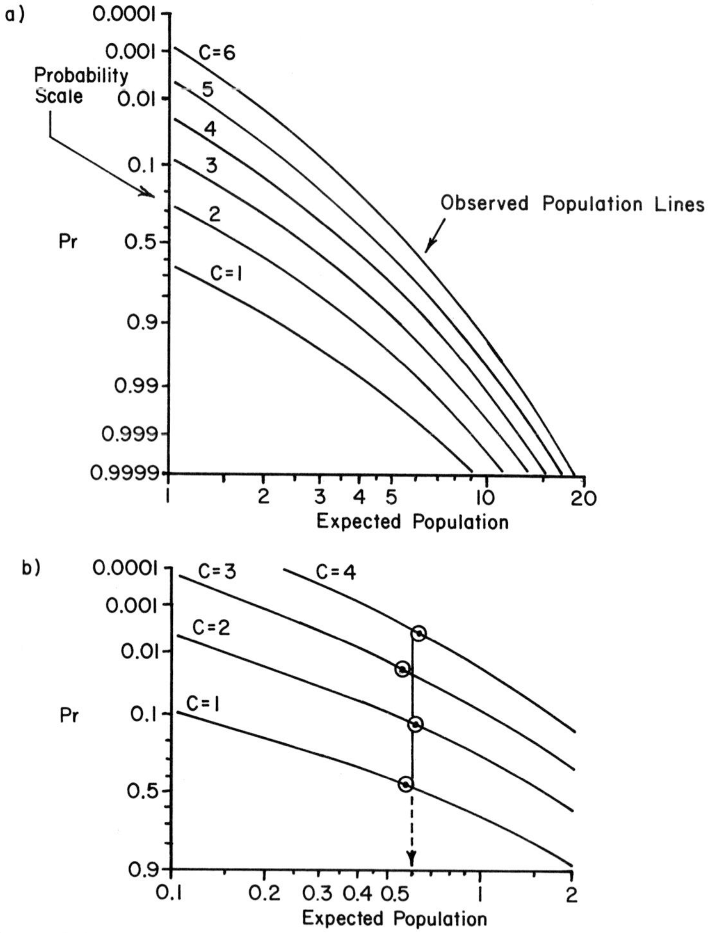

Figure 5.2. A set of observed population lines curving across Poisson probability paper enables the expected population of a variable to be estimated graphically. a) Simplified graph paper emphasizing the structure of the observed population curves. b) Data for "horse-kick death" generates a dataline giving an estimated frequency of occurrence of 0.6 deaths per corp per year.

what are known as "*observed population lines*". To help the reader see the various components of this graph paper more clearly, the observed population curves are shown without the parallel set of lines along with the abscissa and ordinate scales in *Figure 5.2*(a).

To explain how one uses Poisson graph paper, we will use the data from the Prussian horse kick deaths to plot events as datapoints on the graph. First of all, we must transform the kick data into a cumulative distribution in which we calculate the proportion of "corps-years" in which deaths are stated as being at least x. When transformed in this way, we note that at least 1 death occurs with the probability

0.455, at least 2 deaths occur with the probability 0.13, at least 3 deaths occur with the probability 0.02, and at least 4 deaths occur with the probability 0.005. The curved lines crossing the vertical and horizontal lines of the Poisson probability paper represent event lines for observed populations of events equal to the event density linked to the line. The bottom curved line crossing the graph paper is labelled $C = 1$. This corresponds to events of density of at least 1. For the Prussian cavalry events, the event line 1 corresponds to "at least one death from horse kicks in a corps-year". To plot the fact that this observed event has a probability of 0.455, we move down the event line until it crosses the appropriate probability line. In Figure 5.2(b) the bottom data point on the $C = 1$ event line corresponds to the probability of point 0.455. The fact that at least 2 deaths occurred with a probability of 0.13 is plotted on the graph paper by moving down the event density curve $C = 2$ until one meets the probability value of 0.13, where one then locates the datapoint. Similarly, the events for $C = 3$ and $C = 4$ are located on the graph paper.

The structure of the Poisson graph paper is such that if the data being studied can be described by the Poisson probability function, then a straight line can be drawn through the datapoints parallel to the probability axis to cut the logarithmic scale on the abscissa at the expected population governing the Poisson fluctuations observed in the data.

In the case of the Prussian horse-kick deaths, it can be seen that the appropriate straight line through the datapoints cuts the abscissa at an expected average death rate of 0.6. This agrees well with the calculated value of 0.61. In the next few sections we will explore situations where the randomness of several causes generates a variable which can be described by the Poisson probability distribution.

Section 5.2. Using Poisson Graph Paper in the Assessment of Dust Deposition Density in Air Pollution Studies

An important problem that arises in a study of air pollution is the assessment of the concentration of dust fineparticles settling out of the atmosphere. A full discussion of the assessment of any number of dust fineparticles of different sizes settling out on the microscope slide is beyond the scope of this book [2, 3]. However, we can consider the simple problem which arises if all of the dust fineparticles falling out of the air are the same size. This is not an irrelevant over-simplified problem, since it can be shown that the best way to assess the density of dust on a microscope slide is to search for fineparticles of one size at a time, assessing the density of occupancy of that size on the slide, and then repeating the inspection of the slide for

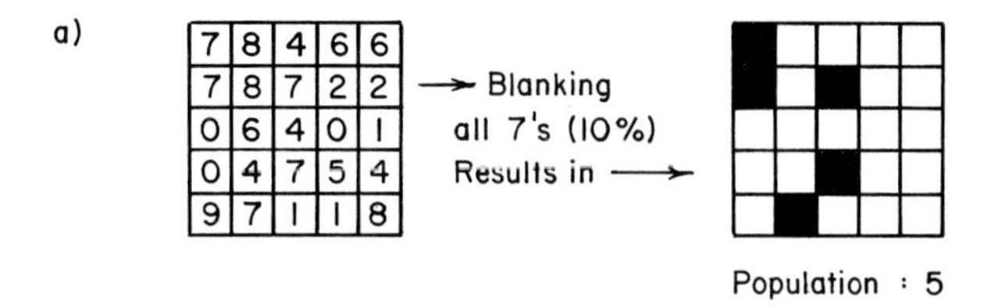

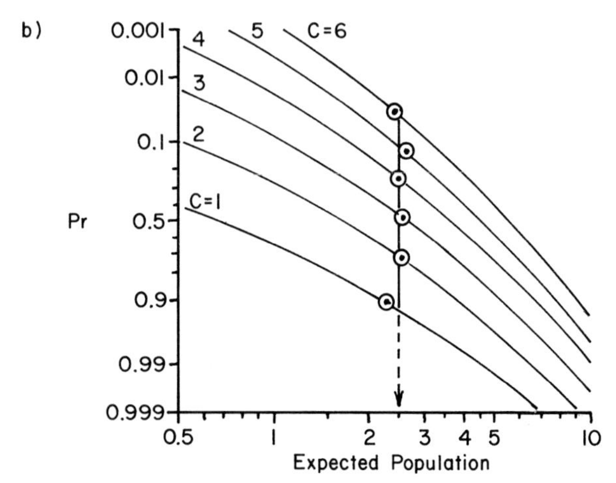

Figure 5.3. The variations in the observed population of dust fineparticles in a series of fields of view fluctuating according to predictions calculated using the Poisson distribution. a) Simplified conversion of a random number table to a 10% field of view. b) Distribution of the populations of 10% simulated fields of view plotted on Poisson graph paper.

a different size. Those interested in this problem should see the discussion given [3] and [4].

The random deposition of monosized dust onto a slide can be simulated by taking blocks of numbers from a random number table and letting any specified digit become a fineparticle. We will assume that the analyst inspecting the slide will be able to resolve and mentally separate contiguous pixels representing dust fineparticles which touch each other. If we were to inspect blocks of 25 random numbers, and if we let any given digit become a dust fineparticle, this would correspond to the inspection of a microscope slide which was 10% covered by deposited fineparticles. A typical simulated field of view with an observed population of 5 in a field of anticipated population of 2.5 is shown in *Figure 5.3*(a). The data for one hundred simulated fields of view are shown in *Table 5.2*, where it is also converted to cumulative populations greater than or equal to a stated value. It can be seen from Figure 5.3(b) that when this data is plotted on the Poisson graph paper, a data line

Table 5.2. Simulated dust deposition data for one hundred "field of view" inspection areas.

Population N	Frequency f	Cumulative Σf	Probability Pr
0	10	100	1.00
1	19	90	0.90
2	24	71	0.71
3	22	47	0.47
4	12	25	0.25
5	10	13	0.13
6	3	3	0.03

is generated which cuts the abscissa at the expected value of 2.5. Student readers can simulate their own dust slides and study the variations in possible populations using the Poisson graph paper (See Exercise 5.1).

Section 5.3. Stingy With the Yellow Buttons?

In Exercise 3.9, we explored the variation in individual weights of a set of candy-covered buttons. In this section we will try to find a pattern in the variations from box to box of the various numbers of each color of button.

Twenty 60-gram boxes of Smarties® buttons were purchased from the store (at the time of writing, Smarties® were not available in the United States; they are similar to M & M® candies; see Exercise 5.5). The number of each color in the boxes is shown in *Table 5.3.* When looking at the data from the first two boxes, one of the participants in the experiment exclaimed; "They are stingy with the yellow ones!" This statement illustrates how people are often willing to reach sweeping conclusions based on small amounts of data. Unfortunately, this quick conclusion was soon shot down in flames by the data from the third box of candy, which turned out to have 12 yellow buttons. All the data for the 20 boxes of candy are summarized in Table 5.3. Although all of the boxes of candy were purchased from the same store at the same time, they differed in that 13 of the boxes contained blue buttons, which appear to have been substituted for light brown buttons in the other batch of candy boxes (we know that the blue buttons were the new ones, since the manufacturer was mounting a major commercial campaign to promote the novelty of blue buttons at the time the boxes were purchased). When I viewed the data of Table 5.3, I suggested to the class of students carrying out the experiment that the manufacturer was actually trying to place the same number of each colored button in each box. Since there were 8 colors and about 64 buttons per box it seemed reasonable to suggest

Table 5.3. Colored button population of 20 Smarties® boxes.

Light Brown	Red	Purple	Blue	Orange	Green	Yellow	Pink	Dark Brown
0	9	7	7	8	10	5	10	10
0	8	10	10	7	6	4	12	9
0	9	10	8	8	10	7	5	7
6	5	5	0	12	6	12	9	7
8	7	9	0	11	9	11	1	6
0	5	9	9	6	16	4	11	5
0	7	9	9	8	12	7	11	6
0	9	5	9	3	13	12	10	4
12	9	5	0	7	3	9	7	13
0	7	8	11	8	6	7	6	12
0	9	9	8	8	9	5	5	10
0	3	8	15	14	8	4	7	5
0	9	7	16	2	7	4	12	7
0	5	8	5	5	12	8	8	13
0	3	4	11	7	12	9	14	4
4	11	5	0	13	5	10	4	7
9	8	8	0	6	10	9	9	6
11	6	6	0	9	5	7	6	7
10	2	6	0	13	8	5	7	12
0	6	8	8	9	7	7	10	7

that the observed population of each color was compatible with attempts to provide 8 of each color per box, with the observed populations fluctuating according to the variations predictable from the Poisson distribution (the 12 students present in the class carrying out this experiment split evenly into two groups, one supporting and the other denying – with scorn – this suggestion). To test the theory of equal color content, the data for the variations in the observed possible intended populations of yellow, green and red buttons are plotted in *Figure 5.4*. When one looks at the datalines of Figure 5.4, one can see that they are compatible with the expected population of approximately 8 of each color per packet.

To test what variations are likely to occur in the population of any one particular color over many boxes of Smarties®, one can simulate the legal variations from box to box using a random number table. A random number table which has only the digits from 1 to 8 in its makeup is particularly useful for simulating variations in the populations of 8 colors in an expected population of 64. If one takes a block of 8 × 8 in this random number table, then any one digit can correspond to a given color and the block of 64 digits becomes a simulated box of candies. In *Figure 5.5*(a) the simulated populations of yellow buttons in 100 boxes of candy are summarized. It can be seen that the actual population per box can vary widely from 2 to 16 as shown.

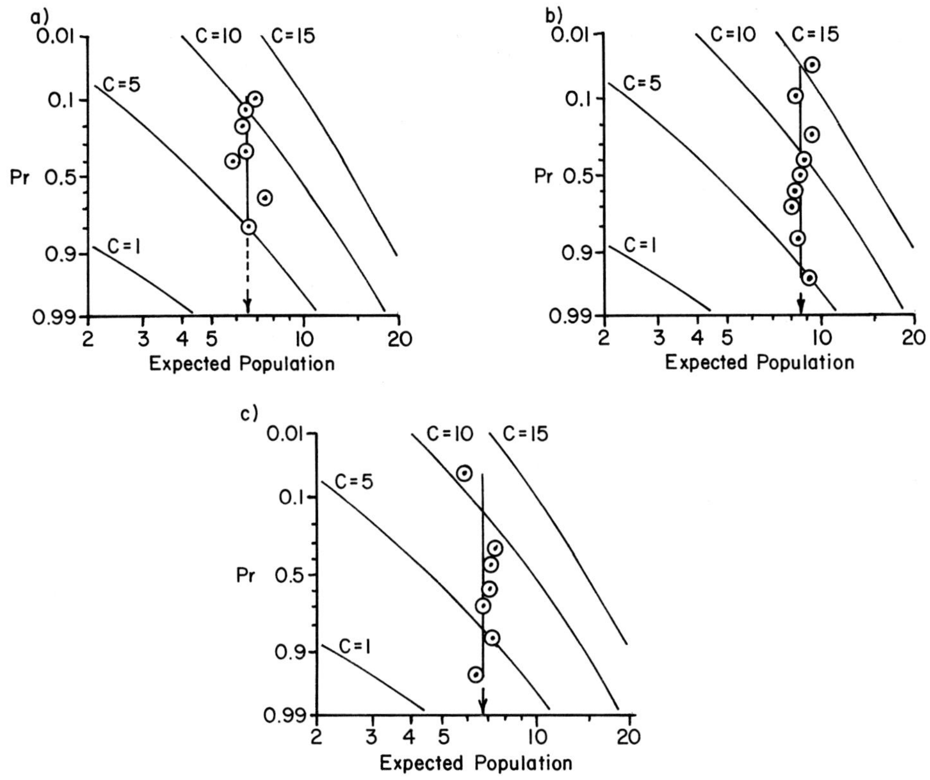

Figure 5.4. The observed fluctuations of the populations of a specific color in a box of Smarties® are consistent with Poisson fluctuations in an anticipated population of eight per box. a) Data for yellow buttons. b) Data for green buttons. c) Data for red buttons.

If one looks at the structure of Poisson graph paper when the expected value is relatively high, such as at the 60 mark, it can be seen that the population lines crossing the frequency lines are almost equally spaced. The ordinate of Poisson graph paper is in fact the Gaussian probability scale, and the fact that the population curves are almost equally spaced on the probability curve at high expected population illustrates the fact that when the expected population is large, the Poisson distribution becomes indistinguishable from the Gaussian probability distribution function. In Figure 5.5(c) the variations in the number of a particular color of candies per box in the 100 simulated boxes is plotted on Gaussian probability paper. It can be seen that the dataline is slightly curved on the Gaussian probability paper. *Figure 5.6* summarizes the data for the simulated variation in the population of one color of "Smarties"® for 1,000 boxes, each containing 160 buttons (an expected population of 20). It can be seen that both the Poisson plots and Gaussian probability paper presentation are close to straight line data plots.

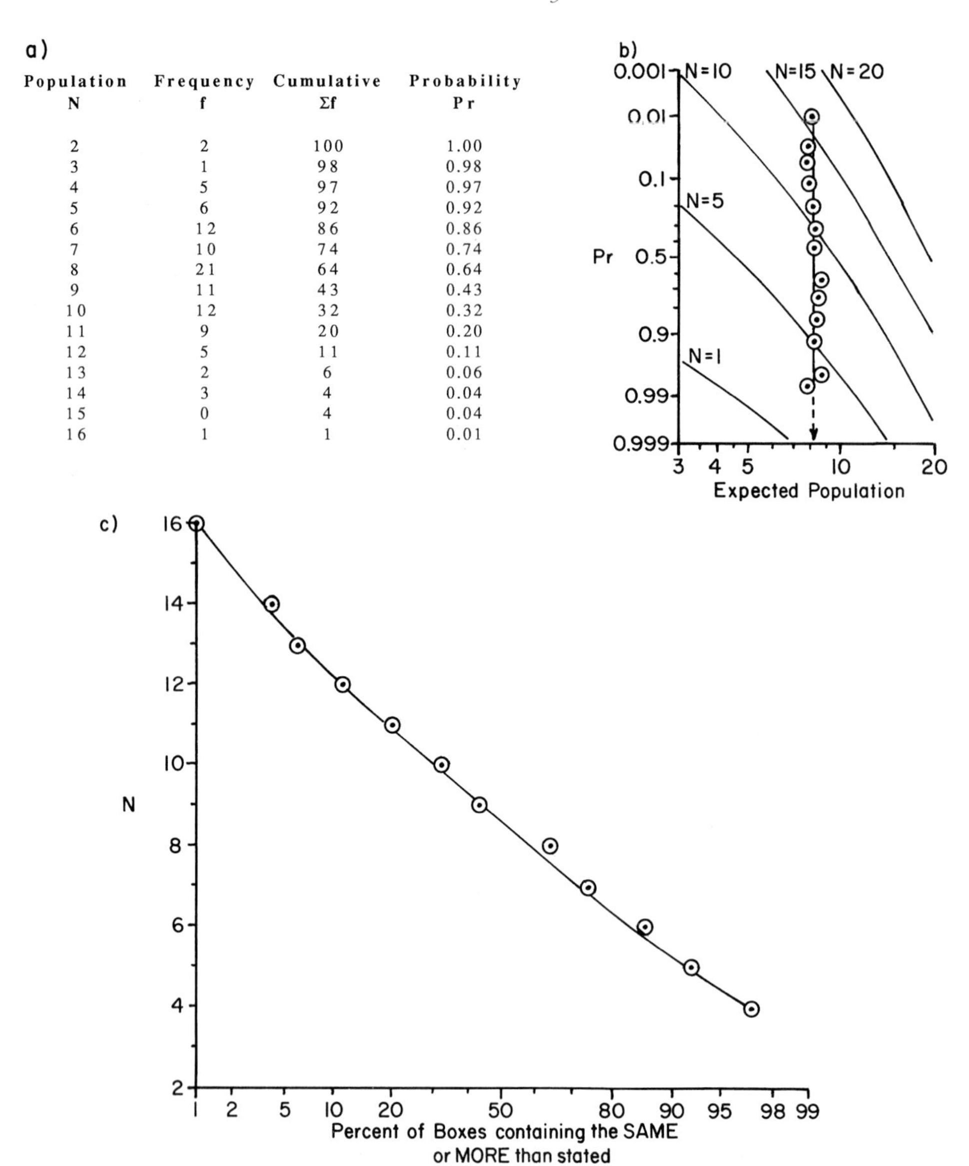

a)

Population N	Frequency f	Cumulative Σf	Probability Pr
2	2	100	1.00
3	1	98	0.98
4	5	97	0.97
5	6	92	0.92
6	12	86	0.86
7	10	74	0.74
8	21	64	0.64
9	11	43	0.43
10	12	32	0.32
11	9	20	0.20
12	5	11	0.11
13	2	6	0.06
14	3	4	0.04
15	0	4	0.04
16	1	1	0.01

Figure 5.5. Data and Poisson graphs for 100 boxes of simulated Smarties® showing the possible fluctuation in any one color. a) Data for 100 simulated boxes of buttons. b) Data for the 100 simulated boxes plotted on Poisson graph paper. c) Data plotted on Gaussian paper.

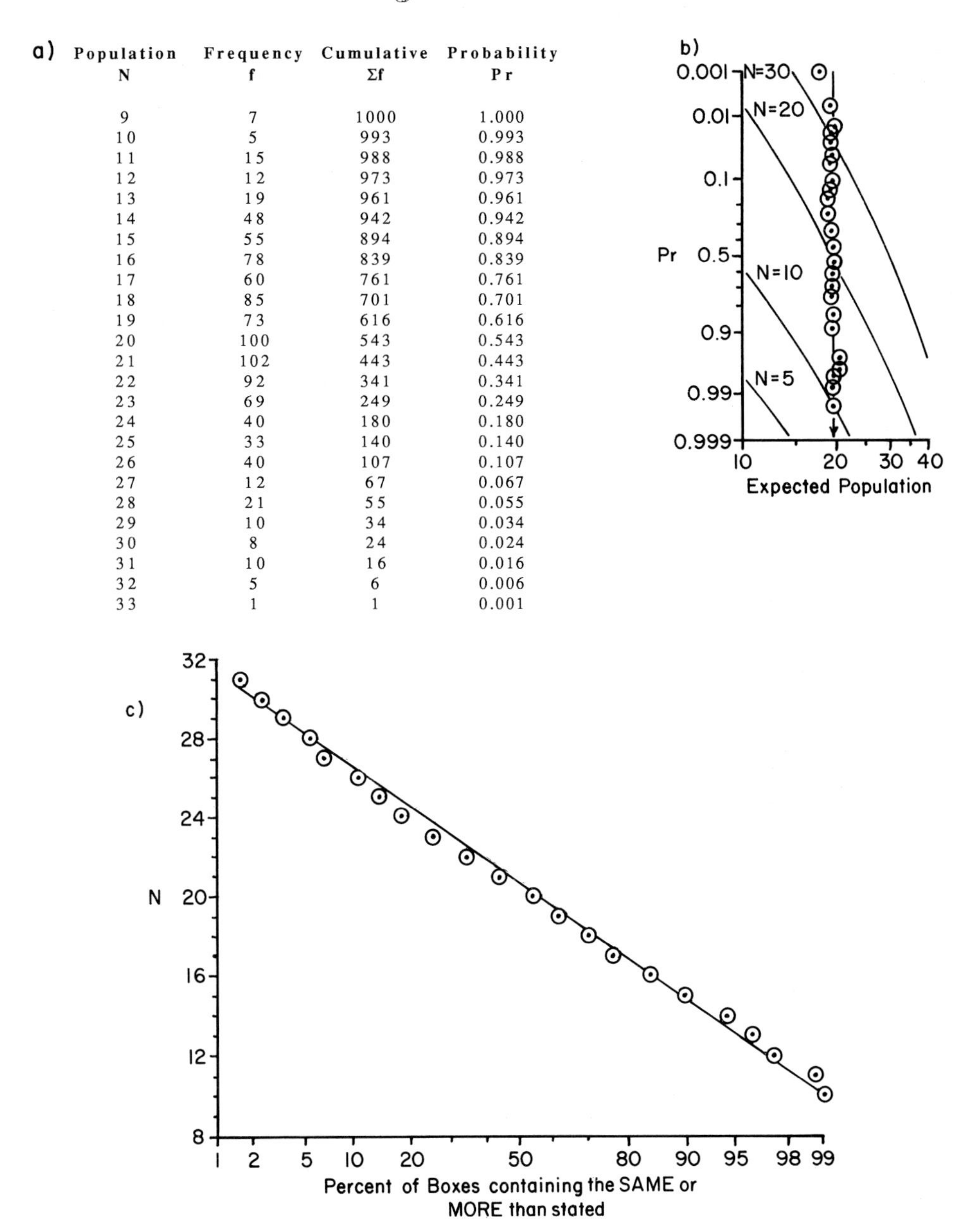

a)

Population N	Frequency f	Cumulative Σf	Probability Pr
9	7	1000	1.000
10	5	993	0.993
11	15	988	0.988
12	12	973	0.973
13	19	961	0.961
14	48	942	0.942
15	55	894	0.894
16	78	839	0.839
17	60	761	0.761
18	85	701	0.701
19	73	616	0.616
20	100	543	0.543
21	102	443	0.443
22	92	341	0.341
23	69	249	0.249
24	40	180	0.180
25	33	140	0.140
26	40	107	0.107
27	12	67	0.067
28	21	55	0.055
29	10	34	0.034
30	8	24	0.024
31	10	16	0.016
32	5	6	0.006
33	1	1	0.001

Figure 5.6. For expected populations of 20 or greater the data plots of permissible fluctuations in observed populations from Poisson or Gaussian probability theory are virtually indistinguishable for many situations. a) Fluctuations in the population of a specified color for simulated data generated for 1,000 boxes of 160 Smarties®. b) Data plotted on Poisson graph paper. c) Gaussian probability plot of data for the simulated 1,000 boxes of 160 Smarties®.

Section 5.4. Using Poisson Trackers to Monitor Chaos in a Powder Mixer

In an earlier section it was pointed out that the process of powder mixing is very important in industry. When one mixes two or more ingredients of a powder mixture, one is attempting to create chaos in that the final position of any particular grain of powder should be independent of its starting point and of the positions of the other grains in the powder mixture. Recently it has been shown that one can check on the chaotic conditions being created in a powder mixture by using what are known as Poisson trackers. To understand what we mean by a Poisson tracker, consider the following situation: The ingredients to be mixed are placed into a small cubical box which can be thrown up into the air and given a random twist at the beginning of each throw. One can imagine that the various rotations of the cube in space would randomize the internal contents so that a good mixture is obtained. To check on whether or not one is achieving a good mixture, one would put a small sampling device inside the cube. Let us assume that something like a thimble is placed on one of the side walls of the cube and that the thimble could contain 1% of the powder mixture when it is full. To enable the grains of powder to tumble freely during the upward and downward movement of the randomly rotating cube, the cube would only be half full of powdered ingredients. At the end of the throw one could retrieve the thimble and measure the contents to see if a good mixture is being obtained. Measuring the structure of the mixture, however, could be quite tedious. Recently Kaye and Clark suggested that if, in such a situation, one were to place 100 plastic beads in the mixture, then on average tha thimble would contain 1 bead (because it contains 1% of the mixture). Then, if one were to carry out a series of experiments of tumbling the cube and examining the bead population in the thimble sampler, the observed population of beads would follow a Poisson distribution fluctuation. Therefore, a series of measurements of the population of tracker beads should generate a straight line on Poisson graph paper generating an expected value on the abscissa. To test their theories, Kaye and Clark built the equipment shown in *Figure 5.7* [5]. They made three different small containers; one was a plastic cylinder, the other a cube and the third a 20 sided geometric figure known as an icosahedron. These different containers were rolled around in a drum which was lined with dimpled foam rubber. In the experiment carried out by Kaye and Clark, one face of each different mixing chamber was equipped with a sampling container which could be taken from the mixer after an episode of tumbling.

Kaye and Clark had hypothesized that the many-sided figure would be the most efficient random tumbling equipment to achieve mixing of the ingredients in a quick and efficient manner. However, they also anticipated that the other simple shapes could be just as efficient in their chaotic tumbling to act as mixing devices.

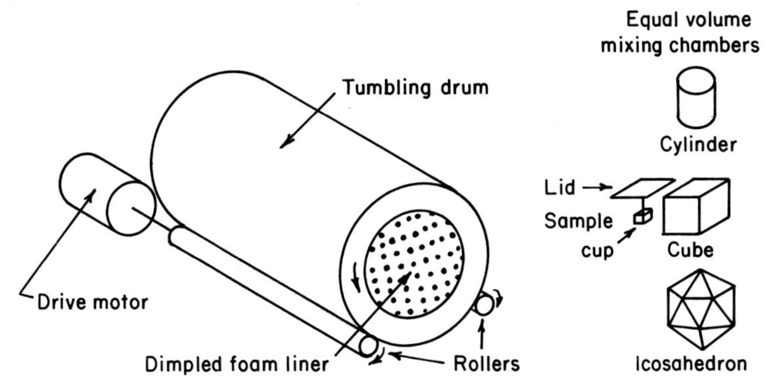

Figure 5.7. Freefall tumbling mixer used to study the usefulness of Poisson trackers in powder mixing equipment.

In their experiments they filled the tumbling chambers one quarter full of sugar mixture and added tracking spheres, which were the tiny silver spheres used to decorate cakes. The mixing chambers had a volume of 140 cubic centimeters and were loaded with 35 cc of powder; 350 tracker spheres were placed in the mixture. The rolling drum had a diameter of 30 cm. When the drum was rotated at a speed of 4 rpm the chambers were picked up by the foam rubber lining of the chamber and moved upwards until they fell down at random, tumbling in a chaotic manner down to the bottom of the chamber. Continued rotation of the drum resulted in a further upward and downward movement of the tumbling chamber. A one-minute rotation of the drum resulted in approximately 12 tumbles of a mixing chamber (one should not tumble the drum too fast, because the tumbling chambers would then tend to be centrifuged against the container and move around without tumbling). After the container had been tumbled for 1 minute, a sample of the mixture was withdrawn and the tracker spheres separated from the fine powder in the mixer with an appropriate sieve. The number of tracker spheres were counted and both the tracker spheres and the fine powder returned to the mixer. The mixing chamber was tumbled again for 1 minute, and then another sample was withdrawn for examination. The fluctuating numbers of tracker spheres were plotted as a cumulative frequency of population equal to or less than a noted population using Poisson probability graph paper. The table of results for the experiment and the appropriate portions of the graph paper are shown in *Figure 5.8*. It can be seen that indeed Poisson fluctuations were generated, and an expected population of 8, 8.8 and 8.2 generated on the axis. These values differ slightly from the 10 expected from the theoretical design of the experiment. This difference was thought not to be significant in view in the relatively small amount of data used to generate the graphs of Figure 5.8. It can be seen that each of the chambers appear to be good mixing equipment and that trackers proved useful in establishing that chaotic conditions

a) Mixing Data for the Cylinder

Population N	Frequency f	Cumulative Σf	Probability Pr
1	0	40	1.000
2	0	40	1.000
3	1	40	1.000
4	2	39	0.975
5	5	37	0.925
6	9	32	0.800
7	2	23	0.575
8	4	21	0.525
9	6	17	0.425
10	3	11	0.275
11	5	8	0.200
12	1	3	0.075
13	1	2	0.050
14	1	1	0.025
15	0	0	0.000

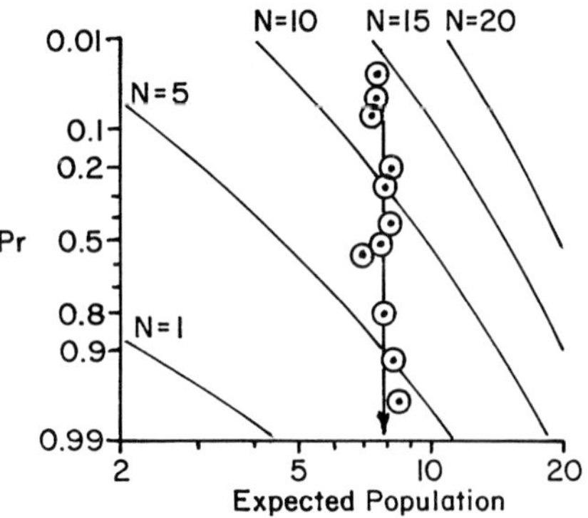

b) Mixing Data for the Cube

Population N	Frequency f	Cumulative Σf	Probability Pr
1	0	40	1.000
2	1	40	1.000
3	0	39	0.975
4	0	39	0.975
5	4	39	0.975
6	2	35	0.875
7	4	33	0.825
8	8	29	0.725
9	1	21	0.525
10	6	20	0.500
11	9	14	0.350
12	2	5	0.125
13	2	3	0.075
14	1	1	0.025
15	0	0	0.000

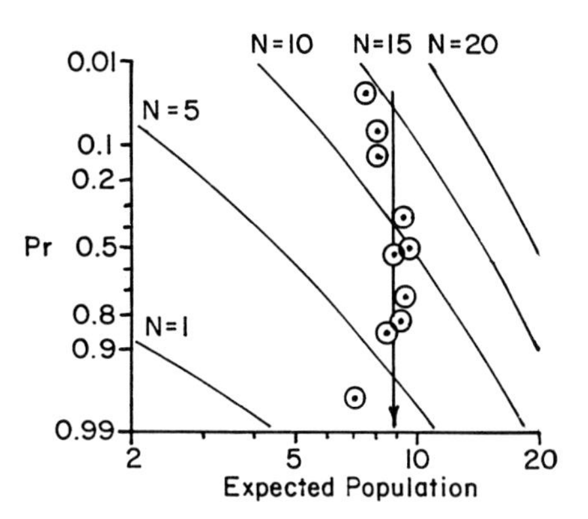

c) Mixing Data for the Icosahedron

Population N	Frequency f	Cumulative Σf	Probability Pr
1	0	40	1.000
2	0	40	1.000
3	2	40	1.000
4	3	38	0.950
5	2	35	0.875
6	7	33	0.825
7	3	26	0.650
8	5	23	0.575
9	3	18	0.450
10	3	15	0.375
11	1	12	0.300
12	4	11	0.275
13	4	7	0.175
14	1	3	0.075
15	2	2	0.050

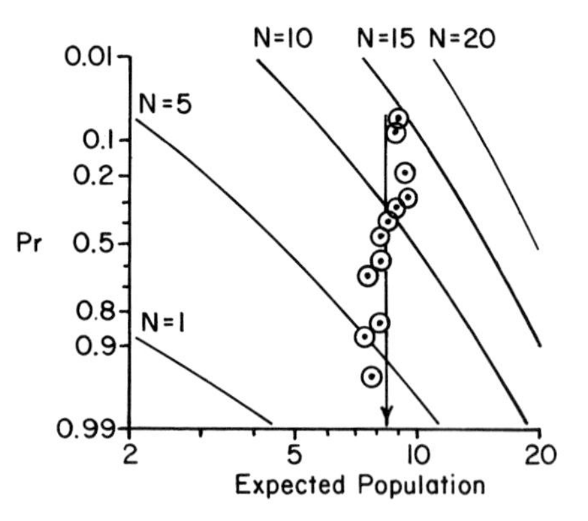

Figure 5.8. Results of Poisson tracker experiments in chaotic mixing investigations. a) Results for the cylindrical container. b) Results for the cubic container. c) Results for the icosahedron.

were being achieved and maintained in the mixers. The equipment shown in Figure 5.7 is relatively inexpensive and easy to build, and students enjoy carrying out experiments with the equipment and various powder mixtures [6].

Section 5.5. Segregated and Tumbled Jelly Beans

Consider a large jar of jelly beans. Presumably, the manufacturer of the jelly beans tries to mix all colors uniformly throughout the mixture. Since the jelly beans are basically of the same size and shape, they should not segregate when they are handled. If one were to purchase a large supply of jelly beans, then one could work one's way down through the supply of beans counting the number of the specific colors to see if the fluctuations in a specific color jelly bean in the jar population follows a Poisson distribution. In *Table 5.4* we show the data for different jelly bean populations from 860 beans, taking a population of 20 beans at a time. In *Figure 5.9* the population fluctuations for the green (expectation 0.7), pink (expectation 2.5), and yellow (expectation approximately 5) are shown. The data variations for the green beans is compatible with the hypothesis that the fluctuations are describable by the Poisson distribution, but the scatter of the data for the pink and yellow would indicate some segregation of these colors in the stack of jelly beans. If one considers the data for the pink beans, one can simulate how a better mixing of the beans can reduce the "bends" and "scatter" in the Poisson graph paper plots.

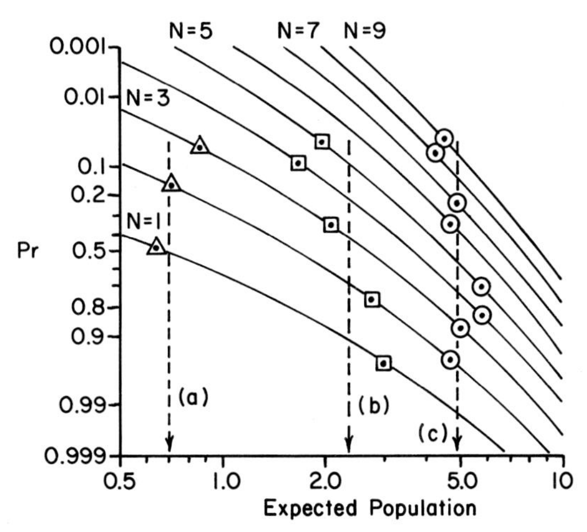

Figure 5.9. Poisson trackers can be used to investigate the possibility of segregation in an assembled mixture. a) Fluctuations summarized for green jelly beans. b) Fluctuations summarized for pink jelly beans. c) Fluctuations summarized for yellow jelly beans.

Table 5.4. Variations in jelly bean colors in sets of 20 beans taken in sequence down through an assembled population of jelly beans.

Sample	Orange	Black	White	Red	Pink	Green	Yellow
1	6	0	1	4	2	2	5
2	2	4	1	2	3	0	7
3	3	5	2	3	1	0	5
4	4	4	1	3	3	0	5
5	4	2	2	3	3	1	5
6	5	5	2	2	2	0	4
7	4	5	2	1	2	1	5
8	3	7	0	2	2	0	6
9	0	3	2	1	4	1	9
10	2	4	0	3	6	1	4
11	2	3	2	4	3	3	3
12	4	3	1	7	3	0	1
13	1	4	2	5	3	0	4
14	5	4	2	0	3	1	4
15	5	7	0	2	1	0	5
16	4	4	4	4	2	0	2
17	6	2	1	1	2	3	5
18	4	4	2	5	1	1	3
19	4	4	1	6	0	0	5
20	2	3	2	3	4	1	5
21	1	3	3	2	2	0	7
22	2	5	0	3	1	1	5
23	8	4	1	4	1	1	1
24	2	4	2	2	2	0	7
25	2	3	2	4	3	1	5
26	2	4	3	0	2	2	6
27	3	2	1	7	1	0	5
28	2	6	0	3	3	0	6
29	5	3	1	3	2	2	4
30	1	5	2	6	1	2	2
31	4	3	1	3	2	2	5
32	3	3	2	2	2	0	8
33	1	4	1	6	1	0	7
34	1	3	0	2	2	1	7
35	4	4	4	1	5	0	4
36	3	2	1	6	2	0	6
37	4	5	2	2	2	0	5
38	5	5	1	2	2	0	5
39	4	1	1	4	0	1	9
40	2	3	1	2	3	1	6
41	2	2	2	2	3	1	7
42	2	3	4	6	2	0	2
43	3	5	0	2	2	0	7

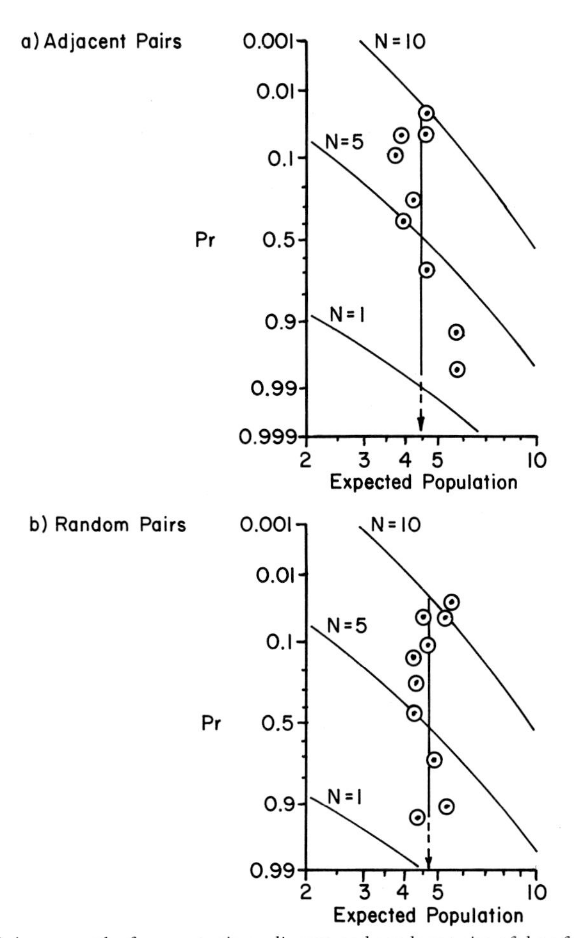

Figure 5.10. Poisson graphs for comparing adjacent and random pairs of data for one color of jelly bean. a) Results obtained by combining adjacent pairs of data. b) Results obtained by combining random pairs of data.

Sequentially, there were 43 sets of beans. If one were to take sets of beans to create populations of 40 beans, then one could create a second set of data by adding two sequential jelly bean samples to generate a set of data. On the other hand, one can take the whole set of data and amalgamate pairs of 20 at random. This is equivalent to randomizing the mixture by some mixing process. In *Table 5.5* the results of the manipulation of the data in sequential pairs and random pairs to generate 2 more sets of data with an expected population of 4.7 are shown. These data are plotted on Poisson graph paper in *Figure 5.10*. It can be seen that in fact when the jelly beans are mixed by randomization of pairs, agreement with the Poisson distribution is achieved. This shows that the extra randomization was sufficient to achieve a good mixture.

Table 5.5. Results for combining the data for adjacent and random pairs of data from Table 5.4 for 1 color of bean.

Adjacent Pairs	Random Pairs	Adjacent Pairs	Random Pairs	Adjacent Pairs	Random Pairs	Adjacent Pairs	Random Pairs
5	4	6	4	3	6	7	5
4	1	6	5	5	10	7	5
4	3	4	6	5	4	4	4
6	3	3	4	3	3	4	4
5	2	4	5	4	3	2	3
4	11	3	6	5	4	3	4
4	6	1	1	3	1	6	6
6	5	4	3	3	5	5	7
10	3	6	6	4	5	4	3
9	4	3	4	3	4	4	8
6	8	2	4	3	8		

Section 5.6. A Cautionary Tale of Tails

It should be noted that most of the distribution functions that we have discussed in previous chapters such as the Gaussian distribution function, the log-Gaussian distribution function and the Poisson distribution often describe most of the data of a set of measurements, but that data in the tails (i.e., data in the low probability of occurrence extremities of the distribution representing rare events) can be "nonconformist" and describable by other distribution functions. West and Shlesinger have pointed out that many data sets which are describable by the log-probability distribution have tails which are describable by other distribution functions [7]. An interesting example they give is that of salaries in the United States. They point out that 97% of all salaries in the United States are describable by the log-Gaussian distribution function, but that when we come to consider rare events represented by very high salaries, the distribution function describing the top 3% of salaries is different type of mathematical function. In fact, these rare, very high, salaries are represented by a relationship which we shall discover as being the appropriate function to describe the frequency of words in the English language, when we come to explore what is known as Zipf's law in a later chapter. The fact that the tails of our distribution differ from the main sweep of the data described by the probability functions does not invalidate our use of the functions to describe stochastic systems, but one should be very cautious when trying to interpret the physical significance of data located in the tails of the distribution functions (see the discussion of elephants in the face powder given in [8]).

Exercises

Exercise 5.1.

Various expected populations for dust content on a slide can be generated using the random number table and the data plotted. In a class project, students may like to determine the expectation of different populations in simulated slides of dust-collecting equipment and compare their results with each other.

Exercise 5.2.

Students can be invited to fill a jar with pennies and then opt to study variations in the population of pennies made in a particular year in sets of coins taken out of the jar. They could find how the population of pennies of one year varies down through the jar in sets of, say, 20 pennies. One could look for the occurrence of the year 1982 in sets of 20 pennies taken from the jar as we worked down through all the pennies in the jar. The data can be plotted on Poisson paper and transformed to a cumulative distribution like that of Table 5.2 shown in Figure 5.3. The further the year selected for study is from the current year, the lower the expected probability. If the study is carried out in 1992, then the expected population of 1982 coins will be less than that of 1985 coins, but both sets will be describable by the Poisson distribution.

Exercise 5.3.

Students can make their own small mixing chambers and load them with different powder mixtures to which one can add plastic beads to serve as Poisson trackers. A novel experiment that the students could try would be to cut a hollow rubber ball, such as a tennis ball, in half and load one half of the ball with the ingredients to be mixed before taping the two halves together. One could throw the reassembled ball around for a short period before investigating to see if the mixture had achieved chaotic conditions. One could also build a small icosahedron which can again be used in the laboratory to throw from one student to the other, giving the icosahedron chamber a twist to achieve chaotic conditions.

Exercise 5.4. Building a Random Number Generator

It is said that some of the random number generators that are built into small computers do not generate truly random sets of data. One way for a class to test a random number generation is to count the number of expected digits in sets of 20 random digits a thousand times and plot the frequency of occurrence of various digits on Poisson graph paper (expected population 2 in 20 for a specific digit).

Exercise 5.5.

The class can purchase several boxes of candy containing different colored items and study the population variations. If the population expectation is low, the variations will be describable by the Poisson distribution when the fluctuations are truly random. The class can then simulate with a random number table the variations to be expected if the fluctuations in observed populations are describable by the Poisson distribution as discussed in Section 5.3. The simulated fluctuations can be compared graphically with the observed fluctuations.

Exercise 5.6. Stingy With the Orange?

At various times when I have lectured on the use of Poisson trackers to study powder mixing, I have used the example of the variation in the colors of Smarties® as an illustration of the potential of the technique. In the United States the inevitable question arises each time I give the lecture; "Can We Use M & M candies instead of Smarties to simulate Poisson fluctuations?" (M & Ms® are a very similar button chocolate covered with sugar candy). Finally, to answer these questions, we repeated the Smarties® "stingy with the yellow" experiment with some M & Ms®. When I checked on the number and color content of a packet of M & Ms® I discovered that there were 2 less colors than in the Smarties packages. In the first package the following distribution of colors was observed; yellow 21, red 12, orange 6, green 10, light brown 6, dark brown 9. This did not seem to fit the hypothesis that the manufacturer was trying to fit the same number of colors into the package therefore one had to work without a hypothesis and study the frequency of the different colored buttons in 20 packets. The table of data generated by a student investigating the color distribution in 20 packets of the M & Ms® are given below.
The reader is invited to transform the data and plot it on Poisson probability paper to discover the frequency of occurrence of the colors from packet to packet. The reader should check for themselves that the data for the yellow buttons is essentially Gaussianly distributed because of the high expected population, whereas the data for the orange is a typical Poisson fluctuation.

Bag	Yellow	Red	Orange	Green	Lt Brwn	Dk Brwn
1	21	12	6	10	6	9
2	16	12	6	5	7	21
3	15	16	4	6	15	11
4	14	16	5	12	9	9
5	20	11	2	8	8	18
6	24	13	5	7	6	11
7	18	16	4	7	5	15
8	20	13	5	10	6	14
9	20	10	6	11	5	15
10	25	12	2	9	5	14
11	18	18	4	5	5	14
12	20	10	2	8	12	13
13	15	18	2	11	3	18
14	21	14	3	4	5	18
15	17	13	8	4	6	18
16	17	12	6	9	8	15
17	10	7	10	6	5	29
18	21	18	3	5	4	15
19	18	17	4	9	9	10
20	19	5	2	8	9	23

References

[1] M. J. Moroney, *Facts from Figures*, 2nd revised edition, Pelican, Harmondsworth, England, 1953.

[2] For a comprehensive study of this problem see B. H. Kaye, *Direct Characterization of Fineparticles*, Wiley, New York, 1981. This book is currently out of print and a revised version entitled *Characterizing Powder, Mists and Other Fineparticle Systems* is in preparation.

[3] B. H. Kaye "Operational Protocols for Efficient Characterization of Arrays of Deposited Fineparticles by Robotic Image Analysis Systems", Chapter 23 in *Particle Size Distribution II, Assessment and Characterization*, ACS Symposium Series # 472, edited by T. Provder, American Chemical Society, Washington D.C., 1991.

[4] B. H. Kaye, G. G. Clark, "Monte Carlo Studies of the Effect of Spatial Coincidence Errors on the Accuracy of the Size Characterization of Respirable Dust", in *Particle and Particle Systems Characterization* 9 (1992) 83–93.

[5] B. H. Kaye, G. G. Clark, "Evaluating The Performance Of Chaotic Powder and Aerosol Sampling Devices Using Tracker Fine Particles", Proc. Nürnberg Conference on Particle Size (Partec), May 1989.

[6] Readers interested in finding out more about powder mixing can consult B. H. Kaye, *Mixing Powders*, Chapman and Hall, to be published.

[7] B. J. West, M. Shlesinger, "The Noise In Natural Phenomena", *American Scientist*, 78, January–February, (1990), pp. 40–45.

[8] B. H. Kaye, *A Random Walk Through Fractal Dimensions*, VCH, Weinheim, Germany, 1989.

Chapter 6

Rubber Number Logic and The Swinging Mouse

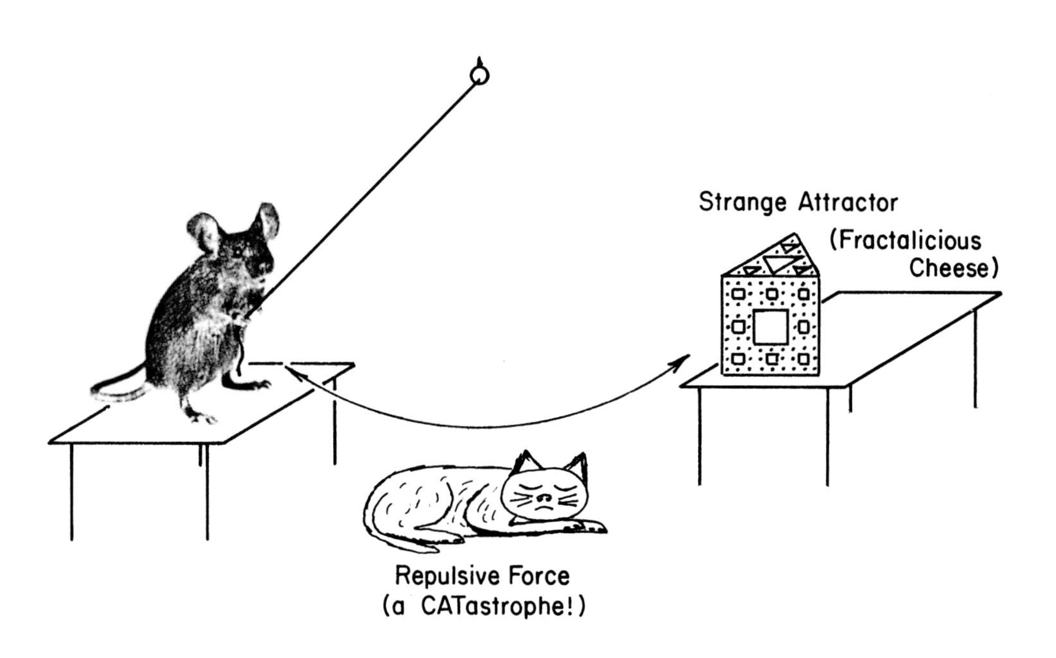

The laboratory mouse of chaos theory is a "swinger"

(Mousus Pendulata Simplicicus)

Chapter 6 Rubber Number Logic and the Swinging Mouse 231

Section 6.1 The Dimensions of Reality 233
Section 6.2 Topolgy, Topography & Stretched Relationships 238
Section 6.3 Stokes' Law Versus Galileo 244
Section 6.4 Rubber Number Logic for Studying the Flow of Viscous Fluid Through Pipes 253
Section 6.5 Dimensionless Numbers as Indicators of Similar Structure and Behavior 256
Exercises 262
References 265

Chapter 6

Rubber Number Logic and the Swinging Mouse

Section 6.1. The Dimensions of Reality

The title of this chapter was inspired by a statement by Gleick that

The laboratory mouse of chaos theory is the simple pendulum [1].

In this colorful statement, Gleick puts forward the idea that, analogous to the way in which many specialists make preliminary explorations of new ideas in biology by carrying out experiments on a mouse population, deterministic chaos specialists can explain their ideas and test concepts by studying modifications in the behavior of a simple pendulum responding to complex stimuli. In this chapter we shall discover how to study the simple pendulum with what many find a surprising form of logic known as dimensional analysis.

In the next few chapters we are going to discover some of the exciting patterns of reality which can be efficiently described using the concepts of fractal geometry. We will discover that fractal geometry, a new branch of mathematics, extends the concepts of dimensional analysis by adding a fractional number to the topological dimension of a system to describe the properties of rugged systems, such as tree bark and porous rocks. The combination of the topological dimension and the fractional *addendum* (a word meaning "something added") constitutes the *fractal dimension* of a system. We will also learn that using the fractal dimension to describe the structure of a rugged system is like using dimensionless numbers to describe the self-similar behavior of complex systems. Obviously, if the reader is to appreciate the role of fractal dimensions in modern science, it is necessary to have a clear understanding of what is meant by topology, dimensional analysis and dimensionless number systems. In this chapter we will explore the surprising power of dimensional analysis to discover fundamental relationships by stretching the truth, a process which I like to call *rubber number logic*. We will discover that some relationships which were deduced by laborious experimental investigation could have been predicted as inevitable patterns of truth if scientists had thought before they investigated.

Students often have their first brief encounter with dimensional analysis when they are told in physics classes to check the *dimensional homogeneity* of their equations.

One of the first sets of equations presented to a beginning student of physics is the group of equations used to describe the speed and distance travelled by an object such as a sphere falling under the influence of gravity. The origin and form of these equations are described in Chapter 14. For example, they are told that for such a falling object

$$s = \frac{1}{2} a t^2$$

where s is the distance travelled by an object, originally at rest, falling freely for t seconds under the influence of gravity (assuming that air friction is negligible) and a is the acceleration caused by gravitational attraction. When students try to remember this formula in the middle of an examination, they may have trouble remembering if the equation is

$$s = \frac{1}{2} a^2 t$$

or

$$s = \frac{1}{2} a t^2$$

The consequences of such a memory lapse can be avoided by carrying out a dimensional homogeneity check of both sides of the two possible versions of the equation. Before the reader who has never studied physics at an advanced level can understand how to carry out a dimensional homogeneity check, it is necessary to explore the concepts of the dimensional description of physical systems.

The word dimension was created by early surveyors as they measured distances along a road or the size of a field. The word comes from two Latin root words: di, meaning "apart", and metiri, "a measure". A *dimension* was originally a measurement of the distance between two posts on a road. However, over the years the word dimension has acquired several different meanings. In everyday speech we describe a computer problem as having many dimensions, not all of which are measurements of length. To the physicist and engineer the dimensions of an object are the number of measurements required to specify the magnitude of an object or the physical properties of a system or physical variable. A length is said to be one-dimensional, an area is said to be two-dimensional and a volume three-dimensional. Scientists have developed a special system of describing the physical dimensionality of an object. The dimensions of an object are represented by capital letters inside square brackets. An exponent is added to the bracket to describe how many times a dimension is used to achieve a dimensional description of a system. Thus $[V]$ represents the dimensions of a volume and

$$[V] = [L]^3$$

is a symbolical relationship summarizing the fact that a volume can be described by three measurements of length. The relationship for describing the dimensionality of area is

$$[A] = [L]^2$$

where $[A]$ denotes the dimensional description of area. Note that the dimensionality of an object is not concerned with magnitude, shape or structure. All of the lines of *Figure 6.1* have the dimension of 1. The dimensions of an elephant, a grasshopper and a cube, which are all topologically equivalent, are all the same: $[V] = [L]^3$.

Students introduced to the concepts of dimensional descriptions of a system often have difficulty accepting the fact that a line is a one-dimensional system or that a piece of paper is a two-dimensional system. To the student, a line always has breadth and so appears to be a two-dimensional system. In some class discussions a teacher will often refer to a thin thread as an ideal one-dimensional system. Again, many students feel that a thread has to have three dimensions to exist as a visible object. This confusion over the reality of the dimensions used to describe a system comes partly from the historic development of ideas in mathematics, and partly from the omission of the adjective "operational" in a discussion of the dimensions of reality. As discussed earlier, the Greeks always saw physical systems as imperfect models of an invisible perfect reality. To a Greek a thread was an imperfect representation of

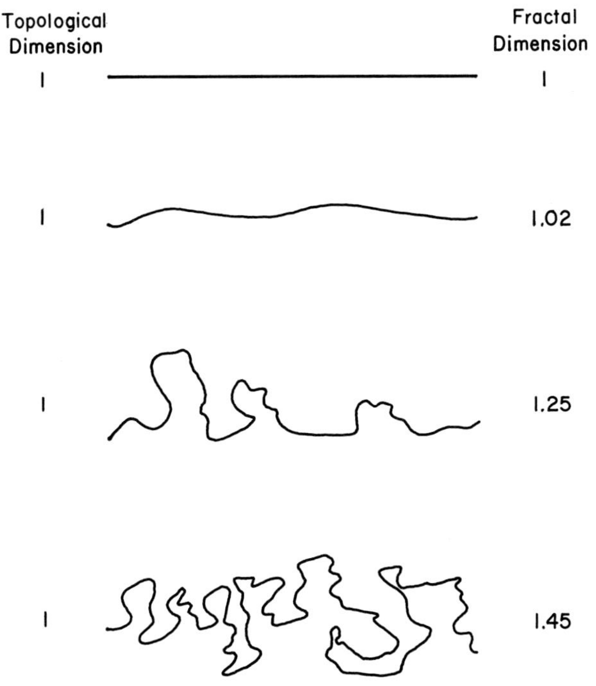

Figure 6.1. The fractional number added to the topological dimension of a line describes its ruggedness. The combination of topological dimensions and the fractional number describing its space filling ability is known as the fractal dimension of the line.

an ideal infinitely thin thread which existed in the realm of the pure mathematics of the imagination. In the same way, a thin sheet of paper represents a crude version of an ideal infinitely thin two-dimensional sheet, which is a perfect model underlying everyday things that we see around us. In his discussions of the basic ideas of geometry, Mandelbrot points out that the difficulty of grasping the dimensions of reality can be simplified by realizing that any dimensional description in a real world always implicitly assumes an operational perspective [2].

If we are interested in constructing a triangle as a mathematical exercise from the perspective of "construct a drawing to represent a triangle", the line is essentially a one-dimensional system. In the same way, a piece of paper can represent a two-dimensional system when one is interested in the properties of sheets of metal that are to be bent to construct a box. However, if one is interested in passing water through a piece of paper to filter out bacteria, then the reality of a piece of paper for the operation "filter out bacteria" is that the paper is a three-dimensional porous body.

Mandelbrot discusses the problem of assigning an *operational dimensionality* to an object in terms of a fly approaching a ball of wool from a great distance. From the operation "navigate towards the ball of wool", the fly can see the ball at a great distance as a point with no dimensions. At an intermediate distance during its journey to the ball of wool, the ball of wool changes its dimensionality from 0 to 2 as it begins to look like a disk. When the fly is close up to the ball of wool, it becomes a three-dimensional structure. If the ball is made of coarse wool and the fly is very tiny, then if the fly gets inside the ball of wool the dimensionality of the wool is a very difficult concept since it involves an apparent maze of ropes occupying three-dimensional space.

Often the dimensionality of an object being discussed in a mathematical or physical situation is not its ideal dimensionality but its operational dimensionality. It is unfortunate that the distinction is not always clearly made when discussing the dimensional structure of a system. At the beginning of any discussion of dimensionality of a system, the perspective from which a given problem is being discussed should be clearly stated.

The reader will have noted that to describe the dimensionalities of volumes, areas and lengths we have used multiples of the one-dimensional measure called length. When discussing the dimensionality of a system in dimensional analysis, scientists distinguish between so called *fundamental dimensions* and *derived dimensions*. Length is described as a fundamental measure of the dimensionality of the universe, whereas the dimensionality of a volume is described as a derived dimension based on the dimension of length. Over the years, scientists have come to agree that a useful system for describing the universe involves the three basic fundamental dimensions of length, time and mass [3, 4]. Most people have no difficulty grasping length and time as dimensions, but mass is a difficult concept. Generations of physics students have struggled to understand the difference between mass and weight. A full discussion of the difference between mass and weight is beyond the scope of this book, but it should be clearly understood that the weight of an object is a force, whereas mass is a measure of the "stuff" inside an object. The *mass* of a body is that property of a body that resists the effect of a force and governs how a body reacts when a given force is applied to that object.

A distance travelled between two points in a given time by a body is called the *displacement* of the body. It is usually represented by the symbol s which has the dimension of $[L]$. The speed of movement of an object is a measure of how long it takes to move a given distance. It has the dimensionality of length divided by time. This can be written in two ways as follows

$$[v] = \frac{[L]}{[T]} = [L]\,[T]^{-1}$$

where v = velocity and $[v]$ means "The dimensions of velocity". When a dimension has to be divided by another dimension to achieve a dimensional description of a given measurement, scientists have agreed to use negative exponentials as shown above.

In everyday speech we tend to use the terms velocity and speed interchangeably. To a scientist, speed is a displacement without a description of direction. Speed is a scalar quantity. Velocity requires a description of the direction in which the displacement is taking place and is a vector quantity. In advanced dimensional analysis, a distinction is made between velocity and speed, but for the purposes of this introductory discussion of dimensional analysis it is acceptable to treat the dimensions of velocity and speed as being the same, that is $[L]\,[T]^{-1}$ [3].

When discussing the movement of bodies subject to a set of forces, an important concept is that of *acceleration*. If a car is moving with speed v_1 at a time 0, and its speed increases to v_2 after a period of time t in which the engine pushes the car faster and faster, then the car is said to have accelerated from v_1 to v_2. The word accelerate comes from two root words: ad, meaning "to", and celer, "swift". When a force accelerates a body, it adds swiftness to the body. Technically, the acceleration of a body is described as the change in speed over the time t during which a force causing the acceleration is applied to the body. In symbolic terms we write

$$a = \frac{v_1 - v_2}{t}$$

where a is the acceleration of the body.

The dimensions of acceleration can be deduced from the formula given above as

$$[a] = \frac{[L]}{[T]}\frac{1}{[T]} = [L]\,[T]^{-2}$$

Now that we have the dimensions of acceleration, we can carry out a dimensional homogeneity check on the two versions of the half-remembered equation discussed earlier.

The half-remembered equation was meant to describe a distance moved and therefore the left-hand side of the equation has the dimensionality of length. For the remembered equation to be correct, the right-hand side of the equation must also have the dimensionality of length. If we take the first half remembered expression we see that (ignoring the 1/2, which is a numerical constant) the dimensionality of the right-hand side of the equation is

$$a^2 t = ([L]\,[T]^{-2})^2\,[T] = [L]^2\,[T]^{-4}\,[T] = [L]^2\,[T]^{-3}$$

This is obviously not a single dimension of length. On the other hand, if we take the second possible version of the equation and check its dimensional structure, we see that it has the dimensions

$$a t^2 = [L]\,[T]^{-2}\,[T]^2 = [L]$$

That is, it has the dimensions of length. Therefore our dimensionality check shows that the second version is the correct version of the equation (note, however, that although dimensional homogeneity checks are valuable they do not help us remember forgotten constants such as the 1/2. If we could not remember whether it was 1/2 or 1/3 in the equation, dimensional homogeneity checks would not help us!). In the next section we explore why the dimensional descriptions of physical variables such as volume, area and acceleration can be described as rubber numbers.

Section 6.2. Topology, Topography & Stretched Relationships

The ancient Greek word topos means "a place", "a location", and the suffix -logy, found in many scientific words such as biology, astrology and geology, comes from the Greek word logos, meaning "a discussion". *Geology* is a discussion about the structure of the Earth, which in Greek is called geo. *Geometry* contains the word geo combined with the Greek root word metron. The first geometers were concerned with making measurements of the extent of the earth. *Astrology* is a discussion about the supposed influence of the stars of the heaven on the future of human beings (from the Greek aster, "a star"; an *asterisk* is a small star placed upon a page to draw the readers attention to an important piece of information). *Biology* is a discussion about living things from the Greek root word bios, "living". Unfortunately, the meaning of scientific words are not always logically linked to the meaning of the root words from which they are formed. *Topology*, which logically should be a discussion of places, is an abstract branch of mathematics which one of my dictionaries defines as

> *"Rubber sheet" geometry; a study of the properties of geometric figures that are not affected by distortions of the figures.*

To understand what this definition means, consider the set of lines shown in Figure 6.1. If these lines were drawn on rubber paper, then in theory we could stretch and squeeze the rubber sheet until they all looked alike. Therefore, to a topologist all the lines of Figure 6.1 are identical. The dimensions of all the lines are 1 from the perspective of the topologist. The varying ruggedness of the individual lines is of no interest to the topologist. The fractal geometer, however, allocates to the lines fractal dimensions which describes their space-filling ability as shown by the values of the fractal dimensions for each line given in Figure 6.1. To a topologist roaming his rubber world, all faces with two eyes, a mouth and a symbolic nose look alike, as illustrated in *Figure 6.2*(a). There are not many invariant relationships that can persist as space is distorted, but the presence of holes in a body over a two-dimensional system does persist as space is distorted. This is illustrated in the outlines of Figure 6.2(b), which show how, to a topologist, a needle, a doughnut and a cup are all the same. They have one hole each. The inside of the mug is not a hole, it is a depression which can disappear as space is distorted, but the hole defined by the handle persists no matter how space is distorted. A topologist counting holes describes the number of holes in a system as the *genus of the system*. A needle is a three-dimensional body of genus 1. Dimensional analysis is really a branch of topology. The theoretical dimension, or ideal dimension, of the Greek mathematicians is essentially the *topological dimension* of the modern mathematician. The dimen-

Figure 6.2. To a topologist all faces look alike. A topologist cannot tell the difference between a doughnut, a mug and a needle. a) Genus four topological "faces". b) Genus one doughnut–coffee mug systems.

sional description of a system such as $[V]$ or $[L]^3$ is a rubber number which describes a property of an object which remains invariant as the space in which it is embedded (i.e., the space which carries it) is distorted. The dimensionality of an elephant or a flea is $[L]^3$. This description remains valid topologically no matter how the support space is compressed, extended, twisted or folded (note, however, the topologist is not allowed to rip or tear the support space) [5].

I am told that mathematics students find their first course in topology a mind-stretching experience. When I introduce the basic concepts of topology to my students studying fractal geometry, I point out that the topological world is one in which there can be no stories of Cinderella and the ugly sisters, because all three ladies in the story are topologically equivalent. Furthermore, since a topological Prince Charming would use a rubber slipper, which would fit all the topologically equivalent beauties of his kingdom, he would have to marry the first lady he encountered after he left the palace.

From a mathematician's point of view, the ugly sisters differ from Cinderella in their basic topography. The word topography contains the suffix -graphy which comes from the Greek word graphein, "to write" or "draw". A *biography* is a set of writings about a living person, and *geography* is a written description of the surface of the earth. A *graph* is a drawing summarizing a lot of information (it is said that a picture is worth a thousand words). One of the aspects of the surface of the earth of interest to the geographer is the rise and fall of the land. The hills and valleys of a country are described by the geographer as the *topography* of the country or land. The geographer will say that the coastal regions of The Netherlands have a flat topography, and that the mountains of Tibet are a good example of a region of the earth with rugged topography. The fractal geometer adds a fractional number to the topological dimensions of a system to create a description of the topography and structure of that system. Topographically The Netherlands will have a fractal dimension of 2.00, and Tibet could be as high as 2.32 (see discussion of topographic fractals in Chapter 10).

In the same way that a topologist stretches and deforms space to seek out any structural feature of a geometric system that remains the same throughout all possible continuous deformations, dimensional analysis can seek out basic behavioral relationships governing a system which arise inevitably as manifestations of the way that the universe is structured no matter how the physical size of the system varies.

In this section we will explore the surprising power of dimensional analysis to discover important physical relationships by looking at a formula which describes the oscillation of a simple pendulum. In Chapter 1 we summarized the swinging behavior of a simple pendulum with the aid of phase space diagrams. In this section we will try and make a statement about the behavior of all pendula by using rubber number logic. Consider the set of pendula shown in *Figure 6.3*(a). To a topologist they are all identical. For the purpose of dimensional analysis, we need to isolate the parameters of the system which will influence its behavior when it is set swinging.

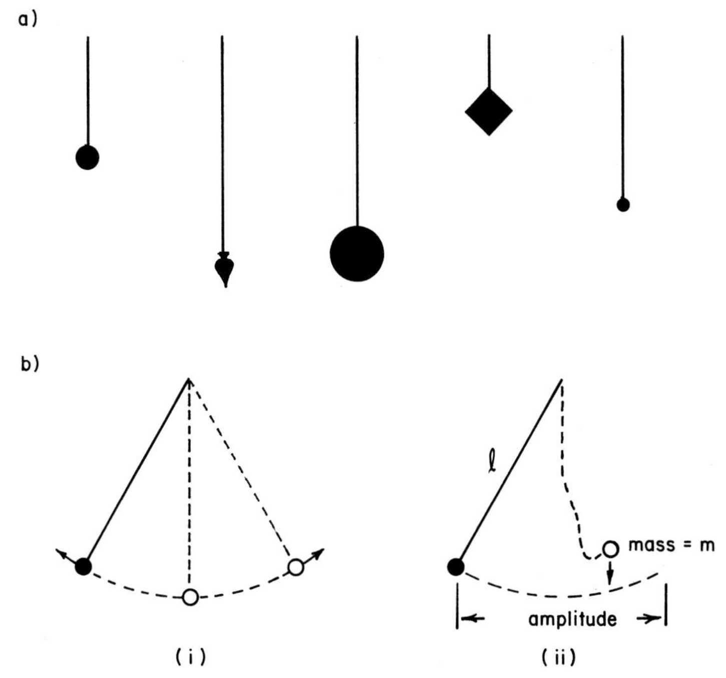

Figure 6.3. Pendula are the swingers of chaos. a) Topologically identical pendula. b) The swing of a pendulum is the result of the competing pull on the bob of the earth and the tension in the support string.

Parameter (not to be confused with perimeter) is a term often used by mathematicians and physicists but not easily located in ordinary dictionaries. One of my dictionaries defines a parameter as "a variable constant", which seems to be another self-contradicting definition! Another defines it as

> *A quantity that has values which are not fixed or absolute but which are chosen to establish a particular set of conditions in a given experiment.*

The word comes from the Greek root words para, "beside" and metron, "a measure". Perhaps we can best illustrate how two different meanings of the word parameter developed by looking at the sketch of the pendulum system given in Figure 6.3(b) (i). Before studying the swinging behavior of the pendulum, the scientist asks "What influences the behavior of the pendulum?" To answer such a question, the early scientists probably wrote down by the side of the sketch of a pendulum:

Variables affecting behavior of pendulum

1. Mass of pendulum bob
2. Length of pendulum string

Over the centuries these variables written down beside the system, which were literally "para metron", came to be called the parameters of the system. Using the language of the mathematician, the parameters we select as possibly governing the behavior of the pendulum are

l, the length of the pendulum string
m, the mass of the pendulum
g, the acceleration due to gravity experienced by a body falling under the influence of the force of gravity.

The word parameter is also used in a slightly different sense to describe the boundary conditions of an experiment. In our search for a pattern of behavior of a pendulum we will assume that (a) the amplitude of oscillation is small and (b) air friction is negligible. These assumptions are known to scientists as the *limiting parameters* of the validity of any relationship discovered linking the *operative parameters* we select as being significant amongst all possible operative parameters governing the behavior of the simple pendulum.

To set our pendulum swinging, let us imagine that initially we held the bob with the string slack, as shown in Figure 6.3(b) (ii). When we let go of the bob, it initially falls under the influence of gravity. It can be shown that all falling bodies, irrespective of their size, accelerate at the same rate under the influence of gravity. This *acceleration due to gravity* is 9.81 meters per second squared (written symbolically ms^{-2}). It is universally represented by the symbol g. Throughout its swing the pendulum bob is trying to fall downwards because of the pull of gravity, and is always being yanked sideways by the thread. The net effect of the sideways pull of the suspending thread and the influence of gravity is the swinging oscillation of the pendulum. Therefore, it seems reasonable to postulate that the acceleration due to gravity should be added to the two behavior-important parameters, length and mass, set out earlier in this section. The swinging back and forth of the pendulum is characterized by its *period*, which is defined by the time taken to complete one oscillation. This is often represented by the symbol τ. To begin the search for a relationship describing the behavior of a pendulum in terms of the selected parameters, we can write the relationship

$$\tau = k l^x g^y m^z \qquad \text{Equation 6.1}$$

In this relationship x, y and z are unknown exponents, the values of which we are going to try and discover by dimensional analysis; k is an unknown constant. In dimensional form this relationship becomes

$$[T] = [L]^x \frac{[L]^y}{[T]^{2y}} [M]^z$$

that is

$$[T] = [L]^{x+y} [T]^{-2y} [M]^z$$

Since the left-hand side of this equation does not contain $[M]$, it follows that $z = 0$. Equating the exponents for $[T]$ for both sides of the relationship

$$1 = -2y$$

Therefore $y = -\frac{1}{2}$

Again, equating the exponents of the dimensional terms for L we have

$$x + y = 0$$

Therefore $x = \frac{1}{2}$

Writing these values for x, y and z in our relationship, remembering that an exponent of $\frac{1}{2}$ means "the square root of", we have

$$\tau = k\sqrt{\frac{l}{g}} \qquad \text{Equation 6.2}$$

A study of the dynamics of a pendulum using the laws of motion and calculus manipulation of the variable generates the formula

$$\tau = 2\pi\sqrt{\frac{l}{g}} \qquad \text{Equation 6.3}$$

If we compare the two equations 6.2 and 6.3, we see that rubber number logic arrived at the basic expression for the period of the pendulum, except for the value of the constant. The reader should note that rubber number logic arrived at the direct relationship without any mathematical analysis and without any use of calculus.

The reader can see what many people find a surprising fact, that the mass of the pendulum bob does not appear in the formula for the period of the pendulum. Individuals without a background in physics will often suggest that a more massive bob on the end of a piece of string will oscillate more slowly than a small bob on the same piece of string. Other people are surprised to find our friend π appearing in the exact analytical version of the same equation. However, a little thought will suggest to the reader it is not unreasonable to find π in the exact equation, since the swinging pendulum traces out an arc of a circle. If the pendulum is given more and more energy it will eventually move around the perimeter of such a circle in a continuous motion.

I remember the day when my physics teacher presented the derivation of Equation 6.2 using dimensional analysis. I was fascinated how such a relatively complex relationship could be arrived at simply by considering the needs for dimensional homogeneity in the structure of a relationship. Since that day I have been fascinated by rubber number logic. Perhaps one day I will be able to write a more extensive essay on the subject. It is worth quoting some interesting comments on the power of rubber number logic made by *Lord Rayleigh*. Lord Rayleigh was an English physicist who lived from 1842–1919. He pioneered the study of black-body radiation, which we discussed earlier in connection with the violet catastrophe and the birth of quantum physics. Lord Rayleigh made the following comments on dimensional analysis:

> *I have often been impressed by the scanty attention paid by even original workers in physics to the great principle of similitude [Dr. Rayleigh's term for dimensional homogeneity]. It happens not infrequently that the results in the form of laws are put forward as novelty on the basis of elaborate experiments which might have been predicted a priore [meaning from first principles and logical considerations] after a few minutes of consideration* [3].

Hopefully, this brief introduction to dimensional analysis as a prelude to the use of dimensions as descriptive parameters in fractal geometry will whet the appetites of the readers to discover for themselves the surprising and fascinating patterns that can be derived by manipulating the rubber numbers described in the dimensional structure of various relationships. Huntley, who has written a book on dimensional analysis, made the following statements:

> *It is strange that the method of dimensions should receive only cursory treatment or none at all in physics teaching, for it is a useful tool and seldom fails to arouse the interest of the student* [3].

Two formulae, known to scientists as Stokes' law for a falling sphere and Poiseuille's flow through a pipe, are two relatively complex formulae put forward after extensive experimental investigations and theoretical studies, when the basic structure of the formulae could have been predicted with one page of dimensional analysis of the type put forward for our study of the simple pendulum. The rubber number logic route to both of these formulae are explored in the next section [6].

Section 6.3. Stokes' Law Versus Galileo

The first day of employment as a graduate scientist was a day of surprises and regrets. Since I was going to be working on the fabrication of parts for the atomic bomb, the first task I was given when I started work at the British Atomic Weapons Research Establishment (in 1955) was that of characterizing the properties of beryllium powder to be used in making parts for the bomb and for nuclear reactors. *Beryllium powder* is nasty stuff. If you inhale the powder it causes an industrial lung disease known as *Berylliosis*. Therefore, when working with beryllium powder in the analytical laboratory it was preferable to keep it suspended in an *inert* liquid (inert comes from a Latin word idte meaning "idle and lazy"; an inert liquid does not react with things placed in it). As a consequence, the preferred method for sizing the grains of the powder was to study the rate at which grains of the powder settled in a viscous fluid. While showing me around the laboratory on that first day of employment, my mentor put it thus:

> *You measure the falling speeds of the grains under low Reynolds number flow conditions and derive the size of the grains from Stokes' equation.*

As he delivered this powerful phrase containing half-remembered technical names, he looked at me anticipating full comprehension of the physics of the situation. After all, he was a chemist and I was supposed to be a physicist. I tried to look intelligent and nodded my head, all the time regretting I had not paid more attention to what is probably one of the most boring laboratory studies inflicted on undergraduate physicists – the measurement of the viscosity of a fluid by studying the movement of a metal sphere down through a column of fluid. I vaguely remembered using Stokes' equation to interpret the data from this experiment. If I had known that I was to spend most of my professional life studying applications of Stokes' equation, you can be sure I would have paid more attention in class when my professor first lectured on the topic. On that first day of employment, as soon as my tour of the laboratory was over, I scurried to the library to begin an intense study of Stokes' equation and Reynolds number. We will explore the history of the Reynolds number of flowing liquids and its physical significance in some detail in Section 6.5, but for the moment it is sufficient to recognize that it is a number which can be used to specify the fluid flow conditions around an object [7]. In *Figure 6.4* the different types of flow around a sphere which we can encounter in the real world is summarized. It should be noted that in Figure 6.4 we have introduced a new mathematical symbol $\approx$ which is used by mathematicians to mean "approximately equal to" or "about the magnitude of". When we use the symbol in Figure 6.4, each of the different systems are what we would observe when the Reynolds number is round about, or approximately equal to, the values stated. The word *approximate* is one that is used very loosely in everyday speech, and it is useful to recall the scientific definition of the word. It means

> *A result in mathematics which is not rigorously exact but so near the truth as to be sufficient for a given purpose.*

It comes from two Latin root words: ad, meaning "to" and proximus, meaning "nearest" or "near". Some scientists write the symbol for approximate in the form $\cong$ (note that the symbol $\equiv$ means not only equal in one particular property but "totally equivalent to").

When the flow of a liquid is slow (at low Reynolds number) the type of motion is described as *laminar flow*. This word comes from the Latin word lamina meaning "a thin plate". *Stokes Law* is an equation which can be used to predict the behavior of a sphere falling in a liquid under low Reynolds number, that is, laminar flow

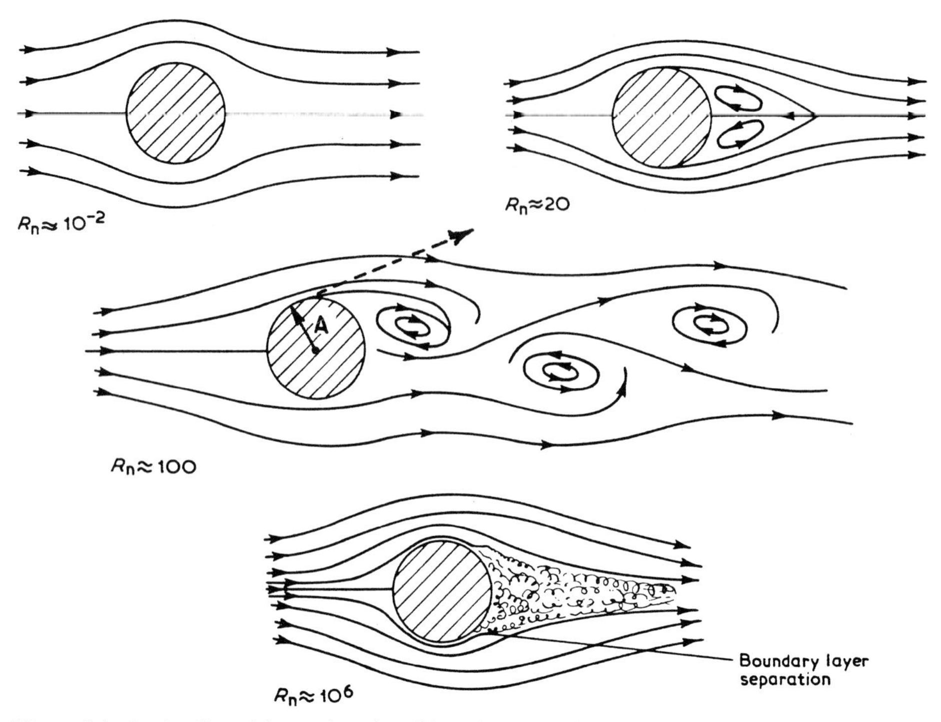

Figure 6.4. As the Reynolds number describing the type of flow past a sphere increases the flow conditions change from laminar to turbulent (R_n = Reynolds number).

conditions. In this section we will show how the basic structure of Stokes Law, or Stokes Equation as it is often described, can be deduced using rubber number logic.

When a fluid is moving in laminar flow it is as if the liquid was composed of many thin sheets which glide over each other like a pack of cards being pushed on the top. This type of motion is illustrated in *Figure 6.5*(a). In such a situation the force opposing motion when the liquid is pushed with a force such as F, on the top lamina of the stack of thin plates is the friction between the layers of the liquid. If the internal friction is low the planes of the fluid slip over each other easily. If the internal friction is high the liquid appears to be sticky and resists motion like *molasses*.

Note that in Europe molasses is often known as "black treacle". *Treacle* is defined in a dictionary as "a dark viscous liquid that drains from sugar juice at various stages in the process of manufacture". The word comes from the Greek word therion meaning "a wild beast". Apparently the Greeks used to put this type of liquid into wounds caused by wild beasts. The word molasses comes from the Latin word for honey – mellis. As the definition of the word treacle given above indicates the scientific term for a sticky fluid is viscous. This word has an interesting history. It comes from the Latin word viscum meaning "*mistletoe*". The mistletoe is a parasitic plant that grows on oak trees. It is used at Christmas as a decoration. Apparently the

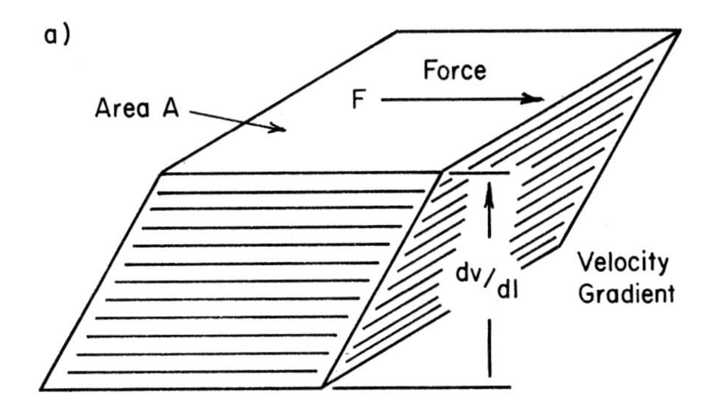

b) Coefficient of Viscosity $(\eta) = \dfrac{\text{Shearing Stress}}{\text{Velocity Gradient}} \left(\dfrac{F/A}{dv/dl}\right)$

$$\begin{aligned}
\text{Force} &= \text{mass} \times \text{acceleration} \\
[F] &= [M]\,[L]\,[T]^{-2} \\
[A] &= [L]^2 \\
\left[\frac{dv}{dl}\right] &= \frac{[L]}{[T]\,[L]} = \frac{1}{[T]} = [T]^{-1} \\
[\eta] &= \frac{[M]\,[L]\,[T]^{-2}}{[L]^2} \div [T]^{-1}
\end{aligned}$$

$$\boxed{[\eta] = [M]\,[L]^{-1}\,[T]^{-1}}$$

Figure 6.5. Frictional forces in a slowly moving fluid are quantified using the concept of the coefficient of viscosity. The dimensions of the coefficient of viscosity are shown above. a) A volume of fluid viewed as an idealized "deck of cards". b) Dimensional analysis of the coefficient of viscosity of a fluid.

Romans used to capture birds by squeezing the berries of the mistletoe to obtain a sticky fluid. They spread this viscous fluid on the ground to attract birds. The birds became trapped as they attempted to walk over the sticky mess. A *viscous fluid* is one that has high internal friction. If in some situations one can ignore the viscosity of the fluid one describes it as being an *inviscid fluid*. The decision to treat a fluid as being viscid or inviscid is an operational decision. As we walk through the air the friction of the air on our bodies can be ignored and we can treat air as an inviscid fluid. However, when attempting to ski at high speeds the athlete must learn to adopt a posture which minimizes the effect of air friction and the athlete cannot consider air to be inviscid. Again when considering the fall of a dandelion seed to the ground, the air must be considered to be viscous.

Scientists characterize the viscosity of fluid by measuring the movement of the fluid when a force is applied to the layers of fluid. Historically scientists have used the *coefficient of viscosity* to describe the behavior of the fluid undergoing change from an applied force. Scientists use the word *stress* to describe force per unit area. If we consider our liquid to be like a pack of cards, and if we stress them by applying a force to the top card, then as the cards slide over each other we say that we are applying a *shearing stress* to the pack of cards. The change produced in any system when a stress is applied to it is known as the *strain* (notice again that words such as stress, shearing and strain are words which are used loosely in everyday speech but which have very specific meanings in science). The shearing stress is defined scientifically as the applied force divided by the area of application. If we were to push the pack of cards the force we would apply would be at a specific location but the next card in the pack would feel the force spread over the area of the top card. As the pack of cards is sheared a velocity gradient is set up in the pack of cards with the top card moving the fastest and the bottom card remaining motionless on the support surface. Again scientists have a special notation for writing a velocity gradient. They use a small "d" to mean "a small change" in any measured quantity. When we write "dv" this means "a small change in a velocity v" with "dx a small change in distance x". The symbolic description of the coefficient of viscosity in terms of the shearing stress and velocity gradient is shown in Figure 6.5(b) (the concept of this type of notation is discussed in more detail in Chapter 14). A study of the various types of liquids that one can meet in the real world becomes very complex. For our present purposes we will confine our discussion to a study of simple liquids such as water, kerosene and glycerine etc. Newton showed that for this type of fluid a graph of the shearing stress against the velocity gradient was a straight line graph the slope of which became known as the coefficient of viscosity. Such simple fluids are known as *Newtonian fluids*. In a later chapter of this book when we start to look at applications of fractal geometry we will find that we can use the fractal dimension of fragments of rock to describe the non-Newtonian behavior of slurry. A *slurry* is defined as a suspension of fineparticles in a fluid. The term is restricted to mud like mixtures. When they have less liquid so they do not move easily they are known as *pastes*. The viscosity of slurries is becoming a very important topic because of the fact that technologists are suggesting that slurries of coal in oil could be a good way to transport crushed coal from British Columbia to Ontario and then these slurries can be used directly to power oil burning electricity generating stations.

Before we can write down the dimensions of the coefficient of viscosity we need to develop the dimensional description of force. Force in the physicists terminology is deduced from the first of Newton's laws of motion which states that

"Force = Mass × Acceleration"

That means if we know the mass of a body and we observe its acceleration when a force is applied to that body then the magnitude of the mass can be calculated from this relationship in dimensional terms as follows

$$[F] = [M]\,([L]/[T]^2) = [M]\,[L]\,[T]^{-2}$$

where F represents the force.

We noted earlier that many students confuse the mass and the weight of a body. The mass of a body has the dimension [M] whereas the weight has the dimension of force. The weight of a body is the force exerted on it by gravity thus

$$[W] = [M]\,[L]\,[T]^{-2}$$

Where W = weight.

When we come to consider the motion of a sphere through a liquid at low Reynolds number we know that the sphere reaches a steady velocity when the viscous forces operating on the sphere match the buoyant weight of the sphere. If we weigh an object submerged in a fluid it appears to weigh less than when weighed in air because of the upthrust of the liquid displaced by the object. This apparent loss of weight of an object supported by a fluid was first studied by the Greek mathematician-engineer *Archimedes*, who was born and died in Syracuse, Sicily (287 BC–212 BC). *Archimedes principle* states that

"The loss of weight caused by buoyant force of a liquid on a body floating on it or submerged in it is equal to the weight of the liquid displaced by the body".

To be able to calculate the buoyant weight of the sphere we must make use of the concept of the *density* of a material. The density of a material is defined as the mass of one cubic centimeter of the material. In these notes we will use the Greek letter ϱ to signify the density of a body. If we have a sphere made of material ϱ_s moving through a liquid of density ϱ_L then the buoyant weight of the sphere is given by the relationship shown in *Figure 6.6*. As the sphere moves down through the liquid the resistance to the motion of the sphere exerted by the liquid on the sphere is known as the *viscous drag*. To use rubber number logic to find the force resisting the motion of the sphere we assume that the resisting force in the fluid will depend upon

1. η: the viscosity of the fluid
2. v: the velocity of the fluid relative to the body
3. d: the size of the body

Based on these assumptions we can write symbolically that the viscous drag experienced by the sphere is

$$F = \mathrm{k}\, v^x \eta^y d^z$$

Where F = viscous drag at low Reynolds number. On the left hand side we have dimensionally

$$[F] = [M]\,[L]\,[T]^{-2}$$

On the right hand side

$$\mathrm{k}([L]\,[T]^{-1})^x \cdot ([M]\,[L]^{-1}\,[T]^{-1})^y \cdot ([L])^z$$

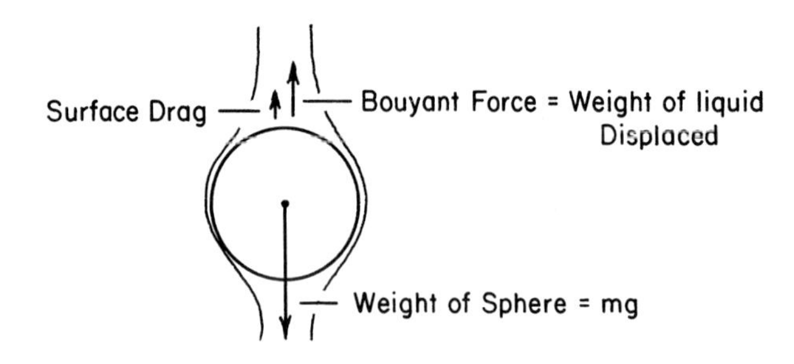

$$(\frac{1}{6}\pi\, d^3\, \rho_s)\, g \quad = \quad \text{weight of sphere}$$

where d is the diameter of the sphere, ρ_s is the density of the sphere and g is the acceleration due to gravity.

$$\frac{1}{6}\pi\, d^3\, (\rho_s - \rho_l)\, g \quad = \quad \text{Bouyant force}$$

where ρ_l is the density of the fluid.

Frictional Force $= 3\,\pi\, v\, \eta\, d$ [expression derived by Stokes]
where v is the velocity of the sphere relative to the fluid and η is the viscosity of the fluid.

$$\therefore \quad 3\,\pi\, v\, \eta\, d = \frac{1}{6}\pi\, d^3\, (\rho_s - \rho_l)\, g$$

$$\therefore \quad \boxed{\eta = \frac{d^2\,(\rho_s - \rho_l)\, g}{18\, v}} \quad \text{Stokes Equation}$$

or since $v = \frac{h}{t}$ where h is the height through which the sphere has fallen in time t :

$$\boxed{d = \sqrt{\frac{18\,\eta\, h}{(\rho_s - \rho_l)\, g\, t}}}$$

Figure 6.6. Stokes' equation for the fall of a sphere moving slowly through a viscous fluid can be derived by equating the frictional forces resisting the fall of the sphere to the buoyant weight of the sphere.

Equating the dimensional exponents for mass, length and time we have

Mass: $1 = y$
Length: $1 = x - y + z$
Time: $-2 = -x - y$

$$\therefore x = y = z = 1 \quad \text{and} \quad F = \mathrm{k}v\eta d$$

The problem of deriving an expression for the viscous drag experienced by a slowly falling body was first studied in detail by the British scientist Sir George *Stokes* (1819–1903). Using complex mathematical reasoning he was able to deduce that the viscous drag experienced by a slowly moving sphere was given by the expression $F = 3\pi v \eta d$. We see that we have been able to derive the same basic expression except for the constant, using half a page of rubber number logic. Obviously this does not detract from Stokes' exact solution. It only hints that, as Lord Rayleigh pointed out, we could have anticipated the basic structure of the equation long before we were able to carry out the exact mathematical analysis. The term 3π could then have been evaluated experimentally.

The relationship between the coefficient of viscosity and the diameter of the sphere known as Stokes' equation is shown in Figure 6.6. The reader will notice that in Figure 6.6 we have introduced another mathematical symbol $\therefore$. The three dots forming a triangle is used universally by mathematicians to represent the word "therefore". I have been unable to track down the origin of this symbol and it does not appear to have any particular logic in its structure but it is useful shorthand.

To complete the task I was given on my first day of employment I was expected to use the rearranged Stokes' formula to calculate the size "d" of a fine particle which fell through a height h in a time t as set out in the second form of Stokes' equation in Figure 6.6.

The title of this section can now be explained. The folklore of science relates that Galileo dropped two objects, a large cannonball and a small cannonball, from the top of the leaning tower of Pisa. We are told that he observed that the two of them fell with the same speed and from this fact he was able to deduce that the acceleration due to gravity was a force independent of the mass of the body. In fact his observations were only true if you could treat the air through which the objects were falling as being inviscid. If he had dropped the equivalent of a ping-pong ball at the same time as he had dropped his cannonball he would have found that the ping-pong ball would have fallen much more slowly than the cannonball. If ping-pong balls had been available to him he may not have been able to discover his universal law of gravitation.

The opposite to laminar flow in a fluid is *turbulent flow*. The word *turbid* meaning "muddy" or "full of sediment" and the word turbulent are related. They both come from a Latin word turbidus, meaning "disturbed" or "disordered". If a stream becomes turbulent the flow stirs up the mud from the bottom and the stream water becomes turbid. The transition from laminar flow to turbulent flow as the flow rate is increased are sketched in the systems of Figure 6.4. We see that the first breakdown from laminar flow is the formation of a small *eddy* behind the sphere. The word eddy is defined in a dictionary as a current of water or air running back contrary to the main stream causing a circular motion such as a small whirlpool or whirl wind.

At a Reynolds number of about 100, eddies start to leave the sphere in a constant stream. This stream of eddies is known as a *Karman street* [7, 8]. The name comes from of one of the pioneers of the study of turbulence – Von Karman. Theodor Von Karman (1881 – 1963) was a Hungarian born physicist and engineer. In 1930 he moved to the United States where he became director of the Guggenheim aeronautics laboratory at the California Institute of Technology. The fluttering of a flag in a stiff breeze is caused by the flag wrapping itself around the eddies of a Karman street of eddies shed by the wind blowing past the flag pole. A complete study of turbulence is a very difficult area of physics. For our purposes in this discussion we will note that the basic difference between laminar flow and turbulent flow is that energy is lost in the swirling of the eddies as turbulence sets in. This means that the sphere has to do more work to move through the fluid under turbulent flow conditions as compared to the situation in laminar flow conditions because it must not only overcome the viscous drag of the fluid but must give energy to the fluid which circulates in the turbulent eddies.

Essentially, as the sphere attempts to move through the fluid at high velocity, the liquid in front of the sphere is moved away from its original position so quickly that by the time it reaches the outer diameter of the sphere, denoted by A in the third element of Figure 6.4, it has kinetic energy and attempts to keep moving in the direction of the arrow shown. However, it is pushed back behind the sphere by the surrounding fluid complete with its energy. Frustrated in its attempt to move away, it must swirl around behind the sphere until it looses its energy. The onset of turbulence in this system of fluid moving past a sphere, causes a jump in the

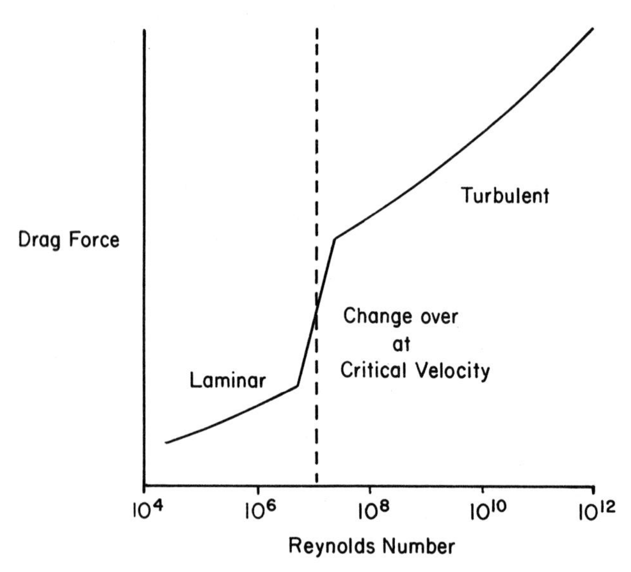

Figure 6.7. At a velocity of fluid motion, known as the critical velocity, flow conditions change from laminar to turbulent conditions.

resistance to motion when the flow conditions change from laminar to turbulent. This is shown by the data of *Figure 6.7*. The Reynolds number of a fluid system is given by the relationship

$$R_n = (v\,d\,\varrho_1/\eta)$$

The velocity at which the change over from laminar to turbulent conditions takes place is known as the *critical velocity* [9]. The magnitude of the critical velocity depends on the surface of the object. For example it takes place at a higher Reynolds number for a golf ball than for a smooth ball of the same size (for a discussion of this surprising fact see Reference 10).

It has been shown that one can assume that Stokes law is applicable to a falling sphere if the Reynolds number is less than 1.0 [8]. In the next section we will use rubber number logic to study the flow of fluid through a cylindrical pipe.

Section 6.4. Rubber Number Logic for Studying the Flow of Viscous Fluid Through Pipes

The first historic record of the various parameters that effect the flow of fluids through a pipe was put forward in 1842 by J. L. M. *Poiseuille*. The relationship that he discovered experimentally is shown in *Figure 6.8*. Poiseuille was interested in the flow of blood in the body. His formula shows how the heart has to work harder when the pipes of the body narrow down through illness. The fact that the radius of the pipe occurs in the equation to the 4th power is often a surprise to students. This 4th power dependence is a very important aspect of the flow of air through filters and other systems. The basic structure of *Poiseuille's formula* could have been devised in half a page of rubber number logic manipulations long before experimental studies were undertaken. This fact endorses Lord Rayleigh's claim quoted earlier that formulae put forward after extensive experimental studies could have been predicted from the beginning using the dimensional homogeneity of the universe.

To establish Poiseuille's formula by rubber number logic we select the following parameters as being relevant to a study of the flow of viscous liquids through a pipe.

1. P_G: the pressure gradient over the length of the pipe
2. η: the coefficient of viscosity of the fluid
3. r: the radius of the pipe

Using these parameters we can write the general relationship

$$[Q] = \mathrm{k}([\eta]^x\,[r]^y\,[P_G]^z)$$

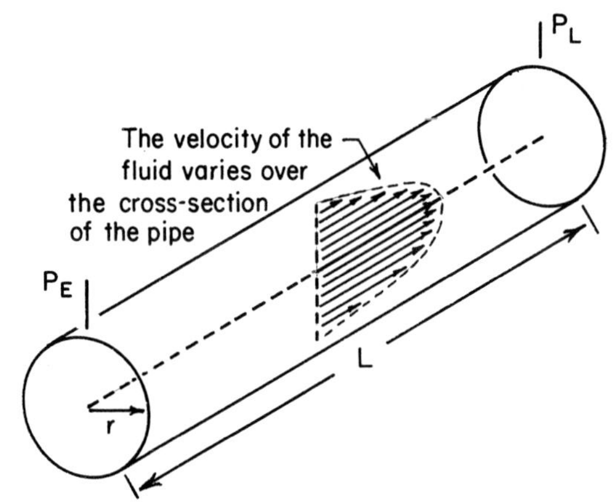

v_a = Average velocity of the fluid = Q/A

v_r = the Velocity of Flow at the centre of the pipe = $2v_a$

where : Q = Volume flowing through the pipe per second
A = Cross-sectional Area of the pipe = πr^2 (r = radius)

The pressure drop due to the work done to drive the fluid through the pipe is :

$$P_G = \frac{P_E - P_L}{L}$$

where : P_E = the Pressure at the Entry of the pipe
P_L = the Pressure at the end of a pipe of length L

Poiseuille's Formula :

$$Q = \frac{\pi (P_E - P_L) r^4}{8 L \eta}$$

where : η = the Coefficient of Viscosity of the fluid

Figure 6.8. The flow of fluids through a pipe was first studied by the French scientist Poiseuille. The formula given above is named after him.

where Q is the amount of liquid flowing per second and k a constant. Transforming these parameters into equivalent forms we can write

$$[Q] = [v_a]\,[A]$$

where v_a = the average velocity and A the cross-sectional area of the pipe

$$A = \pi r^2$$

Therefore, dimensionally, the left side of the equation is:

$$[Q] = ([L]/[T])\,[L]^2 = [L]^3\,[T]^{-1}$$

The right side of the equation contains:

$$[\eta] = [M]\,[L]^{-1}\,[T]^{-1}$$
$$[r] = [L]$$

and

$$[P_G] = [\text{FORCE}]/([\text{AREA}]\,[\text{LENGTH}]) = [M]\,[L]\,[T]^{-2}/[L]^2\,[L])$$

so

$$\therefore [P_G] = [M]\,[T]^{-2}\,[L]^{-2}$$

Thus the right side of the equation becomes:

$$\text{k}\,([M]\,[L]^{-1}\,[T]^{-1})^x\,([L])^y\,([M]\,[T]^{-2}\,[L]^{-2})^z$$

If we then equate the left and right sides we find that:

From the dimensions of M $\quad 0 = x + z \therefore x = -z$
From the dimensions of L $\quad 3 = -x + y - 2z \therefore 3 = x + y$
From the dimensions of T $\quad -1 = -x - 2z \therefore 1 = x + 2z$

$$\therefore z = 1$$
$$\therefore x = -1 \quad \text{and} \quad y = 4$$
$$\therefore Q = \text{k}\,\eta^{-1}\,r^4\,P_G^1$$
$$Q = \text{k}\,(r^4\,P_G)/\eta$$

Exact analysis shows that $\text{k} = \pi/8$. We also know that the pressure gradient is given by

$$P_G = (P_1 - P_2)/L$$

where L is the length of the pipe, P_1 is the pressure at the beginning of the pipe, and P_2 is the pressure at the end of the pipe. The equation then becomes

$$Q = (\pi\,(P_1 - P_2)\,r^4)/8\,L\,\eta$$

To honor the pioneer work of Poiseuille the scientific unit of viscosity is defined as the *Poise*. It should be noted that in the above rubber number manipulations we are working with the average velocity of flow. In reality the velocities range in the fluid form from zero at the wall to the highest velocity in the middle. The non-moving fluid at the wall of the pipe is known as the *boundary layer*. It can be shown that the maximum velocity at the center of the pipe is twice the average velocity [11].

Section 6.5. Dimensionless Numbers as Indicators of Similar Structure and Behavior

Earlier in our journey of discovery we met π the crazy, irrational number. In our original discussion of the fascinating features of π we did not draw attention to the fact that π is a dimensionless indicator of structure. From one point of view π can be described as a rubber number fitting and describing the structure of all circles. In fineparticle science, (which is concerned with the properties of fragmented material such as fibers, powders, mists, sprays etc.), it is often necessary to describe the shape of objects. This is a difficult task in three-dimensional space and often powder specialists have to be satisfied with a two-dimensional description of shape. In *Figure 6.9* some of the dimensionless numbers which have been used to characterize the shape of fine particles are summarized. These dimensionless numbers are called *shape factors* or shape *indices* [12]. The word *index* comes from the Latin word indicare meaning "to show". The index of a book shows you where to find discussions of various topics. A *shape index* indicates to the scientist the basic structure of a profile. The process for creating shape indices is essentially a process of mathematical normalization, a process which was discussed earlier in this book. The first step in creating a shape index for a non-circular profile is to specify what is meant by the diameter of a profile.

The word *diameter* comes from two Greek root words dia meaning "through" and metron "a measure". The diameter of a circle originally meant the through measure of the magnitude of the circle defined by a line moving through the center of the circle. When one specifies the structure of an ellipse then one defines the major and minor diameters of the profile. The major and minor diameters for an ellipse are shown for the set of profiles of Figure 6.9. When studying irregular profiles some scientists have constructed an ellipse of equal area and then specified the structure of the irregular profile as the ratio of the minor and major axes of the equivalent area ellipse. In practice, this is not a useful shape factor because of the difficulty of measuring the area of the profile and the subsequent calculation of the dimensions of the ellipse of equal area (note the word ellipse comes from a Greek word elleipsis meaning "defective", "not perfect". Remember the Greeks admired perfect shapes and to Greek mathematicians an ellipse was a deformed circle).

When dealing with a highly irregular profile such as the diesel exhaust fine particles shown in *Figure 6.10*, scientists often construct an enveloping profile curve around the profile known as the *convex hull* of the profile. The word *envelope* means "to cover by wrapping"; my dictionary states that the origin of the word is obscure. *Convex* means "bulging outwards" and is the opposite of *concave* meaning "inward curving". Features of the soot fine particles which could be regarded as convex and concave are indicated in Figure 6.10. The word *hull* means "an outer covering". It probably comes from an old English word helan meaning "to cover". The convex

Ellipse Number	Semi-Major a (cm)	Semi-Minor b (cm)	Perimeter P (cm)	Diameter D (cm)
1	1.00	1.00	6.29	2.00
2	1.25	1.00	7.09	2.50
3	1.50	0.50	6.68	3.00
4	2.00	0.10	8.03	4.00
5	5.00	0.00	10.00	5.00

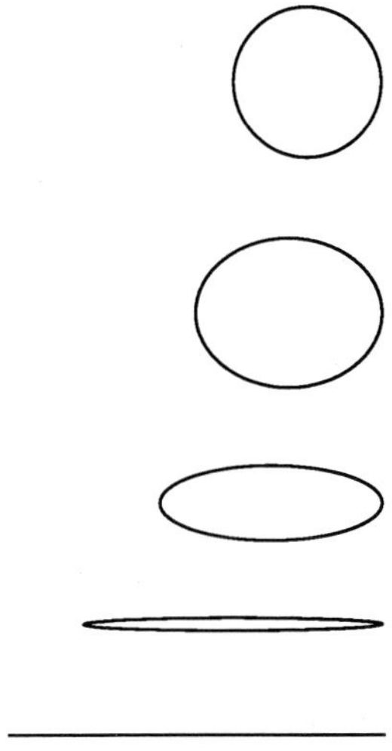

	π' (P/D)	AR (L/B)	Ch (B/L)
1	3.1416	1	1
2	2.8362	1.25	0.80
3	2.2275	(Fiber?) 3.00	0.33
4	2.0086	20.00	0.05
5	2.000	∞	0

L : the maximum Length of the profile, for an ellipse = 2a
B : the Breadth of the profile, for an ellipse = 2b
π' : a dimensionless constant similar to π for a circle
AR : the Aspect Ratio of the profile
Ch : the Chunkiness of the profile

Figure 6.9. Dimensionless ratios are widely used to characterize the structure of fine particles in two-dimensional space.

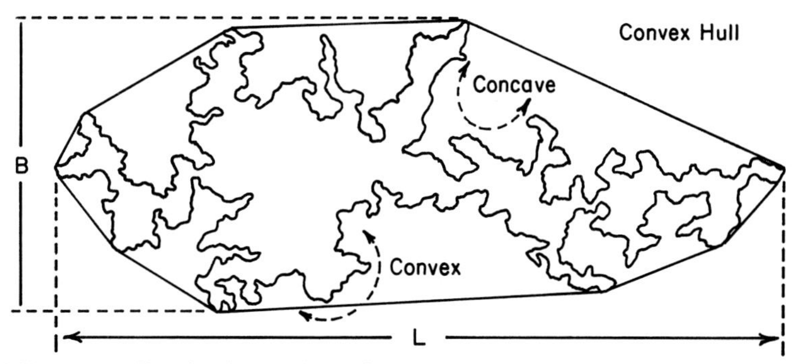

Figure 6.10. The profile of a diesel exhaust fine particle showing concave and convex areas and the convex hull enveloping the profile.

hull is an enveloping curve that forms an outer covering with no concavities. In two-dimensional space the easiest way to think about the convex hull is that it is the shape that would be assumed by an elastic rubber band placed around a model of the profile. The structure of the convex hull of the diesel soot profile is shown in Figure 6.10. For such a profile the shape of the convex hull is used to generate a shape index using the length and breadth of the profile as indicated in the diagram.

One way to specify the structure of the profile for smooth profiles is to modify the concept used to generate π for a circle and quote the perimeter divided by the maximum diameter of the profile. This modified shape index is denoted by the symbol π' in Figure 6.9. It can be seen that this shape factor varies from 3.14 to 2 for a very long fiber. This modified π' shape factor has not been widely used in fine particle science for two reasons. First of all, it is usually quite difficult to measure the perimeter of the profile and secondly it can be shown that for very rugged profiles such as the diesel soot exhaust of Figure 6.10 the real perimeter of the profile approaches infinity and the modified π' shaped factor becomes indeterminate and not a good indication of structure.

A widely used dimensionless shape factor, the *aspect ratio*, is based upon the external dimensions of the profile. The word aspect comes from two Latin root words ad meaning "towards" and specere meaning "to look". When we describe an aspect ratio we are describing what something looks like by quoting the ratio of the dimensions of the profile. Fine particle scientists studying such subjects as air pollution (an important aspect of air pollution is the movement of dirt fine particles in the atmosphere) and industrial lung diseases caused by inhaling dangerous substances such as asbestos dust, need to define what they mean by a *fiber*. From many experimental studies scientists have determined that a working definition of a fiber is a profile that is three times longer than it is wide. The weakness of this definition can be appreciated from the fact that according to this definition the soot profile of Figure 6.10 is a fiber. This obviously does not match our mental picture of a fiber. However, the working definition given here is useful when studying such things as asbestos fibers and other needle-shaped fine particles encountered in industrial hygiene situations. Hygeia was the Greek goddess of health. Therefore, *hygiene* is the name for techniques suitable for preserving and promoting health. Industrial hygiene is the technology for promoting and maintaining healthy working conditions in industry. The subject is also called occupational hygiene.

Another dimensionless number used to describe a useful physical quantity is the specific gravity of a substance. The density of a substance is the mass of a substance contained in 1 cubic centimeter of the substance. Density has the dimensions given by the relationship

$$[\varrho] = [M]/[L]^3 = [M]\,[L]^{-3}$$

The specific gravity of a substance is the density of the substance divided by the density of water at 25 °C and atmospheric pressure. Therefore, the *specific gravity* has zero dimensions and is a pure number. Because the density of water at 25 °C and 1 atmosphere pressure is 1 gram per cubic centimeter the specific gravity and the density of a substance often have the same numerical magnitude. This causes problems when students must work with dimensional analysis. Another dimensionless number used to describe the behavior of physical systems is the Mach Number.

The *Mach number* is defined as the speed of an object in a fluid divided by the speed of sound in that fluid. Sound waves are compressions and rarefactions moving through a fluid. If an object is moving faster than the speed of sound it is moving at such a speed that the noise it makes cannot travel in front of it. If it travels actually at the speed of sound, vibrational energy accumulates around the object and causes serious problems [7]. An aircraft cannot travel safely at the speed of sound. It must move rapidly through the speed of sound; a phenomenon known as breaking the *sound barrier*. The word "phenomenon" comes from a Greek word phainein meaning "to see". Originally a *phenomenon* was anything that could be seen, heard or felt. In modern science it means anything which can be observed. In everyday speech the word *phenomenal* is used to mean something unusual or exceptional. In science however, not all phenomena are phenomenal. A scientist watching a supersonic flight could describe it in terms of its Mach number. If a statement was made such that a missile was travelling at mach 1.2 it would be understood by other scientists that the object was travelling at 1.2 times the speed of sound. The Mach number is named after the Austrian physicist *Ernst Mach*. Mach made many observations of the movement of objects in fluid and pointed out in 1887 that there is a sudden change in the nature of the airflow over a moving object as it reaches the speed of sound. Therefore, when looking at the behavior of a moving object it is more important to know how its speed compares to the velocity of sound in a fluid rather than to know its absolute velocity. The Mach number is a dimensionless number the magnitude of which indicates to the scientist the type of flow that will be occurring as fluid moves past an object.

The British physicist Sir Osborne Reynolds (1842–1912) made a special study of the flow of water in pipes. He discovered that the changeover from laminar to turbulent flow in the pipe could be linked to the magnitude of the dimensionless number known as the Reynolds number which is discussed briefly in the previous section of this chapter. The Reynolds number for flow through a pipe is usually written in the form

$$R_n = (v\, D\, \varrho_L)/\eta$$

where R_n is the Reynolds number, v is the velocity of the moving fluid, ϱ_L is the density of the fluid, D is the diameter of the pipe and η is the coefficient of viscosity of the fluid. The fact that this is a dimensionless ratio can be shown by writing down the dimensions of the various terms to show that they all cancel to produce no dimensions. The Reynolds number can also be written as the ratio

$$R_n = (v\, D)/(\eta/\varrho_L)$$

The ratio of viscosity to density occurs so often in a study of fluid motion that it has been given a separate name. The straightforward coefficient of viscosity is known as the *dynamic viscosity*. As we have already mentioned, it is measured in Poise. The dynamic viscosity coefficient divided by the density of the substance is known as the *kinematic viscosity*. Its units of measurement of kinematic viscosity are stokes named after Dr. Stokes. The symbol often used for the kinematic viscosity is

$$\sigma = \eta / \varrho_1$$

Using this notation the Reynolds number becomes

$$R_n = v\,D / \sigma$$

Reynolds showed that the flow in a pipe would change over from laminar to turbulent when the Reynolds number was approximately 1,000. The reason that one says approximately is that the exact value at which the flow changes over to turbulent depends somewhat on the steadiness with which the fluid is being pumped or driven along the pipe and also on the surface of the pipe. For many purposes however, one can use the value 1150 as the *critical velocity* at which the pipe flow changes over to turbulent flow. The physical significance of the Reynolds number can be appreciated in a more physical sense if we realize that the Reynolds number is actually a ratio of the forces required to move the mass of the fluid through the pipe to the viscous forces experienced by the moving fluid [7]. This helps us to appreciate why the Reynolds number should be a dimensionless number since it is essentially one force divided by another.

The concept of the Reynolds number has been extended from the study of a flow of liquids through a pipe to look at the flow around a moving object. In such a situation the parameter D in the equation becomes some typical dimension of the object. The Reynolds number, by itself, is not an absolute guide to flow conditions past an object since the shape of the object also determines the type of flow conditions that are established around it. If we were to study the flow around the various shapes of Figure 6.9 we would find that the Reynolds number at which turbulent flow began varied from shape to shape but was the same for a given shape no matter how large the object. Scientists studying the flow around wings and ships use a shape factor similar to the aspect ratio to define the shape of an object moving through a fluid. The general effect of changing the shape from a sphere to something more like the outline of a fish is to increase the Reynolds number at which the changeover from laminar to turbulent flow occurs. The outline of something such as a fish is said to be *streamlined*. The fact that the changeover to turbulent flow is delayed for a streamlined body means that the resistance to the motion of a streamlined body from the fluid is less than that of a sphere of the same diameter. The appropriate Reynolds number for various objects are summarized in the tables of *Figure 6.11*. It can be seen from the table of Figure 6.11 that for fluid moving past an object in open water or in an infinite sea of fluid the changeover from laminar to turbulent flow occurs at

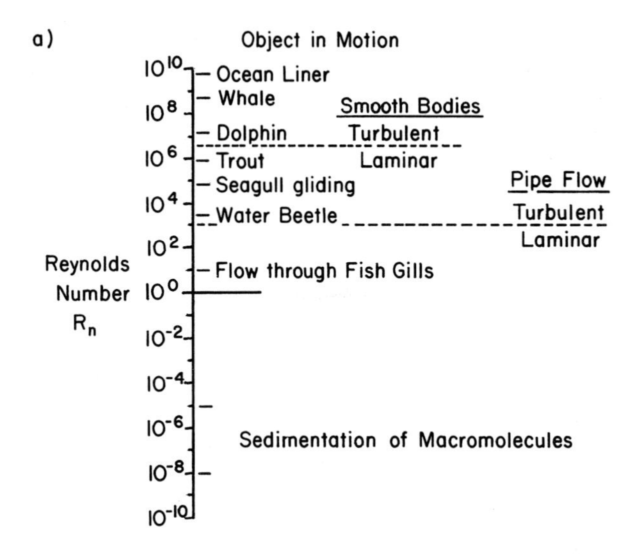

b)

Object in Motion	R_n
A Whale swimming at 10 m/s	300 000 000
A Tuna swimming at 10 m/s	30 000 000
A Duck flying at 20 m/s	300 000
A Dragonfly moving at 7 m/s	30 000
A Crayfish in a burst of 20 cm/s	300
A Small insect flying at speed	30
An Invertebrate larva moving at 1 mm/s	0.3
A Sea Urchin Sperm advancing the species, 0.2 mm/s	0.03

Figure 6.11. The shape of an object, along with the Reynolds number for the flow conditions, indicates the type of flow conditions around the object [13, 14, 15]. a) Reynolds number for different types of flow around different shaped objects as compared to the flow through pipes. b) Educated guesses at various Reynolds numbers governing the flow around large and small objects.

approximately a Reynolds number of one million. Again however, it has to be stated that the Reynolds number is only a guide to what is happening in the fluid because according to the Reynolds number for the flow around a dolphin, the dolphin should be trying to swim in turbulent flow conditions. However, the dolphin's skin which is rough and rubbery, is thought to dampen out any turbulence that starts around its skin. The dolphin is able to swim at high speeds because the structure of its skin maintains laminar flow. It is interesting to note that aeronautical engineers are studying the skins of sharks and dolphins to see if they can design a special type of skin for aircraft which would reduce the frictional energy experienced by the aircraft when flying at high speeds.

The flow of blood in the human body is usually laminar. Some of the early blood circulating machines used to pump blood around the body during heart operations inadvertently created turbulent conditions in some regions of the flow through the heart-lung machines. This turbulence damaged blood cells in the circulating blood. When the flow condition in a fluid are laminar no noise is produced by the moving fluid. However, when the conditions change to turbulent flow the gurgling eddies make noise which can be heard by listening on the outside of the pipe. In some illnesses of the heart the vessels leading to the heart are narrowed down and the flow into and out of the heart can become turbulent and may produce noises. These noises, which can be detected by a stethoscope, can help in the diagnosis of disorders of the human body circulation system [13–15].

From the discussions that have been given in this section it can be seen that as scientists have attempted to tackle complicated systems, many of which qualify for the definition deterministic chaotic systems, they have found dimensionless numbers for self-similar systems to be useful indicators of the type of behavior which can be anticipated in various systems. As our journey of discovery continues we will discover that the fractal dimension of an object can be regarded as a type of dimensionless number which is useful in attempting to describe the self-similar behavior of complex systems.

Exercises

Exercise 6.1

Students should be encouraged to use equation 6.2 to carry out a series of measurements with pendula of various lengths. A rearrangement of equation 6.2 generates the equation

$$\log \tau = 1/2 \log l - 1/2 \log g - \log 2\pi$$

If the student looks up the value of g in a book of tables then the value of π can be obtained from this equation by plotting $\log \tau$ against $\log l$. Alternatively, the student can calculate a whole series of values of τ, the period of the pendulum, from the square root g over l divided by 2 to obtain π as shown by the other arrangement of the equation

$$\pi = (1/(2\tau))\sqrt{(l/g)}$$

The teacher could also lead a discussion as to what would happen to a given pendulum moved to the surface of the moon. Since the gravitational field of the moon is less than that of the earth the pendulum will oscillate more slowly on the surface of the moon and the students can assume a new value for g written symbolically g_m and work out the ratio of the times of oscillation of the pendulum on the earth and on the surface of the moon in terms of g_m. They could look for this value in physics textbooks.

Exercise 6.2

When starting to study physics many students find it surprising that the velocity squared appears in the expression for the kinetic energy of a moving object of mass m. Many textbooks talk about the energy distribution in a swinging pendulum. If the pendulum is released from an original height h then at the start of the swing of the pendulum the *potential energy* of the bob before it starts to move is

$$E_p = mgh$$

By the time the swinging pendulum reaches the bottom of its oscillation all of this potential energy has been transformed into *kinetic energy*

$$E_k = 1/2\, mv^2$$

The students are then often told that the velocity of the moving pendulum can be calculated by equating the potential to the kinetic energy to give the relationship

$$mgh = 1/2\, mv^2$$

so

$$v = \sqrt{2gh}$$

The student should be encouraged to show that the dimensional structure of the term for kinetic energy and potential energy are the same and that $\sqrt{gh}$ has the dimension of velocity. They should be encouraged to check the dimensional homogeneity of Einstein's equation for the interchange of mass and energy:

$$E = mc^2$$

They will find that the velocity of light c has to be squared in this relationship to maintain the dimensional homogeneity of the equation. Rubber number logic (dimensional analysis) does not yield the half in the kinetic energy expression and cannot tell us that the constant we assume in $E = mc^2$ is 1. However, carrying out the exercise helps the students remember the initially surprising fact that velocity is the squared term in energy relationships.

Exercise 6.3

The students should be encouraged to collect dimensional descriptions of various variables such as force using Newton's first law of motion. This law states that

$$\text{force} = \text{mass} \times \text{acceleration}$$

The dimensions of work can be deduced from the fact that

$$\text{work} = \text{force} \times \text{distance}$$

Table E 6.3.1. A short list of the dimensional natures of several quantities.

Quantity	Symbol	Dimensions
Mass	m	$[M]$
Length	l	$[L]$
Time	t	$[T]$
Velocity	v	$[L]\,[T]^{-1}$
Acceleration	a	$[L]\,[T]^{-2}$
Force	F	$[M]\,[L]\,[T]^{-2}$
Work	W	$[M]\,[L]^2\,[T]^{-2}$
Momentum		$[M]\,[L]\,[T]^{-1}$
Energy	E	$[M]\,[L]^2\,[T]^{-2}$
Power		$[M]\,[L]^2\,[T]^{-3}$
Pressure	P	$[M]\,[L]^{-1}\,[T]^{-2}$

Some numerical constants. The students can explore the dimensionality of these constants.

Constant	Symbol	Value
Acceleration due to gravity	g	$9.81\ m/s^2$
Radius of the Earth	R_E	6370 km
Mass of the Earth	M_E	5.98×10^{24} kg
Escape velocity at surface	v_{esc}	11.2 km/sec
Distance from Earth to Moon		3.84×10^8 m
Distance from Earth to Sun		1.50×10^{11} m
Speed of light	c	2.998×10^8 m/s
Speed of sound in dry air	v_s	331 m/s
Density of air	ϱ_A	$1.29\ kg/m^3$
Density of water	ϱ_W	$1000\ kg/m^3$
Standard temperature		$0\,°C = 273.15$ K
Standard pressure		1 atm $= 1.013 \times 10^5\ N/m^2$
Avogadro's number	N_A	6.022×10^{23} particles/mole
Fundamental charge	e	1.602×10^{-19} C
Boltzmann constant	k	1.381×10^{-23} J/K
Planck's constant	h	6.626×10^{-34} Js
Coulomb constant	$1/4\pi e_o$	$8.988 \times 10^9\ Nm^2/C^2$
Gas constant	R	8.314 J/mole K
Gravitational constant	G	$6.672 \times 10^{-11}\ Nm^2/kg^2$
Electron mass	m_e	9.1095×10^{-31} kg
Proton mass	m_p	1.673×10^{-27} kg
Neutron mass	m_n	1.675×10^{-27} kg

References

[1] J. Gleick, *Chaos, Making A New Science*, Viking Penguin, New York, 1987.

[2] See discussion of the dimensional concept in B. B. Mandelbrot, *Fractals, Form Chance and Dimension*, W. Freeman, San Francisco, 1977.

[3] H. E. Huntley, *Dimensional Analysis*, Dover Publications, New York, 1958.

[4] D. C. Ipsen, *Units, Dimensions and Dimensionless Numbers*, McGraw Hill, New York, 1960.

[5] I. Stewart, *Concepts of Modern Mathematics*, Penguin Books, Harmondsworth, Middlesex, England, 1975.

[6] British textbooks of physics in use when the writer attended university used to put more emphasis on dimension analysis than comparable North American textbooks. See for example *General Physics and Sound* by Fender (full reference given in Chapter 15 list) in which both Stokes' law and Poiseiulles' law are derived by dimensional analysis.

[7] More adventurous readers who would like to discover more of the physics of fluid flow and shockwaves could look at the very readable advanced textbook J. E. Allen, *Aerodynamics, The Science Of Air In Motion*, Granada Publishing Company, St. Albans, Herts, England, Second Edition, 1982.

[8] For a non-mathematical introduction to the movement of fluids around various bodies see A. H. Shapiro, *Shape And Flow*, Doubleday, New York, 1961.

[9] A very interesting discussion of the Reynolds number associated with the type of flow that occurs around an insect and other tiny forms of life moving in a liquid is to be found in E. M. Purcell, *Life At Low Reynolds Numbers*, *Am. J. Phys. 45* (1977) 3–11.

[10] P. J. Brancazio, *Sport Science*, Simon & Schuster, New York, 1984.

[11] See discussion of Air Sampling Problems in B. H. Kaye, *Direct Characterization of Fine Particles*, Wiley, New York, 1981.

[12] See discussion of shape characterization in B. H. Kaye, *Direct Characterization of Fine Particles*, Wiley, New York, 1981.

[13] S. Vogel, *Life in Moving Fluids, The Physical Biology of Flow*, Willard Grant Press, Boston, 1981.

[14] D. C. S. White, *Biological Physics*, Chapman and Hall, London, 1974.

[15] P. Davidovits, *Physics In Biology And Medicine*, Prentice Hall, Englewood Cliffs, New Jersey, USA, 1975.

Chapter 7

Congregating Drunks, Soot and Other Pigments

Chapter 7 Congregating Drunks, Soot and Other Pigments 267

Section 7.1 Congregating Drunks Create Surprising Patterns! 269
Section 7.2 Characterizing the Structure of Fractal Agglomerates 275
Section 7.3 Fractal Fingers Generated by Electrolytic Deposition 282
Section 7.4 Creating Fractal Fingers of Moving Fluid in a Hele–Shaw Cell 287
Exercises 289
References 304

Chapter 7

Congregating Drunks, Soot and Other Pigments

Section 7.1. Congregating Drunks Create Surprising Patterns!

The heading for this section reads like a newspaper headline. It probably conjures up in the mind of the reader pictures of swirling drunks creating surprising patterns as they converge to a central point. In Chapter 3 we looked at the pattern created by dispersing drunks. Is there a complementary pattern created by drunks who, giving up the struggle to disperse, attempt to return to their friendly lamp post? A *complementary pattern* in science is one that is the opposite of another pattern which when put together with the original pattern makes something which is a whole. That is, the total assembly is something which is completed. If we had a piece of paper from which a series of holes had been cut, then the pieces of paper taken from the holes mounted on another piece of paper in the same location as the holes on the original piece of paper is a complementary pattern. The two together make a complete whole. The word *complete* was originally created by combining con, meaning "together", with plere, "to fill". A complete pattern is like a cup full to the top with all the necessary information.

To see if the returning drunks create a structured pattern we can simulate the movement of the drunks into a central area containing a lamp post using a Monte Carlo routine. The basic system to be used in such a simulation is illustrated in *Figure 7.1*(a). In this figure the two-dimensional space available to the staggering pixels, representing the drunks is shown. The address of any pixel in this two-dimensional space is given by a combination of the A and B coordinates of the squares. The address of the central lamp post in this space is A7, B7. To start our simulation of the congregating drunks we choose a point of entry at random around the two-dimensional space. To facilitate the readers' selection of an entry point when they carry out an experiment a special random number table giving addresses on each side of the square at random is provided in Figure 7.1(b). If we select our first entry point at the top left hand corner of this entry selection table, the first pixel is placed at the address A4. We now decide which way the pixel will move using the direction

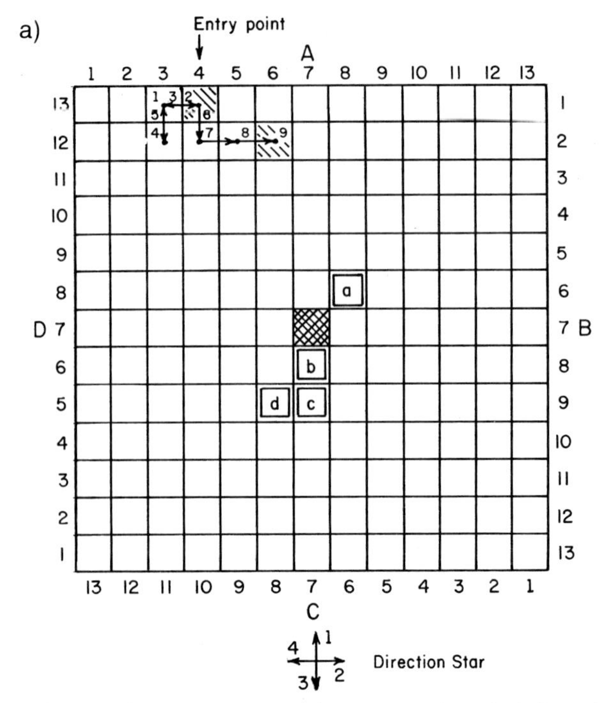

Figure 7.1. We can discover the patterns created by congregating pixels by using a Monte Carlo simulation of the process. a) Two-dimensional staggering space for simulating the staggering progress of drunks by using a Monte Carlo routine. The lamp post is located at the address A7, B7. b) Selection table structured to facilitate the study of returning pixels in the stagger space of part (a). This table may be used to find a random access point to the stagger space. c) This random number table facilitates the selection of four possible *directions of movement* for a pixel undergoing a random walk in the stagger space of (a).

star of Figure 7.1 (a) and the direction random number table of Figure 7.1 (c). If we used the top left hand corner of this table, the first step would be a sideways direction to the address A3, B1. The next two movements selected from the table of Figure 7.1 (c) would be illegal since they would take the pixel off the square, so we ignore these choices. The second legal move is a 2 which moves the pixel back to our starting point. Again the next move, 4, simply moves us to the address A3, B1. Finally with the 6th direction selection we start to move into the body of the square. We allow our pixel to stagger at random around the square until it either hits the lamp post or another pixel which is already fastened itself to the lamp post. Note in scientific terminology our lamp post for the staggering pixels is called a *nucleating center*. The word *nucleus* in Latin means "a small nut". A nucleating center is a tiny object around which another system grows. The growth of a snowflake usually begins when ice molecules attach themselves to a piece of dust which acts as a nucleating center for the snowflake [1, 2]. When considering the growth of an b)

A4	C7	C8	A8	B13	C8	C4	B7	B5	B10	C8
D8	C11	B6	D8	D1	B5	C8	D8	D8	A5	C11
C13	C7	A5	B4	D12	D8	C1	D12	C10	C12	A5
B10	B1	D13	A6	D4	B8	C10	D6	A10	D5	A10
B11	A2	C7	A5	D7	D9	A2	C9	B11	B4	C5
C13	C9	C7	B1	C12	C5	B4	D4	D2	D1	B6
B9	D4	D8	C10	B8	A13	D12	A6	D10	A5	B11
C7	C13	D9	B11	C11	C12	B5	C10	D3	C8	B9
C8	C9	B2	D12	D2	A13	A5	D2	C8	B7	A3
D9	D12	A12	B9	C1	B12	C3	D6	B9	D13	D6
A7	A6	B13	B2	C2	B12	B4	C4	D12	C12	C4
D10	D7	D12	D11	B7	A4	C10	A3	C5	C10	D5
B13	A1	D7	C9	C3	A9	C2	B9	B5	D3	A13
D4	C12	A6	C13	D4	A9	A12	D10	B8	A13	D5
C1	B1	A11	B12	D10	B12	C12	A1	D11	D6	A7
B5	C11	D5	C7	C6	C7	C12	D13	D5	B13	D3
A1	B2	B4	A11	A8	C11	D7	A7	D4	B12	D7
B13	D13	B10	A4	C4	D10	B3	D13	B3	C10	D7
D4	D3	C2	C8	B11	B10	B5	B1	C3	A4	A1
D4	A11	C9	B2	A12	D1	B2	A5	D5	A9	B1
B8	A2	C13	B2	A1	C6	D8	D2	D10	D10	B1
A13	C12	A6	A1	B12	C6	D7	C6	B4	A3	A9
D7	D6	D6	B12	B1	C3	A1	D5	B9	D1	A4
C6	D4	B11	C11	A1	D10	D6	D10	C9	B2	D6
B5	C1	C1	A4	A4	C1	D6	B3	D3	A12	D11
B7	A8	A11	A4	D11	D13	A7	B9	C6	C4	A11
A5	C1	A1	A9	D10	D6	B4	A4	C7	D8	C7
B7	C10	C3	C13	A1	D1	D13	D6	B3	B3	B4
D12	C10	C12	B5	B11	A13	A11	A13	C12	B12	D6
A8	A11	D13	A10	A9	D9	C11	C12	B10	D2	B4
B11	C6	D10	B11	D12	A3	A3	B7	A4	D4	A6
C5	A7	B12	A2	D10	A10	C7	C1	C7	A13	A10
A12	C6	C5	A1	B6	D8	A11	C10	B10	A2	C10
D4	A9	C1	B6	A9	C2	A5	A2	D9	C1	B11
C12	B2	C7	B9	B1	D5	A7	B11	D8	C13	D1
B1	A12	A1	A2	A10	D5	A5	D12	D6	D5	D8
D1	B11	A13	A10	B6	C12	D12	B9	B12	A12	C12
A5	B6	D2	C12	C3	C1	C3	A10	D3	C1	A12
D2	A8	B13	B8	D6	D11	B11	C6	B10	C4	C11
C5	D4	C6	C1	B5	C1	C9	C7	D13	C11	D5
A6	C8	D1	C13	C1	A6	B3	B1	C11	D7	D9
D13	D13	B4	A1	D1	A2	D5	C1	B1	B3	D10
A4	B8	A8	B13	D10	D7	A7	A9	A3	C6	D2
C1	A7	A6	C11	C11	A12	D13	B2	A9	D12	D12
C3	A4	B6	B10	D10	C3	A1	D2	B4	D11	C6
A9	C1	C1	C12	A7	D6	A8	B2	C8	B4	B6
C1	A13	B2	C9	B11	D3	D10	B10	B12	D3	B10
D10	C11	D1	D10	C3	A13	C11	C1	B9	C11	A2
D7	D7	C10	A6	C9	B6	C12	C3	B13	A7	A4
D12	D9	C8	A3	D12	C12	D1	B1	B12	D6	C12
D7	C7	A5	C7	A11	A5	A4	D6	B3	A1	A8
B4	D3	C12	C3	B5	A5	D4	A11	A12	A10	B4
C5	C3	D6	C6	A3	B1	A12	C4	D7	C3	C4
B8	C6	A13	A13	D8	C9	C11	A7	D7	A2	B13
A12	B11	C12	C8	C2	D8	C1	B6	C13	D9	A7

agglomerate of drunks, it should be noted that we do not consider diagonal touching of pixels to constitute an encounter resulting in a bonding of the wandering pixel to the growing agglomerate of pixels around the nucleating center. Only pixels which butt onto the nucleating center or other pixels are considered to have joined the conglomerate. In Figure 7.1 (a) we would consider the pixels b, c and d to be c)

41124 31123 22341 41312 34113 34341 32212 11141 33444 44442
32133 44214 43142 12241 32231 31213 12243 24241 44421 32343
21342 32331 13121 33444 33444 12222 32313 23231 34313 12144
31411 12121 14243 32231 44221 14211 42134 12223 43414 33423
23412 34142 13311 31241 31141 21412 24143 34122 22444 31331

14433 13243 43441 32323 44422 31144 23322 33243 43223 33244
22133 11433 22224 32334 23133 13243 43142 13411 22112 34244
14122 13422 41444 23414 34242 12421 33341 12111 44314 42421
33123 22223 32221 34211 23234 42323 22244 22223 44231 42342
24342 32243 22311 44142 34134 21434 22223 31143 12112 12231

12322 33432 14243 23211 11432 14223 13133 23241 43124 11131
22143 23333 43424 43231 11244 24444 13211 43341 43234 31414
33441 24142 31222 21442 43113 33122 41111 41242 14123 23334
13313 12214 22141 32113 21122 12231 21243 43142 43314 32122
11231 24343 23321 23243 14433 11324 12113 24232 33411 44431

12134 21343 12111 44432 43134 14321 11113 42422 12144 34321
13133 22242 41332 14124 24142 32423 34334 21414 44214 21311
33124 33342 43141 12444 23434 24123 12421 43213 43244 13233
42122 42321 13243 43222 33243 42331 44112 44413 12224 44234
14214 14122 12124 33214 31123 34224 21343 42241 23424 31231

44312 32324 11444 31423 42223 21322 43232 23344 32314 21324
12443 22423 44434 24243 43142 44242 21314 14441 13134 13322
41242 34242 44243 22442 32141 11311 11423 32434 42132 24434
13412 44322 14343 14411 34343 12323 34233 23111 44414 44241
43113 12134 31223 22243 23134 44413 34324 21124 13222 34331

42233 12322 42311 21212 41341 43331 32434 13321 43414 22421
14434 43412 23133 22233 44434 23422 42144 43411 14242 11341
14113 22242 43232 42441 43444 41124 41234 31321 42434 31134
43121 42444 12221 12441 34114 14234 22324 33233 31233 42334
14233 13313 31224 43111 13341 14433 42324 12124 32441 41331

41113 14433 41422 23343 14321 12211 12144 22432 43243 23244
42224 43414 11124 13423 42143 24123 23321 31333 31443 11133
21323 31114 22211 42442 44314 44221 34311 23212 33441 33343
12442 31131 14331 41133 31111 44241 43123 24221 12431 14234
23323 41442 14442 44133 44444 22442 23221 34144 33323 13344

42443 43443 33133 22331 14112 14413 41333 31132 44232 23332
41214 33142 24322 24443 24314 31313 13231 44313 22444 43314
44323 33123 12242 22433 23222 21423 33213 43312 32411 23131
43124 22113 24411 12212 41121 14344 43114 24234 21212 44144
13441 34241 41423 31323 13141 13221 41113 12344 22413 43232

11111 21241 31411 31342 14413 34344 22434 44141 43233 21124
42121 42143 44313 11421 32211 11324 43111 23333 11124 31311
24242 44133 42442 23333 32222 44323 24122 22232 22134 11241
22212 44443 42131 24224 11133 12211 13412 11222 14131 44231
43243 12434 42234 34242 21234 32112 14332 33143 32233 41222

41414 24131 21244 31422 11144 33433 22231 11114 24424 31344
32241 22344 33124 42134 44321 31334 22241 43444 14144 31112
42143 41313 13222 42131 41124 32242 23123 43234 14211 13133
32212 24223 34421 21331 11122 21422 21133 13232 43333 11422
34234 24434 23332 44124 22122 34113 14441 12323 23132 14114

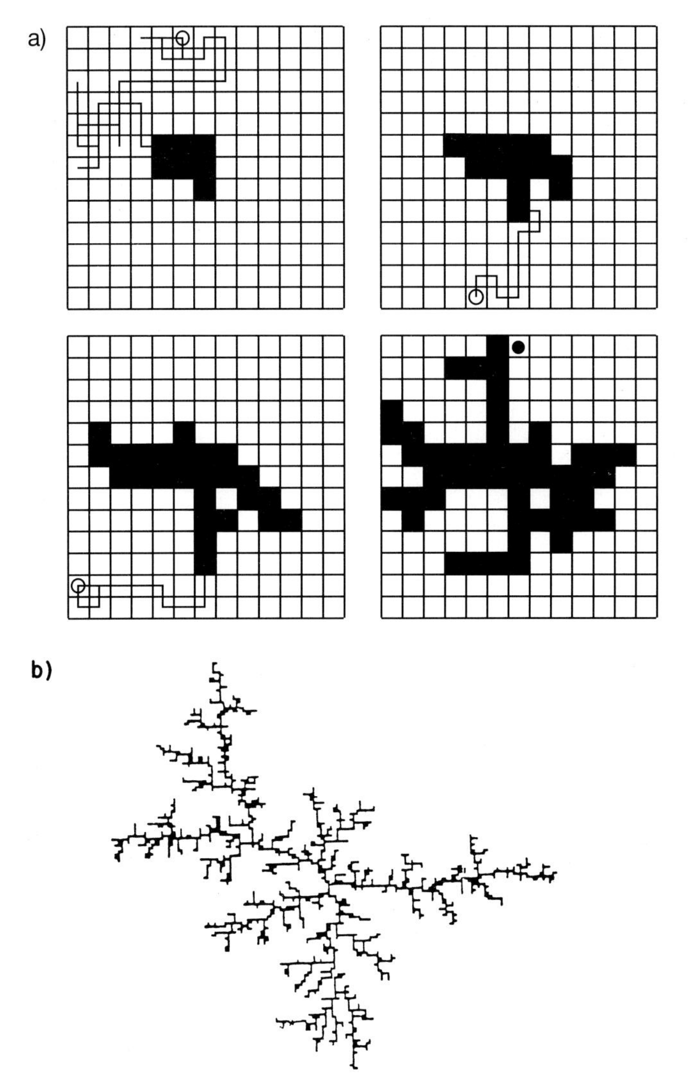

Figure 7.2. The pattern of randomly arriving pixels congregating around the center nucleus is known as a diffusion limited aggregate (often referred to in shorthand as a DLA structure). a) As several pixels stagger towards their friendly nucleating center a spider-like aggregate grows outward from the center. b) A well developed aggregate formed by the staggering pixels is known as a Whitten and Sander fractal aggregate [1].

bonded to the nucleating center but the pixel "a" is free to continue its meanderings. In *Figure 7.2*(a) we show a series of moving pixels and their consequent participation in the growing congregation of pixels.

The word congregation comes from two Latin words con meaning "together" and gregis meaning a "flock of sheep". The term was originally used to describe a group of people coming together to worship at a church. If this seems a strange word to describe such a group of people it should be remembered that the Christian church often uses the imagery of the shepherd and his flock to express religious truths. The pastor of the congregation is literally the shepherd who leads the sheep (in Latin the word pastor means "shepherd"). In scientific textbooks and publications there is considerable confusion between the terms *aggregate* and *agglomerate*. The people who make concrete refer to the gravel that they add to their mixtures as aggregate. However, aggregate is closely related to congregate and literally means to come together like a bunch of sheep. Anyone who has watched a flock of sheep knows that they will disperse the moment that the sheep dog leaves them alone. An aggregate, strictly speaking, is an association of subunits which disperses readily. On the other hand, the word agglomerate means "added to a ball". The word comes from the Latin root words ad meaning "to" and glomus meaning "a ball". This latter word has given us the words *globule* for a small droplet and *globe* for a ball on which the map of the earth is drawn. Looking at the origin of these words it becomes clear that an aggregate is something that is loosely assembled and likely to disintegrate when used, whereas an agglomerate is a group of objects bonded together so strongly that they will not disintegrate when handled. These are the meanings for aggregate and agglomerate that we will use in this book. The reader is warned that in general scientific literature the words are used loosely and interchangeably. In the same way a conglomerate should mean something which is bound together strongly whereas a congregation is free to leave after the sermon!

In Figure 7.2(a) several stages in the buildup of a spider-like aggregate generated by the arriving pixels are shown. This type of structure was first studied by Whitten and Sander who were interested in the growth of soot fine particles [1, 2]. In real soot fine particles the pixels making up the aggregate are tiny spheres of unburnt carbon leaving a burning flame. The turbulence in the flame causes them to move around at random until they collide with each other. The soot fine particles self-nucleate to create spider-type aggregates. The rapid movement of the colliding subunits is referred to technically as *diffusion*. This comes from two Latin root words dis meaning "apart" and fundere meaning "to pour". When something is said to *diffuse* it spreads out in all directions at random like liquid poured from a jug onto rough ground. Since the basic scientific process leading to the buildup of the aggregate of Figure 7.2 is a diffusion process the structure is known as a *diffusion limited aggregate* (note that in this case we do not know if the structure will stay together when used so that we can use the term aggregate or agglomerate). If, however, it was found that the entire structure persisted as the soot was breathed in and out of the lung or presented to a respirator filter it would actually be better terminology to describe it as a diffusion-limited agglomerate. In general the whole question of whether or not a cluster of pixels is an aggregate or an agglomerate can be bypassed for now by referring to it as a *DLA* structure (where the A can stand for either aggregate or agglomerate).

If one grows a Whitten and Sander-type structure several times the structures, although not identical, have essentially the same structure and can be described as being statistically self-similar. One of the properties of a statistically self-similar system is that it is very difficult, from looking at an isolated part of the structure, to know the scale of magnification that one is using to look at the structure. If one was to take a branch of the aggregate of Figure 7.2(b) and look at it in isolation one would not know if one was looking at a magnified branch of a larger structure or at a structure entire in itself. These two properties, statistical self-similarity and scale independent structure, satisfy the requirements for describing the structure as a fractal structure (the meaning of this term will become clearer as we explore the concepts of fractal geometry in more detail in the next several chapters of this book). The fact that the congregating drunks create a fractal structure leads us to an important realization:

If one is studying a system which satisfies the description of a deterministic chaotic system then if the outcome of the various causes interacting to produce chaos is a set of discreet events such as the final position of a dispersing drunk, then the distribution of the events generated by the deterministic chaos process is one of the several probability distributions which were discovered by scientists studying stochastically fluctuating systems. If, however, the outcome of the interacting causes is a structure, then the outcome of the interacting causes is a fractal structure such as the Whitten and Sander fractal aggregate.

West and Shlesinger put this general conclusion forward in a slightly different manner when they said that

"*A fractal structure is the residual effect of interacting causes*" [3].

Section 7.2. Characterizing the Structure of Fractal Agglomerates

In *Figure 7.3*(a) three different Whitten and Sander DLA clusters are shown. It can be seen that although they are not identical they look alike. We now need to discover some method of describing the structure that brings out their statistical self similarity. One way of doing this is to look at the way in which the aggregate occupies space. To characterize the density with which the aggregate fills the space it occupies we can count the number of intersections that the structure makes with a set of concentric rings drawn around the center of the aggregate (*concentric* means "with the same center"). Such a system for characterizing the Whitten–Sander Spiders is illustrated in Figure 7.3(b) (i). In Figure 7.3(b) (ii) the number of intersections with each circle are plotted on log–log graph paper. On the same graph the cumulative

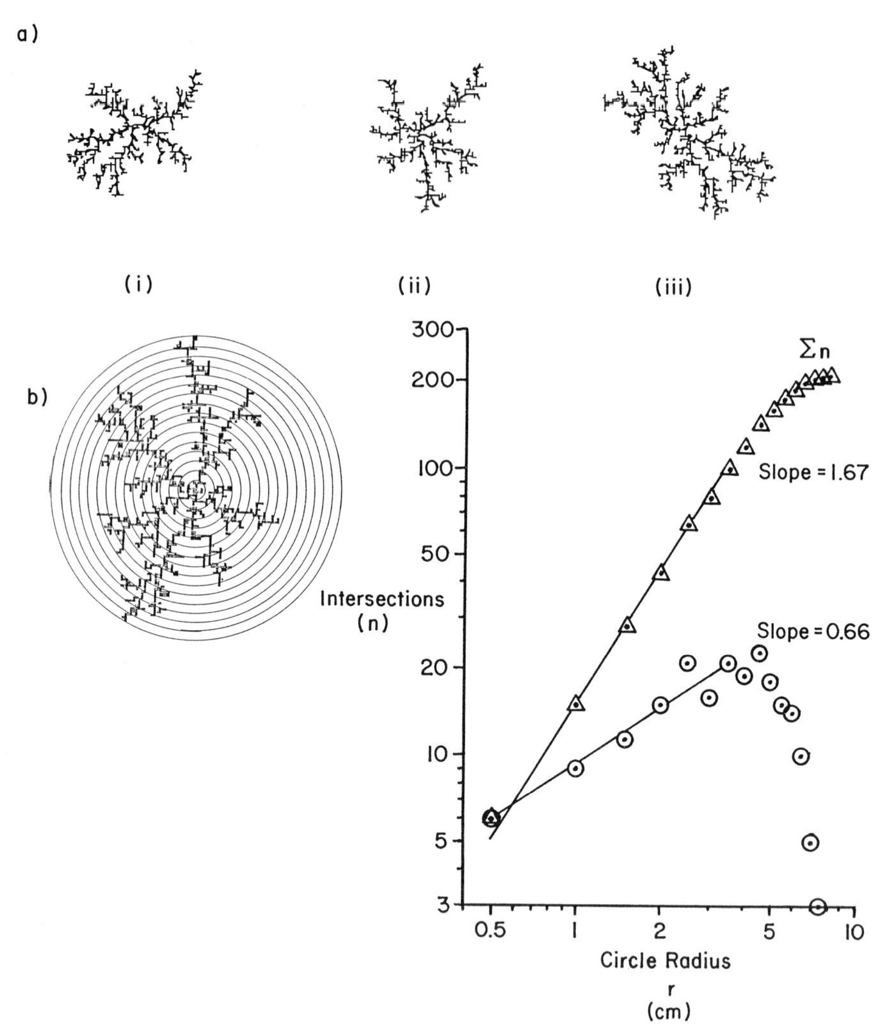

Figure 7.3. The structure of a Whitten and Sander DLA can be characterized by the efficiency with which it occupies space. a) Three DLA of the Whitten and Sander type each grown on a computer using the same construction algorithm. b) Density fractal dimension estimation using search circle intercept frequency data.

number of intersections for all the rings are also shown. It can be shown that one can define a quantity known as the *density fractal dimension* of the aggregate which is related to the data of the graph by the relationship

$$\log N_I(R) = (\delta \varrho - 1) \log R$$

$$\log \sum_{R=1}^{R=n} N_I(R) = (\delta \varrho - 1)\, 2 \log R$$

where $N_I(R)$ is the number of intersections within the search circle of radius R, $\sum N_I(R)$ is the total number of intersections within all the circles of radius R or smaller, and $\delta \varrho$ is the density or mass fractal dimension.

Using the data of Figure 7.3(b) this gives the density fractal dimension for the aggregate as 1.7. The higher this number the more efficiently the aggregate covers the space available. It can be shown that for all Whitten and Sander-type fractal dimensions one always obtains the dimension 1.7. The 1 of this fractal tells us that we are dealing with a branching line and the 0.7 tells us how efficiently the branching line fills the space. Readers can check for themselves that all three DLAs of Figure 7.3(a) have the same fractal dimension. It should be noted that the density fractal of an aggregate is different from its boundary fractal, a parameter which will be discussed in the next chapter.

In *Figure 7.4* a system for generating the intersection data used to measure the fractal dimension of a DLA as the aggregate is generated on the computer is illustrated. For simplicity only a small system is shown. The basic idea is that interrogation squares instead of circles can be drawn around the nucleating center and as pixels move into the nucleating center and join the growing cluster they occupy pixel space in the addresses of the interrogation squares and their number can be tallied automatically to give the equivalent of interception frequencies. For the sake of simplicity only two interrogation squares are shown on a simple matrix in Figure 7.4. In Exercise 7.1 a larger system generated on a computer is presented. In this exercise the readers are encouraged to count the occupied pixels in the interrogation squares and construct the graphs on log–log graph paper leading to the estimates of the density fractal dimension of the structure.

To write a computer program to simulate the buildup of a DLA the major extra work in programming the computer, as compared to following the aggregation process visually on a board game space, is to check the space to which a pixel is asked to move to see if it is occupied or not by the growing aggregates. Checking whether a nearby space is occupied by visual inspection is an easy task whereas checking numerically whether any given address to which a pixel is directed is occupied or not and then to check after a move if it has encountered a valid pixel for joining up to the growing aggregate is tedious but well within the capacity of a small computer. The flow sheet and a typical program written in the Pascal language for construction a DLA on a Macintosh SE computer is given is Exercise 7.2.

Whitten and Sander carried out their original studies of fractal growth as part of their investigations into the structure of soot. *Soot* represents unburned carbon from the fuel of an internal combustion engine. Scientists study the structure of soot to see how it formed in order to be able to work backwards to determine what was wrong with the conditions in the burning chamber when the gasoline and air mixture was burnt to give power to the engine. We have only discussed two-dimensional formation of soot but it is relatively easy, if you have a big enough computer, to grow the DLA in three-dimensional space. A full discussion of this aspect of

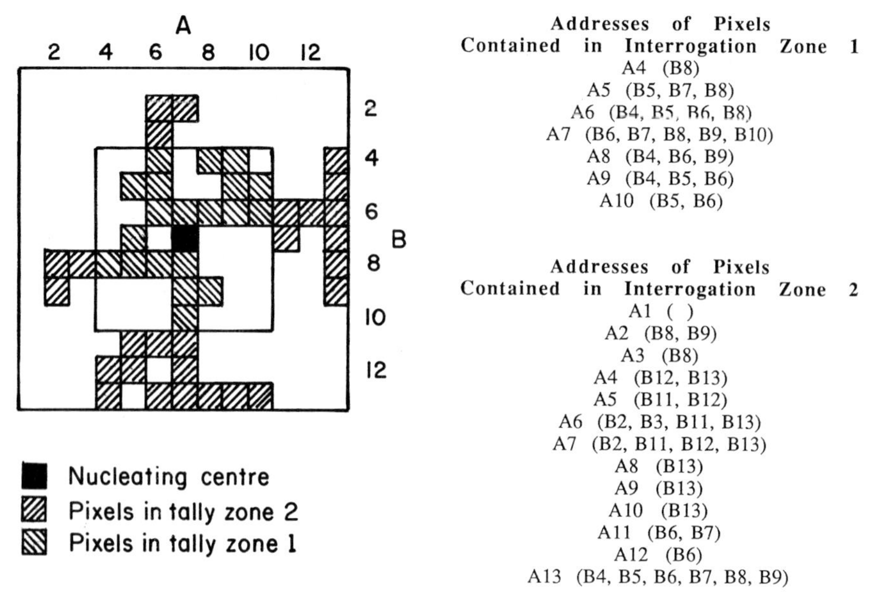

Figure 7.4. The fractal structure of a Whitten and Sander-type DLA can be characterized as it grows by keeping a tally of occupied pixels in a set of interrogation squares drawn around the nucleating center as shown above.

modeling soot is beyond the scope of this book but the interested reader will find details and model soot fine particles in three-dimensional space in Reference 4.

Although the DLA structures of Figure 7.3(a) look like some forms of soot, other soot fine particles have different structures. One can model different types of DLA by altering the probability with which a pixel approaching a growing aggregate sticks to the aggregate. The idea that a pixel may or may not join a growing "spider", models the idea that the staggering soot fine particle on encountering the growing cluster may have sufficient energy for them not to be captured by the growing aggregate at the first encounter (alternatively the bonding between the aggregate and the incoming pixel may be weak so that local flow fluctuations may cause the pixel just to bounce off an aggregate on first encounter). To model the growth of various different types of aggregates in which the capture efficiency of the wandering pixels in the environment of the aggregate are captured with various efficiencies an extra stage is added to the algorithm used to generate the aggregates illustrated in Figure 7.3(a). When a pixel encounters another pixel already forming part of the growing aggregate one consults a random-number table to see if the contacting pixel should be considered to have joined or not according to a prearranged probability schedule. If we wish to simulate a 10% probability of a contact pixel forming a capture bond with the aggregate we would consider that every time we looked up a digit in a random-number table a bond would be considered to be

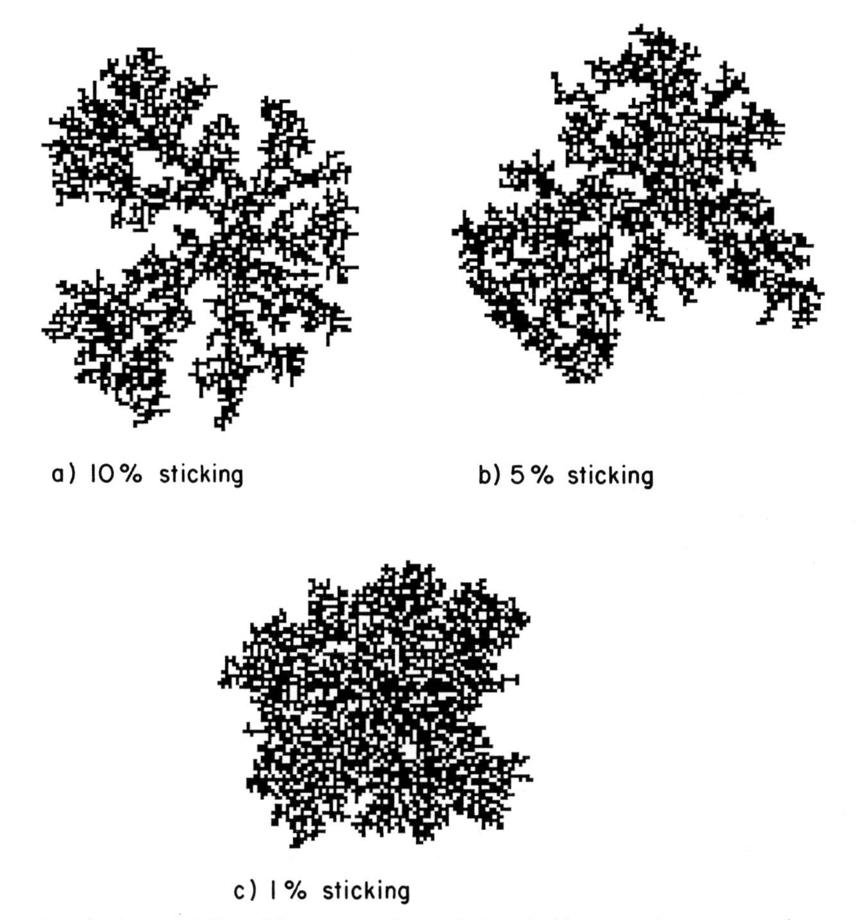

Figure 7.5. As the probability of the capture of a wandering pixel by a growing aggregate decreases the density of the aggregate and hence the density fractal of the aggregate increases.

formed if the digit encountered in the table was 4. Since there are 10 digits in a random-number table the frequency of the action to join the pixel to the spider when encountering the number 4 in a random-number table represents a 10% probability of capture. If we consulted the probability of sticking table and found any other digit we would instruct the computer to let the pixel move away at random until a second potential capture encounter took place when again the probability of capture table would be consulted. In *Figure 7.5* three different DLA spiders grown on a computer using different probability of capture rules are shown. It can be seen that as the probability of capture goes down the aggregate becomes denser. That is its packing density in two-dimensional space increases and its density or mass fractal dimension increases towards the value of two. The reader will also notice that the overall shape of the aggregate begins to look more like a square as the probability of capture falls.

This is quite noticeable for the 1% sticking probability cluster shown in Figure 7.5. It has been shown by Meakin that this shape is due to the square lattice in which the aggregate has been drawn [5]. If one allows the staggering pixel to move in more directions than the 4 of Figure 7.1 then the overall shape of the low probability of capture aggregate will be that of the lattice shape defined by the possible directions of motion of pixel as it approaches the growing aggregate [5]. The important commercial black pigment – *carbon black* is essentially pure soot grown under conditions that generate desirable shape and structure. Many other pigments used in paint, such as titanium dioxide, are produced by a fuming process (a *fume* is a very fine smoke) and the fractal structure of pigments determine their optical properties. The early studies of the fractal structure of commercial pigments has focused on the use of what is known as a boundary fractal. We will delay a discussion of the detailed structure of commercial pigments until Chapter 8 in which we discover the physical nature of structural and textural boundary fractals.

The reader will have noticed that the different DLA we have constructed in this chapter bear a striking resemblance to the frost fingers that decorate windows on a cold day. This is because the growth of precipitated crystals on a surface, or a nucleating center such as a small crystal added to a supersaturated solution, takes place by a mechanism similar to that postulated for the growth of our DLA. In the case of the frost patterns, the nucleating center is often a scratch or a mark on the surface of the pane of glass from which the "frost fingers" spread out. Scientists trying to deposit thin films of metal onto silicon wafers, used to construct the "chips" used in computer memories, discovered that it was easier to grow metal frost fingers on their wafers than it was to deposit uniform films of metal [6]. Scientists studying such systems found that they often grew aggregates like that of *Figure 7.6*(a) in which the rules of capture seemed to change after the aggregates

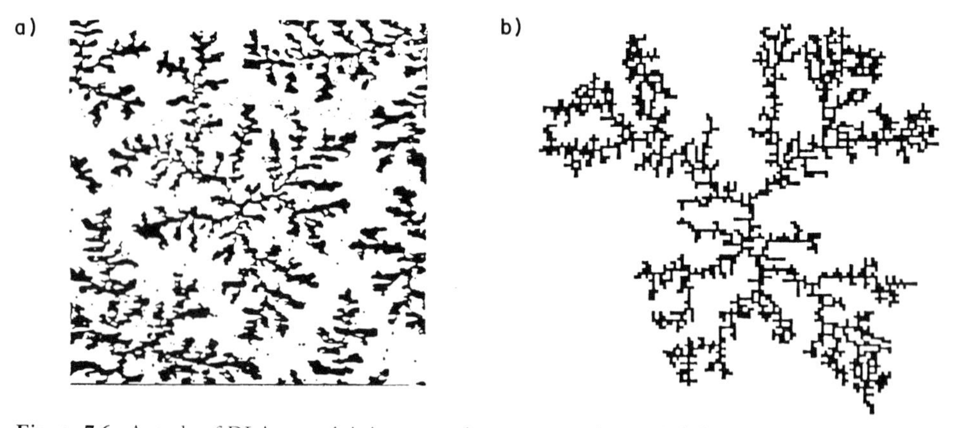

Figure 7.6. A study of DLA growth is important in computer science. a) A fractal aggregate grown on a silicon chip be Elam et al. b) A "bush with leaves" grown by changing the sticking probability in the growth algorithm after the aggregate reaches a predetermined size.

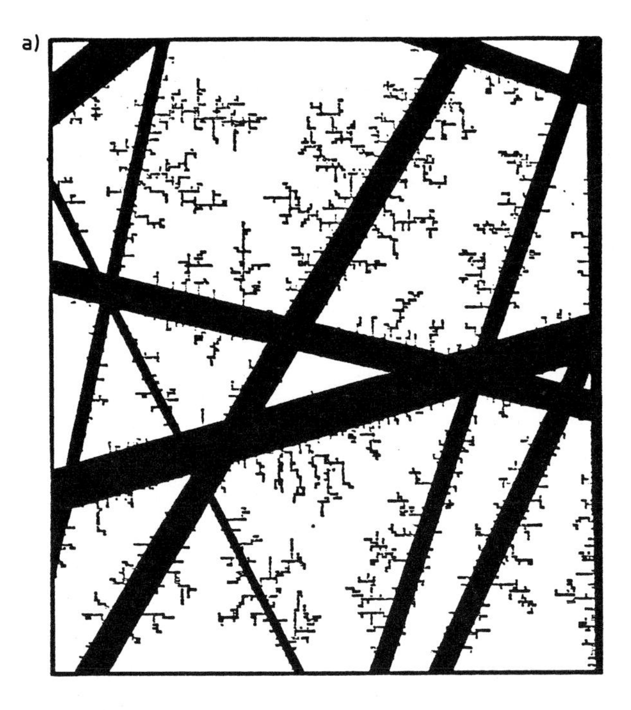

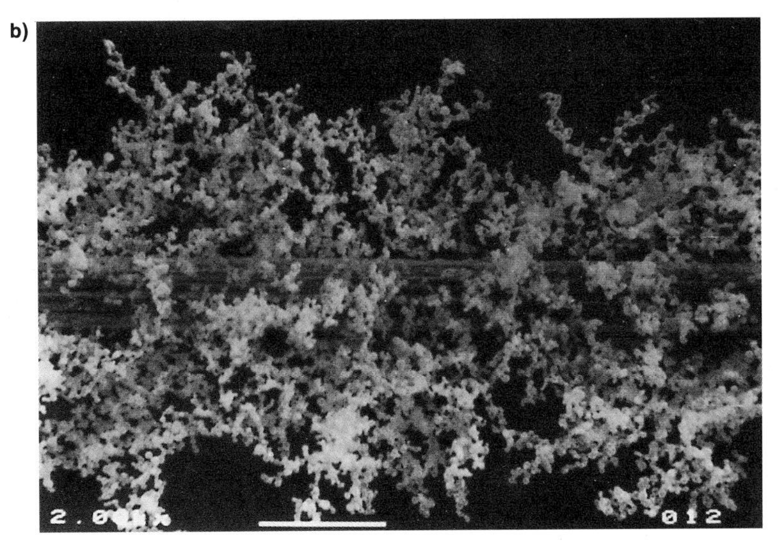

Figure 7.7. Respirable dust captured by a filter generates fractally structured "capture trees". a) Dendritic capture trees generated in a computer simulation [8]. b) Fractally structured deposits reported by Kanoaka [9].

reached a certain size. They described such structures as "bushes that grew leaves". One approach to research in such fields as pigment manufacture, thin-film production and combustion research, is to capture natural aggregates and compare them to DLA structures grown using different constructional algorithms in an attempt to model the natural processes generating fractal aggregates and agglomerates. In Figure 7.6(b) a DLA for which the sticking rules were changed after it had grown to a certain size is shown. It can be seen that it looks like the DLA grown on a surface by Elam et al.

A problem closely related to the growth of DLA in free space is the growth of what is known as capture trees in a respirator or filter when one is attempting to take fine particles smaller than 1 micron out of a moving stream of air (a micron is one millionth of a meter). Dust smaller than 5 microns is particularly dangerous to the lung, fine particles smaller than 1 micron move about with Brownian motion as the air stream carrying them passes through a filter. If we look at the square of *Figure 7.7*(a) it could be considered as an idealized hole in a filter formed by four fibers which lay along the edges of the square to form an idealized hole. If the air stream moving through the square contained respirable dust then from the perspective of the boundary fibers, the dust fine particles would appear at random in the pixel space defined by the fibers. For the system of Figure 7.7(a) a pixel point of materialization (that is the point at which it would appear in the hole formed by the fibers) can be simulated by choosing a pixel address at random and then permitting the pixel to stagger about at random until it hits the wall of the fiber hole or joins a fractal capture tree growing out from the wall. In Figure 7.7(b) a sophisticated version of the growth of several capture trees in a real fibrous filter as reported by Kanoaka is shown [7–9]. In the design of filter systems an important property of the fibers is the attractive force between the fiber and a passing fine particle. If the forces of attraction and capture are strong then randomly moving fine particles will stick to a growing capture tree with 100% probability. This type of growing capture tree will have a fractal dimension of 1.7. If, however, the probability of capture on encounter is less than 100% then the capture tree will look more like a bush and will have a higher fractal dimension. Therefore, by observing the structure of the capture trees in the holes of a filter one can deduce the power of attraction exerted by a specific fiber on a specific dust.

Section 7.3. Fractal Fingers Generated by Electrolytic Deposition

The system shown in *Figure 7.8*(a) looks like the DLA grown by Whitten and Sander on their computers. It is a metal deposit grown in a special two-dimensional

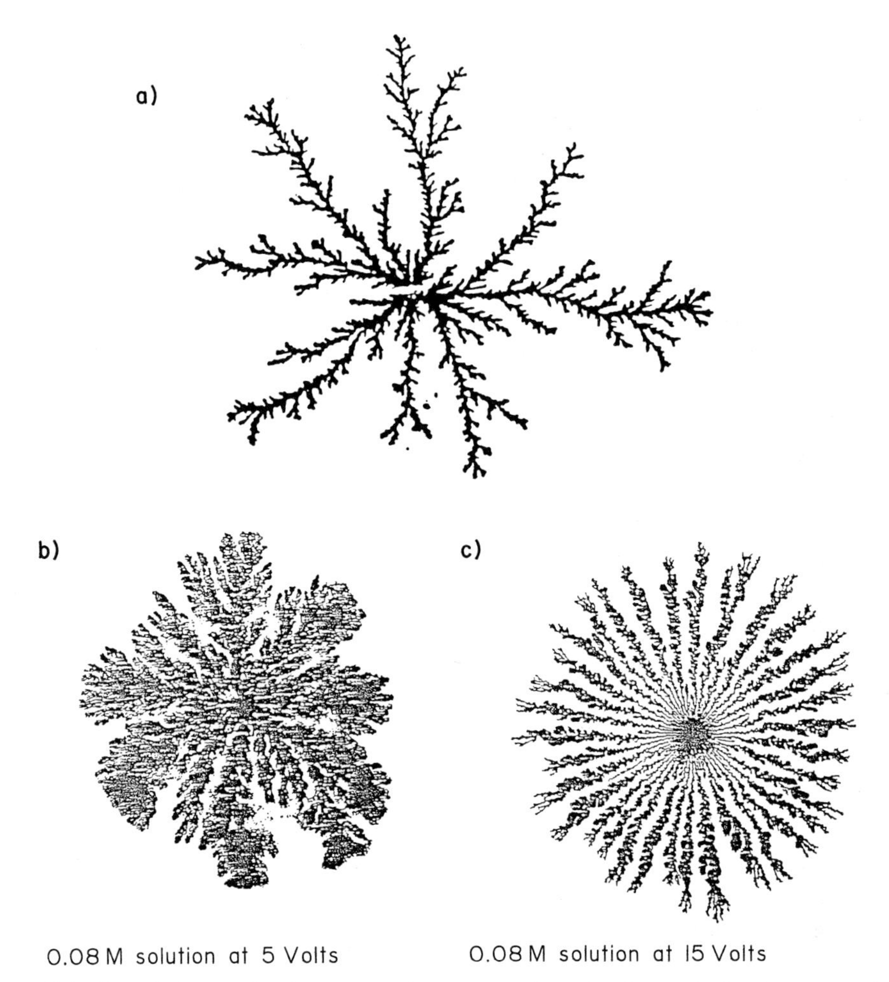

Figure 7.8. Electrolytic deposits of metal growing in a two-dimensional electrolysis cell can grow like Whitten and Sander fractal aggregates under certain conditions of formation. a) Aggregate grown by Matsushita et al. b) Electrolytic deposits for a 0.08 M solution grown at 5 volts. c) Electrolytic deposits for a 0.08 M solution grown at 15 volts.

electrolytic cell designed by Sawada and Matsushita [10]. In order for the reader to be able to appreciate the way in which Sawada and Matsushita fabricated the fractal metal fingers, it is necessary to master the concepts and vocabulary of the process known as electrolysis. The word electrolysis means "freed by means of electricity". The origins of the various words we need to use to describe the techniques for forming the fractal fingers of Figure 7.8 are shown in *Figure 7.9.* The reader can create electrolytically grown fractal fingers such as those shown in Figure 7.8 by using the simple equipment shown in *Figure 7.10.* The shallow dish forming the base of the electrolytic cell of Figure 7.10 is part of an inexpensive plastic Petri dish.

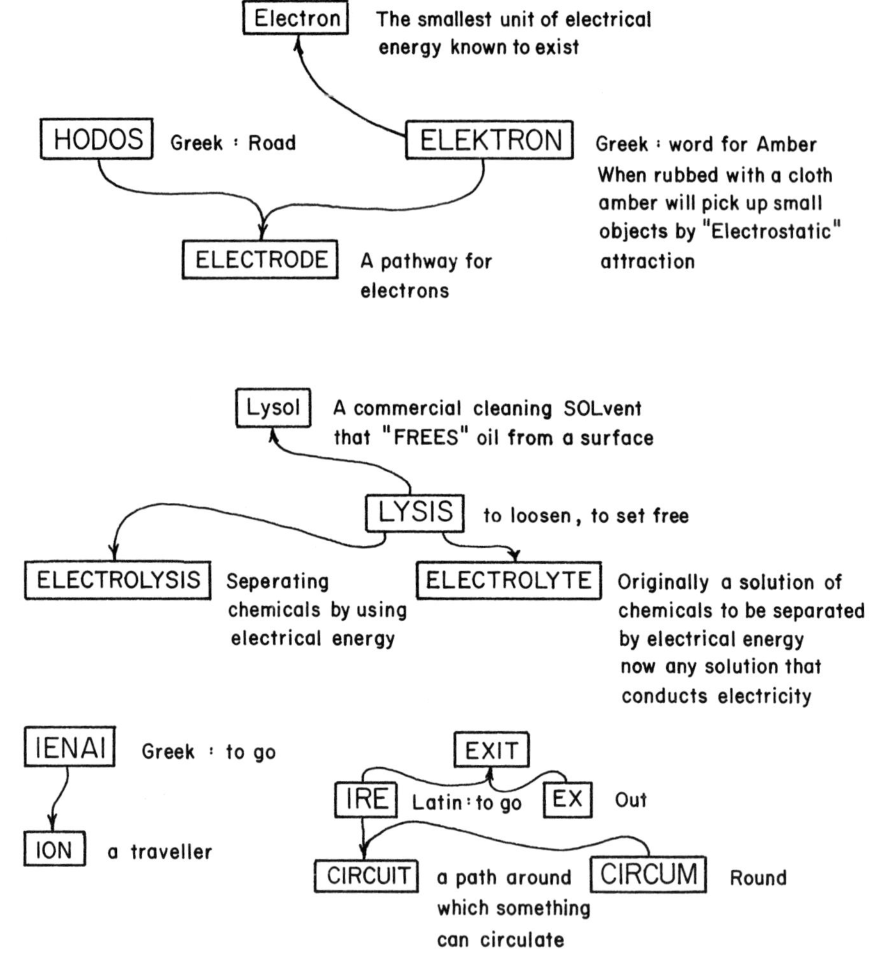

Figure 7.9. Electrolysis is a word created from Greek root words which mean "released by means of electricity".

Richard Petri (1852–1921), a German bacteriologist, invented the set of two round rimmed plates which are put together to form a vessel for growing bacterial cultures. Petri dishes are now used all over the world. The cathode for the two-dimensional electrolytic cell is made by bending a piece of copper wire into a circular hoop. One end of the hoop sticks up above the surface of the Petri dish so that it can be connected to the positive terminal of an electric battery. Students at Laurentian have grown beautiful fractalytic deposits in a cell such as that in Figure 7.10 using a simple six volt electric battery bought from the hardware store. The Petri dish is filled with sufficient solution of copper sulfate to immerse the wire electrode. A thin plastic disc (in our case cut from the top part of the plastic Petri dish) smaller

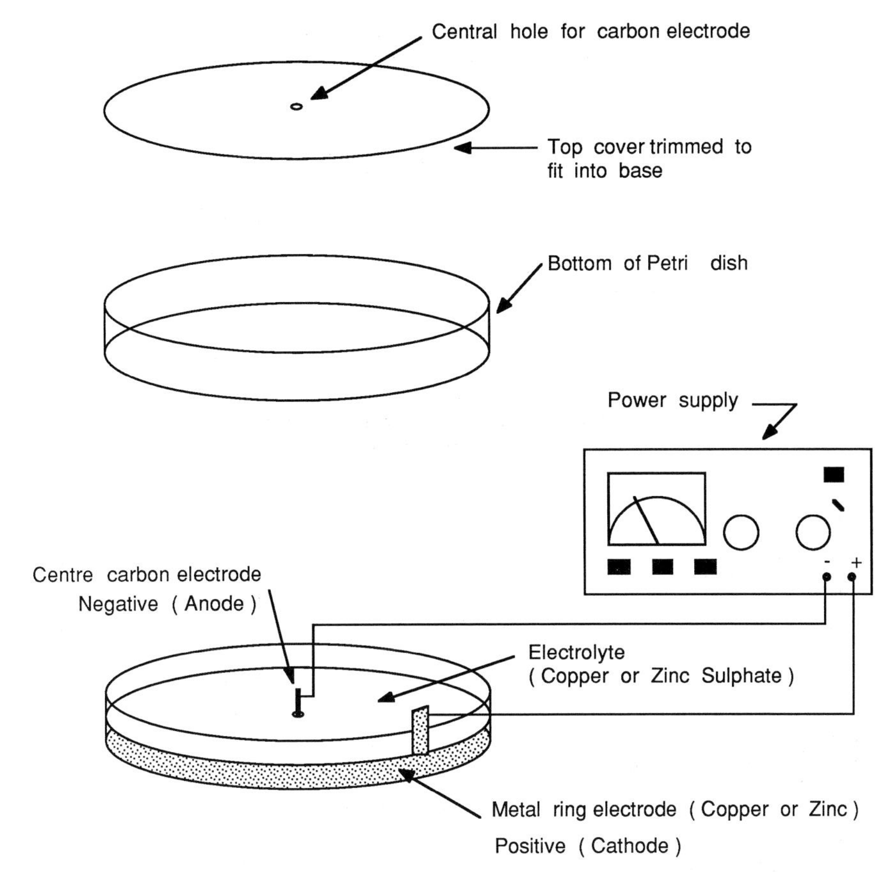

Figure 7.10. The equipment used to grow fractal metal fingers by electrolysis, can be simple and inexpensive.

than the diameter of the Petri dish but slightly larger than the diameter of the copper wire electrode is placed on top of the wire. A hole is drilled at the center of the disc so that a point carbon electrode (made from the center of a pencil) can be placed at the center of the electrolytic cell and can be connected to the negative terminal of the electric battery.

Chemists tell us that when a salt such as copper sulfate is placed in water the force of attraction between the copper cation and the sulfate anion is weakened by the properties of the water. When they are in the water the copper and sulfate part of the salt split up and wander freely about the solution which is called the *electrolyte*. The freely moving parts of the salt are known as *ions* which literally mean "travellers" from the Greek word "to travel". The copper ion is known as the *cation* and the sulfate ion is known as the *anion* because under the driving force of the electric battery the copper ions move toward the cathode and the sulfate ions move toward the anode of the electrolytic cell. To become ions, the copper in the salt donates two electrons to the sulfate unit. When the copper ion reaches the electrode it acquires two electrons from the electrode to become an ordinary copper atom and is deposit ed on the electrode. When this transformation occurs it is said to plate out onto the electrode. When the sulfate ion reaches the copper ring anode it gives up its electric charge as it reacts with the electrode to form more copper sulfate. When the electrolytic cell is operating, electrical energy is carried by the ions between the electrodes and copper is dissolved from the anode and plated out onto the cathode.

If the object of an electrolytic experiment is the coating of an object with material deposited by electrolysis then one should always use low voltage conditions so that the ions approach the electrode slowly with low energy and pack to form a dense coating. In fact, the wandering ions of electrolysis are the staggering pixels of the computer simulation of the growth of aggregates discussed in Section 7.2. In our simulation studies of that section, we discovered that to form a dense agglomerate one must arrange conditions so that the meandering of the pixels takes place with low energy and low sticking probability. This corresponds to using an electrolytic cell with a low driving voltage between the electrodes. The fractal structure of Figure 7.8(b) was made using 5 volts. When higher voltage is used the ions are driven more directly towards the electrode and what is known as *ballistic growth* occurs of the type shown in Figure 7.8(c) [11]. The idea embodied in the term ballistic growth is that the ions no longer meander toward the nucleus but are thrown toward it like a baseball being thrown by a pitcher.

Readers can construct the cell in Figure 7.10 for less than twenty dollars and can carry out many different growth experiments using a range of voltages to drive the ions towards the electrodes. They can also discover the effect of a range of concentrations of the electrolyte. It is also easy to generate this type of fractal growth using a strip of zinc as the cathode and zinc sulfate as the electrolyte. Sometimes, when growing deposits of this kind, an unequal growth will occur with the fractal fingers spreading preferentially towards one side of the cell. Once this imbalance in growth occurs, since the electric field from the tip of the fingers to the metal electrode is shorter than anywhere else in the cell, the electric field is stronger at the point of imbalance in the growth and the fingers race from the unequal growth area to the electrodes. The electrolytic fingers tend to be very fragile but can sometimes be preserved by letting the cell dry out when the fingers are near to the base of the Petri dish.

The fact that the electrolytic deposits on an electrode can be fractally structured or ballistically fingered is the explanation of why it is unwise to charge a car battery at high voltage and high current density. For such a charging condition, the fractal fingers growing out from the electrode can short out the plates. Even if this does not occur, they can fall to the bottom of the battery forming a conductive sludge that will destroy the effectiveness of the battery. Investigation of this problem can form an exciting project. Using the equipment shown in Figure 7.10, a fractal electrolytic deposit can be grown in the dish placed on an overhead projector and the whole class can watch the growth of the fractal fingers.

Section 7.4. Creating Fractal Fingers of Moving Fluid in a Hele-Shaw Cell

In 1898, long before anyone had dreamed of fractal geometry a British engineer Hele-Shaw created the system now known as the Hele-Shaw cell for studying the movement of one liquid into another at an interface [12, 13]. This problem is very important in oil engineering as one studies the movement of oil in a pipeline where often one type of oil must be pushed along the pipe by a different oil. To study what happened at such an interface, Hele-Shaw made a thin film of liquid between two plates held apart by spacers. He injected a liquid under pressure through a central hole in the cell to study the way in which such a pressurized injection could drive the fluid in the cell away from the central hole. Hele-Shaw observed that under many conditions, the fluid injected at the center of the cell did not drive the other fluid away in a regular manner. The invading fluid did not create a uniform circle spreading out from the point of injection. Rather the injected fluid created finger-like patterns penetrating far into the fluid well ahead of any circular front that would have been created by uniform motion of the fluid. From the perspective of fractal geometry, it has been recognized that such fingering is actually a fractal structure describable by means of a fractal dimension. Many interesting studies of fractal fingering at liquid fronts have been carried out in recent years [14]. The reader can create a simple Hele-Shaw cell using the basic system employed in the earlier studies of fractal electrolytic deposits. The copper wire that constituted the electrode now becomes the spacer of the Hele-Shaw cell (the Petri dish base). The top of the Hele-Shaw Cell is the plate that we used to complete the electrolytic cell. It may be necessary to clamp the lid to the Petri dish base to prevent movement when injecting the fluid at the central hole. The hole through which the carbon electrode was placed now becomes the point of injection for the invasion fluid which is loaded into a hypodermic syringe. Gaps must be left in the circular wire separating the upper and lower part of the cell so that the fluid driven out of the cell by the invading fluid can flow out of the cell. Various combinations of fluid can be used [13]. A pair of readily available fluids used by Laurentian University students to study fractal fingering at viscous fluid interfaces are latex resin glue (white office glue available at office supply stores) and black drawing ink. The space in the Hele-Shaw cell made from the Petri dish is filled with the white resin glue and the black ink is injected using a hypodermic syringe at the center of the disc. A viscous fingering pattern

Figure 7.11. An inexpensive Hele-Shaw cell can be fabricated from a plastic Petri dish set. The upper plate is trimmed to create an upper retaining plate. The pattern shown above was created by injecting black ink into the center of a layer a white resin glue.

Figure 7.12. Patterson studied the fingering of fluid penetrating a random array of cylinders packed into a Hele-Shaw cell.

obtained in this way by a student at a first attempt to create a fractal fingered interface is shown in *Figure 7.11*.

In 1983, Lincoln Paterson, who was interested in studying the movement of water being used to drive out oil from a sandstone reservoir, modified the basic Hele-Shaw cell by filling it with plastic cylinders. He then studied the movement of fluid into the simulated sandstone and found that often he created fractal fronts. In *Figure 7.12* one of his photographs is shown [15]. In the next chapter, we will discover how to measure the fractal dimension of this moving front. Note that this type of fractal front differs from that of Figure 7.11 in that now the progress of the fluid into the Hele-Shaw cell is governed by the availability of random paths through the porous medium rather than the physical properties of two fluids.

Exercises

Exercise 7.1. Intersection Count Technique for Characterizing the Fractal Structure of a DLA

Students will generally be unfamiliar with the plotting of data on log–log graph paper and they should use the large scale display of *Figure E 7.1.1* (p. 290) to count the intersection frequency of the Whitten and Sander aggregate with the interrogation squares drawn on the pixel space. The teacher could discuss with the students how this square interrogation pattern is not significantly different from the use of circles as long as one varies the size of the interrogation unit systematically. The students should plot the two lines representing the single count for each interrogation square and the cumulative intersection count. The teacher could discuss how the data per square is more scattered whereas the scatter for the cumulative intersection is less because the procedure involves self cancelling variations as the totals are accumulated from square to square.

Advanced classes may wish to grow their own clusters with subsequent evaluation of the fractal structure using the basic technique outlined in this section.

Exercise 7.2. Flow Chart and Sample Program for Growing a DLA

In *Figure E 7.2.1* (p. 291) the flow chart for growing a DLA in square pixel space (four possible directions of motion are used for the random walk). The computer program for use with a Macintosh personal computer to implement this flow chart is given below.

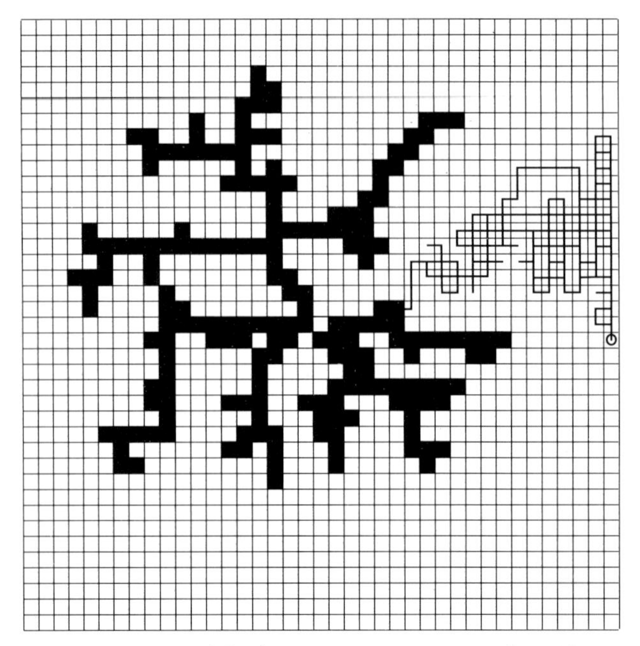

Figure E 7.1.1. Is a large Whitten and Sander aggregate on a rectangular grid.

Exercise 7.3. How Long Does a Wandering Pixel Take to Arrive at a Growing Cluster?

The stagger space of Figure 7.1 can be used to look at the question "How long does it take for a pixel to reach a growing aggregate at the center of the stagger space?". This exercise can be set up on a computer for the students to study the movement of the pixels or alternatively it can be played as a board game using a board constructed as shown in Figure 7.1. The pixels can be made out of black cardboard and a small piece of magnetic material attached to the underside of the square. The stagger space can be made of a metal sheet. In *Figure E 7.3.1* data showing the number of steps required to reach the growing agglomerate for the first three pixels joining the aggregate are shown for two separate experiments. Students will usually be surprised at how many steps can be involved in generating this type of random walk. In each case the actual path of the third pixel is shown. The data of Figure E 7.3.1 (a) can be used to discuss the fact that a tiny piece of material undergoing Brownian motion will eventually cover an area entirely if given an infinite wandering period. The teacher could also discuss with the class techniques for shortening the period required to grow a fractal agglomerate. In the early stages of the study there is no reason to introduce the wandering pixel at the perimeter of the stagger space. It could be allowed to materialize at any unoccupied area around

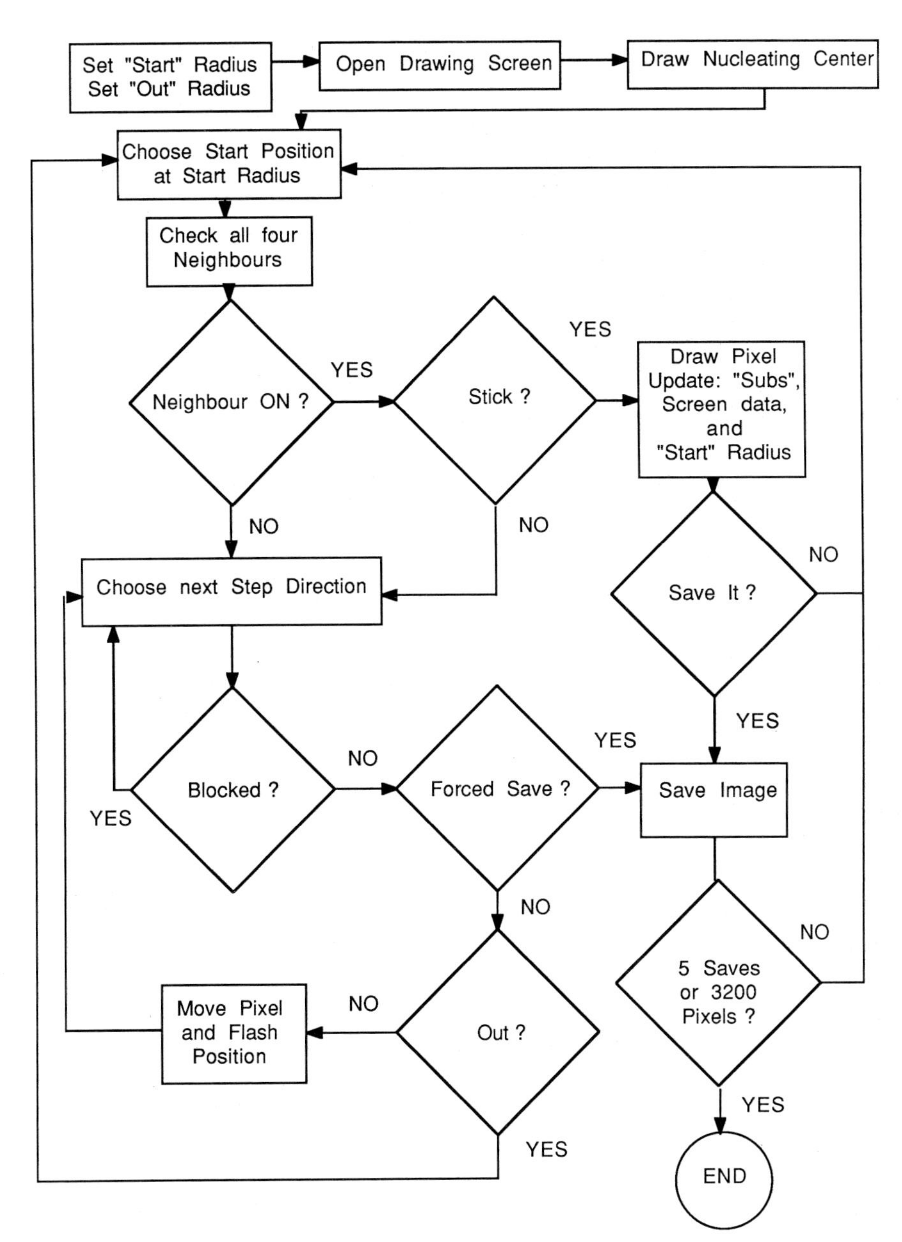

Figure E 7.2.1. Flow chart of a pixel–DLA calculation.

```
program DLAsimulation;
{Pascal program written for the Macintosh SE}
{By Garry G. Clark and Remi A. Trottier Laurentian University}

  type
    NG = array[1..4] of integer;
    drw = array[1..5] of string;

{Variable declarations}
  var
    x, y: integer;
    x1, x2, y1, y2: real;
    fx, fy, lx, ly: integer;
    Radius, MaxRad, Dist: integer;
    fullScreen: rect;
    i, j, k, l: integer;
    step, Stick, Stuck: integer;
    dir, ckdir, SQuad: integer;
    m, ag, StAg, b: real;
    ltype, width: integer;
    Px1, Py1, Px2, Py2: real;
    sx, sy, ex, ey, sp, ep: integer;
    err: OSErr;
    ok: boolean;
    myPict: picHandle;
    DrawingRect, ScratchRect: Rect;
    tempPoint: Point;
    NoGo: NG;
    name: drw;
    change, newper, subs, psubs, saves: integer;



  label
    1, 2, 3, 4, 5;

{Set up drawing screen}
begin
  SetRect(DrawingRect, 5, 35, 505, 335);
  SetDrawingRect(DrawingRect);
  ShowDrawing;
  myPict := OpenPicture(DrawingRect);
  ShowPen;

  begin

{Seed random number generator}
   randSeed := integer(TickCount);

{Draw frame around drawing screen}
   Drawline(5, 5, 480, 5);
   Drawline(480, 5, 480, 280);
   Drawline(480, 280, 5, 280);
```

```
   Drawline(5, 280, 5, 5);

{Draw nucleating centre}
   DrawLine(240, 140, 240, 140);

{Define image names}
   name[1] := 'first';
   name[2] := 'second';
   name[3] := 'third';
   name[4] := 'forth';
   name[5] := 'fifth';

{Define growth parameters}
   Stick := 100; {initial sticking probability}
   change := 1500; {number of pixels until new percent sticking employed}
   newper := 10; {percent sticking after change}

{Set up initial conditions}
   subs := 1; {number of subunits ... pixels}
   saves := 1; {number of images saved}
   step := 1; {step size in pixels}
   Radius := 3; {starting radius from nucleating centre}
   MaxRad := 5; {maximum radius from centre before terminating walk}
   fx := 240;
   fy := 140;

   PenMode(PatXor);
   PenSize(1, 1);

{Beginning of main loop}
1: {Check cursor position ... jump to SAVE if cursor at TOP-LEFT}
   GetMouse(tempPoint);


   while (tempPoint.h > 10) or (tempPoint.v > 10) do

     begin
       GetMouse(tempPoint);

{Choose Starting Angle and Starting Quadrant for random walker}
       StAg := (abs(Random) mod 91) * pi / 180;
       SQuad := (abs(Random) mod 4) + 1;

{Calculate starting point for the walk relative to centre}
       case SQuad of

         1:  {first quadrant}
          begin
            x2 := Radius * sin(StAg);
            y2 := -Radius * cos(StAg);
          end;

         2: {second quadrant}
```

```
          begin
            x2 := Radius * cos(StAg);
            y2 := Radius * sin(StAg);
          end;

        3:  {third quadrant}
          begin
            x2 := -Radius * sin(StAg);
            y2 := Radius * cos(StAg);
          end;

        4:  {fourth quadrant}
          begin
            x2 := -Radius * cos(StAg);
            y2 := -Radius * sin(StAg);
          end;
      end;

{Change walker position to screen coordinates}
      x := round(fx + x2);
      y := round(fy + y2);

      k := 0;

{If starting pixel is on then choose a new starting position}
      if GetPixel(x, y) then
        goto 4;

{Check neighbouring pixels}
{Neighbour is on if k=100}
      while k < 50 do
        begin
          lx := x;
          ly := y;

          k := 0; {set detector to zero}
          for l := 1 to 4 do
            begin
              NoGo[l] := 0;
              ckdir := 0;
            end;

{Check Right}
          if GetPixel(lx + 1, ly) then
            begin
              k := 100;
              NoGo[1] := 100;
              ckdir := ckdir + 1;
            end;

{Check Down}
          if GetPixel(lx, ly + 1) then
            begin
```

```
            k := 100;
            NoGo[2] := 100;
            ckdir := ckdir + 1;
          end;

{Check Left}
        if GetPixel(lx - 1, ly) then
          begin
            k := 100;
            NoGo[3] := 100;
            ckdir := ckdir + 1;
          end;

{Check Above}
        if GetPixel(lx, ly - 1) then
          begin
            k := 100;
            NoGo[4] := 100;
            ckdir := ckdir + 1;
          end;

{If neighbour is on then decide if the walker should stick}
        if k > 50 then
          begin

{Generate sticking number}
            Stuck := round(abs(Random) mod 100) + 1;

{If not stuck then proceed with walk}
            if Stuck > Stick then
              begin
                k := 0;
                goto 2;
              end;

{if stuck then draw the pixel and update screen info}
            begin
              DrawLine(lx, ly, lx, ly); {draw pixel}
              subs := subs + 1; {update count}
              PenMode(PatBic);
              PenSize(20, 20);
              DrawLine(10, 10, 250, 10); {clear data line}
              PenMode(PatXor);
              PenSize(1, 1);
              DrawLine(20, 20, 20, 20); {position pen for writing}
              WriteDraw(Stick, newper, change, subs); {write data to screen}

{if subunits are more than change then select new sticking probability}
              if (subs >= change) then
                begin
                  stick := newper;
                end;
```

```
{Update starting radius to keep it outside the cluster}
                if (abs(lx - 240) >= Radius - 2) or (abs(ly - 140) >= Radius - 2) then
                  begin
                    Radius := Radius + 2;
                    MaxRad := MaxRad + 2;
                  end;


                goto 3; {Start a new walker}
              end;
            end;

2: {if not stuck select next step direction}

          if ckdir = 4 then
            goto 3; {if trapped abort walk}

          dir := (abs(Random) mod 4) + 1; {choose direction}
          if NoGo[dir] > 50 then
            goto 2; {if blocked choose a new direction}

{Update walker position}
          case dir of
            1:
              x := x + step;
            2:
              y := y + step;
            3:
              x := x - step;
            4:
              y := y - step;
          end;

{Check cursor position again}
          GetMouse(tempPoint);
          if (tempPoint.h < 10) and (tempPoint.v < 10) then
            Leave;

{if maximum position is beyond the screen border bail out}
          if (abs(x - 240) > MaxRad) or (abs(y - 140) > MaxRad) then
            Leave;

{Blink the pixel at the walker position}
          DrawLine(x, y, x, y);
          DrawLine(x, y, x, y);
3:
        end;
4: {Save the image}
      if (tempPoint.h < 10) and (tempPoint.v < 10) then
        begin
          SaveDrawing(name[saves]);
          saves := saves + 1;
        end;
```

```
{if maximum position is beyond the screen border save the image and quit}
        if MaxRad >= 140 then
          SaveDrawing(name[saves]);

{if five images have been saved ... quit}
        if saves > 5 then
          Leave;
        if MaxRad >= 140 then
          Leave;
        goto 1;
      end;


5 :
    PenNormal;
    HidePen;
    ClosePicture;


  end;

end.
```

the growing nucleus. For example if we looked at Figure E 7.3.1 (a) we could choose the starting point of an entering pixel at any address other than (A7, B6), (A7, B7), (A7, B8), (A6, B8). To be even more efficient we could choose the next pixel from the set of addresses contiguous with the central nucleus that is from the set of pixels which are shown lightly hatched in Figure E 7.3.1 (a). The teacher could use this simple example to discuss the efficiency of any computer algorithms. The data of Figure E 7.3.1 (b) and *Table 7.3.1* can be used to illustrate how the number of steps required to join the agglomerate decrease when the distance of the starting pixel from the growing agglomerate decreases. When the exercise is available on a computer, the class can do a study of a repeat of a given pixel joining the agglomerate. One could make 50 runs to measure the variation in the number of steps required for the third pixel to join the nucleating center similar to the runs of Figure E 7.3.1 (a). The students should be able to discover from this data that the variation in the number of steps taken by a given pixel is log-normally distributed. On the other hand, if the total number of steps for 50 runs were to be averaged and the experiment carried out 20 times (a total of 10,000 pixel wanderings!) they would find that the total number of times for 50 runs were distributed Gaussianly. The significance of the log-normal structure for the repeated tracks toward the nucleating center versus the Gaussianly distributed average for 50 pixelated unions at the same stage of growth can be discussed by the teacher.

In *Table E 7.3.2*, the number of steps taken by a sequence of 48 pixels joining a growing agglomerate, as shown in *Figure E 7.3.2*, along with the starting position

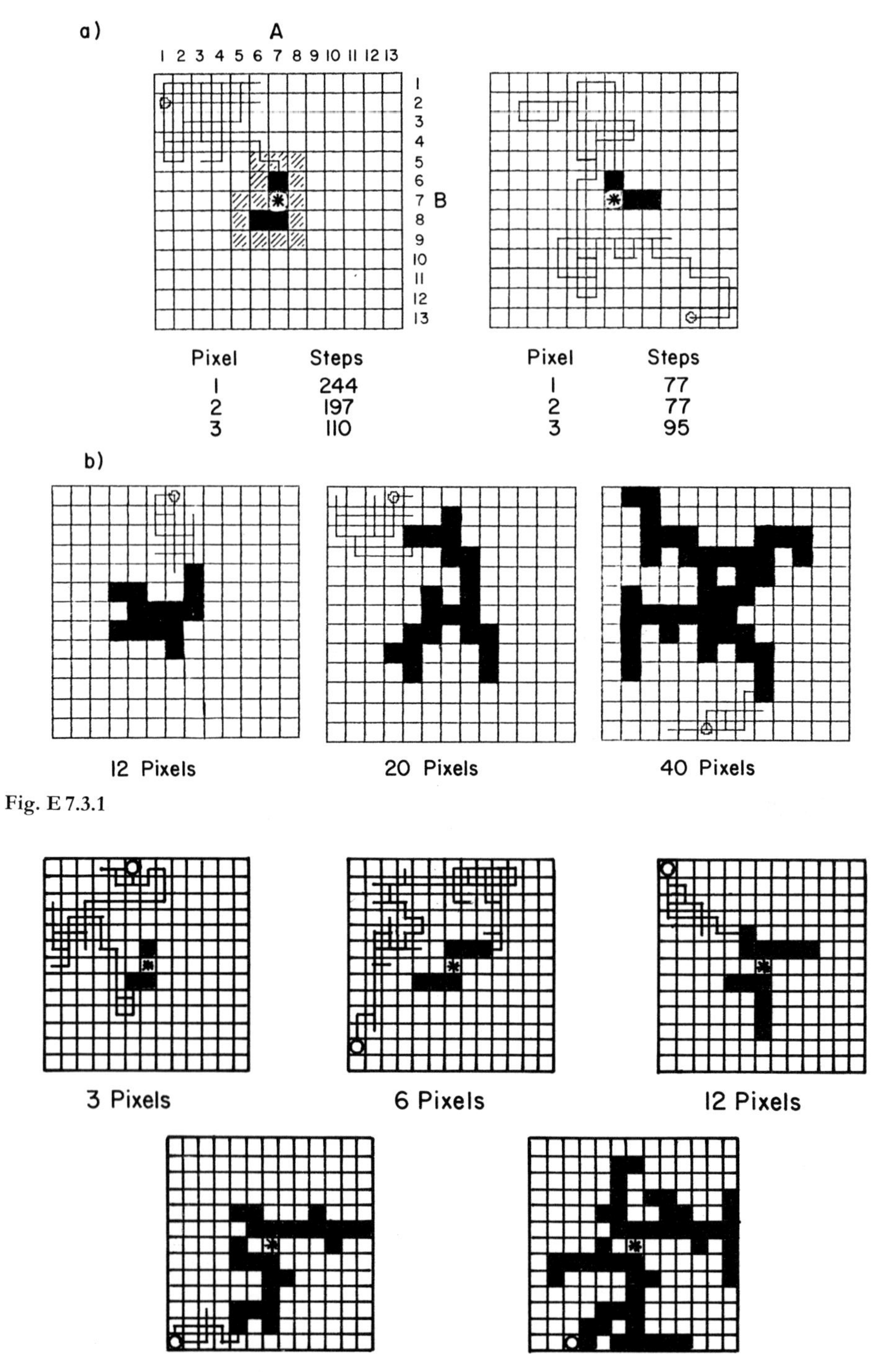

Fig. E 7.3.1

Fig. E 7.3.2

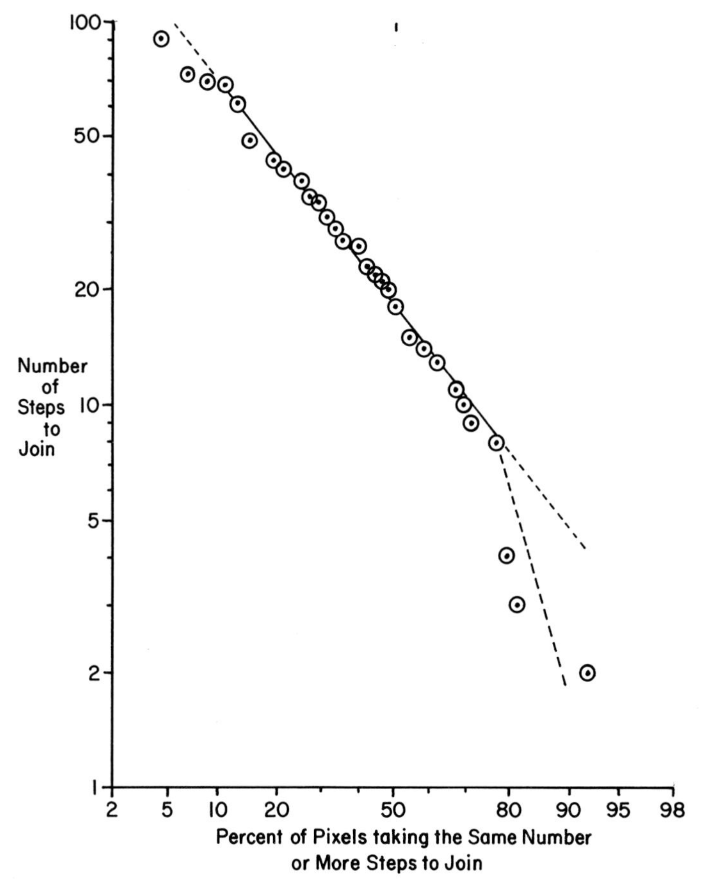

Figure E 7.3.3. The number of steps required to join the growing cluster undergoes a log-probability decay until the wall effects of the limited simulation space distorts the data.

and joining position of the pixel are shown. Students can discuss the fact that, if the random entry process is working, the pixels should just as often start from one side as from the other. There should also be no bias in which side the joining pixel encounters the growing agglomerate. By the time 48 pixels have joined the agglomerate in a space of 169 pixels it can be seen that the available space is too small to

Figure E 7.3.1. Growth stages in the development of a DLA cluster. a) Two separate attempts to grow the first three pixels of an agglomerate showing the number of steps required for each pixel to join. b) The growth of three further clusters to various sizes.

Figure E 7.3.2. In the small space used in this Monte Carlo routine the cluster starts to grow along the "wall" of the available space. This series shows the stages of the growth of one aggregate toward the boundary. At each stage the random walk of the last pixel joining the aggregate is shown.

Table 7.3.1. The number of steps taken by each randomly moving pixel before joining the growing aggregates of Figure E 7.3.1 (b).

Pixel	Steps	Pixel	Steps	Pixel	Steps
1	107	1	36	1	194
2	105	2	212	2	293
3	102	3	241	3	78
4	79	4	68	4	87
5	188	5	432	5	41
6	46	6	27	6	104
7	175	7	42	7	107
8	64	8	39	8	59
9	12	9	52	9	60
10	15	10	54	10	132
11	13	11	19	11	26
12	35	12	7	12	184
		13	13	13	6
		14	52	14	28
		15	6	15	17
		16	158	16	41
		17	41	17	4
		18	72	18	19
		19	37	19	11
		20	52	20	2
				21	13
				22	8
				23	162
				24	24
				25	5
				26	2
				27	32
				28	4
				29	9
				30	20
				31	2
				32	9
				33	11
				34	27
				35	12
				36	15
				37	1
				38	41
				39	12
				40	26

Table 7.3.2. For the aggregate of Figure E 7.3.2, the number of steps required to join the growing aggregate decreases as the growth approaches the boundary of the stagger space.

Pixel	Enters	at	Steps	Joins	at
	x	y		x	y
1	2	1	39	7	6
2	13	6	217	7	8
3	6	1	70	6	8
4	3	13	8	5	8
5	13	6	34	8	6
6	1	12	91	9	6
7	11	1	61	10	6
8	13	12	22	7	9
9	5	13	8	7	10
10	13	2	68	6	6
11	7	13	73	7	11
12	1	1	26	6	5
13	8	1	29	10	5
14	2	13	26	5	7
15	13	6	3	11	6
16	13	12	11	7	12
17	13	10	43	8	9
18	13	13	27	11	7
19	1	12	23	6	11
20	13	3	9	12	6
21	1	11	13	5	11
22	10	1	49	13	6
23	5	1	31	5	5
24	1	13	18	5	12
25	6	13	2	7	13
26	7	1	35	9	5
27	13	4	43	9	4
28	1	7	15	4	8
29	6	13	1	6	13
30	13	8	2	13	7
31	13	13	14	13	8
32	1	12	21	3	8
33	11	13	4	8	13
34	3	1	39	6	4
35	1	12	8	2	8
36	13	4	2	13	5
37	3	1	14	6	3
38	11	13	13	13	9
39	13	1	10	13	4
40	1	9	2	2	9
41	7	1	15	8	4
42	2	13	20	4	12
43	13	11	41	9	13
44	4	13	1	4	13
45	9	1	11	6	2
46	10	13	1	10	13
47	7	1	2	7	2
48	3	13	1	3	13

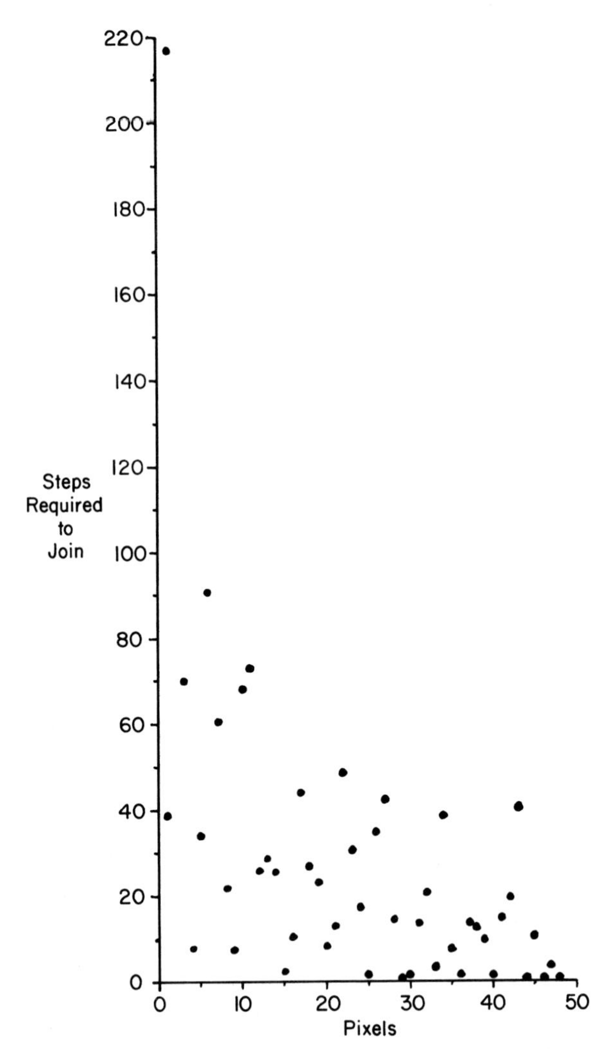

Figure E 7.3.4. The scatter of the pixel steps required to hit the growing agglomerate orthogonally is high.

let the agglomerate grow freely and the agglomerate is beginning to grow along the confining walls of the available space as shown in Figure E 7.3.2. In *Figure E 7.3.3* the distribution of the number of steps taken by all of the pixels joining the agglomerate from the first to the last pixel is shown. It can be seen that for 80% of the data, the variation in the number of steps taken is describable by the log-normal distribution. The number of steps per pixel is obviously declining as the agglomerate grows. The data of *Figure E 7.3.4* shows the decay in the number of steps required to join the agglomerate decays logarithmically as the agglomerate grows. The data

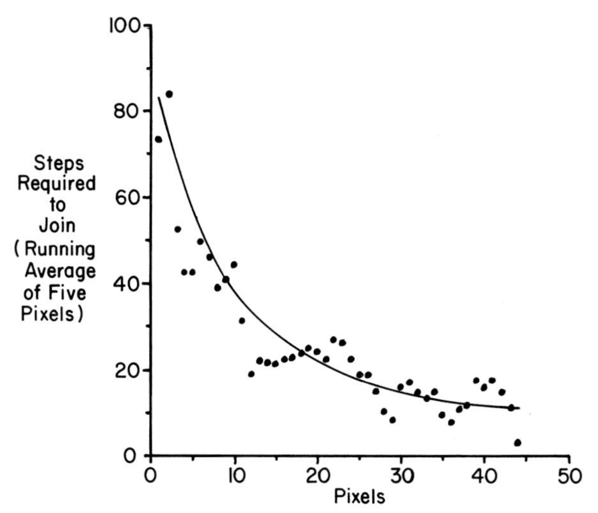

Figure E 7.3.5. The technique of using a running average can help bring out significant trends in scattered data.

obviously has a wide scatter. In such situations trends can sometimes be more easily shown by using what is known as a *running average*. One can average the number of steps taken by pixels 1 to 5 then pixels 2 to 6, 3 to 7 and so on. In *Figure E 7.3.5* the decay in the running average of five adjacent data points is shown. The word adjacent comes from two Latin root words, ad meaning "to or toward" and jocere meaning "to throw" (the words "javelin" for a throwing spear and "reject" for something thrown away have the same source). Students can discuss why in Figure E 7.3.3 the short number paths are more rare than expected and how this discrepancy would be corrected as one extended the size of the experimental space (the word *discrepancy* originally meant "not sounding right" from dis meaning "not" and crepare meaning "to sound"). It was used to describe a musical note which did not fit with the other sounds. Now it is used to describe anything which does not fit properly into a set of things such as data or facts.

Exercise 7.4. Studying the Effect of Electrode Configuration on the Structure of Fractal Fingers Grown in a Two-Dimensional Sawada–Matsushita Electrolytic Cell

The student can put a kink in the circular electrode and study what effect this has on the growth of the fractal system deposited in the cell. The student can also create a modified cell in which the cathode is a straight piece of copper wire and the anode is a graphite pencil core at the other side of the cell. This should lead to multi-fern type growth of fractal deposits on the line carbon electrode. Deposits grown this way can be compared to those which Voss simulated for the electrolytic processes [16].

References

[1] T. A. Whitten, L. M. Sander, *Phys. Rev. Lett. 47* (1981) 1400.

[2] The basic theory and development of diffusion-limited aggregation is reviewed in B. H. Kaye, *A Random Walk Through Fractal Dimensions*, VCH, Weinheim, Germany, 1989.

[3] B. J. West, M. Shlesinger, "The Noise in Natural Phenomena", *Am. Sci.* January-February 1990, *78*, 40–45.

[4] R. Richter, L. M. Sander, Z. Cheng, "Computer Simulations of Soot Aggregation", *J. Colloid Interface Sci. 100* (1984) 203–209.

[5] P. Meakin, "Formation of Fractal Clusters And Networks by Irreversible Diffusion Limited Aggregation", *Phys. Rev. Lett. 51* (1989) 1019–1022.

[6] W. T. Elam, S. A. Wolf, J. Sprague, D. U. Gubster, D. Van Vechten, G. L. Barz, Jr., P. Meakin, "Fractal Aggregates In Sputter Deposited Niobium Germanium Films", *Phys. Rev. Lett. 54* (1985) 701–703.

[7] R. Trottier, G. Clark, A. Hoffman, "Fractal Characterization of Experimental And Simulated Dendritic Capture Tree Growths in Filter Systems", paper presented at the 8th annual meeting of the American Association for Aerosol Research, Reno, Nevada, October, 1989.

[8] B. H. Kaye, "Describing Filtration Dynamics from the Perspective of Fractal Geometry", KONA, 9 (1991) 218–236, published by the Hosokawa Foundation, 780 Third Avenue, New York, NY, USA, 10017.

[9] C. Kanaoka, H. Emi, S. Hiragi, T. Myojo, in *Proc. 2. Int. Conf. Aerosols: Formation and Reactivity*, Pergamon, Oxford 1986.

[10] M. Matsushita, M. Sano, Y. Hayakawa, H. Honjo, Y. Sawada, "Fractal Structures of Zinc Metal Leaves Grown by Electro-Deposition", *Phys. Rev. Lett. 53* (1984) 286–289.

[11] L. M. Sander, "Theory of Ballistic Aggregation and Deposition", Reprint provided by Dr. Sander, Physics Department, University of Michigan, Ann Arbor, MI, USA, May, 1986.

[12] J. Walker, "Fluid Interfaces, Including Fractal Flows, can be Studied in a Hele-Shaw Cell", *The Amateur Scientist, Sci. Am.*, November, 1987, pp. 134–138.

[13] H. S. Hele-Shaw, *Nature* 58 (1898) 34. See also [14].

[14] See review of these experiments in Chapter 8 of B. H. Kaye, *A Random Walk Through Fractal Dimensions*, VCH, Weinheim, Germany, 1989.

[15] L. Paterson, "Fingering With Miscible Fluids in a Hele-Shaw Cell", *Phys. Fluids 28* (1985) 26–30.

[16] R. F. Voss "On 2-D Percolation Clusters and on Multi-Particle Fractal Aggregation", in F. Family, D. P. Landau (Eds.), *Kinetics of Aggregation and Gelation*, pp. 8–9, Elsevier, Amsterdam 1984.

Chapter 8

Infinite Coastlines and Other Wiggly Lines

Chapter 8 Infinite Coastlines and Other Wiggly Lines 305

Section 8.1 Easy Questions and Impossible Answers 307
Section 8.2 Beware! Richardson Plots May Have More Than One Straight Line Data Relationship Lurking in the Scatter of the Data Points 311
Section 8.3 Characterizing Profiles Which Manifest Various Fractal Structures Around Their Perimeter 314
Section 8.4 Estimating Fractal Dimensions by Penny Plating Procedures . 324
Section 8.5 Mosaic Amalgamation – Another Variation of the Minkowski Sausage Method for Characterizing Rugged Curves 331
Section 8.6 Fractal Rabbits and Manitoulin Island 334
Section 8.7 How to Tell a Vulcan from Another Carbon Black 340
Section 8.8 Putting Fractal Dimensions to Work in Applied Science 344
Section 8.9 More Ideas for Putting Fractals to Work 359
Exercises 363
References 383

Chapter 8

Infinite Coastlines and Other Wiggly Lines

Section 8.1. Easy Questions and Impossible Answers

The story "Alice In Wonderland" was written by a serious mathematician under the assumed name of Lewis Carrol (The author's real name was Ludwig Dodson). In Alice, Lewis Carrol pokes fun at the logical tangle one can become entrapped in if words are allowed to dominate ones thinking. The strict logic of "Alice In Wonderland", and it's sequel "Alice Through the Looking Glass" should warn us of the need to check with reality when words seduce us into paradoxes and impossible conclusions [1]. Lewis Carrol would have wallowed gleefully in the paradoxes of what we now know as fractal geometry. Consider for example the kind of dialogue Alice could have had with a fractal geometer.
Fractal Geometer: "What is the length of a coastline?"
Alice: "Which coastline?"
Fractal Geometer: "Any coastline"
Alice: "Don't be ridiculous, I must know which coastline you are interested in before I can answer your question."
Fractal Geometer: "Not so."
Alice: "Well, you answer your own question."
Fractal Geometer: "That's easy; all coastlines are the same, they are infinite in extent."

Strictly speaking, the Fractal Geometer is correct. The answer that all coastlines are infinite is true in a strict mathematical sense but this fact is of no practical use to the geographer or the geologist. If a geographer answers a question, "How long is a coastline?", the question always implicitly assumes some measurement procedure by which one is estimating the coastline. However, as geographers' measurement techniques become more and more refined, the estimate becomes larger and larger. In fact, the question disappears once one is down to the size of the coastline which is affected by the movement of the tides and the to and fro movement of the waves. If one attempts to give an answer to within a meter to the question, "how long is a coastline?", then not only is the coastline indeterminate but its physical extent changes minute by minute as the tides ebb and flow. Not only are they practically infinite but coastlines change permanently as they erode under the action of the waves and weather (note that the word *erode* comes from a Latin root word rodere "to nibble" and "eat away". This root word has also given us *rodent* for animals that nibble at objects and *corrosion* meaning "eaten away by chemicals and other agents"). Mandelbrot, when discussing the structure of coastlines, points out that there is no answer to the question, "how long is the coastline of Great Britain?". One can only answer this question by giving an estimate based on a particular method of estimation [2]. Furthermore, the geographer who attempts to improve his estimates of the coastline length faces the problem that the estimates increase quickly as the scale of scrutiny decreases. How can one usefully summarize all the information on the estimates of the coastline at various scales of scrutiny?

Mandelbrot, in his book "The Fractal Geometry of Nature", suggested that we can usefully describe the difference in the ruggedness between two coastlines, both of which have infinitely long boundaries, by switching our attention from the question of, "how long is the boundary?", to "how quickly do the estimates of a boundary tend to infinity as we change the resolution of inspection of the boundaries?". The physical basis of the way in which Mandelbrot used the rate of increase of the estimates of boundary length to characterize the ruggedness of a boundary can be appreciated from the data available on two theoretical curves, the Koch triadic and quadric islands. We met these two curves briefly in Exercise E 1.3. We discovered that these interesting curves have infinite area and infinite perimeter and that the quadric island has a perimeter of higher infinity than the triadic island. The appearance of these two curves after three iterations of the construction algorithm are shown in *Figure 8.1* (a). These curves are described by Mandelbrot as *ideal fractal curves* because of the fact that they are physically self-similar at all levels of inspection resolution.

A physical scientist looking at the two curves of Figure 8.1 (a) immediately sees in them the possibility that they are ideal representations of natural systems such as agglomerated precipitates, fumes and fractured rocks. To distinguish between the mathematical curves with ideal properties and natural systems which look like fractal curves, such as soot agglomerates, terminology has been developed in which systems such as soot profiles and other naturally occurring systems are referred to as *natural fractal systems*. When we investigate the perimeter properties of the triadic and quadric islands by means of a compass walk of the type illustrated in Figure 8.1 (b) we generate data like that of Figure 8.1 (c). The step size represents our resolution of inspection and the total length of the polygons constructed by walking around the profile with the compasses represents our perimeter estimate as illustrated in the Figure 8.1 (b).

The terms *resolve* and *resolution* are ones that cause problems to the novice because they have very different meanings in everyday language and in technical English. The root word re at the beginning of a word can indicate "doing something in an aggressive or intensive manner" or "coming back". The word solvere means "to

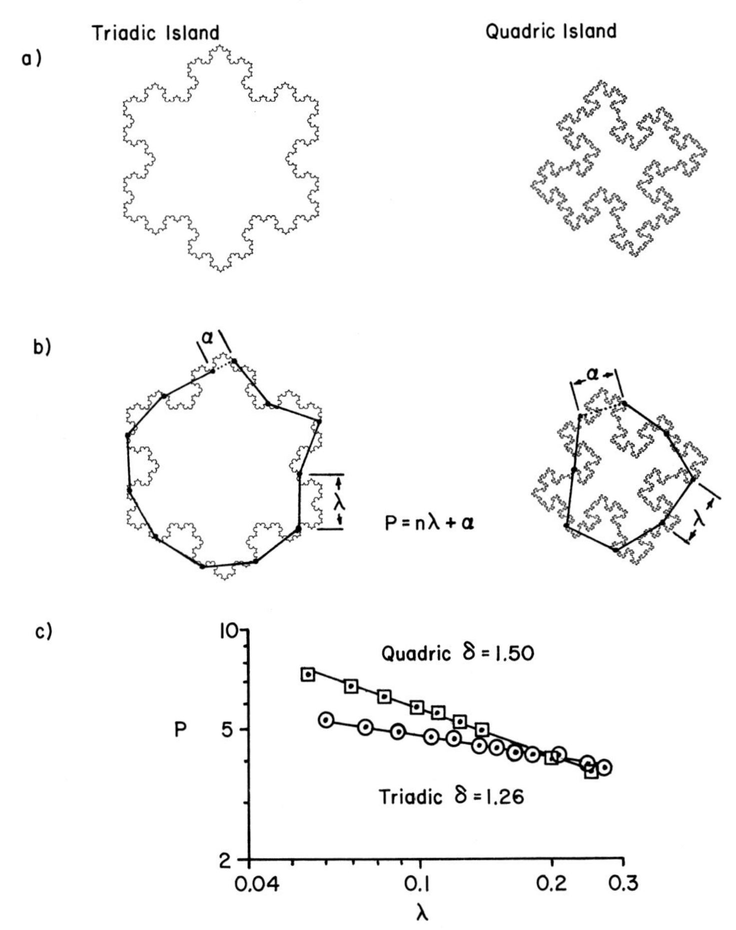

Figure 8.1. The rate of the increase of perimeter estimates as one looks at the Koch islands at increasing levels of scrutiny is a useful measure of the ruggedness of their profiles. a) Depiction of third order islands. b) Basic procedure used to estimate the profile by stepping out along the profile with step size λ. c) Richardson plot of normalized perimeter estimates versus the normalized stride length for both islands.

loosen something" thus a solution is something in which we loosen the parts of a molecule from each other and thus a copper sulfate solution contains the copper and sulfate ions split up from each other. On the other hand, in everyday English, a solution to a problem means that we have been able to break it down into its parts in order to be able to gain a complete understanding of the problem. Originally, a "resolve" was an intense desire to reach a solution to a problem and hence our "New Years resolutions" are intended to reach solutions to our personal problems. To resolve a conflict was to break it up into its different parts.

The scientific use of the word resolution came about from the need to develop words to describe what one saw through a microscope or in any examination of something like a map. If we look at a small map of England we would be able to make out the broad outlines of the country and we could put dots in place for the major cities. However, we could not see any useful detail on such a low magnification map which would help us to travel by car across the country. In scientific terms, one is not able to see separate parts on the map such as roads so that the scientist says one has not resolved the details of the structure of the road system. If one increases the magnification one can now see things separately which were fused in the lower magnification picture. The scientist says that by increasing the magnification he is able to resolve the separate parts of detail in the structure of the map. The *scale of magnification* then became known as the *resolution of inspection* and the phrase began to be used that in a low-resolution inspection one cannot see the detail that one can see at high resolution. A high-resolution inspection of a system enables one to see details which are not visible at a lower resolution. The term scrutiny is also used to describe the magnification that one uses to look at a picture. It is said that a low-resolution inspection corresponds to a low scale of scrutiny and that at a high scale of scrutiny one can resolve details. The term scrutiny comes from a Latin word meaning "rags". It can be traced all the way back to the searches carried out of a citizen's house by Roman officials. A high-intensity search of the house even looked at the rags and was said to be a detailed scrutiny of the residence. When we discuss the measurements of the boundary of Great Britain by striding around the boundary, the large steps are said to be low-level resolution – low-scrutiny-level inspection of the boundary. As we decrease the step size of the exploration of the boundary, we are said to increase our resolution of inspection, since at small step size we can discover more details in our exploration of the boundary than when we use a large step size.

The data of Figure 8.1 (c) illustrates how the perimeter estimates for the quadric island, as we take smaller and smaller steps, move to infinity much more quickly than those for the triadic island. A series of careful measurements soon establishes the fact that the perimeter data for the triadic island on log–log graph paper has a slope of -0.26. Some consideration of the formation algorithm for the triadic island shows that at each stage of the construction of the island, as illustrated in Figure E 1.3.1, the perimeter of the island increases by the ratio 4/3 so that the slope of the dataline on log–log paper should be related to $\log 4/\log 3 = 1.26$. In the mathematical sense, the slope of the perimeter increase with increasingly resolution of inspection is related to the formation dynamics of the mathematical curve. This general truth is also illustrated for the quadric curve data in that its slope is -0.5 which is related to $\log 8/\log 4 = 1.50$ which is the rate at which the perimeter increases for each increase of construction algorithm.

Mandelbrot suggested that the slope of the graph of perimeter estimates against the inspection resolution parameter on log–log graph paper, a display of data now called a *Richardson Plot,* was very characteristic of an ideal curve and represented the ability of the fractal curve to occupy space. The triadic island perimeter does not occupy space as efficiently (as densely) as that of the quadric island. This is shown by the slope of the Richardson data plot for the two curves. Mandelbrot also suggested that the scientific community could start to describe the space filling ability of a curve by adding the slope of the Richardson plot data to the topological dimension of a curve. The combination of the topological dimension and the slope information he described as the fractal dimension of the curve. The Koch triadic and the Koch quadric islands are said to have fractal dimensions of 1.26 and 1.50 respectively. When describing the structure of the boundary of a curve this type of fractal dimension is called a *boundary fractal dimension* to distinguish it from the mass or density fractal dimension of a system of the type discussed in the previous chapter.

Very often in discussions of fractal dimensions the short form "fractal" is used for physically different types of fractal dimensions. The reader sometimes may have difficulty in deciding exactly what a writer means by "fractal" especially in some of the earlier publications on the subject. When one explores the possibility that a natural boundary has a boundary fractal dimension which can be used to describe its structure one has to investigate the possibilities that a Richardson plot of the perimeter estimates against resolution parameter generates a straight line. The slope of such a line, should it appear on the Richardson plot, is then taken to be the empirically discovered boundary fractal of the system. In an operational sense, a curve which manifests a fractal region on a Richardson plot is described as a fractal because it behaves like a fractal. Since the fractal dimension of the boundary of the Koch islands is related to their formation dynamics it is not unreasonable to assume that the empirical fractal dimension of the boundary of a rugged system will contain information not only on its structure but on the formation dynamics which generated the boundary. This is one of the reasons why scientists are so interested in characterizing the boundary fractals of many complex systems.

In later sections of this Chapter I will indicate various areas of research where fractal investigations are flourishing, but first we will explore some different methods for characterizing boundary and line fractals and give examples of situations to be avoided when exploring empirically the possibility that a natural system has a fractal structure.

Section 8.2. Beware! Richardson Plots May Have More Than One Straight Line Data Relationship Lurking in the Scatter of the Data Points

As I have discussed in detail elsewhere, my first efforts to characterize natural fractal systems focused on the problems of characterizing the structure of the carbon black profile shown in *Figure 8.2*(a) [3]. The first experimental data curve that I drew

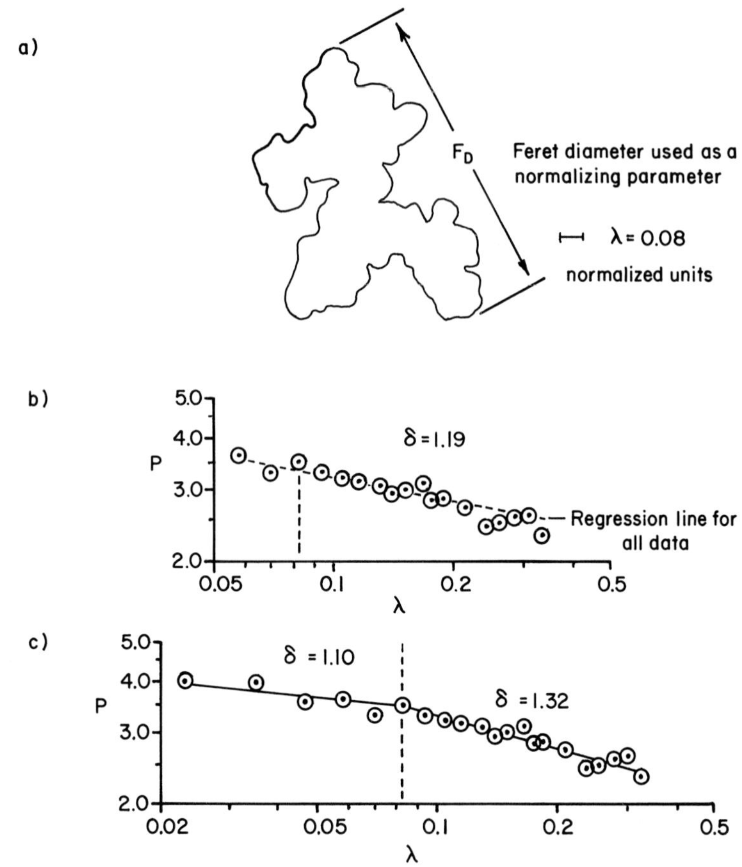

Figure 8.2. Detailed exploration of a real fine particle profile revealed that natural boundaries can be more complex than ideal fractal curves such as the Koch islands of Figure 8.1. a) Carbon black profile subject of the first exploration of a fine particle profile. b) First data plot of a walk exploration of the perimeter of carbon black profile. c) Upon further exploration of the profile it was found to manifest two data lines on the Richardson plot.

in my investigations is shown in Figure 8.2(b). As I looked at this curve in the summer of 1977, a time when fractal geometry was a very new subject, I was very pleased with the part of the data curve up to a stride length of 0.08 normalized units since with some scatter it definitely showed a straight-line relationship between perimeter estimates and inspection resolution parameter. However, I was somewhat troubled by the last two points of the curve which definitely seemed to be falling away from the data curve. I wrestled with this problem of data interpretation for some time and then decided that, even though the compass points were set at a small distance and the estimate of the perimeter was an extremely tedious operation, I would carry out further measurements to generate the more extensive data set of Figure 8.2(c).

Initially, when contemplating Figure 8.2(c), I felt that by obtaining more data I had not solved the problem; rather it had increased the complexity of the situation by providing two lines instead of one. However, the explanation of the two data relationships became possible once I realized that the break point between the two straight line data relationships represented a change-over point between two different aspects of the carbon black structure. This can be appreciated from the sketch shown in Figure 8.2(a). For steps larger than 0.08, one is exploring the basic structural features of the profile whereas for step sizes smaller than 0.08 one is looking at the texture of the agglomerate. Pursuing the idea that the interaction of causes creates the structure of the fractal system it is now apparent that for the high-resolution inspection of the carbon black agglomerate one was looking at the way in which tiny spheres of carbon produced by incomplete combustion of a hydrocarbon in a burning flame had packed together to form primary units of soot agglomeration. These subunits had collided in the cooler part of the flame to produce agglomerates made up of obvious subunits.

Thus, the carbon black profile of Figure 8.2(a) had probably been formed by the collision of at least three clearly separate subunits to produce the overall agglomerate. When the line changed its slope with increasing resolution it was because at the higher resolution inspection one was looking at the packing of the subunits of the agglomerate not the structure of the agglomerate. This explanation of the physical significance of the line at high resolution is now well established. In current terminology, it is said that the slope of the line at high resolution generates the *textural fractal dimension* of the agglomerate and that the slope of the line at a low resolution of inspection defines the *structural fractal dimension* of the agglomerate (both fractal dimensions are boundary fractals). This example illustrates how the fractal structure of a natural system may only exhibit a given fractal structure over a short range of inspection resolutions and that there may be more than one line on a Richardson plot representing different formation dynamics for different aspects of the fractal boundary. The range of resolution used when making and reporting measurements of boundary fractal dimensions should always be quoted at least once in the data report.

Many scientists in their initial attempts to use fractal dimensions left the computer to draw the best straight line through available data. However, the computer, unless it is specifically instructed to look out for more than one line, is likely to draw a line such as the dotted line of Figure 8.2(b) which suppresses the fact that there are really two lines. This example should warn the student that computers are electronic robots that may suppress detail in ines if they are left to draw their own conclusions without the benefit of human experience.

Section 8.3. Characterizing Profiles Which Manifest Various Fractal Structures Around Their Perimeter

Before fractal geometry was invented it was very difficult to characterize the structure of coastlines. One often had to have a vivid imagination to be able to describe the structure of island boundaries. Generations of British students grew up learning that the profile of Great Britain looked like an old lady with a Scottish bonnet and a Cornwall leg riding a welsh pig (see cartoon at the front of this chapter). The coast of Ireland when rotated by 90° looks like a very rugged terrier dog or a shaggy buffalo depending on ones viewpoint.

Again, generations of British school children learned that, "long legged Italy kicked poor Sicily into the Mediterranean Sea." With fractal dimensions, we can describe structural features of the coast but note however, that we still have to have some overall shape factor to describe the long leggedness of Italy or the height of the old lady of Great Britain.

Fractal geometry not only enables us to move away from picturesque descriptions of coastlines but it also enables us to interpret the structural features of a coastline in terms of possible formation dynamics. However, any exploration of real coastlines soon teaches the student that one must be careful not to blend data from different structures around the coastline in the same Richardson plot.

In the previous section, the reader was warned that computers will sometimes draw straight lines through data which in reality manifest more than one line. In the same way, a blind approach to the characterization of a coastline or other natural fractal can result in ruggedness information which merges data from regions of different ruggedness to generate a parameter which is not directly related to any specific section of the structure of the profile. For example, a visual inspection of the coastline of Great Britain immediately suggests that the ruggedness of the Atlantic coastline (the west coast) is more rugged than the east coastline. This difference becomes vividly apparent if one constructs synthetic islands by splitting the coastline into two regions and using reverse or inverted images of the coastline to create a synthetic island as illustrated in *Figure 8.3*(b, c) [4]. In Figure 8.3, the Richardson plots for data generated by exploring the structure of the natural coastline of Great Britain and the two synthetic islands formed from synthetic assembly of the west coast (SCOWAC) and the east coast (SCENG) are summarized.

The data of Figure 8.3 was generated using an experimental measurement technique pioneered by Schwartz and Exner [5]. The method is called the *equipaced polygon estimation technique*. The method is often referred to briefly as the *equipaced method*. The basic concept of this method can be appreciated from the data of *Figure 8.4*. The line to be characterized is first digitized as in Figure 8.4(a). Its length is then estimated by stepping along the line a stated number of steps and subsequently drawing a chord between the starting and ending point of the paced out distance as shown in

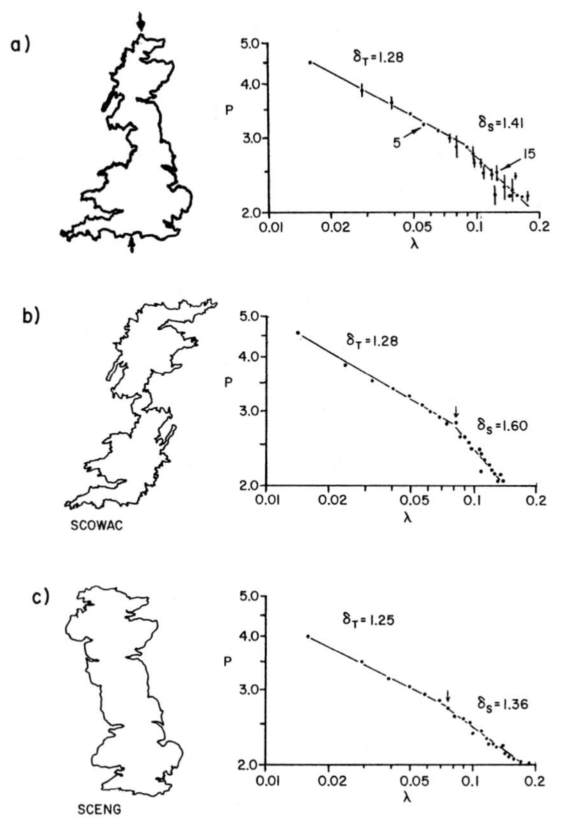

Figure 8.3. Coastlines may have regions of different ruggedness. This fact is illustrated for the coastline of Great Britain which is more rugged on the west side than the east side. This difference in ruggedness can be made visually apparent by creating synthetic islands using the two sides of Great Britain separately. SCOWAC: two copies of Scotland, Wales and Cornwall. SCENG: two copies of the east coastline of Scotland and England. a) Data for the whole coastline of Great Britain. b) Image made by splitting Great Britain at the arrows shown on the profile of (a) and combining the line of the west coast with an inverted image of the same part of the coastline. c) Image made by splitting Great Britain at the arrows shown on the profile of (a) and combining the line of the east coast with an inverted image of the same part of the coastline.

Figure 8.4(b), (a *chord* is defined in a dictionary as "the string of musical instrument" or "a straight line joining any two points on a curve". The two meanings grew out of the practice of making musical instruments out of curved pieces of wood with strings stretched between two points on the support bow to form the chord which is plucked to make a musical sound).

We can imagine that if one were to pace out a wiggly line on the ground, after a given number of paces the surveyor would drive a peg into the ground and create a chord by tying a piece of string from the peg located at his starting point to the

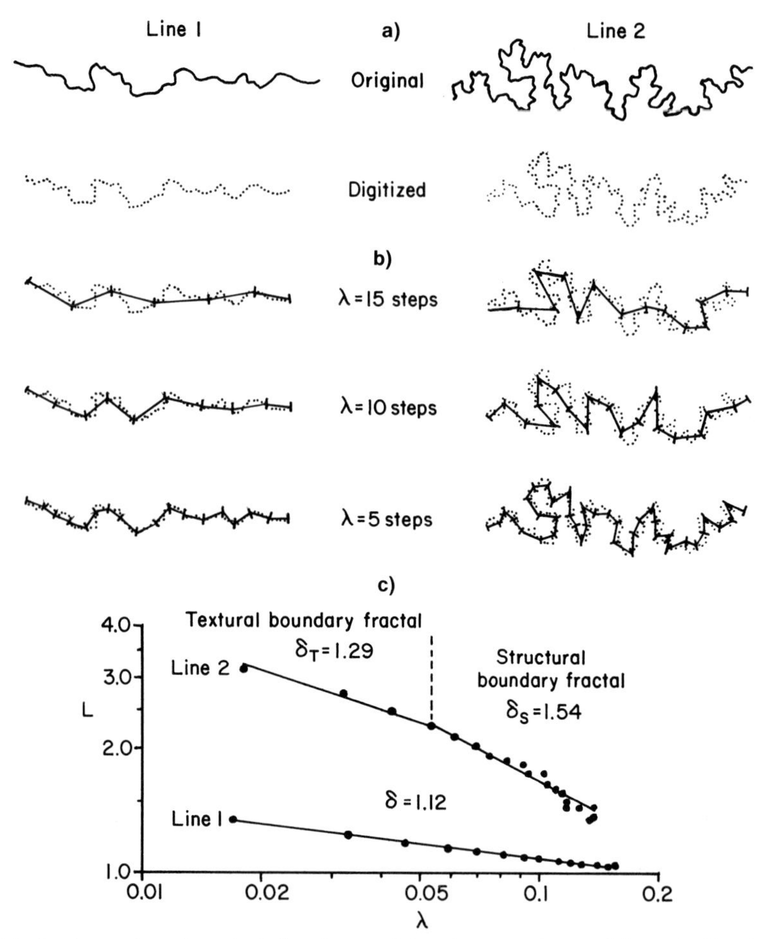

Figure 8.4. In the equipaced method for characterizing the ruggedness of lines the line lengths are estimated using sets of chords representing various "stepped out" distances along the digitized lines. a) Original and digitized profiles of the rugged lines to be characterized. b) Length estimates based on various "stepped out" distances. c) Richardson plots for the equipaced explorations of the two rugged lines of (a).

new peg. The overall length of the wiggly line is estimated by adding up the lengths of the chords created by the equipaced exploration of the line. In Figure 8.4(b), estimates of the length of the two different lines are shown for equipaced chords of 5, 10 and 15 steps. The line-length estimates and the paced-out step size can be normalized by using the straight line drawn between the beginning and end point of the rugged line. When the data is summarized on a Richardson plot, the number of steps used to generate the chord is the measurement of the resolution inspection of the line. The Richardson plot for the exploration of the rugged lines of Figure 8.4(b) are presented in Figure 8.4(c).

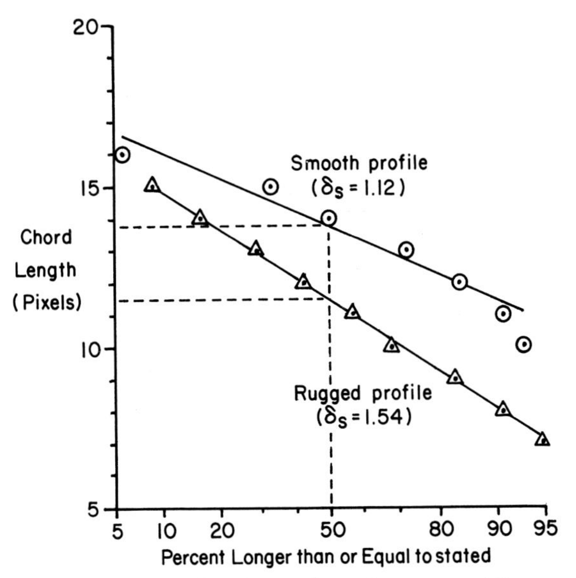

Figure 8.5. The variations in the chord lengths paced out along a rugged line may be describable by the Gaussian distribution with the mean chord length proportioned to the fractal dimension of the line. The data shown are for five step chords on the lines of Figure 8.4.

When estimating the ruggedness of a profile or line using the compass striding technique all the steps are of the same size. In the equipaced method, the chord lengths vary but the technique for estimating the length of the line is unbiased and systematic. The variations in the chord length for a given number of steps taken along the rugged boundary are often Gaussianly distributed with the standard distribution of the chord length being related to the fractal dimension of the boundary. This is illustrated by the data of *Figure 8.5*. In *Figure 8.6*(b), the way in which the average chord length varies with ruggedness for the exploration of the lines of Figure 8.6(a) at a resolution of five steps, is shown.

The data of Figure 8.6(b) demonstrates the relationship between the average chord length and the fractal dimension of the lines. It has been shown that this fact can be used to teach the computer how to split a rugged boundary manifesting more than one fractal dimension into separate regions for separate inspection [4].

If we look at the Richardson plot for the east and west side of Great Britain (Figure 8.3) we see that the structural fractal of the coastline is much higher for the west side than it is for the east. This is probably a function of the fact that the west coast of Great Britain is generally hilly whereas the east side tends to be made up of coastal plains of clay or low-strength chalk deposits. This again points to the fact that the fractal dimension contains information on the formation and/or structural strength of the rocks forming the coastline.

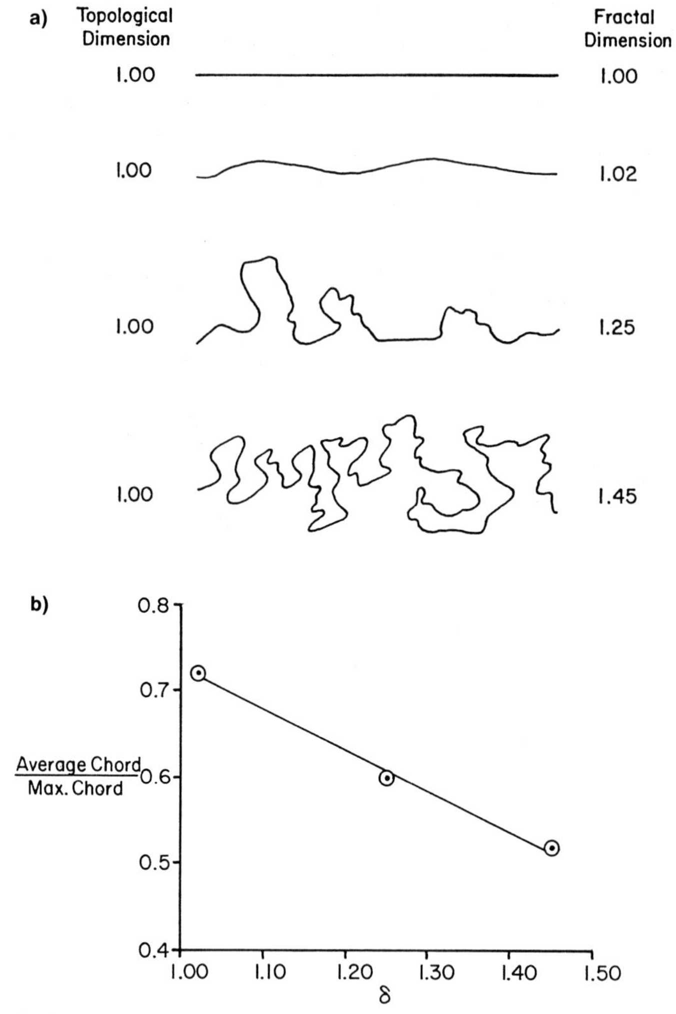

Figure 8.6. As the ruggedness of the lines increases, the average chord length in an equipaced exploration at the same level of resolution decreases. a) Lines of various ruggedness. b) Ratio of average chord to maximum chord for exploration of the three rugged lines with five step chords.

It should be noted that the value 1.26 is widely reported for the coastline of Great Britain. From the data of Figure 8.3, it is now apparent this is the average value of the textural fractal of the coastline since it is intermediate between the 1.25 and the 1.28 for the textural fractal dimension estimated for the two portions of the coastline.

Everyone should do at least one fixed stride exploration of a rugged boundary using a pair of compasses (see Exercise E 8.1) to gain a physical appreciation of the basis of the Richardson plot method of evaluating a fractal dimension. However, the explorer of rugged lines soon discovers that the equipaced method has the big advantage that the calculations and explorations of the lines can be handed over to a computer once the digitized version of the boundary line has been stored in the computer memory. This is because the chord length between two points paced out on the line can be calculated using *Pythagoras' theorem* and the co-ordinate addresses of each point on the digitized version of the boundary.

Pythagoras was a Greek mathematician who lived between 582 BC and 497 BC. The use of Pythagoras' theorem to automate fractal calculations is so important that it is worthwhile making a short diversion to make sure that everyone knows the meaning of the words and ideas wrapped up in Pythagoras' theorem. Many students gain an intense dislike of Pythagoras' theorem early in their study of geometry. This is because they often meet it as an irrelevant idea in a strange subject known as trigonometry.

A student can be forgiven sometimes for wondering if scientists deliberately switch words on them to make a subject difficult. For example the word *triangle* means "three angles" from the Latin word tres meaning "three" and the word angle which in Latin means "to bend". An angle is formed by bending a line. However, when mathematicians needed a name for the systematic study of triangles, they chose the Greek word *trigonometry* from three Greek root words tri meaning "three", gonia meaning "angle" and metron "to measure". The word gonia has given us *diagonal* for a line drawn through the middle of an angle and the word *polygon* for a many sided angle from poly meaning "many".

As a student, I remember struggling with many trigonometric relationships wondering how they could possibly be of any use in a real world. Trigonometry bored me to tears! I particularly remember having great difficulty spelling *isosceles* for a triangle with two equal sides. I now know that isosceles comes from Greek roots meaning "having two equal legs". I wish I had known that when I was trying to study an isosceles triangle.

When one first meets Pythagoras' famous theorem in trigonometry, it is usually stated in the following way:

> "*For a right-angled triangle the sum of the squares of the two sides forming the right angle is equal to the square of the hypotenuse*"

The meaning of this statement is illustrated by the information summarized in *Figure 8.7*(a) where the squares represented by the capital letters A, B, and C have been drawn on the sides of the triangle with sides a, b and c. Our major interest in the theorem of Pythagoras' is that we often know b and c and wish to calculate a. The word *hypotenuse* has probably caused more grief to novice geometry students than any other word. We can understand the meaning of this word if we appreciate that many geometric figures began life as instruments for surveyors and builders. In particular when one wanted to build a vertical wall, one often needed to have devices to help one to make sure the wall did not lean one way or the other. A simple device used by the early builders was the plumb line; still a familiar device on today's

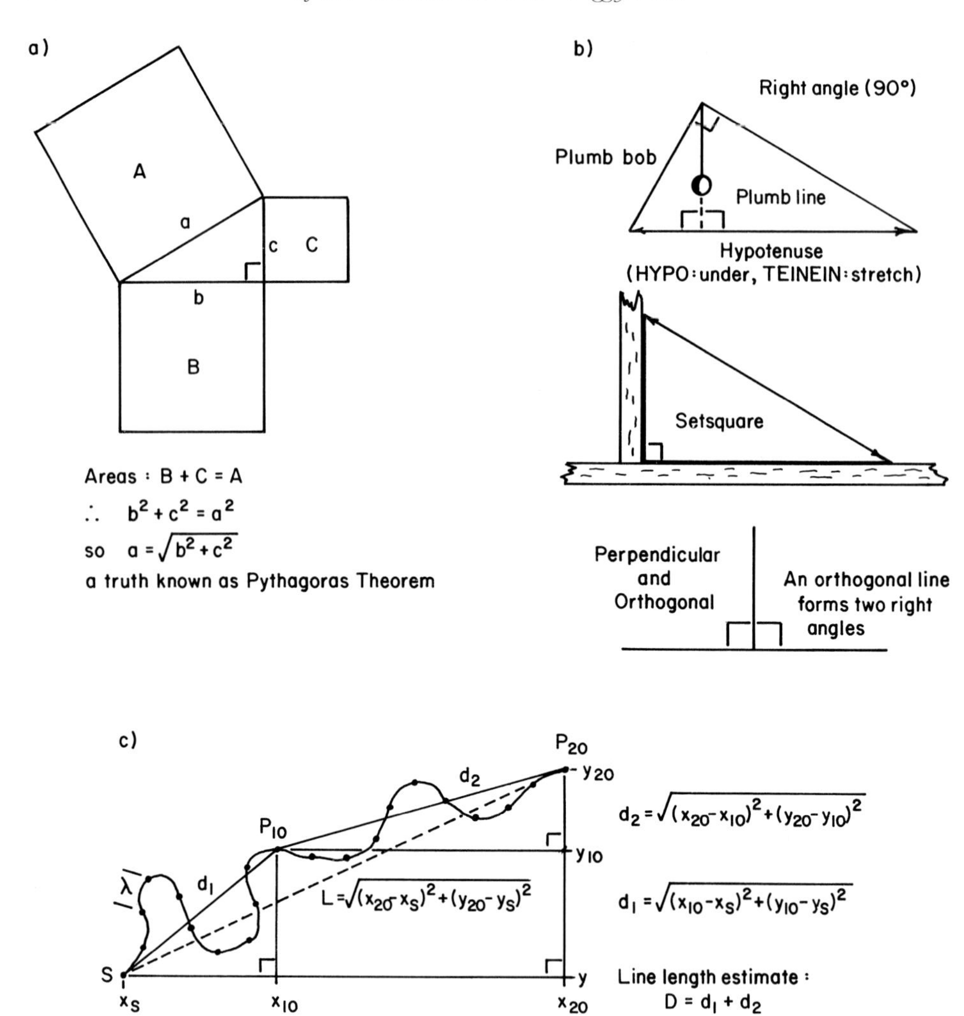

Figure 8.7. Pythagoras' Theorem can be used to manipulate the data from a digitized version of a rugged line to estimate the perimeter from equipaced exploration of the boundary. a) Basic representation of Pythagoras' theorem algebracially and geometrically. b) The plumb line and set squares were tools of Greek surveyors. c) Putting Pythagoras' theorem to work in the equipaced method.

building sites. The *plumb line* is a weight on a string hanging from a support. It took its name from the fact that the small bob of heavy material making the string hang vertically was made of lead (the word for lead in Latin was "plumbus" hence, the name *plumber* for a man who used to work mainly with lead pipes but who now often works with plastic pipes). The plumb line would often be suspended from the junction of two pieces of wood which were joined together at the bottom by a base. This frame usually had a right angle at the top since it could also be sat on its side and used to make a visual check to see if a wall being built was going straight up. This is illustrated in Figure 8.7(b). The builder, when trying to check on the straightness of a wall, uses a large wooden triangle with a right angle base line. The wooden triangle, used in the way shown in the sketch, is called a "*set square*". This is a strange name for a triangle until one realizes that it is what it does, not what it looks like, that has given it its name.

The third line of the frame for supporting the plumb line was called the hypotenuse by the Greeks because the third line was "stretched out" beneath the support point of the plumb line. The word hypotenuse comes from hypo meaning "under" and teinnein meaning "to stretch". One of the reasons for the obscurity of this word for geometry students is the fact that modern textbooks always show right-angled triangles with the base of the triangle sitting firmly on the ground and the hypotenuse leaning like a ladder from the ground to the wall. If it is drawn the way the Greeks used to draw it the hypotenuse is a more meaningful name.

The line drawn from the tip of the frame to the base of the triangle using the plumb line is known as the *perpendicular* line from the Latin words per meaning "through" and pendare "to hang" (this word has also given us *pendant* and *pendulum* for things that hang down around the neck and in a grandfather clock). Mathematicians describe the perpendicular line joining a horizontal base line as being *orthogonal* to the base line. The Greek word orthos means "straight" (an *orthodontist* is someone who makes the teeth straight). When the perpendicular meets the base line orthogonally it is said to form two right angles. The word right is one of those words which has so many meanings that its scientific meaning sometimes escapes the novice student. It comes from the Latin word rectus which has several meanings including "upright". A right angle is formed when one line stands upright on another. In Figure 8.7(c) the use of Pythagoras' theorem to work out the chord lengths created by stepping out along a digitized line is illustrated. When we use the theorem of Pythagoras to calculate the chord lengths of an equipaced exploration of a rugged line the entire process of calculating perimeter estimates of a digitized coastline or any other profile can take place in the computer processing part of a machine without having to view the boundary throughout the measurement process.

In *Figure 8.8* two profiles which have different ruggedness located around different parts of the perimeter are shown. The first profile, in Figure 8.8(a), is that of a piece of *debris* found within some lubricating oil. The word debris is a recent arrival in the English language from the French word brieser meaning "to break". The fine debris in the lubricating oil are the pieces of material which have broken away from the surfaces being lubricated. Engineers can look at the structure of the fine particles of debris in the oil and discover what is going wrong within an engine being lubricated. When working with very complex engines, such as modern aircraft jet engines, taking the engine apart to see if it needs maintenance causes more problems than leaving the engine running. By looking at the debris the engineer can decide if maintenance is required and in this way can avoid unnecessary stripping down of

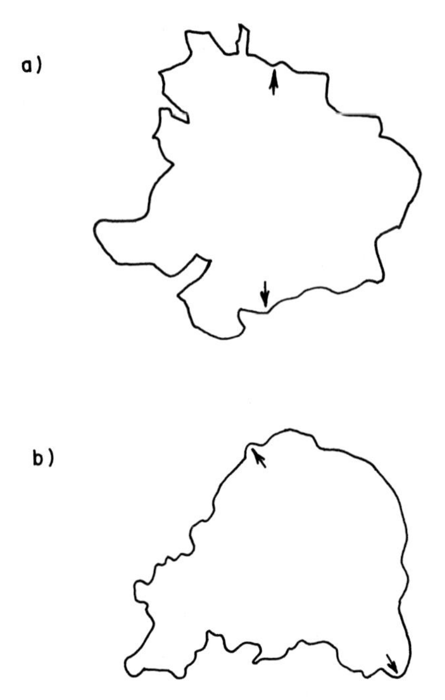

Figure 8.8. Many naturally rugged systems manifest different levels of ruggedness around the periphery of the profile. The ruggedness of the two systems shown above differs to the left and the right of the two arrows shown. a) Debris fragment from lubricating oil [9]. b) Activated leukocyte [4, 10].

the engine [6, 7]. The scientific study of the wear caused by the rubbing of surfaces together is known as *tribology* from the Greek word tribos meaning "to rub".

For historical reasons the detailed technology for looking at the lubricating oil debris to gain tribologically important information is known as *ferrography*. This is an unfortunate name because it means literally "drawings of iron" from the the Latin word ferrum meaning "iron". The name arose because the first type of debris studied was either made of iron or was contaminated with iron which enabled the debris to be separated from the oil by a strong magnetic field. More advanced techniques of ferrography sometimes use other methods to look at the debris and the study has been extended from the wear of engines to such interesting problems as the study of the debris to be found in a human synthetic joint such as a hip or knee replacement. Note that the medical term for a replacement part such as a metal knee joint is *prosthesis*, another word derived from the Greek word "to place" (see the word web for synthetic words, *Figure 8.9*). By studying the shape and size of the debris washed out of a human prosthetic device, the surgeon is able to decide what is causing the wear inside the joint and whether it needs to be replaced [8].

The piece of debris shown in Figure 8.8(a) was described by Beddows and came from oil used to lubricate a gear [4, 9]. The different roughness of the two sides of

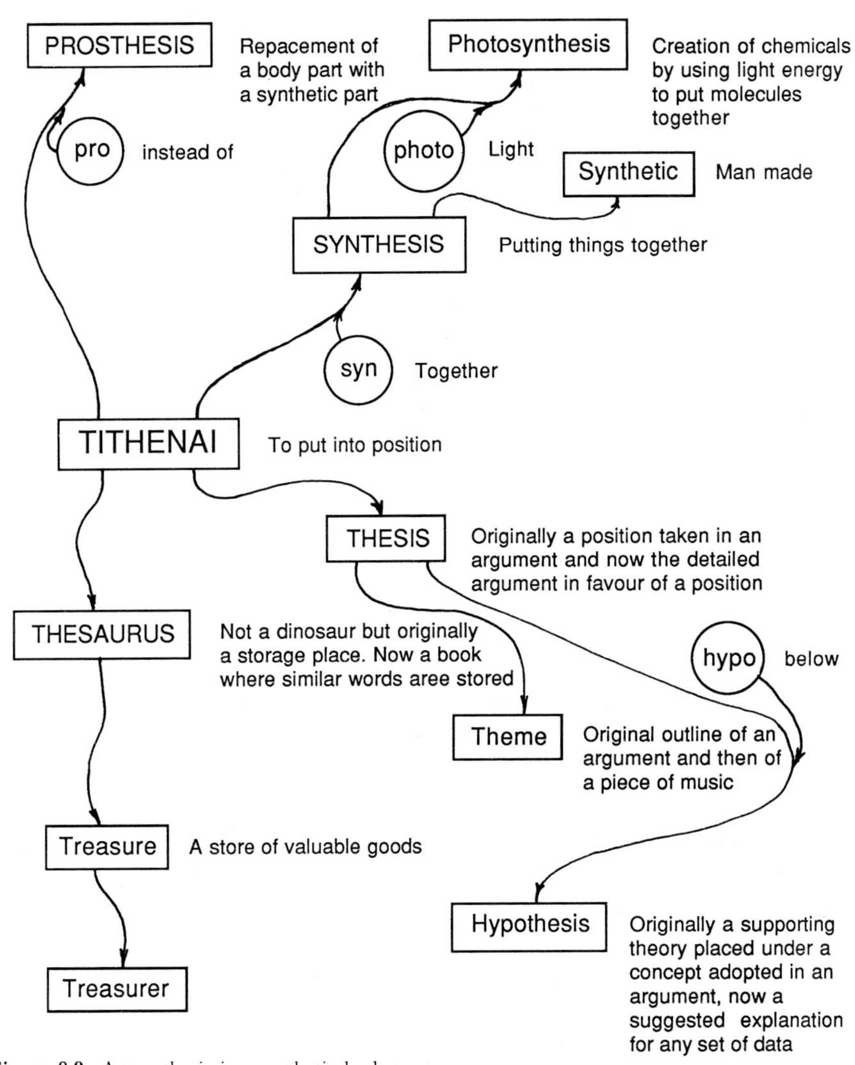

Figure 8.9. A prosthesis is a synthetic body part.

the piece of debris could represent important information with regard to the wear going on inside the gear.

The other profile of Figure 8.8(b) is an activated leukocyte. The word leuko in Greek meant "white". The Greek word kytos was the name for a hollow space and has given us the word for the basic unit of construction of many biological systems. The specialist who studies living cells is known as a *cytologist.* A dictionary defines *cytology* as, "a branch of biology dealing with the structure, function, multiplication, pathology and life history of cells." *Leucocytes* are otherwise known as white blood cells. Note *leukemia* is a disease of unknown cause in man and other warm blooded animals. It involves the blood-forming organs and is characterized by an abnormal increase in the number of leucocytes in the tissue of the body with or without corresponding increase of those in the circulating blood. Leucocytes are part of the body's defense against foreign materials in the blood. When leucocytes sense the presence of foreign bodies, they change shape to get ready for their attack on the foreign bodies. The cytologist and other medical experts are interested in this change of structure and also in the amount of the cell perimeter which becomes activated. If we look at the profile of Figure 8.8(b) the portion to the right of the two small arrows represents unactivated cell perimeter whereas that to the left represents the activated state. Fractal geometry is proving useful in a study of the activated state of leucocytes. The ratio of perimeter which is activated to unactivated state is important information. Scientists at Laurentian University are cooperating with Dr. Don McIver of Western University, London, Ontario, in a study of the fractal structure of activated leucocytes [10].

Section 8.4. Estimating Fractal Dimensions by Penny-Plating Procedures

What we call the penny-plating procedure in this book is an adaptation of a technique for measuring the length of a rugged line by the *Minkowski Sausage* method. Herman *Minkowski* was a Russian–German mathematician who lived from 1864 to 1909. He is famous for his work on the mathematical basis of Einstein's theory of special relativity. The basic idea employed in Minkowski's sausage technique for measuring the length of a wiggly curve such as a coastline is to assume that someone walks around the coastline dropping tires or other discs onto the coastline [11]. This disc dropper does not need to know exactly where the coastline he is covering with discs is located; he only needs to know if a disc he has just dropped has covered part of the coastline and touches the previous disc dropped on the coastline. A giant walking around the coastline dropping rather large discs would plate the coastline of Great Britain as shown in *Figure 8.10*(a). The diameter of the discs used in the coast-covering experiment is the inspection resolution used in the estimation of the coastlines length.

To estimate the coastline length from the number of discs placed around a coastline one imagines that one straightens out the discs to form a sausage string, as shown in Figure 8.10(b), the ribbon area of which is

$$A_R = nd^2$$

where n is the number of discs around the coastline and d is the diameter of the discs. To convert this into a linear estimate of coastline length one divides the measured

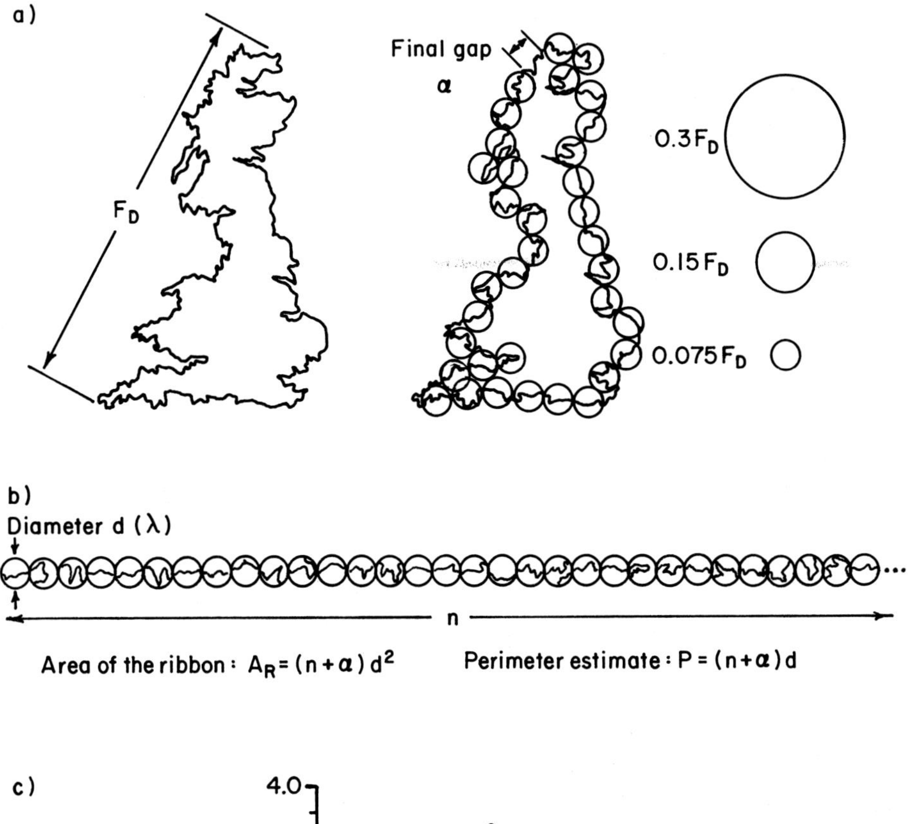

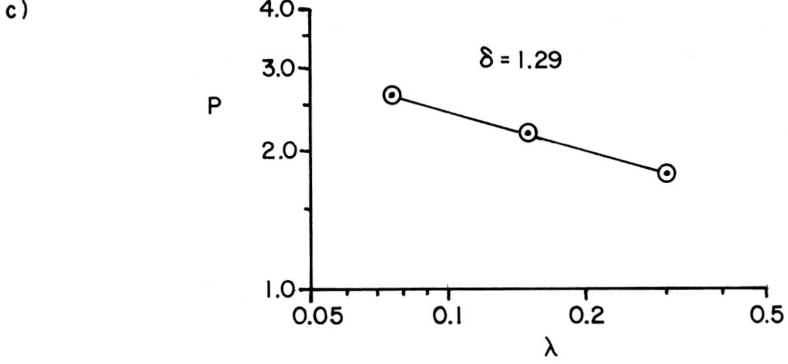

Figure 8.10. In the Minkowski Sausage technique for characterizing fractal lines contiguous discs are placed on the rugged line to be characterized to estimate its length. In the penny-plating modification of the procedure the discs are placed outside the line to be characterized creating a rather larger ribbon of discs with a gap. a) Discs dropped around the profile of Great Britain. b) Ribbon of discs of (a) opened up to form a line. c) Plot of the data obtained from the method part (b).

area by d. Therefore, the estimate of the coastline length is nd (a variation of this Minkowski sausage method is to plate the coastline with discs, a procedure in which the base of the disc is allowed to touch the coastline but not cover it). When working with this procedure, one ends up with a small gap between the beginning and the end of the ribbon of discs. One therefore has to estimate the width of a fractional disc which completes the ribbon.

The plating of the line with discs can be adapted to make a classroom method for estimating the length of rugged lines by plating the line with coins. The first time we tried this in the classroom, we used the Canadian penny, a readily available disc to generate the data, hence our name for the technique "penny plating". To obtain estimates of the length of the wiggly lines at different levels of resolution, one can use other coins such as the Canadian quarter and the Canadian dollar. To obtain a wider range of data for constructing a Richardson plot, we changed the magnitude of the basic curves using a photocopier and used the same coins again. A rudimentary Richardson plot based on Minkowski Sausage techniques for three different discs is shown in Figure 8.10(c). The reader is invited to extend the data of Figure 8.10 using different discs and coastline magnifications.

We have found that carrying out estimates of wiggly lines in this manner helps the student to gain a physical appreciation of the Minkowski sausage method for measuring the length of the irregular line. It also lays a basis for the understanding of the profilometer-based method for measuring the fractal dimension of a wiggly line and the fractal dimension of a rough surface by gas adsorption as described in the following section.

Before discussing the profilometer method of measuring the surface, it is useful to develop some vocabulary for use when describing the structure of rugged lines. First of all, we note that one of the words used to describe the features of a rugged line is that it is said to be *indented*. Individual features of the rugged line are described as indentations. This word literally means "bitten into" from the Latin word for "tooth", dentis. The dictionary definition of indent is "to cut into zig zags, to divide a document along a zig zag line". This latter definition may be surprising to modern readers. It came from the fact that when illiterate people were involved in contractual arrangements such as an agreement to serve an apprenticeship to become a carpenter they used to sign a document which was read to them. The document was then torn down the middle and the apprentice kept one half and the master the other. The pattern of the tear was so unique that the apprentice knew that the master could not change his half of the agreement without tampering with the actual original document. If the master attempted to forge a new document it would not match the tear of the half of the document kept by the apprentice. This led to the term *indentured servant* meaning "one who had signed a contract, which was torn in half, to serve a given period of time". This legal use of the term indentured bears witness to the fact that throughout history people have recognized that fractures and tears generate unique patterns. It also reminds us that fractal dimensions describe the average ruggedness of a surface but not the exact structure of a particular run of a specific short rugged line.

The other word that bears witness to the practice of matching fractal surfaces as a means of recognition of the qualifications of a person is the word *symbol.* This word comes from two Greek root words syn meaning "together" and ballein meaning "to throw". Greek generals would break a ceramic token or a piece of pottery into two halves. One half would be given to a messenger and the other to the future recipient of a message from the general. Later if a messenger arrived with news of a battle the credentials of the messenger could be established by asking him to produce the piece of pot the fractal surface of which had to match that of the other part of the token or pottery. This origin of the term also warns us that a symbol is only meaningful if both parties to the use of the symbol share a common experience. To inflict a mathematical symbol on a student who has never met that symbol is a false message from the battle field of knowledge acquisition.

Another term used to describe the structure of a rugged surface is to say that it is *convoluted.* This word originally meant "rolled together" from the Latin root word volvere meaning "to roll". This root word has given us the term revolver from the design of a gun that has a rotating ammunition chamber. The use of the term convoluted to describe a line such as those of Figure 8.6(a) probably arose from the fact that something which has been rolled up tightly for any given length of time usually does not straighten out immediately when unrolled but forms a profile such as those of the lines of Figure 8.6(a). If we take a section through a surface, we often refer to the boundary of the section as being a *profile.* This is an appropriate name since the word profile is defined as "a portrait and side view as if outlined by a thread", from the Latin word filum meaning "thread".

One way to measure the length of a wiggly line is to lay a thread along it. Another procedure is to make a little device used by geographers in which a small wheel with a rubber tire or tiny teeth (to stop it from slipping when used) is mounted on a screw thread between the jaws of a fork as illustrated in *Figure 8.11*(a). Usually the geographer uses this device orthogonal to the surface of the map and he can follow the convolutions of a river or a road. As it is wheeled forward, the wheel moves along the screw. At the end of the exploration of the river or road, one can move the wheel back along a straight line to measure the length of the convoluted line. When using this type of device to explore the fractal structure of a rugged line, one has to use it in a different mode. One could transfer the wiggly lines of Figure 8.6(a) to a piece of plywood with the edge of the plywood being cut to form a replica of the line to be characterized (this makes a useful class demonstration). One can then measure the effective length of the line by moving the wheel along replica boundary with different diameter wheels. For a given wheel the estimate of the line length will depend on whether the wheel enters or glides over an indentation in the profile of the line. The Richardson plot is constructed by plotting the profile length estimate

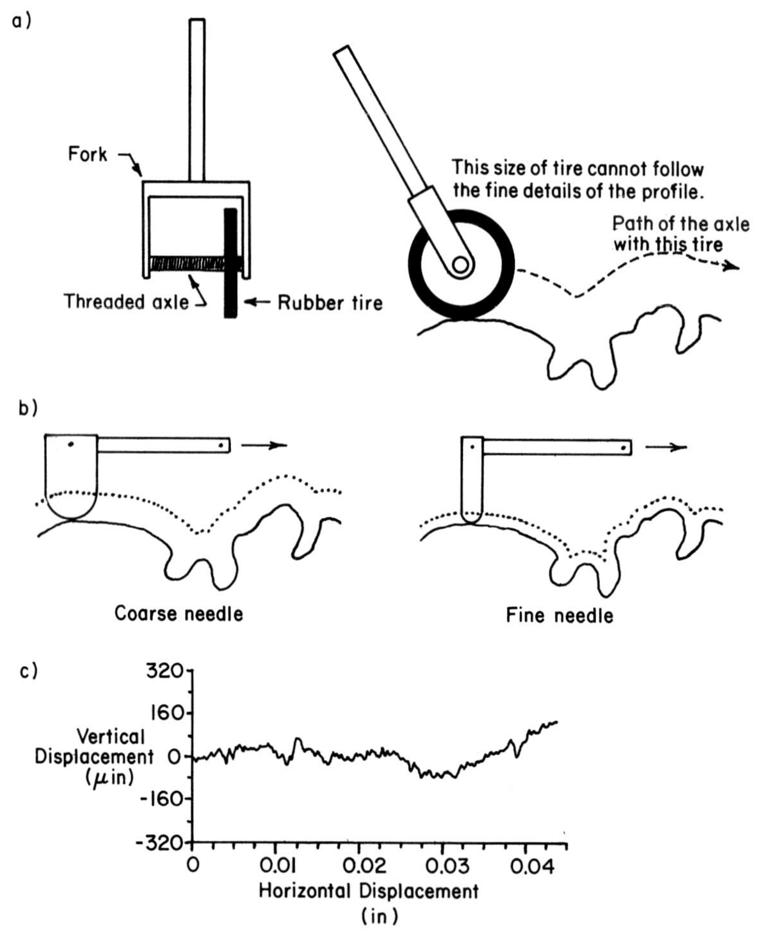

Figure 8.11. Profilometer data can be used to characterize the roughness of a surface. a) Rubber wheeled device for measuring distance can be used to explore fractal structure. b) The fractal structure of a rough surface can be explored using a profilometer with needles of various radii. c) Typical profilometer trace generated in an industrial study of the roughness of a surface.

against the wheel diameter. By carrying out this experiment in the class, the teacher develops the background for understanding the use of industrial profilometer equipment for measuring the fractal structure of a surface (a very important area of applied fractal geometry) [12–15]. When working with the photograph of a surface, one is not usually working with a profile of the surface but a *silhouette*. A silhouette is defined as "a representation of the outline of something usually colored black". The name comes from that of a French minister of finance who attempted to reform the French economy in 1757. He introduced so many taxes that the people used to joke that they could no longer afford to have their portrait printed; all they could do was have an artist sketch their outline.

The basic concept of an industrial profilometer based on needle scanning used to measure the profile of a rough surface is illustrated in Figure 8.11 (b). The surface to be explored is moved along under a needle which rides upon a pivoted arm. As the needle rides up and down on the rugged surface, its motions are amplified and recorded to produce a trace such as that shown in Figure 8.11 (c). The faithfulness with which the needle follows the wiggles in the surface being investigated depends upon the diameter of the needle tip being used to explore the roughness. This fact is illustrated in Figure 8.11 (b) in which it can be seen that a broad needle would not enter into some of the smaller cavities of the surface (the sketches of Figure 8.11 (b) are not intended to be drawn to engineering scale of design but are only sketches intended to demonstrate principles).

The immediate thought is that to get the best trace of a surface one must use the finest possible needle. However, it must be remembered that the weight of the arm and the needle is pressing down on the tip of the needle. Therefore, the smaller the needle, the higher the profilometer pressure at the tip of the needle created by the weight of the profilometer needle head assembly (remember that pressure is weight divided by surface area). If one were to use a very fine diamond needle on a surface such as leather, a very fine needle would probably deform and/or rip the surface to create new roughness. For this reason commercial profilometers using needle heads to trace the structure of the surface usually come equipped with several needles of different roundness so that the scientist can select an appropriate needle for his investigation. This is fortunate for the scientist who wishes to study the fractal structure of a surface since he can now create a Richardson plot, from which he can deduce the fractal dimension of a surface, by carrying out repeat measurements of the surface roughness using needles of different roundness. A graph of the length of the trace recorded by the needle against the radius of the tip of the needle generates the data for measuring the fractal dimension of the surface. Industrially this could be a very good technique for measuring fractal roughness since the data coming from a profilometer can be sent straight to a computer for processing. Some profilometers have more than one needle arm so that one could carry out three measurements side by side with the data being used to give immediate readout of the fractal roughness of the surface. Some profilometers use optical methods to explore the roughness of the surfaces. This optical profilometer can also be used to generate a surface profile which can be characterized by its fractal dimension and this instrument avoids any danger of pressure deformation when exploring the surface.

The fractal structure of a rough surface can be measured using *gas-adsorption studies.* Techniques for measuring the surface area of a powdered solid by studying how gas is adsorbed onto the surface at low temperatures have been in wide spread use for over 50 years. Pioneer studies of this technique were carried out by three scientists *Brunauer*, *Emmett* and *Teller* [16].

In gas-adsorption studies of surface area, the powdered substance to be characterized is placed in a glass vessel and the air above it is pumped out as the powder is being heated to drive off any adsorbed gas molecules on the surface. When describing this technique for measuring surface areas, it is important to distinguish between *adsorbed* gases and *absorbed* gases. Adsorbed gases are only on the surface of the material, whereas, absorbed gas can go into the material and be very difficult to remove. The process of pumping off any adsorbed gases from the surface of the powder is known as *outgassing*. After the sample has been outgassed for several hours, a small amount of a pure unreactive gas is introduced into the container holding the powder. Next, the volume of the gas-powder container is reduced by applying pressure to the system with a device such as a mercury manometer. This step is repeated several times and the volume–pressure data recorded. From the pressure and volume measurements, the scientists can discover when the gas molecules are adsorbed onto the clean surfaces and can work out when the surface of the powder is just completely covered by a monolayer of the gas. We do not need to be too concerned with the actual mechanics of the process. The interested reader can consult a textbook on surface-area measurement by gas adsorption. For the purposes of this book, we simply note that the theory used to interpret gas-adsorption studies of this type is widely known as the *BET method* or theory. These letters stand for the initials of the three scientists who developed the theory.

There is an interesting story told of how Brunauer and Emmett were having some difficulty working out the detailed theory of the gas-adsorption studies so they discussed their problems at a luncheon meeting with Edward Teller. Edward Teller was one of the scientists who developed the atomic and the hydrogen bomb. His work was so important that he is often called the father of the hydrogen bomb. A widely circulated story concerning the BET theory is that Dr. Teller worked out the details of the relevant theory for Drs. Braunauer and Emmett on the table cloth of the restaurant where they had lunch. Dr. Brunauer had to buy the cloth from the restaurant before he could return to his laboratory and write the necessary scientific papers!

Many different types of commercial equipment are available for measuring the surface areas of powders by gas adsorption [18, 19]. It is widely stated in the literature on gas-adsorption studies that the measured values of the surface area of powders vary when using different gases because of uncertainties in the gas molecule size. Fractal geometry is causing most of the gas-adsorption textbooks to be rewritten because we now realize that using gas molecules of increasing size will result in different packing patterns of the molecules along the surface of the powder. Therefore, surface-area measurements made with different sized gas molecules will give different surface area estimates because of the different molecular accessibility to the surface contours of the different gas molecules. It follows that a Richardson plot of the measured surface area of a powder against the size of the molecule used in the gas adsorption study will generate a Richardson-type plot, the slope of which is the fractal dimension of the rough surface of the powder. What has been considered to be error or uncertainty for 50 years turns out to be hidden information on the roughness of the surface. Dr. David *Avnir* of the Hebrew University pioneered this re-evaluation of gas-adsorption studies from the perspective of fractal geometry [17]. He reports that every study he has re-examined has shown a fractal dimension lurking amongst the data. A scientist can now measure the fractal dimension of a rough surface by doing gas-adsorption studies with a series of gases such as helium, neon, argon, and krypton which represent increasing molecular sizes of gas adsorbent; a technique that can be simulated in two-dimensional space using coin-plating studies.

Dye molecules have also been used to measure surface areas by packing dyes of different molecular sizes on the surface of the powder. We can expect many scientists to be investigating the use of dye adsorption to generate fractal surface-area information since by using dyes of different molecular size they can generate fractal information from their adsorption studies [16].

Section 8.5. Mosaic Amalgamation – Another Variation of the Minkowski Sausage Method for Characterizing Rugged Curves

In *Figure 8.12*, the basic concept used in the mosaic amalgamation technique for studying the fractal structure of rugged lines is illustrated. The three parts of Figure 8.12(a) were generated by placing three different transparent grids over the profile of Lake Ramsey (see Figure 3.19). Wherever the boundary crossed a grid square it was colored black and then the perimeter of the lake was estimated by counting the number of squares which were located on the perimeter. This is the same as the Minkowski sausage method for measuring the perimeter of the profile in which one is using squares instead of discs. By placing the square grid over the profile, one is in fact converting the lake into a mosaic with a subsequent estimation of the perimeter of the lake by counting the pixels of the mosaic that lie along the shore of the mosaic. If one were to be carrying out this experiment with a computer limited to an imaging system, one could use a device known as a *charge coupled device camera* which converts the entire image into a mosaic which can be transferred to a computer memory system [20]. Then one can measure the area of the profile and the fractal structure of the boundary by plotting the number of squares on the boundary against the size of the square on log–log graph paper [21].

The Richardson plot of the data generated by the three grids used in Figure 8.12(a) is shown in Figure 8.12(b). Doing such a pixel count using different grids is a relatively tedious way of measuring fractal structure in the laboratory but is probably the best way in which to measure fractal structure when processing images

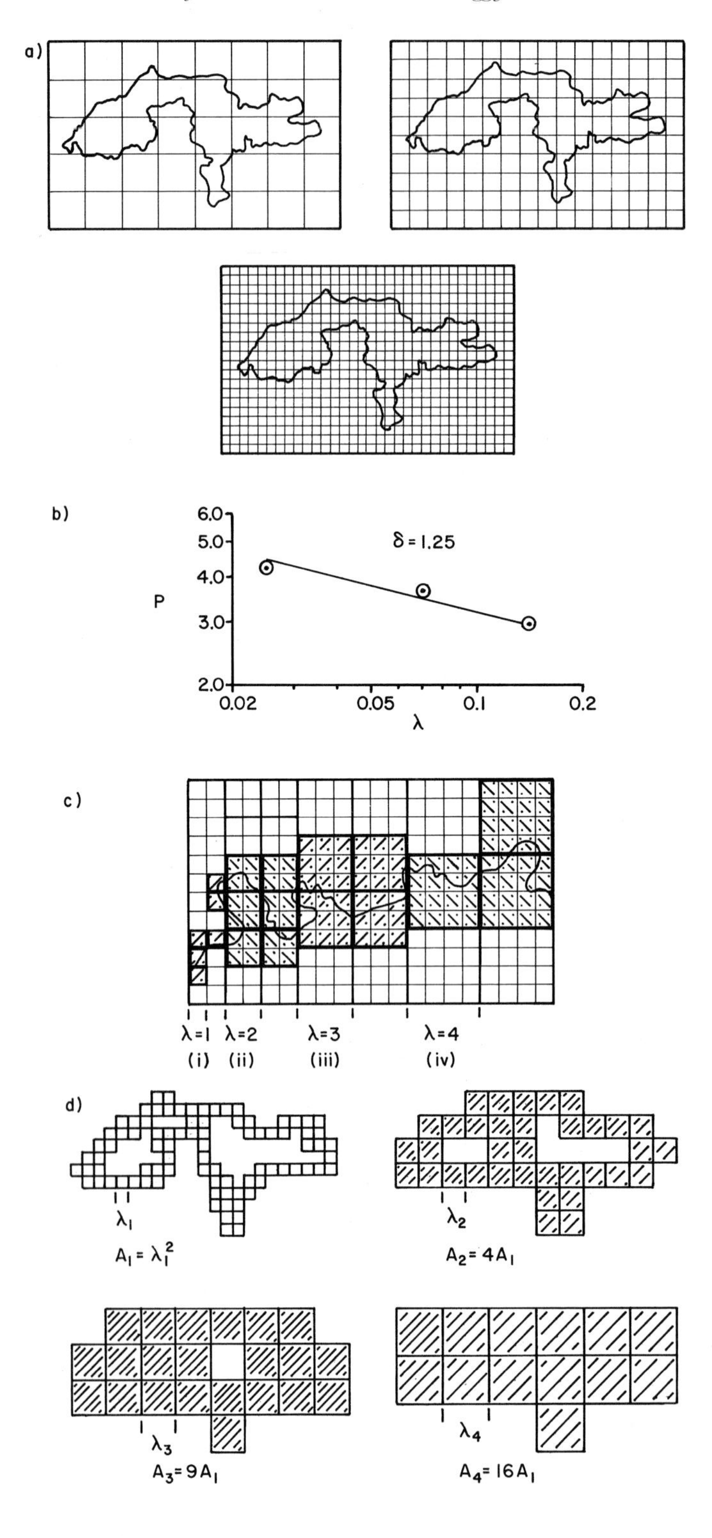

a)
b)
6.0
5.0
4.0
3.0
2.0
P
δ = 1.25
0.02
0.05
0.1
0.2
λ
c)
λ = 1
λ = 2
λ = 3
λ = 4
(i)
(ii)
(iii)
(iv)
d)
λ1
A1 = λ1²
λ2
A2 = 4A1
λ3
A3 = 9A1
λ4
A4 = 16A1

generated using television cameras and/or scanning electron microscope pictures. As in the case of the equipaced method discussed in Section 8.3, the automated procedure for generating mosaics using different pixel sizes is carried out in a different way to the procedure one would use when using different sized grids to generate the data. In the electronic processing, one generates the highest resolution image of the line that can be achieved with the given electronic camera. One then proceeds to create mosaics in the computer of different pixel size by mathematically amalgamating data from adjacent pixels. This procedure is illustrated in Figure 8.12(c).

In part (i) of this sketch, the outline being studied is transformed into single element pixel units, then, moving to the right, the appearance of a coarser mosaic with pixel size equal to four pixels of the original mosaic is shown in part (ii). A computer would group the pixels in sets of four and make them black if any one of the previous separate pixels had been on the boundary line. The result for one mosaic tile of four pixels size is as shown in the sketch. The next downgrades of the mosaic, in which the pixels are grouped in sets of nine and then sixteen pixels, are shown as parts (iii) and (iv) of this sketch. To calculate the length of the perimeter at a given mosaic tile inspection, one counts the number of mosaic tiles and divides them by the size of the tile. The appearance of the mosaic as it is continuously degenerated is shown for illustrative purposes in Figure 8.12(d).

To understand the technical term for the mosaic technique, we must explore the meaning of the word, "amalgamate". The word *amalgam* entered the English language through the art of metallurgy. A dictionary defines *metallurgy* as

> "*The art and science applied to metal extractions from ores, refining etc.*"

The word comes from two Greek root words metallon "a mine" and ergon "to work". A metallurgist was originally someone who worked in a mine to obtain metals such as tin, lead, and copper. The metallurgist then widened his skills and started to make alloys such as bronze and eventually began to work with such alloys to produce armor etc. The word ergon combined with the word en meaning "within" has given us the word *energy* which originally meant "the ability of someone to achieve or carry out work". It is interesting to note that the word *metal* itself comes from the Greek word for a mine. This is an example of what is known as "transfer" in the development of words. For instance the word *coin* originally meant "wedge" and then when wedge-shaped tools were used to make metal discs,

◀

Figure 8.12. In the mosaic amalgamation method for characterizing the structure of a rugged profile, the number of "tiles" of the mosaic lying on the boundary of the structure are counted as the structure is converted into a mosaic using smaller and smaller tiles (pixels). a) Three typical mosaic transformations of Lake Ramsey. b) Richardson plot for the boundary estimates based on the mosaics of (a). c) Illustration of the steps used to create larger mosaic tiles by amalgamation data from adjacent squares. i) Pixels of unit size 1. ii) Pixels of unit size 2. iii) Pixels of unit size 3. iv) Pixels of unit size 4. d) Amalgamation versions of the mosaic of Lake Ramsey created mathematically.

the name was transferred from the tool to the object and we started to describe the metal discs used as money as *coins*.

One of the metals that the metallurgist learned to make alloys with was mercury. When mercury is mixed with another metal the initial stage of the alloy is a soft dough type material similar in structure to the material used to make bread. The Greek word for "dough" was malgam so that the combination of mercury with silver was known as *silver amalgam*. This dough-like material quickly hardened to make a very tough substance which is widely used to make "silver" fillings for teeth (there is no silver in the tooth fillings, they just look silvery). Like so many other words, what started life as a technical term started to be used in society to describe social happenings. When two tribes joined together to become one they were said to be amalgamated. In modern English the amalgamation of two companies results in a new company under a new management structure. The separate parts of the original companies become indistinguishable from each other. It is this modern sense of amalgamation which is used to describe our technique for estimating boundaries by merging pixels together in groups as "*mosaic amalgamation*".

Section 8.6. Fractal Rabbits and Manitoulin Island

In the early days of the application of fractal geometry to the study of rugged systems, people sometimes drew straight lines through curving data points which had nothing to do with fractal dimensions but actually were coarse resolution data points beginning to home in on the structure of a euclidean profile. This enthusiastic discovery of fractals that did not exist was described by David Avnir, of the Hebrew University, as the discovery of fractal rabbits by analogy to the fact that magicians could produce unexpected rabbits out of black hats [22, 23]. Therefore, even though it may be a disappointment to some of the readers, this section is not going to discuss how infinitely rough haired rabbits with infinite appetites are destroying the vegetation of an island. It is a discussion of the problems of characterizing the profile of Manitoulin Island which hopefully will demonstrate to the reader how to avoid being deluded by *fractal rabbits*.

The profile of Manitoulin Island is shown in *Figure 8.13*(a). This island is the largest island in a fresh water lake in the world. It is over 100 miles in length. It is located in the northern part of Lake Huron in the Great Lakes area of the North American continent (approximately 70 miles from Laurentian University). It was chosen for this particular study because its north and south shores are exposed to very different erosive conditions, the profile obviously has different ruggedness on the north and the south side of the island. The resolution with which the perimeter of the island was stored in the memory of the computer as a digitized profile is

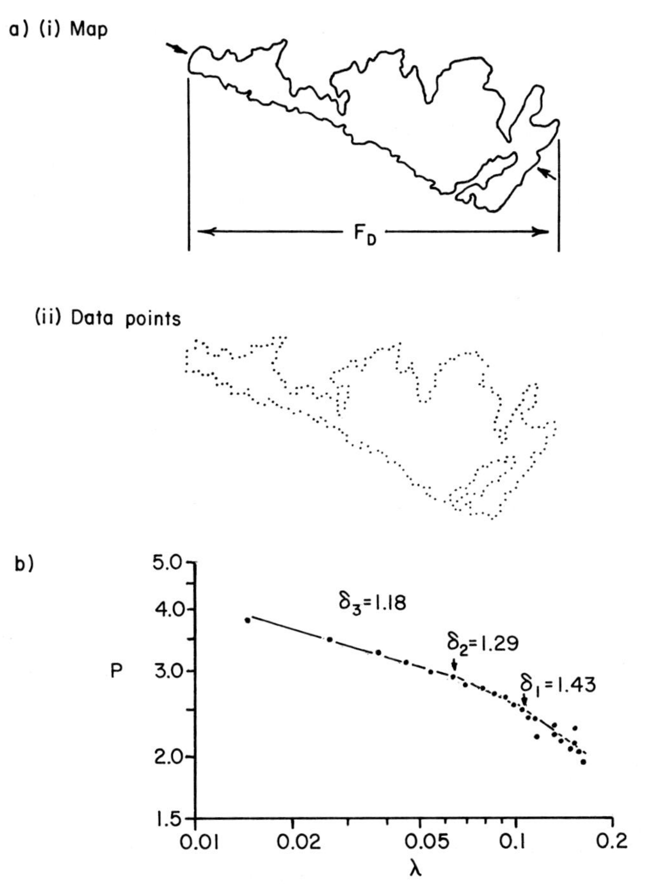

Figure 8.13. An uncritical amalgamation of data from an equipaced exploration of a convoluted profile can generate "average values" for fractal dimensions that are difficult to interpret. a) Continuous and digitized maps of the structure of Manitoulin Island. b) Richardson plot for overall exploration of the ruggedness of Manitoulin Island.

illustrated below the original profile in Figure 8.13(a). The Richardson plot for a wide ranging exploration of the ruggedness of the island profile is shown in part (b) of the diagram. An uncritical acceptance of the data points of Figure 8.13(b) would indicate that there were three regions of different fractal dimension of values shown in the figure. However, a visual examination of the profile suggests that in fact the island has only two different roughness. To demonstrate this fact, two synthetic islands representing the north and the south shore were created by splitting the island profiles at the arrows shown in Figure 8.13(a). Two new synthetic islands created from each shore line and its complimentary image along with the Richardson plots generated by the equipaced exploration of the new profiles are shown in *Figure 8.14* and *Figure 8.15*. When this data was first generated, the information for

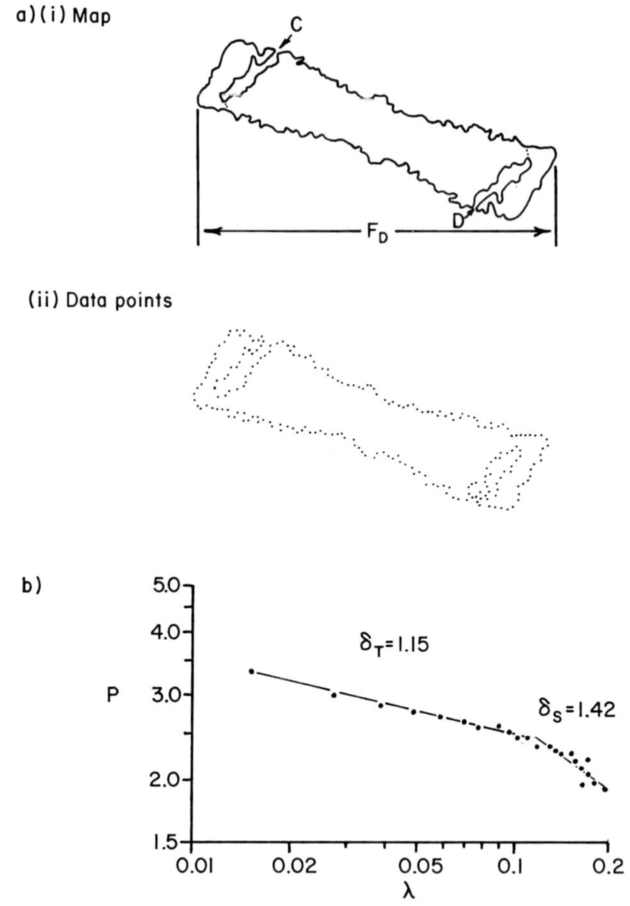

Figure 8.14. The lower ruggedness of the south shore can be demonstrated by creating a synthetic island from two copies of the southern shoreline. a) Synthetic south shore island continuous and digitized maps. b) Richardson plot of the synthetic south shore island exploration.

the north shore seemed satisfactory but we (this work was carried out in cooperation with G. Clark) did not think that the data for the island fabricated from the south shore seemed correct [4]. First of all, the 1.15 textural fractal seemed too high for the visual appearance of the long stretches of almost straight coastline forming the two long sides of the synthetic island. Secondly, you will note that 1.42 is almost the same as that obtained for the whole island for the coarse resolution inspection of the synthetic island of Figure 8.14. We suspected that the two indentations "C" and "D" were completely masking the true structure of the south shore and should have been included with the north shore data. Therefore, we made an alternate split of the profiles at the points at the end of the deep bays to create the synthetic profile of *Figure 8.16.*

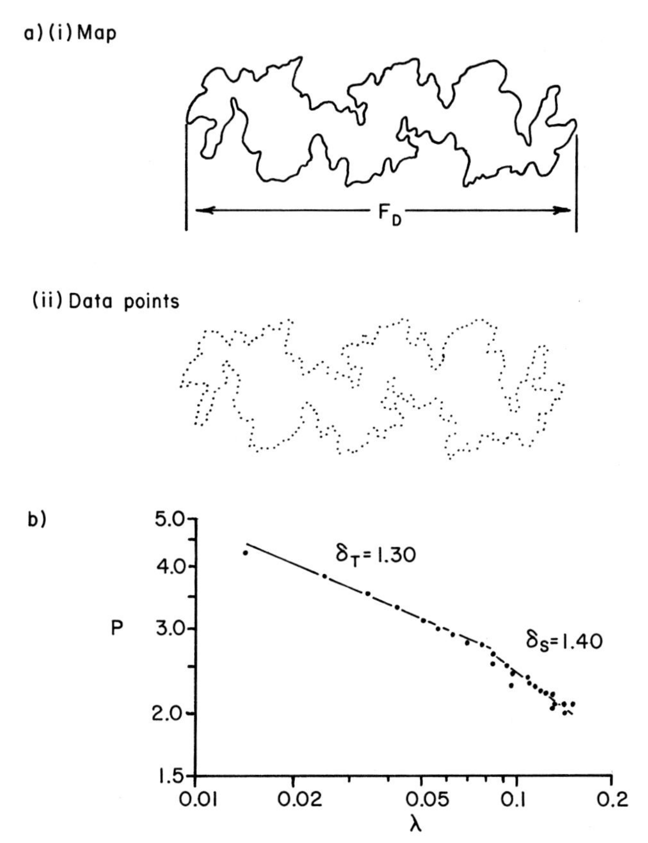

Figure 8.15. The synthetic island produced from two copies of the north shore of Manitoulin Island demonstrates the ruggedness of the north shore of the island. a) Synthetic island and digitized points for the northshore of Manitoulin Island. b) Richardson plot for the exploration of the synthetic island.

The textural fractal for this new island of 1.09 was more compatible with our expectations from a visual inspection of the ruggedness of the south shore. However, the unexpected short slope of 1.14 at a coarse resolution did not match any visually apparent structure of the island. We began to suspect that it was a fractal rabbit. To test this hypothesis, we explored the figures shown in *Figure 8.17*. First of all, a polygon with approximately the same shape as the outline of the synthetic island of Figure 8.16 was drawn and this was found to give a data slope of 1.04. This was obviously not a fractal shape and the data slope only came about from the difficulty of drawing a polygon to fit the original shape exactly. The 1.04 line is a description of the difficulty of fitting polygons to a narrow elongated Euclidian shape and not a fractal dimension. If we now draw a smooth-sided figure shown in Figure 8.17(b) which is an even better approximation of the profile of Figure 8.16, we find that the slight bulging at the corners is sufficient to give an extended deviation from the

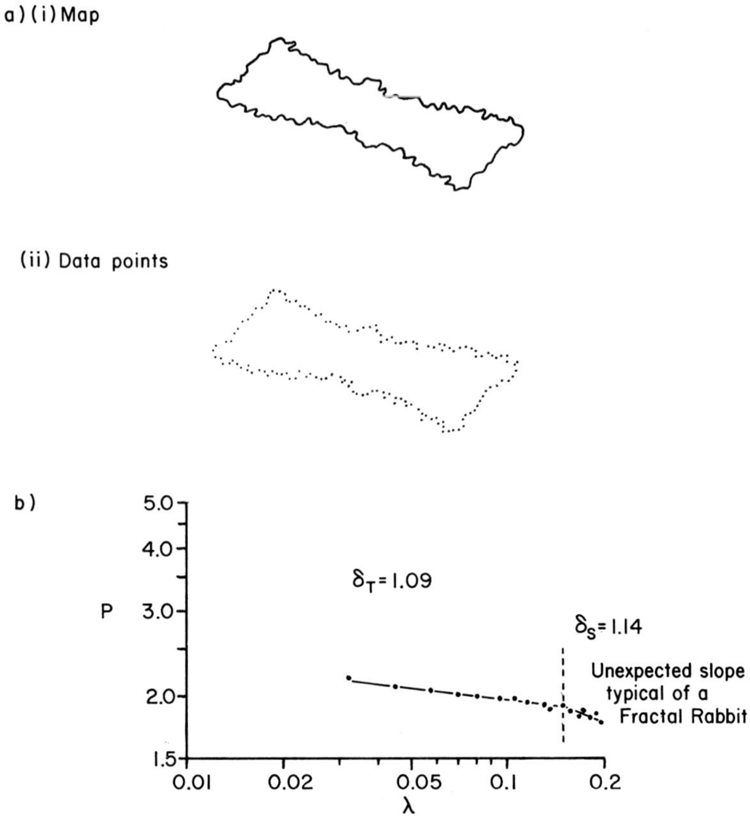

Figure 8.16. The point at which one divides a rugged line into regions of different ruggedness can effect the physical significance of the regions of a Richardson plot generated by an equipaced exploration of the profile. a) New island created from a more restricted range of shoreline and its digitized points. b) Richardson plot of the exploration data for the new shoreline of (a).

upper curve. If one did not look at the curve to see if it had fractal structure, the computer would draw a curve through the first part of the line and tell the investigator that it had discovered a fractal dimension higher than 1.03. The data of Figure 8.17 convinced us that the 1.14 line of Figure 8.16 was a delusion produced by the geometry of the situation and was properly called a fractal rabbit rather than a fractal dimension. To complete the presentation of the problems of drawing polygons within euclidean figures, the two extra diagrams of Figure 8.17(d) are presented. An inexperienced computer exploring a circle would designate its experimental uncertainty as a low fractal dimension of 1.01.

There are two lessons to be learned from the data of Figure 8.17. In our exploration of fractal boundaries at Laurentian University, we have developed a *rule of thumb* that one should not use a normalized stride length of greater than 0.3 of the

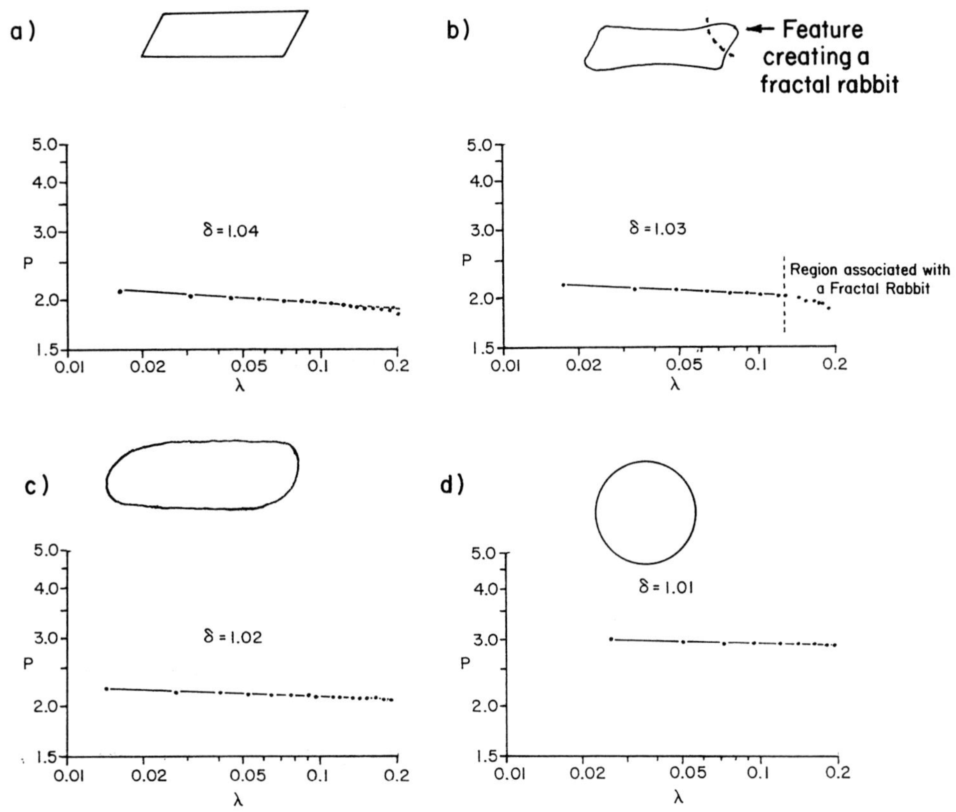

Figure 8.17. Experience in the difficulties of interpreting the equipaced exploration of a boundary can be gained by exploring some Euclidean figures of high aspect ratio. a) Regular polygon approximation. b) Bone-shaped approximation. c) Ellipsoid approximation. d) Simple circle for comparison.

projected maximum length. Brewers Dictionary of Phrase and Fable describes "a rule of thumb" as

> "*A rough guess work measure; practice or experiences as distinguished from theory; an allusion to the use of the thumb for rough measurements. The average human thumb being approximately one inch across the top*" [24].

The data of Figure 8.17 shows that even this rule of thumb is not sufficient guidance if one inadvertently has some smooth bounded figures amongst a set of fractal profiles being evaluated by computer-aided image analysis. The computer could tell the difference between the euclidean figures of Figure 8.17 and a real fractal system by looking at the variation in the chord lengths of the equipaced steps at a given step length. However, for the amount of work that one would carry out in nonspecialist studies of fractal systems visual inspection as a check on the reality of a measured fractal dimension is a useful precaution.

Section 8.7. How to Tell a Vulcan from Another Carbon Black

Devotees of the popular science fiction television series Star Trek, might suspect that they are about to encounter Mr. Spock in fractal land. However, the problem that we are going to discuss in this section is much more down to earth. In the previous section of this chapter, we have discussed how to characterize the fractal structure of a boundary but we have avoided the question as to how useful such a measurement is if the fractal boundary is varying within a group of profiles. There is no general answer to such a question. In this section, we will review a research investigation into the structural difference between two sets of carbon black profiles and show that there are indeed indications that this type of study could provide useful information on the differences between populations of profiles each of which have fractal structure [25].

The Cabot Corporation manufactures many different grades of carbon black. Two of them are called Vulcan 7H and Sterling NS1 (Vulcan was the Roman god of fire, hence the name *volcano* for a firey mountain which smokes and from time to time throws out rocks and lava. Volcano was the name used originally for Mount Etna in Sicily. The Romans used to regard this volcano as being the blacksmith's shop of Vulcan where he hammered out the weapons of the Gods).

In *Figure 8.18* two sets of profiles of carbon blacks, Vulcan 7H and Sterling NS1, are shown. These profiles are taken from a set of highly magnified electron micrographs kindly provided by N. Mace of the Cabot Corporation [26]. A visual inspec-

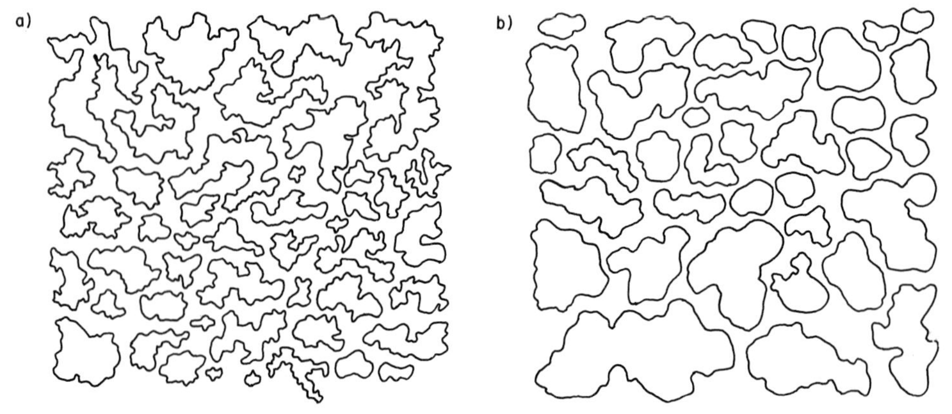

Figure 8.18. An array of profiles of two different grades of carbon black pigment fine particles suggests that fractal dimensions may be used to quantify the difference between them. a) Vulcan 7H carbon black fine particles. b) Stirling NS1 carbon black profiles.

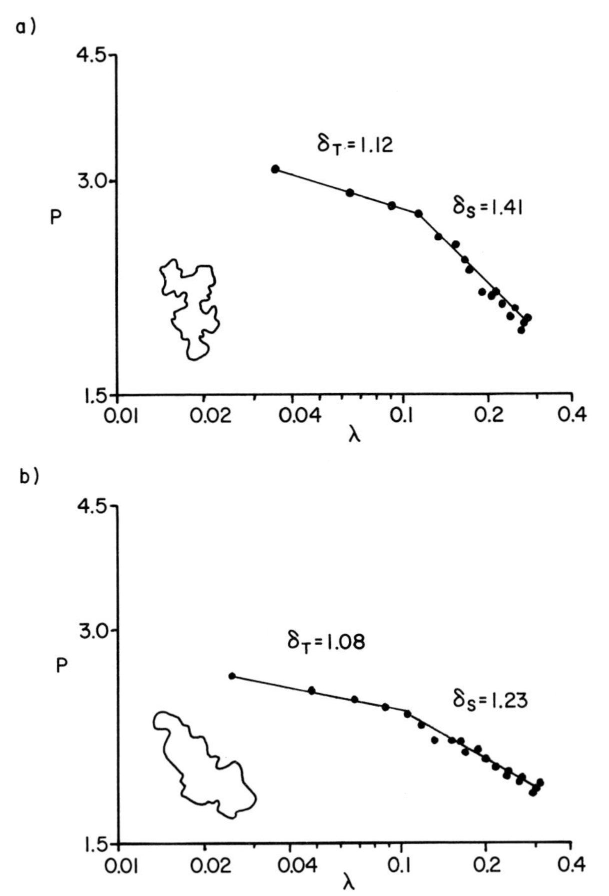

Figure 8.19. Fractal characterization of typical carbon black profiles from the two pigment populations of Figure 8.18 demonstrate the structural differences between the two pigment groups. a) Richardson plot for a randomly selected Vulcan 7H carbon black pigment profile. b) Richardson plot for a randomly selected Stirling NS1 carbon black pigment profile.

tion of these sets of profiles suggests that the Vulcans are more wiggly than the Sterlings. The question we asked at the beginning of the research project was "could this visual impression be converted into a numerical index of structure for each set using the concepts of fractal geometry?". As a first step in our investigations, the fractal structure of each profile of the sets were characterized using the equipaced exploration technique. In *Figure 8.19* two Richardson plots typical of the data generated for the two different sets of profiles are summarized. It can be seen that for each profile we obtained a structural and a textural fractal. From our previous experience of the two-dimensional images of aerosol profiles, we knew that the structural fractal of 1.41 for the Vulcan profile of Figure 8.19(a) was typical of an agglomerate that had been formed by the collision of sub-agglomerates. The lower fractal dimension of 1.23 was typical of the growth of an agglomerate from subunits without the subsequent collision of agglomerates with each other. The 1.08 fractal dimension of the texture of the Sterling profile of Figure 8.19(b) was typical of the "bunch of grapes" texture of a system formed from packed spheres [27]. Therefore, when looking at the two profiles of Figure 8.19 experience suggests the hypothesis that the Vulcan carbon black consisted of more agglomerated profiles, hinting that in the manufacturing process the turbulent mixing of primary agglomerates was allowed to continue longer than in the manufacturing of the Sterling agglomerates.

When a process such as the turbulent collision of primary fine particles in a flame is terminated quickly the terminology used is that the reaction is quenched (*quench* means "to put out or to extinguish a flame". When one talks about quenching ones thirst the idea is that your throat is burning and you put out the burning sensation with a drink of water. The word comes from an old English word cwencan meaning "to cause to disappear"). In technical terms our data of Figure 8.19 suggested that the turbulent interaction of the spheres of carbon produced by incomplete combustion was quenched earlier in the case of the Sterling product than for the Vulcan carbon blacks. If all of the carbon black profiles had identical ruggedness then the data of Figure 8.19 would have been sufficient to establish our hypothesis. However, we were now faced with the problem of coordinating the data for all of the profiles to see if one could distinguish between the sets of data.

If one considers the kind of chaotic conditions under which turbulent collisions of the agglomerates are occurring it would seem to be a situation in which multiple collisions did not favor any particular growth statistics. One could therefore reasonably anticipate that the range of structural fractals and the range of textural fractals for each set of data could be Gaussianly distributed (it should be admitted that in fact this was only inspired guess-work and that if the hypothesis had not been established by the subsequent coordination of the data, one could probably give an equally reasonable background that the collision process was producing a log–normal distribution of fractal dimensions!). In *Figure 8.20* the distribution data for the variation of structural and textural boundary fractal dimensions of the two populations of carbon black profiles are displayed. Although this data is not entirely convincing to the specialist, the data does suggest very strongly that all four distributions are Gaussianly distributed and that the Sterling carbon blacks have a range of fractal dimensions characterized by the following information.

Vulcan carbon black has a mean structural boundary fractal of 1.31 (standard deviation $\pm$ 0.1), and a textural fractal dimension with a mean fractal structure of 1.16 (standard deviation $\pm$ 0.05 units). Comparable data for the Sterling carbon black is 1.26 $\pm$ 0.06 for the structural fractal and 1.10 $\pm$ 0.03 for the textural fractal.

The data of Figure 8.20 indicates that the scientists at Cabot Corporation could use this type of study to control the quality of their two different products by specifying the acceptable range of fractal dimensions to be permitted in the profiles of the

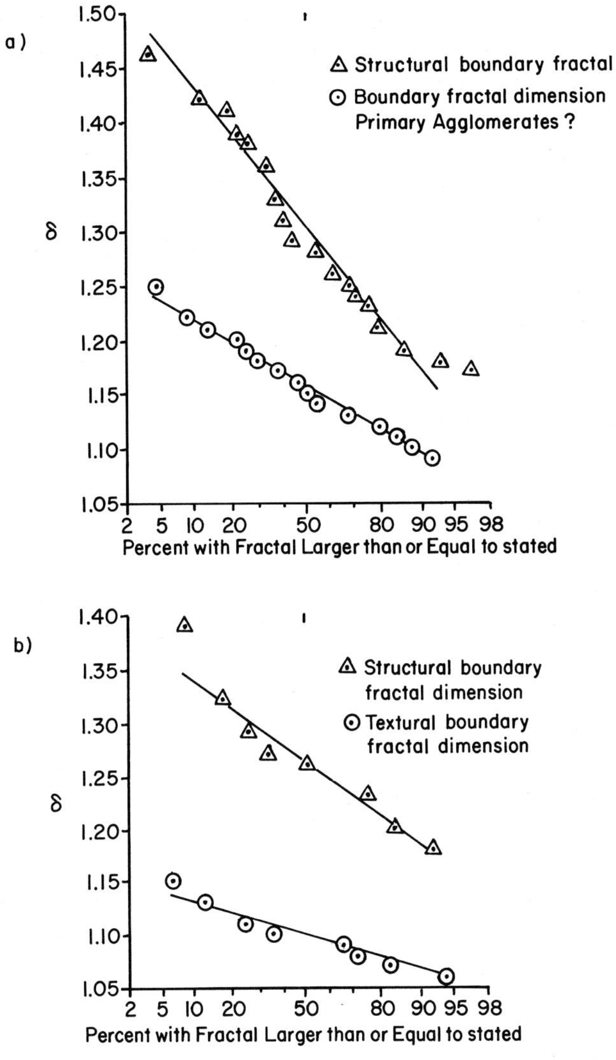

Figure 8.20. The distribution of the various fractal dimensions of the two sets of carbon black profiles of Figure 8.18 appear to be describable by the Gaussian probability function. a) Fractal dimension distribution for Vulcan 7H carbon black profiles. b) Fractal dimension distribution for Stirling NS1 carbon black profiles.

individual carbon blacks. They could start to correlate the performance of the carbon black in terms of the fractal dimension of the population of profiles. The fact that the distribution of fractal dimensions present in the population could be described by the Gaussian distribution is useful information on the formation kinetics prevalent in the burning flame. This information could be used by research scientists to predict the effect of changes in the flame conditions on their product. The data also suggests that when studying fine particle systems, the population range of fractal dimensions may constitute very useful information for the working scientist. Again, it should be noted that when interpreting data such as that of Figure 8.20, one should not use ones experience of scatter in data plotted on arithmetic scales to interpret scatter in a cumulative presentation of the data on Gaussian probability paper. In fact, the central tendency of the data is much more important than scatter at the tails. The two or three points in the low probability regions of the graph should not worry the working scientist unduly.

Section 8.8. Putting Fractal Dimensions to Work in Applied Science

In this section we will not be concerned with direct laboratory experiments. The intent of this section is to stimulate ideas for students who would like to do extensive science projects, and/or to suggest laboratory projects to more senior students at university level. Some suggested explorations could perhaps generate ideas and projects suitable for M.Sc. theses.

An important aspect of occupational health and hygiene studies is to determine the risk posed to a worker by the dust that he might inhale during his work endeavors. Respirable dust hazards range from direct exposure to dust in a working environment where powdered material is being produced to the hazard from solid fumes generated in operations such as welding [28, 29]. On a larger scale there is concern as to the type of fume hazard that would be created if a nuclear reactor was to melt down and produce fumes. Dr. Bolsaitis of the Massachusetts Institute of Technology has also pointed out that we should be concerned for the fact that incineration of city garbage could produce fumed metal oxide which could pose considerable health hazards if it is not removed by cleaning equipment installed at the exit of the combustion chamber of an incinerator.

In occupational health and safety, it has been common practice to assess the size of inhaled dust fine particles by means of a parameter known as the *aerodynamic diameter* of the fine particles [30]. This diameter is the size of a sphere made of material of unit density which would have the same falling speed in the same fluid as the dust fine particle of interest. Since the density of water is unity the use of the aerodynamic diameter to characterize dust dynamics is the same as comparing the behavior of dust fine particles to the behavior of their equivalent water drops. The aerodynamic diameter is usually measured experimentally by measuring the falling speed of the dust fine particle and using the Stokes equation (which was discussed in Chapter 6) to calculate the magnitude of the aerodynamic diameter from the measured aerodynamic diameter.

The focus of the occupational health specialist on the aerodynamic diameter of a fine particle arises from the fact that very often the density of the material being inhaled is not known. The dynamic behavior of the fine particles in a fluid have to be observed experimentally. One of the most modern methods for measuring the aerodynamic diameter of dust fine particles involves the movement of dusty air to a converging nozzle. As the dusty air emerges from the nozzle, the dust fine particles are still being accelerated by the fluid in which they are suspended. The time they take to travel across the distance between two laser light beams is measured. The instrument is calibrated using known aerosols [31].

The occupational health specialist is often concerned with studying how far into the lung a dust fine particle will penetrate during the act of breathing. For this purpose, the aerodynamic diameter is often a sufficient parameter. However, aerodynamic diameters can be very misleading when looking at the probability of a fine particle lodging in a respirator filter or when trying to assess how much poisonous material is piggy backing on the surface of the dust fine particle into the lung. It is known for example, that people who smoke and work with asbestos have 80 times the cancer rate of those who work with asbestos and do not smoke [32]. It is thought that the reasons for this are that the poisonous chemicals that stimulate the development of cancer are adsorbed onto the asbestos dust fine particles and carried far into the lung where they attack the surface of the lung when the asbestos dust is deposited in the lung. To understand the potential impact of fractal geometry on the study of the health hazards posed by dust fine particles consider the profiles of dust fine particles shown in *Figure 8.21*. These profiles were studied by Dr. Kotrappa [33].

The three sets of profiles of Figure 8.21 have been fractionated from other fine particles in a device known as a *Stober Disc Centrifuge* (named for its inventor Dr. Stober). The profiles within any one set all have the same aerodynamic diameter and are said to be *isoaerodynamic* [34]. A short word placed in front of another is called a *prefix* from the Latin word pre meaning "in front". The prefix iso placed at the beginning of a word means "the same". *Isotherms* on a weather map link all the places having the same temperature. *Isobars* on the weather map link all the places having the same barometric pressure (baros in Greek means "heavy" – a *barometer* measures the heaviness of the air, that is the pressure of the air at the given location). Something stuck on the end of a word is described as a *suffix*. This word comes from the Latin words sub meaning "under" and figer meaning "to fix". The root word sub has given us *submarine* for a boat that goes underwater. A suffix that is widely used in dust studies is osis which means "inflamed or diseased". *Pneumoconiosis* is a lung disease caused by the inhalation of dust. This word is formed by taking the core word konio, which is Greek for "dust", and combining it with the prefix pneumo meaning "to do with breathing" and the suffix osois. Pneumoconiosis is caused by the inhalation of coal dust. The disease is often described as *black lung*.

In the vocabulary used by occupation health specialists coal dust is described as a simple dust because the grains of dust are relatively regularly shaped. The shape

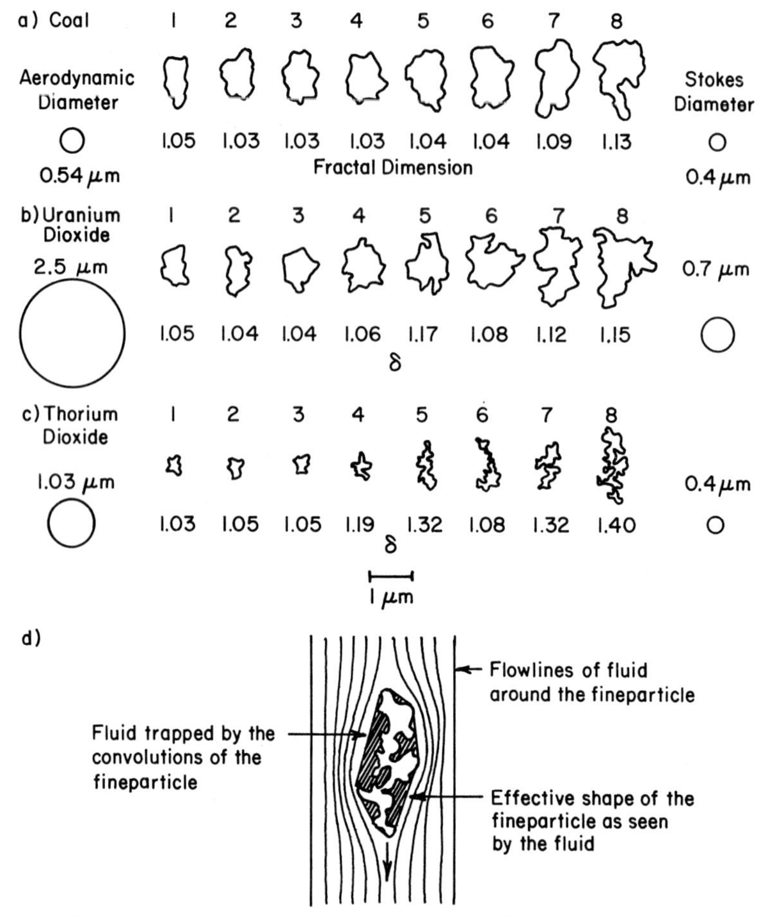

Figure 8.21. The physical size of a piece of dust can be quite different from its aerodynamic size and the physical behavior of a piece of dust may depend on its fractal structure and overall morphology. a) Coal dust fine particle. b) Uranium dioxide fine particle. c) Thorium dioxide fine particle. d) A highly fractal fine particle in fluid flow.

does not vary widely within an isoaerodynamic group of dust grains. The coal profiles of Figure 8.21(a) have an aerodynamic diameter of 0.54 μm whereas the physical size of the profiles ranges up to more than one micrometer as can be seen from the scale bar in the diagram.

The density of coal can range from 1.8 grams/cm^3 up to 2.0 grams/cm^3 depending on the type of coal. Therefore, it is somewhat surprising to see the physical dimensions being larger than the aerodynamic diameter. However, coal can be porous (containing closed air pores) and it would appear that the particular type of coal used by Kotrappa must have had a relatively low density because of air inside the fragments of coal. Even for the coal profiles one can begin to see that the larger profiles within an isoaerodynamic group have a more convoluted structure tending towards a fractal outline. The reason for this is that the more convoluted and open the structure of a fine particle the more air resistance it meets as it attempts to fall. Also, because of the stagnant fluid trapped by the fine particle as it attempts to fall through the fluid, the effective density of the profile is lower than the density of the material from which the dust grain is made. Profiles 2 and 3 of the coal dust would yield a fractal dimension if explored using an equipaced exploration technique but one can see that one probably does not make a serious error in estimating the surface area of the dust and the dynamics of the coal dust fine particles from a knowledge of the aerodynamic diameter.

For a more rugged set of profiles such as the isoaerodynamic group of uranium dioxide fine particles of Figure 8.21 (b), the effect of shape and ruggedness is most important. Uranium dioxide (also called *Urania*) is a dense material. We see that the physical size of the profiles is considerably less than the aerodynamic diameter of the set. Again it is obvious that the larger profiles within the isoaerodynamic set are more fractal in structure. This becomes even more obvious when one comes to the profiles of the thorium dioxide (also called *Thoria*) set of Figure 8.21 (c). Here the physical dimensions of the dense grains is much smaller than those that are obviously agglomerated material such as profile 8 of the third set of Figure 8.21.

It is interesting to note that the uranium dioxide material was made by a chemical precipitation process followed by ball milling. A *ball mill* consists of a cylindrical container partially filled with hard balls and the material to be turned into a powder. The container is tumbled so that the material is turned into a powder as the balls collide with each other and the trapped material. In other words the material is *pulverized* by the balls (a word coming from the Latin pulver meaning "powder"). The thorium dioxide was also made by a precipitation process but was not subjected to ball milling. The precipitation process is similar to the fuming process in that crystals grow by the material being deposited on a nucleating crystal from the solution being used to create the precipitate. The growing crystals join up with each other and grow in a random manner in a process similar to that in which the DLA are grown in a fume. If a precipitated material such as that of the thoria set of profiles is subjected to ball milling, one would expect that agglomerates such as 6, 7 and 8 would break up easily and the fragments would look very similar to the remaining rugged profiles of the uranium dioxide set.

If one were to be looking at the health hazard of a dust such as thorium dioxide, the estimates of the amount of material inhaled based on the assumption that all of the dust was like 1 or 2 of the isoaerodynamic set would be grossly in error when estimating both the amount of material entering the lung and in the reactivity of the material. A profile such as 8 would be much more chemically reactive than a profile such as 2. Furthermore, if there was adsorbed material on the dust the amount of material entering the lung, such as carcinogenic tar from a burning process, would be orders of magnitude bigger than were estimated if one was assuming that the grains were all like number 2. It may be that future studies of the health hazard of fractally structured dust could make use of the fractal dimension of a convoluted dust fine particle to give a more accurate estimate of its health hazard and also to correlate its physical dimension with its aerodynamic diameter.

Figure 8.21 (d) shows that as a profile such as 8 of the thorium dioxide set falls in air the fractal dimension will give some indication of how much fluid is trapped in the convolutions of the profile as the liquid flows around it. It may be that one could multiply the physical dimensions of the profile by its fractal dimension to deduce a characteristic parameter that would help to predict the dynamics of the dust in the lung. A fractally structured dust grain moves much more slowly than one with a similar mass but a more densely constructed profile. However, it is much easier to filter a fractally structured dust fine particle in a respirator or some other device than one would anticipate since profiles such as 8 are more extensive in space than a fine particle such as number 2 of the thorium dioxide set. Therefore, when assessing the "filterability" of a dust perhaps we need a parameter which is the aerodynamic diameter multiplied by a function of the fractal dimension of the profile and its aspect ratio. The reader might like to try to see if there is a relationship between the physical dimensions, the fractal structure and the area of the profiles of the sets of the profiles of uranium dioxide and thorium dioxide given in Figure 8.21. One could measure the area and the structural boundary fractal dimension and see how the two are related. A simple relationship to try is the area divided by the fractal dimension or some power of the fractal dimension such as the square root or the square of the fractal dimension. Parker Reist and his colleagues have applied fractal geometry to the characterization of aerosol fine particles [35].

In *Figure 8.22* we show more outlines of profiles which illustrate how fractal geometry is of potential use by scientists who are active in materials research. The first profile is from a greatly enlarged view of a lead oxide fume of the type generated by the welding of lead. Japuntnich characterized this material by comparing the area and the aerodynamic diameter of the profile [36]. He showed that the aerodynamic size could be half the physical size. He also confirmed that this type of profile is quite easy to filter. Therefore, when studying welding fumes, the bad news is that this type of fume is often underrated as a health hazard but the good news is that it is often much easier to filter than one would have predicted from a knowledge of the aerodynamic diameter of the profile without knowing its physical structure.

Figure 8.22(b) shows a greatly enlarged zinc oxide profile described by Bolsaitis. It can be seen that if such fumes were to be present in the gases leaving an incinerator they would pose a considerable health hazard [37]. Another area where scientists are creating fractal systems is in the search for new ceramic materials. The ceramicist traditionally was the maker of pottery and other products such as bricks and tiles. Today, ceramicists are working on the development of ceramic materials which can be used to make car engines and other things such as aircraft engines. To make such

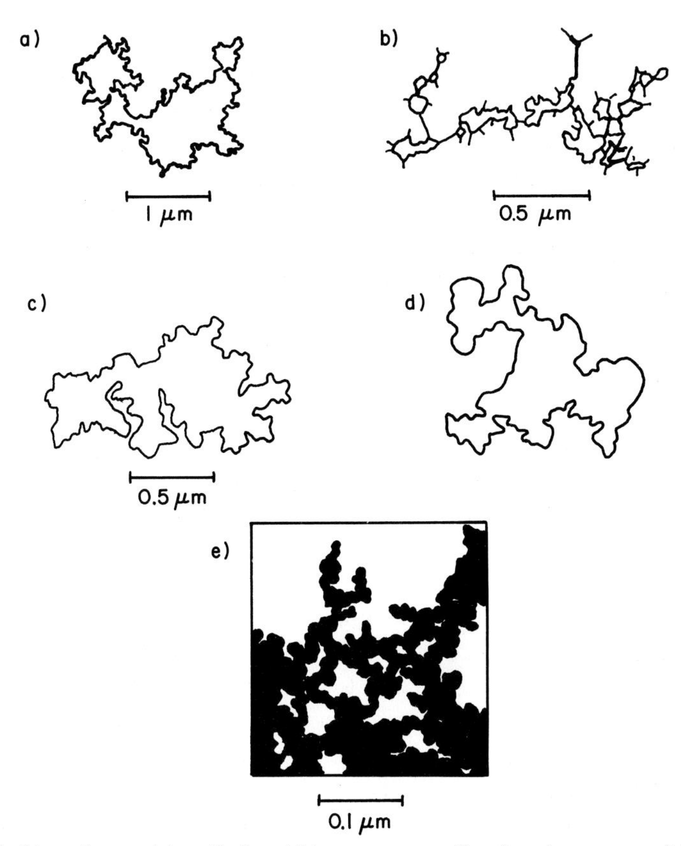

Figure 8.22. Many fine particles of industrial importance manifest fractal structures. a) Lead oxide [36]. b) Zinc oxide [37]. c) Barium titanate [39]. d) Ceramic grade titanium dioxide [40]. e) Sintered titanium dioxide.

sophisticated equipment, the ceramicist is seeking to make highly reactive new ceramic powders as raw materials for their processes [38].

Figure 8.22(c) shows a highly magnified image of a grain of barium titanate generated by Vivekananden and Kutty [39]. This profile has both a density fractal and a boundary fractal. The very high surface area of this grain of powder means that when the ceramicist starts to fuse the material using heat it consolidates relatively rapidly. However, creating fractal agglomerates to trap high energy in the structure of the profile has a negative side to the performance of the powder when attempting to form dense compacts, since powder made of grains such as those shown in Figure 8.22(c) will probably be very difficult to pour into the container, i.e. the powder will have a difficult rheology.

In Figure 8.22(d), a fine particle of what is known as ceramic-grade titanium dioxide is shown [40]. Notice the very open and rugged structure indicating high surface energy. As ceramicists become more familiar with the techniques of fractal geometry, we can expect to see many publications in which the fractal dimension of the profiles is related to the sinterability of the powder and to the flow properties of the powder [41].

The technical term for the process of consolidating powders such as those shown in Figure 8.22 by means of heating them below their melting point is known as *sintering*. This word was coined by German scientists and is related to the English cinder. Raw coal contains many tiny silicate fine particles which when burnt in a conventional furnace or fireplace turn into glass. A *cinder* from a coal fire is a sintered conglomerate of glassy ash fine particles. As the ceramic fine particles fuse with each other during a sintering process, the result is often a complex structure which can be described using the concepts of fractal geometry. In Figure 8.22(e), a micrograph of sintered titanium dioxide is shown.

Several new words are used in relation to this diagram. First of all, the highly magnified pictures similar to those shown in Figure 8.22 can be taken with the aid of a device that passes a beam of electrons through the material. This type of device is described as a *transmission electron microscope* because the electrons go right through the material. In short form this type of equipment is referred to as *TEM*. In another major method for creating highly magnified pictures, the electrons are bounced off the object to be imaged and the reflected electrons captured by a special device. The images are created by moving a beam of electrons across the surface of the object to be imaged. The equipment is described as a *scanning electron microscope* often referred to as an *SEM*. Scanning electron microscope pictures give an impression of depth and ruggedness of the surface whereas transmission electron microscopes look as if they are sections taken through the material. Methods for measuring the roughness of a surface by studying the variation in brightness across a scanning electron microscope picture have been described in the scientific literature [42]. A unit of measurement used by technologists when characterizing very fine systems is the *nanometer*. This unit is one billionth of a meter, 10^{-9} meter (1,000 nanometers = 1 micron). The scale bar of Figure 8.22(e) is 100 nanometers. It can be seen that the constituent spheres of the material are of the order of 10 nanometers. The ceramicists have started to refer to such very finely divided material as *nanophase material* [43]. Techniques for describing the internal structure and state of subdivision of materials such as those in Figure 8.22(c, d, e) using the concepts of fractal geometry is beyond the scope of this book but there has been some excellent work on this topic carried out by Schafer and colleagues [44, 45].

In *Figure 8.23*(a) some profiles drawn from highly magnified scanning electron microscope images of soot generated by free burning diesel oil are shown [46]. Two of the clearly distinct agglomerates visible in this SEM photograph are shown in Figure 8.23(b, c). Scientists are interested in studying the structure of such soot agglomerates for several reasons. Let us consider first of all, the problems faced by people concerned with health hazards to workers in underground mines where the

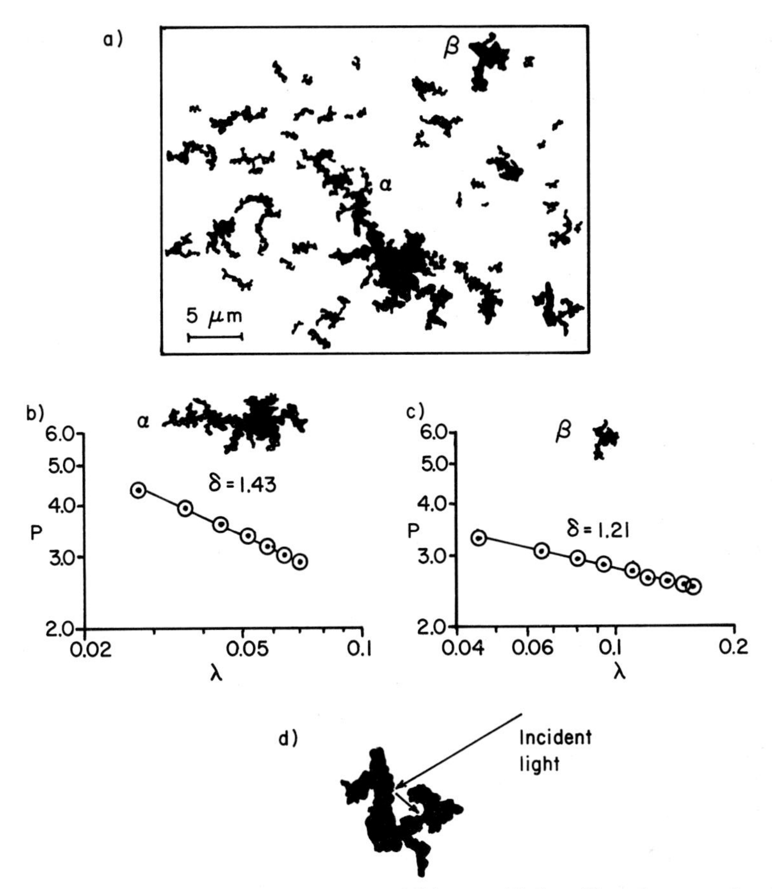

Figure 8.23. Diesel soot has a fractal structure. a) Highly magnified profiles of soot produced by free burning diesel oil. b) The largest agglomerate of part (a) has a structural boundary fractal dimension of 1.43. c) Smaller, denser soot agglomerates typically have a structural fractal dimension of 1.21. d) Fractally structured carbon black agglomerates are very efficient light absorbers because the convoluted cavities of the profiles constitute effective light traps.

production of diesel fumes is a major problem for ventilation engineers and health workers. For a period in the 1970s and 1980s there was wide spread use of diesel equipment in mines; since then the use of diesel engines underground has tended to decline because of the cost of providing air to allow the burning of the diesel fuel. I am told by workers underground that one of the major problems with such diesel engines occurs when they are not properly tuned. When operating in this way they can produce a smoke which nucleates fog and creates a persistent underground fog which hinders visual communications.

If one studies an agglomerate such as that of Figure 8.23(b) by measuring its aerodynamic diameter, it appears to be very small due to its open convoluted structure. This structure also causes it to settle very slowly (measuring the aerodynamic diameter of such agglomerates is like assessing the size of a dandelion seed floating in the wind by its falling speed. The seed is large enough to be seen clearly with the naked eye but measurements would indicate that it's aerodynamic diameter was no more than one or two microns). In Figure 8.23(b) the fractal dimension of the large agglomerate is shown. It can be seen that it has a boundary fractal dimension of 1.43. This is typical of large agglomerates produced by collision of smaller more dense agglomerates which typically have the type of fractal dimension shown in Figure 8.23(c). In fact, the large agglomerate of Figure 8.23(b) has physical dimensions greater than 10 microns. For this reason, it is very likely to be caught in the upper respiratory tract and moved into the digestive tract by the defence mechanism of the lungs. It is interesting to note that scientists discovered empirically that diesel fumes were very easy to filter even though they were apparently of a very small aerodynamic size. There are also growing indications that workers exposed to diesel fumes are not developing lung problems but are developing cancer of the bowels since the carcinogenic chemical produced by the incomplete combustion of diesel oil is being carried into the digestive tract, rather than the lung, by the large fluffy agglomerates [47]. Once one is aware of the effect of fractal structure on the fine particles of inhaled fumes, the ease of filtering such fumes and the threat to the digestive tract become obvious. Hopefully more work on the fractal structure of diesel fumes plus the use of a shape factor such as the aspect ratio will enable the industrial hygienist to give better descriptions of the real threat to the health of the worker from large fluffy diesel agglomerates.

The fractal structure of such soot fine particles can also be a key issue in any law suit governing the emission of smoke from a diesel engine. Very often the pollution hazard is measured in terms of the optical opacity of the fume, that is the blackness of the plume. However, the blackness of the plume is not entirely related to the amount of unburnt carbon in the emission from the diesel engine. The fractal structure of the soot fine particles governs their ability to absorb light. This is demonstrated in Figure 8.23(d) which shows schematically how light energy hitting the profile is trapped by the many convolutions of the surface. Engineers who try to design smoke screens for military defence would obviously like to give the highest possible fractal structure to the soot agglomerates that they generate by burning the fuel to create the smoke screen. The higher the fractal dimension of the agglomerate, the greater the ability of the agglomerate to absorb radiation (that is to hide the object being defended). Also, the high fractal dimension will create a fluffy agglomerate with lower falling speed and hence a more stable smoke screen. One can imagine that those scientists concerned with creating smoke screens are busy trying to create combustion conditions which produce highly fractal agglomerated soot.

The fractal structure of soot is also of great interest to scientists concerned with what is known as the *nuclear winter* effect [48]. It was pointed out in the early 1980s that if the world was foolish enough to have a nuclear war, fires started by the use of nuclear weapons would probably be more devastating than the direct effect of the weapons. This is because the fierce fires would produce enormous clouds of smoke which would drift around the world and cut off the sunlight causing crop destruction and famine. Therefore, scientists began to be very active in calculating how much radiation from the sun would be turned away by the soot in the atmosphere. However, in their calculations, many of them assumed that the soot would be spherical. This is because it is easier to calculate theoretical consequences with spheres as models rather than for real soot fine particles (see discussion of spherical chickens in reference 3). In the late 1980s, J. Nelson of the University of Bristol, England pointed out that when one takes the fractal structure of soot into account when calculating the effect of the nuclear winter, the calculated temperature can drop by several (up to five) degrees Celsius [49].

In the coming years, we can expect to see many experimental studies of the actual structure of soot and light opacity studies of diesel soot created in the laboratory with subsequent study of the fractal nature of the soot [50, 51]. Soot which gets high into the atmosphere can persist for a long time. It is known that smoke from forest fires in Canada can float across the Atlantic to create what is known as a blue-moon effect in Scotland. The *blue moon* is created because the soot fine particles reflect away the red light and only the blue light of shorter wave lengths get through to the Scottish observer. However, such drifting clouds of soot are a relatively rare event as indicated by the saying that something only happens "once in a blue moon".

The fractal structure of soot is also being studied by scientists intent on improving the efficiency of internal combustion engines. The fractal structure of the soot contains information on the burning conditions within the engine. Such information that can be used by the engineer to make better engines with less pollution-causing emissions [52, 53]. The fractal structure of emergent soot is frozen information on the combustion dynamics in the engine.

Another area of active research in modern science is the study of the catastrophic disappearance of the dinosaurs about 65 million years ago. One theory claims that a massive asteroid, or a cluster of asteroids hit the earth and created an enormous cloud of dust. This dust circled the earth and by cutting off the sun's energy caused the death of plant life and hence, by starvation, the death of the dinosaurs. In support of this claim, some scientists study a deposit of iridium rich dust to be found at various places around the earth. Recently, scientists have claimed to be able to isolate ancient soot fine particles from the layers caused by the fires initiated by the asteroid collision and the soot fine particles they have discovered have fractal structure. Therefore, anyone wishing to calculate the radiation flux effects of a cloud of dust generated by an asteroid collision will have to take into account the fractal structure of the soot content of the global clouds [54].

In *Figure 8.24*(a) a set of profiles typical of what is known as rock tailings are shown. *Rock tailings* are the virtually worthless fragments of the original ore which once formed the matrix around the grains of the valuable material in the ore. The

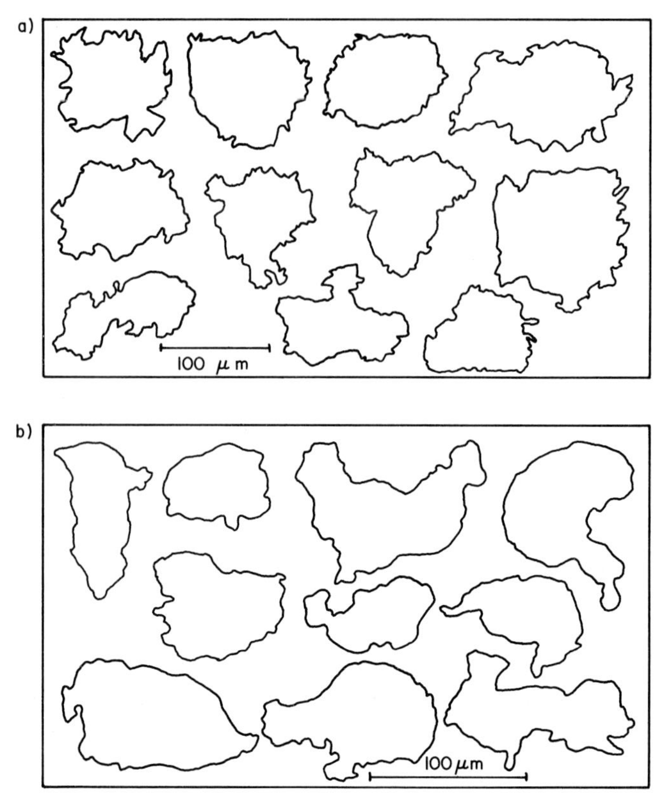

Figure 8.24. A typical selection of the fractally structured fine particles used in rheological investigations. a) Rock dust tailings from nickel mining operations. b) Aluminum shot powder.

rock tailings of Figure 8.24(a) are the fragments of silicate-type rocks left after the valuable nickel-bearing material has been taken out. They are called tailings because they are the "tail end" of the mining process. They are often used in a process known as *back filling*. In this process a slurry (a dense suspension) of the rock tailings is pumped into the tunnels and work areas from which the ore has been produced to fill up the cavities underground. This prevents the settling of the rock or the undesirable collapse of rock working areas. The volume of the rock tailings however is always greater than the volume originally occupied by the rocks and the excess material is allowed to gradually accumulate in what are known as *tailing ponds*. Ultimately, the tailing ponds fill with sediment. They are then usually covered with soil and planted with trees and grass. Sometimes the tailing dust is potentially dangerous as in the case of the tailings from uranium mines which still contain a small amount of radium. The way in which water soaks down through old tailing ponds is a very important problem which we will look at briefly in Chapter 13.

Industry must pump many other slurries in which the suspended fine particles have fractal structure. For example, in a power station very often the material collected by air cleaning devices is washed down from the equipment into a slurry which is then sent to a settling pond. This material is called *flyash*. It is often fractally structured [55]. Again, technologists are considering pumping slurries of ore from one location to another sometimes over large distances and the power industry is considering using crushed coal in oil as a fuel in electricity-generating power stations. When a slurry is made up of fine particles which are nonspherical and which have fractal boundaries, the material exhibits what is known as *thixotropic behavior*. *Thixotropy* is defined in the dictionary as "the property of becoming fluid when shaken or disturbed". It is the opposite of what is known as *dilatancy*. Anyone who has had the opportunity to walk across the surface of wet sand at the beach has had an opportunity to study dilatancy. When the wet sand is undisturbed, the grains of sand have had time to line up with each other and the space between the grains is at a minimum. When the foot presses down on the wet sand it disturbs the grains and they take up random orientation. In this random packing, the pore space between them increases and so the sand appears to dry out because there is not enough water from the original closed packed situation to fill all of the spaces between the grains of sand. The effect does not disappear immediately after walking over the sand and if one looks backwards at ones tracks one can see the footprints apparently raised above the level of the wet sand and they appear to be dry. However after a short time, they will normally slump back into their wet condition. The word *dilate* comes from a Latin root word dilatatus which means "to become wider". When stepped upon, the amount of sand appears to be wider. The word thixotropic comes from two Greek words thixis meaning "a touch" and trope meaning "a turn or a turning". Anyone who has mixed a batch of concrete is aware of the problems of thixotrophy. As one turns a mixture of sand, cement and water with a shovel it appears to be quite fluid and then if one stops turning it apparently drys out. This is because as one turns the material with the shovel, the grains of sand and concrete etc. all line up with the movement of the spade and like the quiescent wet sand the gap between the grains is at a minimum. Once one stops stirring the grains they take up random positions and the material appears to dry out because there is not enough water to fill the spaces. Beginners in the art of concrete making very often add too much water to the mixture because of its apparently dry state when sat waiting to be used.

Putty used to fix the glass of windows in place, is made from a suspension of crushed limestone in linseed oil. If putty is allowed to stand for a time it will appear to weep oil as the grains of the limestone take up an optimum packing position. To make the putty useable after such weeping, one must randomize the grains of limestone by manipulating the oily mixture. When stirring a concrete mix, the worker soon knows that one has to start off the movement of the mixture by stirring and that in fact the viscosity of the mix drops as one starts to move it. This becomes a very important problem when transporting coal slurries and in fact, the design engineers dread a power failure because if a coal oil slurry or coal water slurry starts to settle in the storage tank the amount of power required to start it moving again can actually break the shaft of the stirring device. When using industrial slurries, the golden rule is never to let them settle until they reach their final destination. Thixotropic and dilatant suspensions are known as *non-Newtonian fluids*. A study of the fractal nature of fine particles in non-Newtonian slurries being moved by industry will probably become the major area of applied research [56].

Molten chocolate and other food mixtures are often thixotropic. This fact can cause problems in the factory manufacturing prepackaged foods. The thixotropic behavior of molten chocolate comes from the fact that there are sugar and cocoa grains in the milk matrix forming the chocolate. I discovered the thixotropic nature of chocolate one day when I placed a snack of chocolate near a Bunsen burner. The heat from the Bunsen burner transformed the chocolate into a static molten mass retaining the shape of the chocolate but when I attempted to move the bar it flowed everywhere.

In *Figure 8.25* some data describing how the viscosity of a slurry of rock tailings of the type shown in Figure 8.24(a) varies with the volume concentration of the rock grains in the slurry. This data was obtained using mineral oil as the supporting liquid. This material was chosen for its relatively high viscosity to avoid problems connected with the settling of the rock tailings in the suspension during the making of viscosity measurements. The equipment that was used to generate the data of Figure 8.25 was a relatively unsophisticated viscometer of the type available in the undergraduate section of the university laboratories [57]. We will discuss this type of viscometer in some detail since it can be used to do research for school science projects of interest to industry and also because a study of turbulence in this type of viscometer was one of the key experiments which led to the development of the subject known as deterministic chaos. This type of equipment is widely discussed in books on deterministic chaos [57, 58].

The equipment consists of a cylinder placed co-axially inside another cylinder as shown in Figure 8.25(a). The inner cylinder can rotate and create a velocity gradient between the wall of the fixed, outer, cylinder and the rotating inner cylinder. The equipment can be used at several levels of sophistication. In the simplest mode equal weights are placed upon the two pans which cause the inner cylinder to rotate. The rotational speed of the inner cylinder can be calculated from the falling speeds of the pan and the rate of rotation can obviously be varied by using different weights in the pans. When used in this simple mode, one is obviously working at low speeds of revolution. In a more sophisticated version, the outer cylinder is rotated by means of a pulley added to the outside cylinder. In this instrument, the force required to keep the inner cylinder stationary is measured. Using this configuration higher speeds of rotation can be used. For the size of powders used in our experiments, and by using a relatively viscous clear mineral oil, one could carry out experiments with

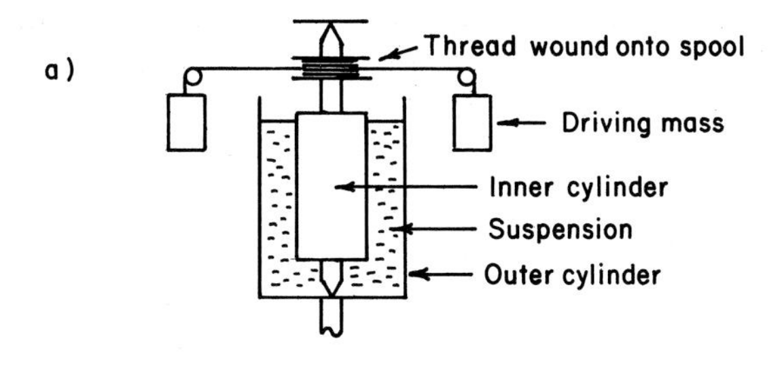

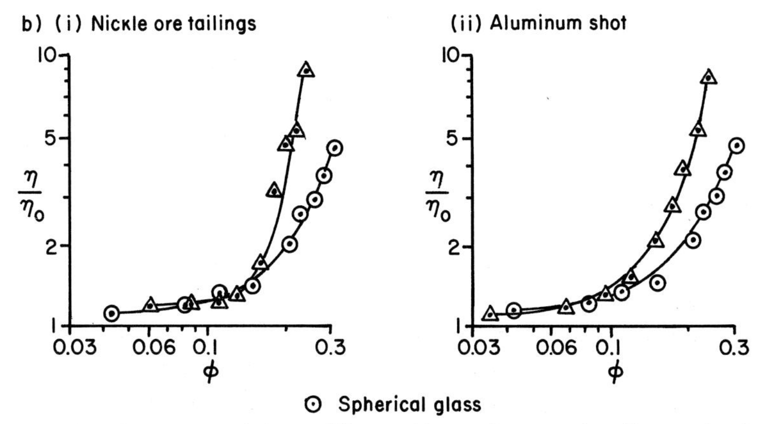

Figure 8.25. The fractal structure and shape of fine particles can have a major effect on the viscosity of a suspension. a) Simple viscometer for measuring the viscosities of suspensions. b) i) Normalized viscosities of suspensions of glass beads and nickel ore tailings in mineral oil. ii) Normalized viscosity of suspensions of aluminum shot fine particles (see Figure 8.24(b)) v/v_0: ratio of the viscosity of the slurry to the viscosity of the mineral oil. φ: volume fraction of the powder in the slurry.

various slurries of different solids concentration to generate the type of information shown in Figure 8.25(b).

To enable one to compare viscosity data for different slurries of varying solids content, it is more useful to express the solids content of the slurry in terms of the volume of the slurry occupied by the fine particles than it is to use weight measurements. This is because it is the volume of the fine particles that is important in determining the structure of the spaces between the fine particles which is the parameter that determines the viscosity of the slurry. All the viscosity data of Figure 8.25 is converted to a dimensionless form by dividing a measured viscosity of a slurry by the viscosity of the oil without fine particles. In Figure 8.25(b) (i) the change in viscosity of the slurry of glass beads as the solids content is increased is expressed as a graph of dimensionless relative viscosity against the volume fraction of the suspension occupied by glass beads. In this same graph similar data for the nickel ore tailings fine particles of Figure 8.24(a) are summarized [59]. It can be seen that for solids concentrations less than 5% by volume solids there is little difference between the viscosity of the glass sphere and nickel tailing slurries. However, at a concentration of 0.2 volume fraction solids (20% by volume solids) the nickel ore tailings suspension is five times as viscous as the slurry of glass beads. This big increase in viscosity is essentially due to the entrapment of fluid in the fractal cavities of the nickel ore tailings fine particles. In Figure 8.25(b) (ii) data for the viscosity of slurries of glass beads and suspensions of the aluminum shot profiles of Figure 8.24(b) are shown. Again it can be seen that due to trapped fluid, the aluminum shot slurries are much more viscous than the glass bead slurries for any given volume fraction of solids. If one superimposes the glass bead, nickel tailings and aluminum data for the region between approximately 0.15 and 0.2 solids volume fraction an interesting difference appears between the aluminum shot and the nickel tailing slurries. It appears that the aluminum shot viscosity starts to increase sooner than the nickel ore tailings fine particles. A visual inspection of the aluminum shot fine particle shows that although there is a wide range of structure amongst the profiles, some hardly have any fractal aspects whereas others are extremely convoluted, in general, the structural fractal of the aluminum shot is more pronounced than for the nickel ore tailings. This gross fractal structure traps significant amounts of stagnant fluid which moves with the fine particles at lower solids concentrations than for the nickel ore tailings fine particles. However, once the solids concentration becomes fairly high, the larger textural fractals of the nickel ore tailings fine particles starts to give the same order of magnitude of increase in the viscosity of the slurry as compared to that of the pure liquid. Perhaps the concentrations where the aluminum shot and the nickel ore tailings fine particle slurry start to deviate significantly from that of the glass beads are a measure of the average structural texture of the profiles. Again, the ratios of the viscosity at the 0.2% solids concentration may be a very good measure of the average fractal structure of the textures of the profiles.

The measurement of the difference between the viscosities of slurries of solids with fractal structure and equivalent suspensions without fractal structure may prove to be a very quick and inexpensive measure of the fractal structure of powder systems. It may be that the ratio of the dimensionless viscosities at some arbitrary chosen volume fractions of solids such as the 0.2 solids content may be a useful shape factor to describe the behavior of a population of irregularly structured – fractally textured fine particles. The reader is invited to explore the problems of the effect of size and shape of industrially important powders on the viscosities of suspensions of the powder.

Readers may want to obtain a sample of pulverized powder which has fractal structure and carry out both fractal characterization of the profiles and the viscosity of the suspension. Students could then carry out ball milling experiments of material such as rock tailings to see if they could reduce the fractal structure of the profile and explore the effect of fractal surface reduction on the decrease in slurry viscosity at a given solids concentration. Readers may also like to practice their fractal measuring skills with the profiles of Figure 8.24. A class project could be undertaken in which each student took a different profile to investigate the distribution of characteristic fractal dimensions amongst the profiles of the figure.

Section 8.9. More Ideas for Putting Fractals to Work

After my book "A Random Walk Through Fractal Dimensions" was published, I received interesting letters from all over the world from people who were starting to use fractal geometry to study various subjects. In this section we will look at some of their ideas and from their activities the reader will probably think of many laboratory projects for themselves.

In 1989, Miss Jennifer Newbury of Washington D.C (U.S.A.) wrote to tell me about her successful science project in which she applied fractal geometry to the characterization of *sun spots* or *solar flares*. Sun spots are eruptions on the surface of the sun that have been observed by astronomers for centuries. They have been very difficult to characterize. It may be that their fractal structure is related to the formation forces and the strength of their effect on the weather and the earth's climate [60].

Again in 1989, a colleague, Dr. Frank Mallory of the Biology Department at Laurentian University, raised the interesting possibility that the line present at the top of a wolf's skull (known technically as the *sagittal suture*) might be characteristic of the species of wolf. He initiated experiments to study the fractal dimension of the wiggly line on the top of the skull of the animals. The trained biologist can tell the difference between wolf species by looking at the structure of the skull but what was needed was a computer based automated recognition procedure for taking some feature of the skull to determine the species of the wolf to enable population densities to be estimated accurately for given areas of the forest. Later Dr. Hartwig of the University of California, Berkley wrote to tell me he was applying fractal characterization to the sagital suture of the human skull [61–62].

A friend who wishes to remain anonymous, has told me that after studying fractal geometry he was able to initiate quality control procedures for evaluating the quality of a paint finish on a surface by measuring the ruggedness of the surface of the paint with a profilometer to measure the fractal dimension of the surface. The quality of the paint could be quantified directly by the fractal dimension of the line contour of the paint finish. The company does not wish to publish any data on their measurements because they would like their customers to believe that all paint surfaces are perfectly Euclidean with a fractal dimension of 1.0000!

In January 1990, Dr. Christopher Davey of the Department of Biological Sciences of the University College of Wales wrote to tell me that he was applying fractal geometry to a study of animal cell walls. He also told me that in cooperation with a fellow research worker and fractal lover (Gerad Marx) he was studying the shape of microbiological colonies grown in Petri dishes. They were looking at the shape of the perimeter of the bacterial colonies grown on agar jelly. A dictionary tells us that

> "*Agar is a polysaccharide gel (sugar loaded jelly) made from seaweed and used by bacteriologists to grow colonies of bacteria called "cultures". Bacteria are deposited on the gelatinous substance and grow into colonies that can be identified by the bacteriologist.*"

Dr. Davey points out that "many colonies grown on agar have slight indentations while others resemble very rugged islands and even swirling feathers." Perhaps, he suggests "the measurement of the fractal dimension of such colony edges may enable us to better understand the development of such colonies. It could be that under defined conditions the structural fractal dimension could be very characteristic of a given species".

After reading his letter, I wondered if this might be a very sensitive method to look for the effect of bactericide and disinfectants on bacteria. The root word cide means "to kill" and hence bactericide kills bacteria; *pesticide* kill pests and *herbicides* kill weeds and growing plants. Perhaps the interface between the growing colony and a drop of bactericide placed on the jelly might give us some information about the potency of the interaction between the growing colony and the bactericide. There are obviously many fractalicious science fair projects waiting to be grown in Petri dishes.

In early 1990, John Graf wrote to me from Michigan Technological University in Houghton, Michigan to tell me he was working on the possible relationship between the fractal structure of moon dust fine particles and the flow of such powders on the moon. The general study of the flow of powders and liquids is known by the scientific name of *rheology*. The word rheology comes from the Greek root word rheos meaning "a flow of water or other substances". Scientists are interested in the rheology of moon dust because it is an important aspect of any attempt to sample moon dust by robotic explorers. Also, the rheological properties of the moon dust determine how the powder under the feet of a lunar module will compact and/or move when the space module lands. In *Figure 8.26*(a) two lunar fine particles are shown which exhibit fractal structure. The reader is invited to carry out measurements on the profiles using the equipaced exploration technique.

Following a presentation at a seminar on the characterization of a rugged line by its fractal dimension, I was asked by a scientist in the audience if it would be possible to characterize the appearance of a broken cookie (biscuit). Apparently consumers of cookies come to expect that their cookies, when broken at the commencement of a munch session, will have an appearance characteristic of the type of cookie. If the ruggedness of the break of the cookie is not to their expectations, they tend to complain that the cookie has not been cooked properly. From the brief conversation

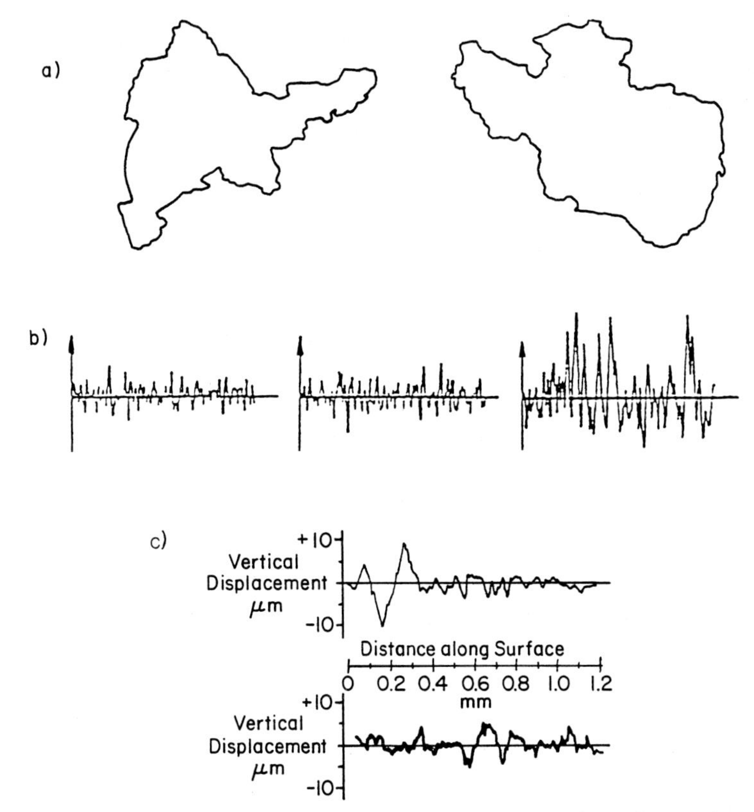

Figure 8.26. Fractal problems of interest to scientists range from the flow of moon dust to the wear of pipes used to convey powders from one part of an industrial process to another. a) Two lunar dust profiles from scanning electric micrographs taken by John Graf. b) Profilometer traces of steel surfaces used in conveying coal. By A. W. Roberts [63]. c) Profilometer traces of metal alloys studied by Upadhyaya [66].

I had with this scientist, I gained the distinct impression that he was rushing back to his laboratory to characterize the broken structure of cookies to give a mathematical definition to the common phrase: "That's the way the cookie crumbles!". Students can try breaking various types and brands of cookies to see how one could characterize the edge of the cookie. One could photograph or photocopy the fracture edge and subject it to fractal characterization study. In my discussion with the food scientist, he indicated to me that he hoped the technique could become a quality control technique for characterizing batches of cookies made by the same process. One of the unfortunate aspects of successful science in industrial applications of new scientific technology is that the scientists are not usually permitted to publish their results. If the cookie scientist with whom I had a conversation does develop a new quality control method for his company's cookies he will probably keep the information as a secret within the company's files.

People who weave cloth like their fibers to have a crimp in them to help the fibers hang together in the woven cloth. The lines of Figure 8.6 look like crimped fibers to the textile chemist. Again I know from discussions I have had with several textile scientists that they are busy using fractal dimensions to characterize their fibers. Manufacturers of beauty supplies and appliances sell kits to enable the consumer to put a permanent crimp into hair fibers. I have been told that studies are under way in the cosmetics industry to measure the fractal structure of hair before and after the application of "permanent wave chemicals." A class of students can study this kind of problem by sampling the hair (on a voluntary basis!) of different members of the class. The students will soon discover that mounting the hair to be studied onto a microscope slide is a major problem since it is all too easy to straighten out the hair by pulling on both ends as one places it in position. One has to devise a natural way of letting the hair fibers drop down onto a sticky surface. A good way to do this is to take a microscope slide and coat it with "double sided sticky" office tape used to mount documents. One then lets the hair float down and sit on top of the coated slide before applying another glass slide and examining the system under a microscope. A class could study such things as the curly hair from animals and the structure of pieces of fiber taken from woven cloth.

The set of profiles of Figure 8.26(b) are those reported in a scientific paper by Dr. A. W. Roberts of the University of Newcastle in Newcastle, New South Wales, Australia [63]. Dr. Roberts and his colleagues are interested in the *pneumatic conveying* of powdered solids (the Greek word pneuma meaning "breath" has given us many words in English; a *pneumatic drill* system is one that is operated with compressed air). To transport solids pneumatically, air or another gas is added to the powder and it is blown along a tube. At several points in a pneumatic conveying system, the powder must move over the surface of a pipe and the roughness of that surface is an important property that determines how much power is required to move the material. During the operating life of material, the surfaces are altered by the movement of the powder. In some cases, the powder can polish the surface, in others it can wear it away at a very fast pace. Therefore, the scientist must continually monitor the roughness of the surfaces in his pneumatic conveying system. This is particularly important when devising a new system for a powder which has not been conveyed pneumatically in previous industrial situations [63–65]. If the student has access to a profilometer at a science teaching institution or in a neighboring industrial research laboratory, a very interesting science fair project would be to measure the rate of corrosion of material exposed to acid rain or to industrial abrasion. In the school laboratory, one could also make strips of metal which are then corroded in mild acid and studied with a profilometer or alternatively a stone surface such as a piece of polished marble could be eroded by a mild acid such as vinegar and the rate of penetration evaluated from the fractal dimension of the surface traces.

The pair of rugged lines shown in Figure 8.26(c) were presented in a scientific paper by G. S. Upadhyaya and colleagues [66]. They are profilometer traces of a metal alloy and a composite alloy containing 14% by volume of talc (an industrial chemical). An *alloy* is any combination of two or more metals. The word comes from two Latin root words a meaning "too" and ligare meaning "to bind". The latter word has given us the word *ligament* which was originally the name of anything used to tie things together in a bundle. It is also used to describe string-like parts of the body. In an alloy the two or more components are bound to each other. *Brass* is an alloy of zinc with copper, and *bronze* is an alloy of tin and copper.

By measuring the roughness of the surface of the alloy and the composite material, the scientists were able to deduce important information about the structure of the alloy and the composite material. Again, one can now measure the fractal dimensions of such profiles and use them to describe the properties of the alloys and composite materials.

Exercises

Exercise 8.1. Measuring the Fractal Dimension of Rugged Lines

As suggested earlier in the text, everyone should do a fractal estimation by striding around a perimeter using a pair of compasses. In *Figure E 8.1.1* (a) the two rugged lines that we met in Section 8.3 are shown formed into a triangle. We have found it very convenient to set up our computers to examine coastlines where the coastlines form a complete circuit and to use such programs to characterize wiggly lines by forming synthetic islands using several images of a wiggly line. Working on a synthetic island instead of directly on the wiggly line helps to minimize problems created by end effects as one strides around a profile. For this reason, it is convenient to put our profiles into a triangular set. When one strides around such a synthetic island, the size of the strides should be related to the size of the convolutions being explored otherwise, the data for the perimeter estimates will not lie on a fractal portion of the Richardson plot. The suggested stride lengths for the setting of the compass legs shown in Figure E 8.1.1 (b) range from the largest to the smallest of the convolutions to be explored in the two synthetic islands. If one were to use a stride length much larger than the one shown it would not really be exploring the structure of the convolutions.

Students left to themselves to choose stride lengths for exploring a rugged system such as that of Figure E 8.1.1 (a) tend to use a series of larger stride lengths which do not collect useful information on the fractal structure of the line. For example, when the lines of Figure E 8.1.1 (a) are enlarged so that the length of the line is 17.7 centimeters, the students will often choose to explore it using a series of stride lengths such as 4 centimeters, 3 centimeters, 2 centimeters etc. They have to be

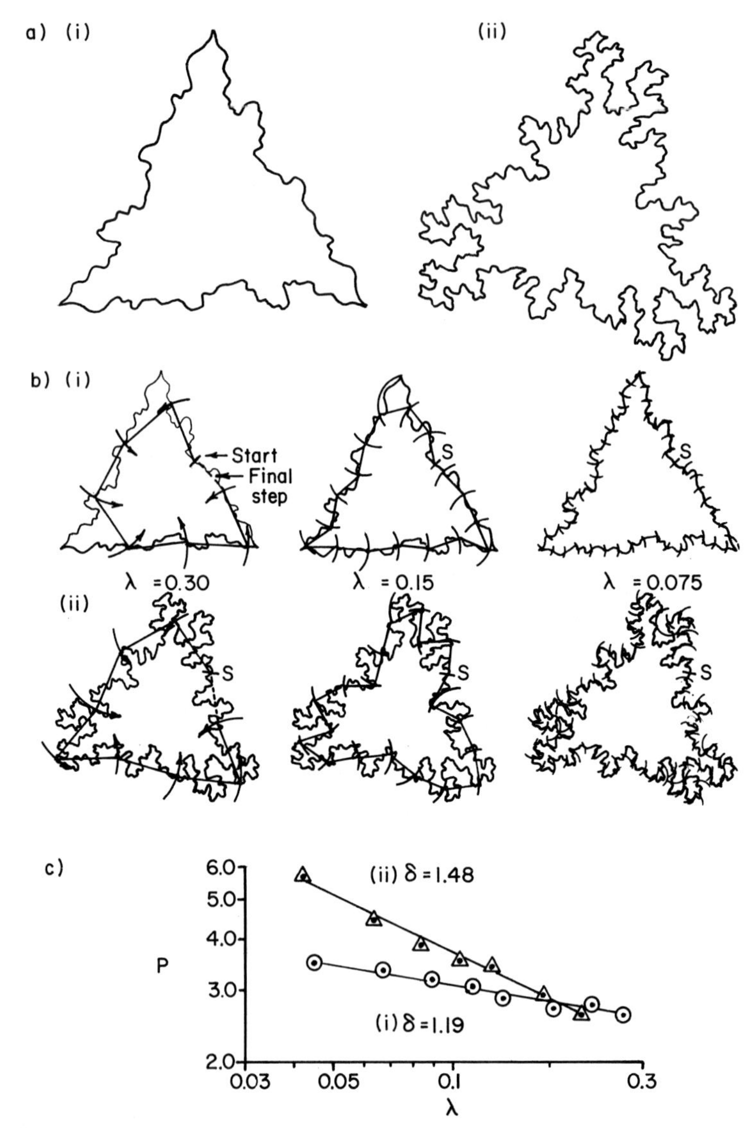

Figure E 8.1.1. Stride around data from a characterization study of wiggly lines transformed for convenience into island profiles. a) Connecting three copies of a wiggly line to be characterized by its fractal dimension to form a synthetic island can facilitate the measurement procedure. b) The perimeter estimate of an irregular profile based on the polygon constructed between the points of contact for three different stride lengths c) Richardson plots for the two synthetic islands.

taught to appreciate the fact that since data points on the Richardson plot will be spaced according to the logarithm of the step size they use they should choose a series of steps which are half the value of the preceding step. When the curve is such that the line is 17.7 cm a useful set of exploration steps are 5 cm, 2.5 cm, 1.3 cm, and 0.6 cm. These four exploration strides give four well separated points on the Richardson plot. The student can see from Figure E 8.1.1 (b) that these stride magnitudes represent the largest to the smallest feature of the fractal structure of the line being explored.

When carrying out an exploration of a profile such as that of Figure E 8.1.1 (a), the answer that one gets for a particular perimeter estimate depends upon the starting point. In Figure E 8.1.1 (b) three different stride polygons generated by walking around the profile are illustrated. Also illustrated in this sketch is the fact that the completion of one complete trip around the profile usually requires a partial stride to complete the polygon. An unbiased and efficient method of doing this is to join the last complete stride to the starting point as illustrated. The teacher should discuss with the students that the variation of the perimeter estimates at a given stride length is not an error but is an uncertainty because of the lack of precision in the concept "polygon estimate" of the perimeter by striding around the profile with a stride of given magnitude. The teacher should discuss how a point to be plotted on the Richardson plot should be an average of several explorations of the profile and that a double headed arrow, could be placed on the Richardson plot to show the range of values represented for the average value shown.

If a class is given the profiles of Figure E 8.1.1 (a) then each of them can do several walks at different starting points and the class data amalgamated on one Richardson plot. The students should also be encouraged to make several measurements at the same starting point and at a fixed stride magnitude so that they can estimate their experimental uncertainty (their measurement error) in making a perimeter estimate. They will discover that their error of measurement is much smaller than their uncertainty due to the fluctuations depending upon the starting point. The teacher could point out this means that it is better to do several measurements from different starting points rather than to take too much effort to make any one individual perimeter estimate accurate as they swing around the profile. In Figure E 8.1.1 (c) the Richardson plots are shown for the two synthetic triangular islands of Figure E 8.1.1 (a).

An experimental strategy that avoids recording many polygon lengths with subsequent addition and normalization is to use a graphical technique known as strip integration, for finding the length of a stride around a polygon. In this method, one takes a strip of paper which is longer than the perimeter of the polygon. Then one places the top of the strip of the paper at the starting point of the polygon constructed by striding around the profile. The edge of the strip is laid along the polygon side. A mark is made at the end of the first side. Next the paper strip is realigned along the next polygon side with the first mark representing the first polygon side at the beginning of the next side. A second mark is made on the paper strip at the end of the polygon side. In this way as one moves around the polygon the lengths are automatically added together and the total length of the strip marked off during the exploration is the total of the polygon sides. This is much quicker and easier than writing down the measured value of each side and the students will find that the method is as accurate as detailed measurement plus summation. It is not so easy to use the method when using small stride exploration but for the suggested search stride magnitudes of Figure E 8.1.1 (a) when the length of the triangle is magnified to the order of 20 cm this technique is very quick and easy. Using this method also encourages the student to devise different experimental strategies when faced with an experimental task.

The synthetic island profiles of Figure E 8.1.1 (a) have been digitized in *Figure E 8.1.2*(a) so that the equipaced exploration technique can be used to characterize the lines. Again these profiles can be enlarged using a photocopier and used directly in class. In Figure E 8.1.2 (b) of this diagram typical polygons constructed on the synthetic profiles for 4, 8 and 16 steps are shown. Again students should appreciate the fact that there is nothing to be gained by increasing the equipaced side one step at a time because of the fact that logarithmic plotting is going to be used to summarize the data. It is better to double the steps at each exploration in order to obtain the equipaced points on the Richardson plot. Without this guidance, students tend to feel that they are making big gains in accuracy by carrying out a series of explorations increasing the pace number one by one. They are disappointed to find such data crowds together on the Richardson plot and that they are short of data points to obtain a good data line to calculate the fractal dimensions.

As in the case of the fixed stride exploration using the compasses there is often a small distance left after a complete equipaced polygon has been drawn and again an unbiased and efficient method of completing the polygon is to join the end of the last complete chord to the beginning point with a short chord as illustrated in the figure. Again the polygon perimeter estimate depends upon the starting point and students should demonstrate the variability of their estimates for a given equipaced stride. Typical data is shown in Figure E 8.1.2 (c). The students should be cautioned against claiming too many decimal points in their estimated fractal dimension. Normally graphical methods are only accurate to within 1% so that the fractal dimension should be quoted to two decimal places. In class discussion it should be pointed out that to avoid fractal rabbits it is not normally useful to explore a perimeter with chord lengths of stride magnitudes of greater than 0.3 of the projected length of the profile.

The fractal dimension of the two synthetic islands can also be measured by a mosaic examination procedure. Grids equivalent to the step sizes shown in Figure E 8.1.1 (a) should be drawn. We have found it convenient to draw the grids and copy them onto overhead transparency sheets. The transparency sheets are then laid over copies of the profile which has been magnified so that the side length is

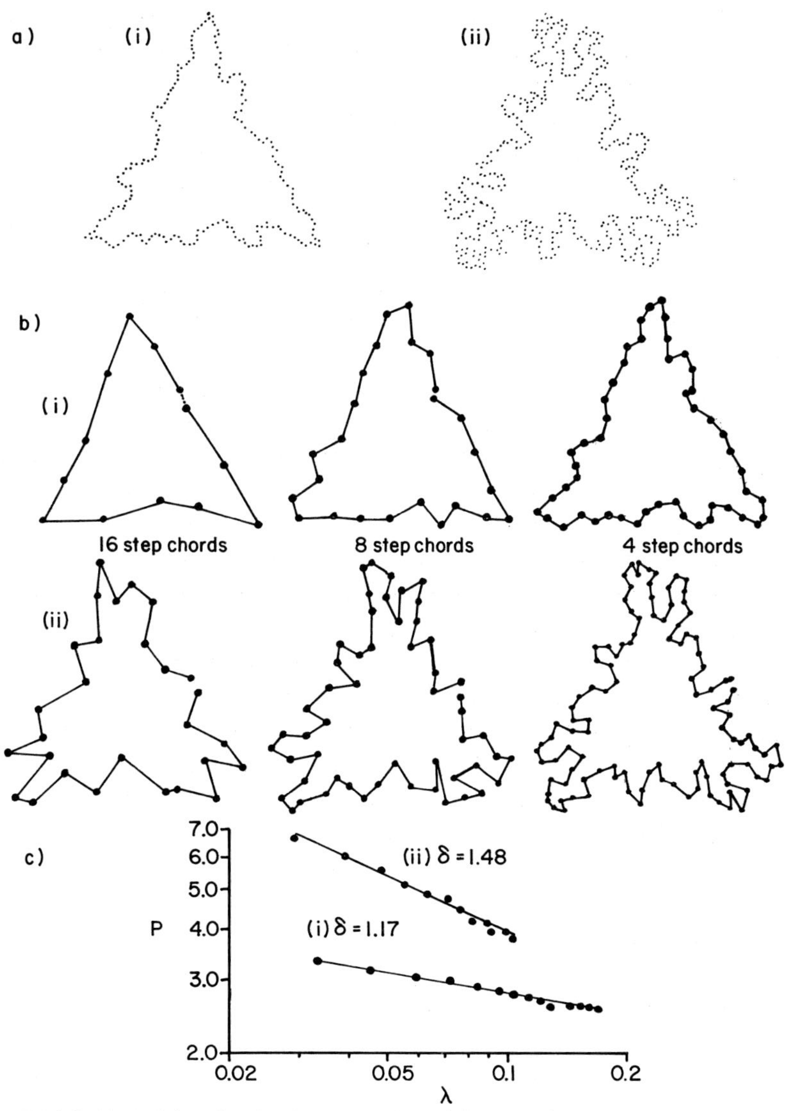

Figure E 8.1.2. Typical data for the characterization of the ruggedness of the two synthetic islands by the equipaced exploration technique. a) Points obtained by digitizing the two synthetic islands. b) 4, 8, 16 step polygons showing end joining for both profiles. c) Richardson plot and fractal dimensions for the two islands.

approximately 20 cm. The students then color in the squares which lie on the perimeter of the synthetic islands with removeable ink marker pens. After counting the squares, the marker can be removed with a damp cloth. This can also be used to let the whole class work with one set of materials using an overhead projector. There will normally be some squares which are doubtful candidates for being

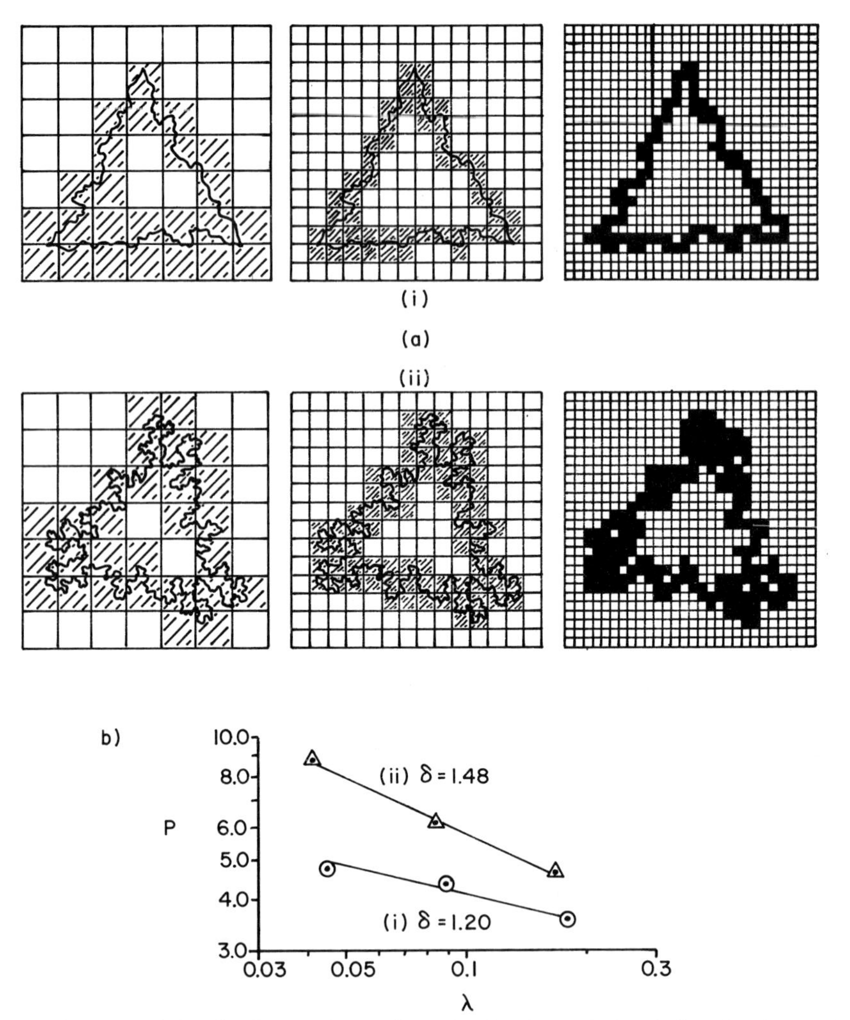

Figure E 8.1.3. Typical data for the mosaic transformation (amalgamation) technique for characterizing the fractal dimension of the synthetic islands.

included in the Minkowski ribbon of squares. The class should discuss how an efficient procedure is to color in every other doubtful square on the basis that half of the doubtful ones would really be in the profile and the other half would be out. Typical sets of data for the mosaic evaluation of the fractal dimension of the synthetic islands are summarized in *Figure E 8.1.3.* The students can also use the transparent overlay to try for themselves the idea of tracing the profile onto the transparency with the smallest set of squares. Then using a washable color pen to block our larger blocks of pixels to illustrate the basic technique of mosaic amalgamation which would be used in a computer automation of the measurement technique.

Exercise 8.2. Wiggly Rivers and Their Topographical Determinants

In *Figure E 8.2.1* two sections of the Vermillion river – a river that flows near Sudbury, are shown. The two portions obviously differ in their ruggedness. If one characterizes their fractal structure the Richardson plots of the two sections of the river are as shown in Figure E 8.2.1. The fractal structure of the sections of the river can be related to the topography of the regions through which they are flowing. The straighter portion of the river is located in a portion of the river basin where the height above sea level is dropping rapidly. Over this region, shown in Figure E 8.2.1 (a), the water of the river gains energy as it travels because potential energy is converted into kinetic energy. This energetic water will not tolerate obstacles in its path and cuts a relatively straight path through the local terrain. On the other hand, the part of the river shown in Figure E 8.2.1 (b) is moving through a short flood plain where the difference in height between the beginning and the end of the section is not that great. Therefore, over this section of the river, the water

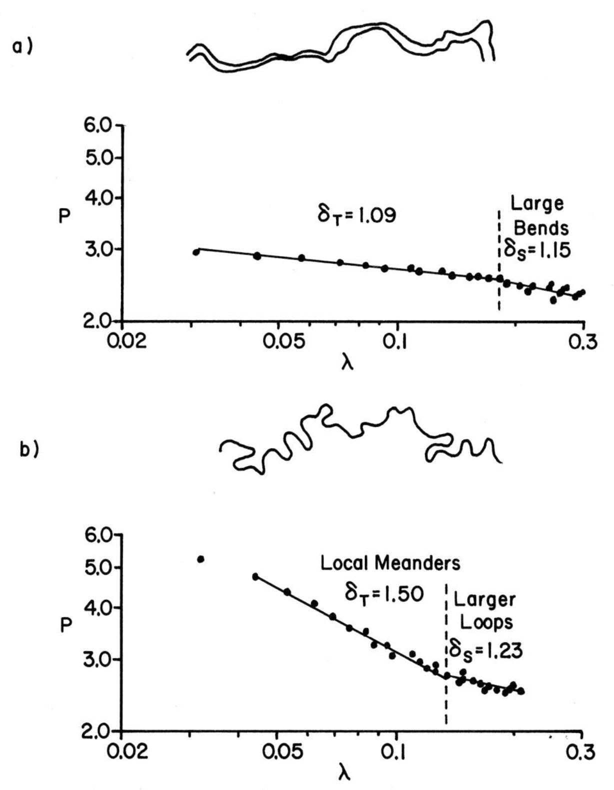

Figure E 8.2.1. The fractal dimension of a river flow path can be related to the local topography. a) In a region of falling topography the fractal dimension of a river bed is low. b) In a relatively flat area (flood plain) the meandering of the river bed can be described by a high fractal dimension.

is moving more sluggishly. As a consequence, boulders and trees etc. in its path pose a more difficult problem and the river chooses to go around them rather than attempt to demolish them. Therefore in the flood plain, the river requires a higher fractal dimension to describe its convolutions than when it is passing through rapid land fall. The student can discover many rivers in an atlas and measure the ruggedness of the river and relate it to the topography of the region. The branching structure of an overall river drainage basin is also an interesting fractal problem but a discussion of the measurement of such drainage basin fractals is beyond the scope of this book. The interested reader will find discussion of river networks in Mandlebrot and in Fedder [2, 67].

Exercise 8.3. Creating Fractals with Paper Chromatography

In *Figure E 8.3.1* a fractal pattern created on a filter paper using a drop of brown drawing ink is shown. To create this pattern, a drop of ink was placed at the centre of a disc of filter paper which had been immersed in water and then placed on the top of the rim of a shallow dish. After the ink blob was placed at the center of the filter paper a steady stream of clean water drops were dripped onto the center of the filter paper. This caused water to migrate out through the wet filter paper toward the edge of the disc and eventually drop off the edge. The different dyes present in the ink moved out at different speeds with the moving of the fluid. The process of dye movement with moving fluid is described as *elutriation*. The word to *elute* meaning "to wash out" comes from the Latin root word e meaning "out" and luere meaning "to wash". The term *polluted* water originally referred to the state of water after dirty garments had been washed in the water using the root word pro meaning "thoroughly". The polluted water represents the state of water after all the dirt has

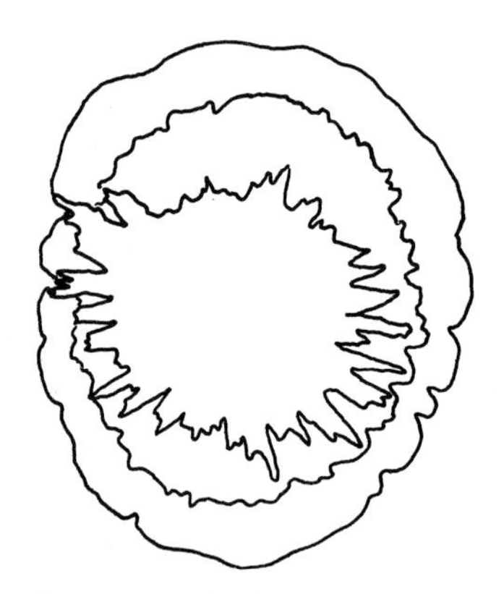

Figure E 8.3.1. In thin layer paper chromatography the components of a mixed solution such as drawing ink can be fractionated by an elutriating fluid to form a set of colorful rings (reminding us that the name chromatography means "colored writing") some of which have fractal structures.

been washed out of a location into the river. The different types of dye molecules in a complex substance such as drawing ink elute at different rates and also have different accessibilities to the pores of the paper. The drawing ink that we used appeared to have three major components. The outer ring shown in the figure was a pink color, the inner ring was an orange color and the rugged line was obviously a more slowly moving larger dye molecule which had a dark brown color. It is probable that the fractal structure of the slow moving brown ring has information on the pore structure and hence of the fibrous network making the paper. The fractionation technique illustrated in Figure E 8.3.1 is known as *thin layer paper chromatography* (TLPC for short). The name chromatography was given to the technique by the Russian scientist Tswett who first fractionated colored compounds from chopped up leaves using this technique. He dissolved the colored material in petroleum ether and then poured the solution through a glass tube tightly packed with powdered limestone. As he kept pouring fluid down through the column, the molecules present in the leaves separated and produced a rainbow display of colors down the tube. To Tswett the separated bands of color seemed to be like colored writing. Therefore he called the fractionating system *chromatography* from the Greek word chroma meaning "color". When scientists were studying cells from living creatures, they found that certain tiny parts at the center of the cell could be made visible with colored dyes. In 1879 a German scientist Walter Fleming called these tiny features of the center of the cell *chromosomes* from the Greek word soma meaning "body". *Chromium*, the metal, is so called because it gives rise to many colored compounds even though it itself is colorless. Today, modern chromatography often has colorless chemicals coming out of the column but the name has stuck and all methods of fractionation of this type are called chromatographic techniques.

An important branch of applied science is *forensic science* defined as the use of science in the support of the administration of law and justice. The name comes from the fact that the Romans used to administer their justice system in the market place of a town. The Latin word for "marketplace" is forum. Forensic scientists use thin layer paper chromatography to analyze ink from documents and the components of many different types of fluid. The student can simulate forensic science studies and at the same time generate some fascinating fractal lines by using the same filter paper to look at the fractal patterns using different drawing and writing inks. They could also investigate the effect of different paper structures by carrying out experiments with the same ink and different discs of paper. Many school laboratories have different grades of filter paper and the student can also try using writing paper and blotting paper as a basis for the fractionation technique. The teacher could prepare diluted samples of different inks in a set of small bottles and the students could discover which bottles contained different samples of the same ink by studying the patterns generated in thin film chromatography.

Students interested in looking into the use of thin layer paper chromatography in forensic science could look at the discussion given in the book "Science and Criminal

Detection" by John Broad, one in a series called "Dimensions of Science" published by MacMillan Education (1988) London England.

Exercise 8.4. The Ruggedness of Mountain Ranges

A look through a modern magazine often shows scenic portraits of mountain ranges from countries you are invited to consider as the location of your next vacation. The jagged peaks obviously have a fractal structure. However when using a photograph of the ruggedness of the outline of a mountain range to deduce a characteristic fractal parameter, one must be careful to take into account the fact that the real ruggedness is usually greater than the apparent ruggedness because of something which is described technically as *projection occlusion*. This property is illustrated in the sketch shown in *Figure E 8.4.1*. The term *occlusion* comes from the Latin word claudere meaning "to shut" and ob means "in the way of" (often when words like "ob" are combined with a word that begins with "c" the "b" changes to a "c" so that the combination is easier to pronounce). Occlusion is the hiding of something by shutting out the view of the details of an object by placing something in front of it. In this case the front range of hills. The front row of hills form a "door" which blocks a clear vision of the back hills. Technically one states that when looking at a range of mountains the skyline of the mountains is a combination of the real outlines of the back range of the mountains plus the projected profile of the mountains nearer to the observer which occlude features of the back range with the general effect of reducing the visible ruggedness. The real and observed ruggedness of the mountain sketched in Figure E 8.4.1 are illustrated in the diagram. Digitized profiles have been given so that the students can enlarge the picture and work directly on the digitized lines to generate the data required for constructing the Richardson plot for the mountain ranges. Occlusion effects are very obvious for measurements made on mountain ranges but the same effect reduces the apparent ruggedness of the system to be characterized if one is looking at a rock fragment or a fracture of a fairly thick specimen of a material. In general the projected fractal dimension may be considerably smaller than the same fractal dimension made by a profilometer trace on a surface which faithfully follows the up and downs of the surface and does not include hidden valleys closed off by the projected hills between the observer and the rugged terrain to be characterized (see discussion of Korcak Fractal for fragmented plastic in Chapter 9 of Reference 11). The readers can look through magazines and select photographs of hilly terrain to be characterized. If they look at mountain ranges in places like South Africa, the ranges tended to be rounded and eroded because they are older and they have lower fractal dimensions. More recent mountain systems produced in places such as the Himalayas in India have higher fractal dimensions. This fact can be used in a geography class concerning mountain formation and erosion.

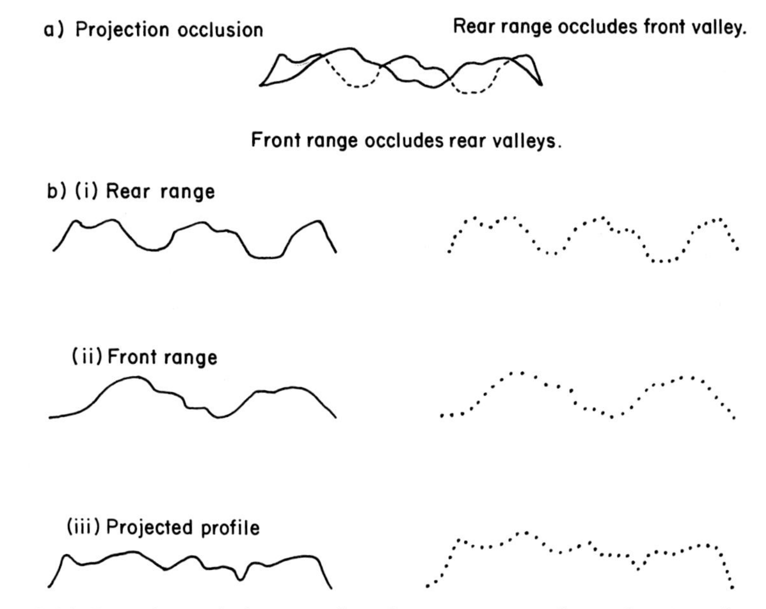

Figure E 8.4.1. Projection occlusion can reduce the apparent ruggedness of a range of mountains or a rugged fracture line. a) The projected appearance of a mountain range. b) Lines representing the true profiles of the two mountain ranges as well as their projected appearance.

Exercise 8.5. Characterizing the Structure of Islands

The reader will probably find it useful to start off a study of real islands by looking at the fractal structure of the three profiles shown in *Figure E 8.5.1* (a). These three profiles can be used as a first study of islands to alert the students to the need to pay attention to details of their experimental technique. Data on the three profiles by fixed stride exploration with a pair of compasses in Figure E 8.5.1 (b) and equipaced exploration using the digitized lines are shown in *Figure E 8.5.2.* Students soon appreciate the fact that working with the compasses is not usually feasible to work over a wide range of stride lengths. There is a lack of certainty in the interpretation of the sparse data points that one uses when basing ones calculations on the fixed stride length. This is shown by the data presented in the figure. On the other hand, the computation of the perimeter estimates based on the equipaced method usually leads to a much greater amount of data. In Figure E 8.5.1 (b) we show the detailed graphs for the data generated by the fixed strick method. It can be seen that for profile (i) one has definitely reached the point of exploration where one is essentially studying the smooth line tracing the rugged outline showing that after an inspection size of 0.85, the surface is essentially euclidean (smooth). For profile (iii), it is obvious that the profile can be split into two regions of different ruggedness. The data of the figure shows that if this separation is not made an average roughness

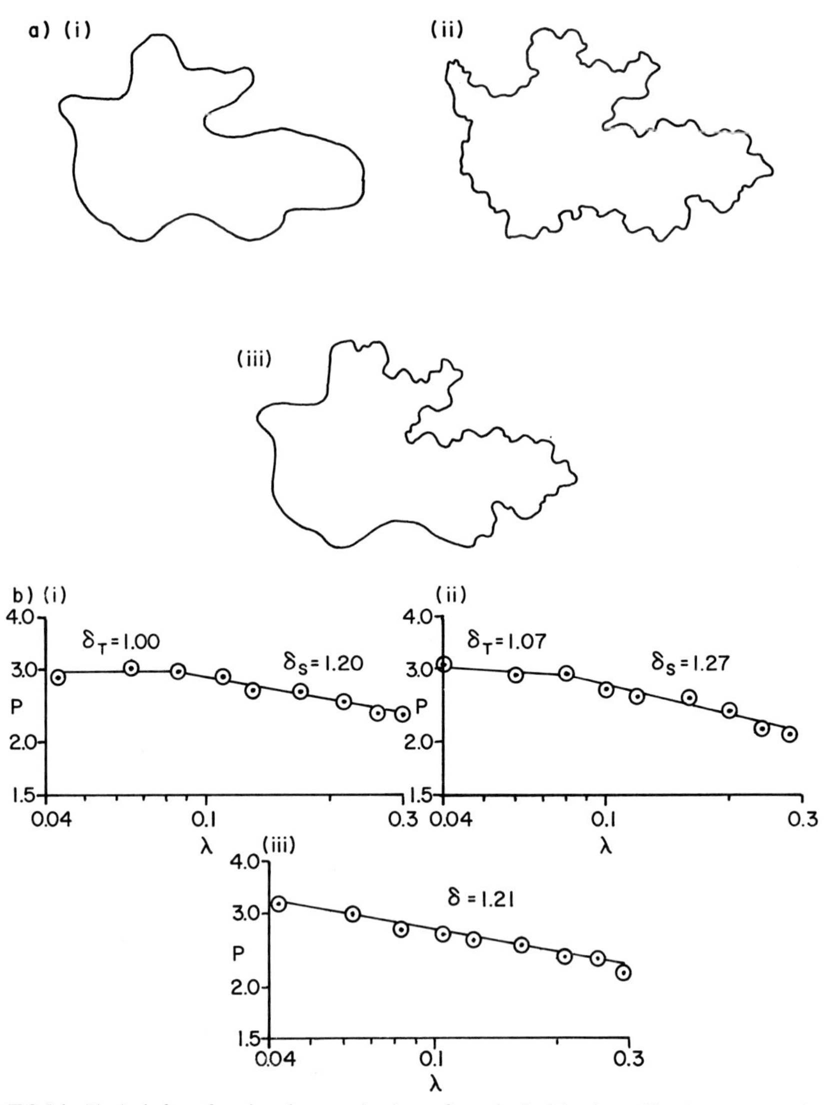

Figure E 8.5.1. Typical data for the characterization of synthetic island profiles by structured walk procedures. a) Three synthetic profiles. b) Richardson plots for the three synthetic profiles

dominates the Richardson plot and information is suppressed. In Figure E 8.5.2 data for the same islands explored by the equipaced method is shown. It is left as an exercise for the student to split profile 3 into two parts to be explored separately. The two figures show that the 2 methods of measuring fractals, stride and equipaced, do not yield identical results. The class can discuss why (Hint the equipaced method is a more intensive evaluation which always explores all nooks and crannies where as the stride method steps over details in the early examinations).

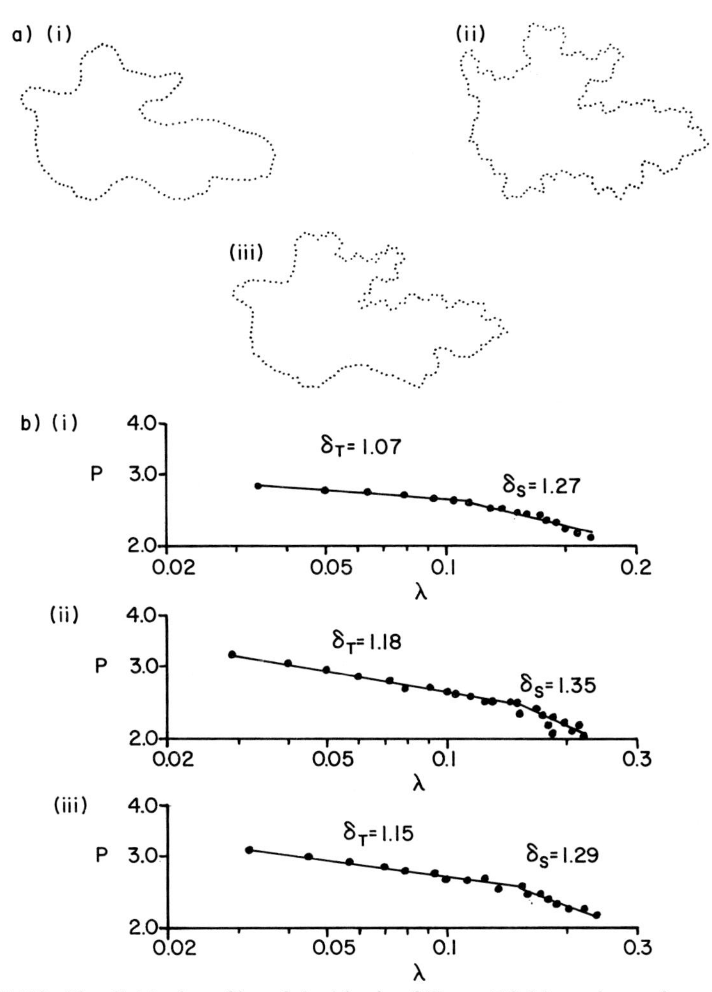

Figure E 8.5.2. The digitized profiles of the islands of Figure E 8.5.1 can be used to characterize the ruggedness of the profiles. a) Digitized profiles. b) Richardson plots for the three profiles of (a) explored by the equipaced method.

After gaining experience with the artificial islands of Figure E 8.5.1, the reader is ready to move on to the study of a real island *Archipelago*. Originally the Archipelago was the name of the chief sea around the Greek islands. The word comes from Greek root words archi meaning "chief" and pelagos meaning "sea". It then came to refer to any group of islands (the root word erchi is to be found in many words such as *archbishop* – the chief bishop, not one who lives under an arch!). In Figure 3.23 we saw an archipelago of islands to be found off the tip off the Cornish peninsula of Great Britain – *the Isles of Scilly*. Only the major islands are shown in that map. The question that immediately arises is "is the fractal dimension of the coastline of the islands independent of the size of the islands?" Obviously to gain an accurate answer to this question, one would need to have a more detailed structure of the smaller islands but certainly for the five largest islands one can make measurements and attempt to give an answer to the question. One only needs to browse through an atlas to see many different archipelagos which invite fractal characterization.

Exercise 8.6. Draw Your Own Clusters

In Section 8.7 we discussed how an aerosol agglomerate structure can be related to the formation dynamics of the cluster. A group of students could discover this information for themselves by drawing their own clusters and exploring the boundary structure of the clusters. In *Figure E 8.6.1* three clusters of balls have been drawn. The first one is the type of cluster that one sees when spheres have been placed in a liquid suspension and then the suspension turned into a spray. As the spray droplet drys out the surface tension of the evaporating liquid is the dominant agglomerate forming force (it is assumed that the spheres were separate from each other in the original suspension) and the dried out agglomerate of Figure E 8.6.1 (a) looks like a raspberry, or a bunch of grapes. If one explores the boundary structure of such an agglomerate one finds that one has an unusual Richardson plot in that the perimeter estimates do not increase very much until the stride size being used to explore the agglomerate is of the same order as the size of the individual spheres forming the agglomerate. Once this exploration size is reached, the perimeter starts

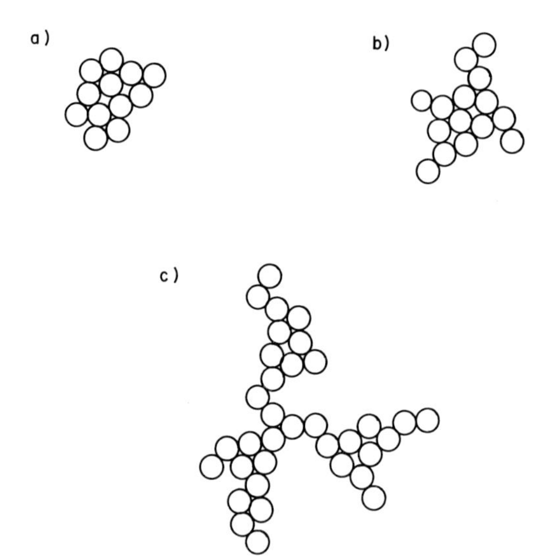

Figure E 8.6.1. The fractal dimension of a cluster can indicate the way in which it was formed. a) Dried out bunch of grapes. b) Simple agglomerate. c) Agglomerate of agglomerates.

to increase and one obtains an increasing data line on the Richardson plot. To investigate cluster structure of this kind, one needs to have an enlarged agglomerate containing a relatively small number of constituents spheres as illustrated in the figures. In fact, the start of the increasing linear region is related to the size of the subunits and the students could demonstrate this for themselves by using clusters with increasing sphere sizes. A human inspector can quickly see the difference between the bunches of grapes in Figures E 8.6.1 (a), (b), (c) but the computer finds it hard to recognize such a difference. The appropriate Richardson plot can be the basis of computer learning process by which the computer can learn to recognize what type of an agglomerate it is looking at and what size of spheres have been incorporated in the agglomerate. In other words, one could program the computer heuristically to discover what kind of subunit had become agglomerated in the cluster being investigated.

In Figure E 8.6.1 (b) a branched type dense cluster of the kind that forms as a primary agglomerate in a flame is shown. The students could look at this cluster and show how the structural fractal of the profile is approximately 1.30 which is typical of a free space grown agglomerate. The cluster of Figure E 8.6.1 (c) has been formed by the random collision of clusters such as those shown in diagram number two and it can be shown that this agglomerate formed by the collision of agglomerates has the fractal dimension of about 1.42 which is almost identical with the fractal dimension of 1.43 that we found experimentally for the diesel soot agglomerate in Section 8.8. In general a fractal dimension of the order of 1.38–1.43 indicates an open structured cluster formed by collision of many subsiduary clusters. Obviously the variations of this experiment are endless and classes can draw all sorts of agglomerates for characterization studies.

Exercise 8.7. Rugged Political Boundaries and Watershed Boundaries

In *Figure E 8.7.1* are two rugged lines of interest to the geographer. The first, of Figure E 8.7.1 (a), is of historic interest since it is the boundary between Portugal and Spain. Richardson tells us that in his original study of boundaries he discovered that this particular boundary was quoted as having different lengths in a Portuguese reference book than in Spanish reference books. This illustrated how the length of a political boundary depends on how one measures that boundary. The students can explore the fractal structure of these two boundaries. The students can check in their atlas to find how much of this boundary is made up of rivers. They can also discover that if they make a rectangle by using a second drawing of the line, joining it to the first, they can create fractal rabbits with such a rectangular shape if they use too coarse an exploration procedure.

The line of Figure E 8.7.1 (b) in the diagram is the boundary between France and Italy in the region of Mount Blanc. It is a *watershed boundary* in which the political

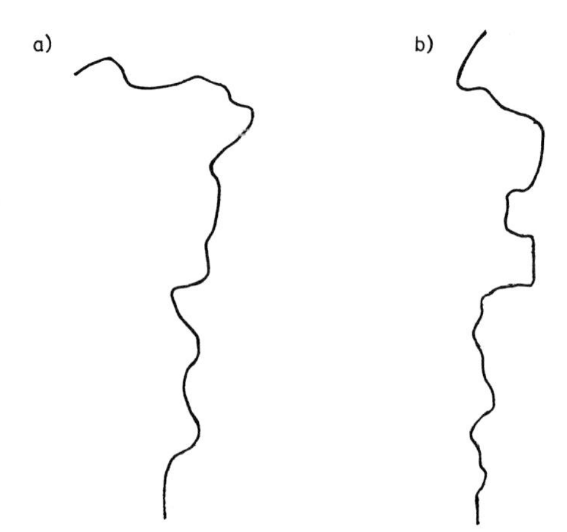

Figure E 8.7.1. Political boundaries and mountain watersheds are sometimes fractal lines. a) A line representing the border between Spain and Portugal. b) A line representing the border between France and Italy.

boundary follows the tip of the mountains with rivers running downwards on each side of the line away from the line. The students can measure the ruggedness of the watershed line and look for other watershed boundaries in different parts of the world. The boundaries of the early American states tend to follow river boundaries but then as they moved further west the Americans became positively euclidian. They started to draw straight lines for boundaries even if this created small areas completely separated from the rest of the state. There is a small part of Washington state contiguous with the Vancouver peninsula with no land communication to the rest of Washington state. The state boundaries following rivers make ideal problems for fractal studies for the exploring student.

Exercise 8.8. Characterizing an Alumina Profile Described by Schäfer and Pfeifer

In *Figure E 8.8.1* two interesting drawings of alumina dust studied by two German scientists Schäfer and Pfeifer are shown [68]. These two scientists were interested in studying how one looked at the detailed distribution of sizes in a powder which was said to be "one" micron alumina powder. *Alumina* is the short form name for *aluminum oxide*. Alumina is used in ceramic fabrication and also as an abrasive polishing powder. Figure E 8.8.1 (a) shows a magnified picture of the dust fine particles captured by a random network of very fine fibers. The fibers are 0.2 microns in diameter. This compares with 100 microns for a coarse human hair. When looking at the photograph, the scientist has to decide whether conglomerates such as the one shown in Figure E 8.8.1 (b) have been formed by many smaller fine-

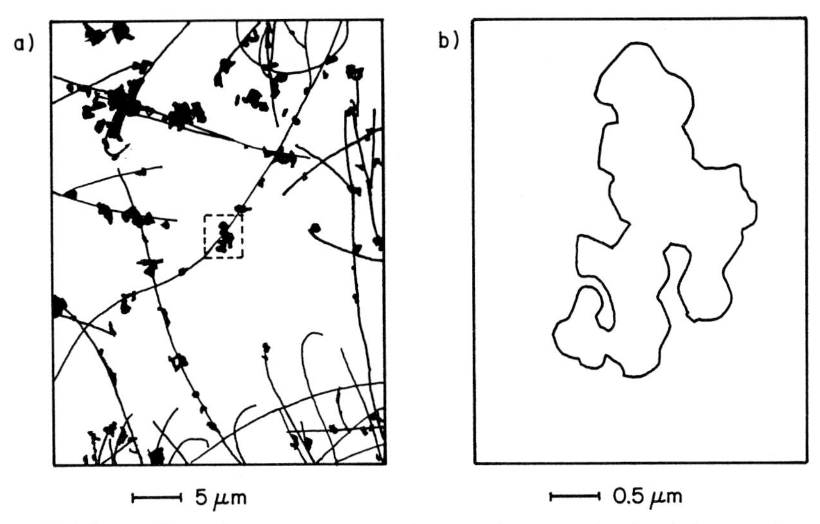

Figure E 8.8.1. Deciding when an apparent agglomerate is a operational unit in aerosol studies is a difficult problem. Fractal characterization of aerosol agglomerates is an interesting study. a) Alumina captured in a network of 0.2 mm fibers. b) An isolated conglomerate of alumina magnified from (a).

particles arriving at the whiskers of the collector or if they constitute a large agglomerate which existed in the powder as a fused concrete entity before becoming airborne. As discussed earlier in Section 8.8 deciding exactly what constitutes an operative aerosol fine particle is a crucial decision when estimating the health hazard of a powder. A resolution of the question as to how many of the profiles of the photograph represent a single entity is beyond the scope of this book but the profiles can be used in a discussion session in which the class discusses how difficult it is to measure a dust hazard to which industrial workers are exposed. Many of the profiles of the photograph obviously invite fractal characterization. The profile of Figure E 8.8.1 (b) is magnified 10 times with respect to the first photograph. Remembering that this is a projected profile and that the ruggedness of the agglomerate is probably less than it would be in three dimensions. One can definitely measure its projected fractal structure and attempt to guess how many subunits are present in the agglomerate. Students should be encouraged to discuss the fact that the agglomerate has a much greater surface area than would be assumed if one has simply measured its aerodynamic diameter and hence, would be more chemically active than a sphere of equal volume.

Exercise 8.9. Fractal Structure of Wear Debris in Lubricating Oil

One could remove some lubricating oil from a piece of moving machinery such as a bicycle gear chain. This could be washed off the chain using solvent and the diluted fluid filtered to leave the debris behind on the surface. It is almost certain that some of the debris found in the oil will have a fractal structure that can be photographed under a microscope and then subjected to fractal characterization. This type of study could be made into a science fair project by looking at what happens in the oil between two surfaces rubbed together. The student could study the effect of lubricating oils and the effect of deliberately placing grit between the moving surfaces. The presence of the grit in the lubricating oil would also create a fractal surface on the material being rubbed together. If profilometers or high-resolution magnifications were available to the student, the structure of the fractal surfaces could be studied (see discussions of rough surfaces in Chapter 10). This is also an ideal project for science students to work on in cooperation with local industrial scientists to give them access to the specialized machinery to look at industrial wear and tribological problems. It has been estimated that many millions of dollars could be saved by industry if they became more aware of the information they could gather by the systematic application of tribological knowledge and ferrography to their lubrication problems.

Exercise 8.10. Fractal Assessment of Damage to the Central Nervous System

Sometimes damage to the central nervous system of the human body is manifest in deterioration of the motor skills of the body involved in carrying out tasks such as being able to draw a line or to walk along a straight line. Sometimes this damage is temporary as manifest by the inability of a drunk to walk along a straight line when his motor skills are impaired by the alcohol. Prolonged exposure to alcohol can however cause permanent damage to the nervous system of the body as demonstrated by "the shakes" that afflict alcoholics. Damage from mercury exposure to the central nervous system can also be detected when the worker exposed to the mecury is asked to draw a line following a curved or straight line. In *Figure E 8.10.1* a line being used to study the recovery of motor skills in people receiving therapy after a stroke is shown. The top line of this figure represents the track to be followed by the patient receiving the therapy. The two lower traces show the ability of the patient to follow the line 77 weeks and 93 weeks after the stroke. The word *stroke* has many meanings in English. The meanings vary from "one movement of a piston of an engine" to "the mark made by a paintbrush". In medical terms a stroke is a sudden attack of weakness or paralysis which is the consequence of an interuption of the flow of blood to the brain. The link between the illness and the word stroke comes about from the fact that in primitive times people used to believe that illnesses were caused by the act of God and paralysis occured when God struck you in revenge for some act you had performed. It is obvious from the perspective of fractal geometry that one can state that the attempt made by the patient after 77 weeks has a lower fractal dimension than the curve drawn after 93 weeks. In fact, the fractal dimension

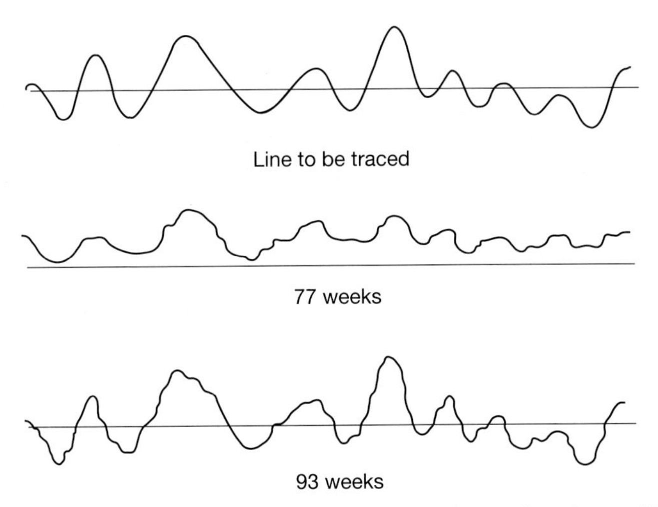

Figure E 8.10.1. Attempts by a stroke victim to trace the upper line made at the specified times after the stroke as the patient recovered.

of the top and bottom line of the diagram are probably identical. The use of fractal dimensions to describe the attempts of patients to draw lines of various structures probably could become an important way of quantifying neurological damage (the word *neuron* in Greek meant "a nerve").

Exercise 8.11. The Fractilicious Structure of the Stock Market Record

A wiggly line that appears daily in many newspapers is the record of the stock market value. In our particular example shown in *Figure E 8.11.1* (a), the record of the movement of what is known as the *Dow Jones Industrial Index* average over a period of several months is shown (the teacher developing this exercise would have to explain to the students how people and institutions invest in stocks and shares and how the Dow Jones averages is the value of certain shares traded at the New York stock exchange). It is obvious that the value of the Dow Jones industrial average increased over the 120 trading days represented by the graph. On the graph the trend line as judged by eye has been drawn. The average line should have as much area above the line as below. The variation of the Dow Jones average above this line obviously has a fractal structure. The student can measure the fractal dimension of the rugged line either by means of a "stride walk" with a pair of compasses or using the equipaced method described in this chapter. The teacher can collect information from the financial pages of the newspaper to show that the variation within hourly trends on one day, are very similar to the trends in variations between days and recent years. This fact illustrates how a fractal system is a statistically self-similar

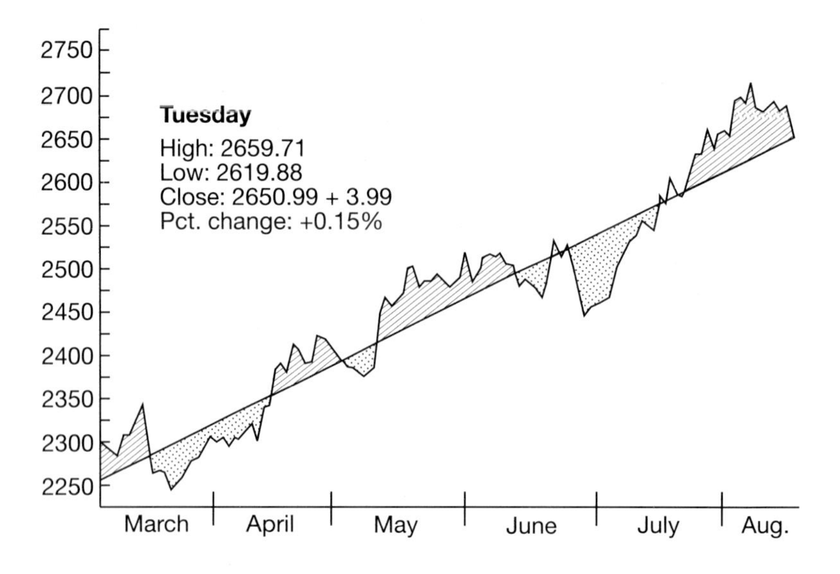

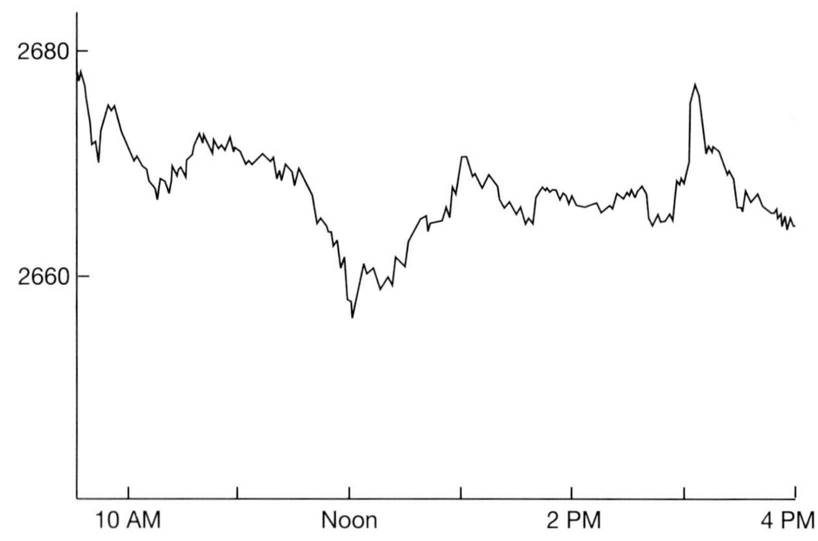

Figure E 8.11.1. A record of the fluctuations in the Dow Jones Industrial Index over several months demonstrates the concept of statistical self-similarity if compared to weekly or daily records.

system. Students can look up the variations in the stockmarket in newpapers from 20, 30, and 40 years earlier using the archives of the local libraries. They could measure the fractal dimension over different historic time periods and compare them. The students could then discuss if the fractal structure had changed over the years which would indicate a variation in the buying habits of investors.

More advanced class studies could look at the fact that the Dow Jones averages over a period of 12 years is known to follow a cyclical pattern of a period of 4 years linked to the presidential election in the United States. It actually peaks about a year after a new president is elected when the value of promises made during the election have had their full effect. The fractal structure represents the day-to-day variation of sales and investment behavior whereas the underlying structure of the curve is related to fundamental aspects of North American lifestyle (see discussion of the stock market in Chapter 15).

A basic theme developed in this book is that fractal structures represent the outcome of the interaction of many causes and that fractal description does not represent an abandonment of determinism but rather a retreat of determinism to a lower level of resolution inspection. The study of the stock market is an ideal example of this theory because the individual investors and institutional finance managers do not buy and sell at random. They are always trying to make the best deterministic assessment of when to buy and sell. The pattern of the stock market is the fractal structure coming out of the interaction of these many causes which contribute to the structure of the Dow Jones average. The teacher could discuss how the determinism of the individual investors illustrates how fractal structures are the outcome of the complex interaction of many causes.

The students could also consult international newspapers to see if the fractal structure of the stock markets of London, Japan etc. varies from the stock market of North America.

References

[1] There are many editions of "Alice in Wonderland" and its companion "Alice Through the Looking Glass", by Lewis Carroll (the pen name of Ludwig Dodgson, an Anglican (Episcopalian) Minister and Professor of Mathematics at the University of Oxford). The well-worn copy of these two books that I have on my desk is the combined issue, Companion Library Edition, published by Grosset and Dunlop, New York, 1965.

[2] B. B. Mandelbrot published an updated and expanded version of his first book, "Fractals Form Chance and Dimension", under the title "The Fractal Geometry of Nature", W. Freeman, San Francisco, 1983. This book is considered by Dr. Mandelbrot to be the definitive book on the subject (Personal Communication).

[3] See discussion given in B. H. Kaye "Direct Characterization of Fineparticles", J. Wiley & Sons, New York, 1981.

[4] The discussion of coastlines with different ruggedness in different parts of the perimeter of the coastline is discussed in more detail in the scientific publication, B. H. Kaye, and G. G. Clark, "Experimental Characterization of Fineparticle Profiles Exhibiting Regions of Various Ruggedness", *Particle and Particle Systems Characterization 6* (1989) 1–12.

[5] H. Schwarz, H. E. Exner, "The Implementation of the Concept of Fractal Dimensions on a Semi-Automatic Image Analyzer", *Powder Technology 27* (1980) 207–213.

[6] D. Scott, W. W. Seifert, V. C. Westcott, "The Particles of Wear", *Sci. Am. 230* (1974) 88–97.

[7] R. Bowen, D. Scott, W. Seifert, V. C. Westcott, "Ferrography", *Tribology Int. 9* (1976) 109.

[8] C. Evans, "How Human Joints Wear – Learning From Machine Methods", *New Scientist* (November 1978) 444–445.

[9] J. K. Beddows, University of Iowa, personal communication.

[10] D. J. L. McIver, B. J. Rogers, R. Trottier, B. McFarlane, B. H. Kaye, "A Multifractal Analysis of Cell Activation", in A. J. Hurd, D. A. Writz, B. B. Mandelbrot (Eds.), *Fractal Aspects of Materials: Disordered Systems*, Materials Research Society, Pittsburg, Pennsylvania 1987.

[11] For a discussion of the basic ideas embodied in the Minkowski sausage technique for measuring coastlines see the discussion given in Chapter 2, of B. H. Kaye, *A Random Walk Through Fractal Dimensions*, VCH, Weinheim, Germany, 1989.

[12] TOPO optical interferometry based profilometer available from WYKO Corporation, 2990 East Fort Lowell, Tucson, Arizona, USA, 80716.

[13] Stylus-based profilometer available from Rank Taylor Hobson Limited, P.O. Box 36, New Star Road, Thurmaston Lane, Leicester, England, LE4 7JQ.

[14] K. C. Clarke, "Computation of the Fractal Dimension of Topographic Surfaces using the Triangular Prism Surface Area Method", *Computers and Geosciences 12*, (1986) 713–722.

[15] Laser interferometry based profilometer available from Zygo Corporation, Lavrel Brook Road, P.O. Box 448, Middlefield, CT, USA, 06455-0448.

[16] Readers interested in a general survey of the gas adsorption techniques for measuring surface area can consult a general textbook such as T. Allen, *Particle Size Analysis*, 4th edition, 1990, Chapman and Hall, London.

[17] Dr. D. Avnir, of the Hebrew University pioneered the fractal reinterpretation of gas adsorption studies. The research scientist will find an excellent review of this work in *The Fractal Approach to Heterogeneous Chemistry*, D. Avnir (Ed), John Wiley & Sons, Chichester 1989.

[18] Commercially available equipment for gas adsorption studies is sold by Quantachrome Corp., 6 Aerial Way, Syosset, New York, 11791.

[19] Gas adsorption equipment for measuring surface areas is available from Micromeritics Instrument Corporation, 800 Goshen Springs Road, Norcross, GA, 30071.

[20] *Textbook of Electronics for the Charge Couple Device Camera*, EG & G Reticon, 345 Potrero Avenue, Sunnyvale, California, USA, 94086.

[21] For a discussion of the basic method employed in the mosaic amalgamation technique for characterizing a fractal dimension see Chapter 2 of B. H. Kaye, *A Random Walk Through Fractal Dimensions*, VCH, Weinheim, Germany, 1989.

[22] An introduction to the general problem of fractal rabbits is given on Pages 23 and 24 of B. H. Kaye, *A Random Walk Through Fractal Dimensions*", VCH, Weinheim, Germany, 1989.

[23] Dr. Avnir's observations on the reality of fractal rabbits formed part of his presentation at the conference "Fractal Aspects of Materials: Metals and Catalyst Surfaces, Powder and Aggregates", organized as part of the annual meeting of the Materials Research Society, November 26–27, 1984. The book of extended abstracts of this Symposium edited by B. B. Mandelbrot and D. Passoja is available from the Materials Research Society, 9800 McKnight Road, Suite 327, Pittsburgh, PA, USA, 15237.

[24] *Brewer's Dictionary of Phrase and Fable*, Centenary Edition, revised by Ivor Evans, Cassell, London 1970.

[25] The data in this section is taken from the scientific publication, B. H. Kaye, G. G. Clark, "Formation Dynamics Information, Can it be Derived from the Fractal Structure of Fumed Fineparticles" Chapter 24 in *Particle Size Distribution II; Assessment and Characterization*, T. Provder (Ed.), proceedings of a symposium organized by the American Chemical Society, Boston MA, April 22–27, 1990. published by The American Chemical Society, Washington, DC 1991.

[26] Cabot Carbon Corporation, Concord Road, Billerica, MA, USA.

[27] Readers wishing to pursue the physical interpretation of agglomerate structure at the research level should consult B. H. Kaye, "Characterizing the Structure of Fumed Pigments Using the Concepts of Fractal Geometry", *Particle & Particle Systems Characterization 8* (1991) 63–71.
[28] F. P. Perera, A. K. Ahmed, *Respirable Particles*, Ballinger Publishing Company, 1979.
[29] T. Mercer, E. Morrow, W. Stober, "Assessment of Airborne Particles", proceedings of the Third Rochester International Conference on Environmental Toxicity, published by Charles C. Thomas, Springfield, Illinois, 1972.
[30] P. C. Reist, *Introduction to Aerosol Science*, Macmillan, New York 1984.
[31] G. J. Sem. "Aerodynamic Particle Size: Why is it Important", *TSI Quarterly 10* No. 3 (1984) 3–11. TSI Quarterly is a house journal published by the TSI Incorporated, 500 Cardigan Road, P.O. Box 43394, St. Paul, MN, USA, 55164.
[32] C. Wagner, "Disputes on the Safety of Asbestos", *New Scientist* (7 March 1974) 606–609.
[33] P. Kotrappa, "Shape Factors for Aerosols of Coal, Uranium Dioxide and Thorium Dioxide in Respirable Size Range", in *Assessment of Airborne Particles*, T. T. Mercer, P. E. Morrow and W. Stober (Eds.), proceedings of the third Rochester International Conference on Environmental Toxicity, published by Charles C. Thomas, Springfield, IL, 1972.
[34] W. Stober, "Dynamic Shape Factors of Non-Spherical Aerosol Fine particles", in "Assessment of Airborne Particles", T. T. Mercer, P. E. Morrow and W. Stober (eds.), proceedings of the third Rochester International Conference on Environmental Toxicity, published by Charles C. Thomas, Springfield, IL, 1972.
[35] P. C. Reist, M. T. Hiseieh, B. A. Lawless, "Fractal Characterization of the Structure of Aerosol Agglomerates Grown at Reduced Pressure", *Aerosol Science & Technology 11* (1989) 91–99.
[36] D. Japuntich, "Fume Filter Efficiency", Article in *Job Health Highlights*, published by the 3M Corporation, Vol. 3, No. 1, Occupational Health Safety products Division of 3M, 220-7 W, 3M Centre, St. Paul, MN, USA, 55144.
[37] P. P. Bolsaitis, J. F. McCarthy, G. Mohiuddin, J. F. Elliott, "Formation of Metaloxide Aerosol for Conditions of High Supersaturation", *Aerosol Science and Technology 6* (1987) 225–246.
[38] G. C. E. Olds, "New Ceramics", *Science Journal*, August 1966.
[39] The profile of the barium titanate fine particle is presented in the publication by R. Vivekananden, T. R. N. Kutty: Characterization of Barium Titanate Fine Powders formed from Hydro Thermal Crystallization, *Powder Technology 57* (1989) 181–192.
[40] Ceramic grade titanium dioxide was photographed by K. Kendall in the paper: "Influence of Powder Structure on Processing and Properties of Advanced Ceramics", *Powder Technology 58* (1989) 151–161.
[41] T. A. Wheat, B. H. Kaye, R. Trottier, "Quantitative Characterization of Ceramics by Fractal Geometry", *J. Canadian Ceramic Society 55* (1986) 57–66.
[42] J. C. Russ, J. C. Russ, "Feature-Specific Measurement of Surface Roughness In SEM Images", *Particle Characterization 4* (1987) 22–25.
[43] J. Eastman, R. W. Siegel, "Nanophased Synthesis Assembles Material From Atomic Clusters", *Res. Develop.* (1989) 56–60.
[44] D. W. Schaeffer, "Fractal Models and the Structure of Materials", *MRS Bulletin 13* (1988) 22–27.
[45] D. W. Shafer, "Polymers, Fractal and Ceramic", *J. Mater. Sci. 243* (1989) 1023–1027.
[46] This diesel soot profile was photographed by R. G. Pinnick, Department of the Army, White Sands Missile Range, New Mexico, 88002-5501. Use with permission of Dr. Pinnick.
[47] See discussion of the health hazards of fractally structured dust in, B. H. Kaye, "Assessing the Health Hazards of some Fractally Structured Respirable Dusts", in preparation for *Particle and Particle Systems Characterization.*
[48] W. R. Cotton, "Atmospheric Convection and Nuclear Winter", *American Scientist 73* (May–June, 1985) 275–280.
[49] J. Nelson, *Nature 339* (1989) 611, See also News Story, Fractal Winter, New Scientist, July 1, 1989.

[50] I. Colbeck, E. J. Hardman, R. M. Harrison, "Optical and Dynamical Properties of Clusters of Carbonatious Smoke", *J. Aerosol Sci. 20* (1989) 765–774.

[51] I. Colbeck, L. Appleby, E. J. Hardman, R. M. Harrison, "The Optical Properties and Morphology of Cloud Processed Carbonaceous Smoke", *J. Aerosol Sci. 21* (1990) 527–538.

[52] D. B. Kittelson, D. F. Donan, "Diesel Exhaust Aerosols", Particle Technology Laboratory, Mechanical Engineering Department, University of Minnesota, Minneapolis, publication No. 387, 1978.

[53] C. A. Amann, D. C. Siegla, "Diesel Particulates: What They Are and Why", *Aerosol Science Technology 1* (1982) 73–101.

[54] W. Glen, "What Killed the Dinosaurs?", *American Scientist 74*, (July–August 1990) 354–370.

[55] For a discussion of the fractal structure of "Flyash" see B. B. Mandelbrot, *The Fractal Geometry of Nature*, W. Freeman, San Francisco, 1983.

[56] A more extensive discussion of the effect of fractal structure on the rheology of suspensions will be presented in the forthcoming book B. H. Kaye, *Powder Mixing*, Chapman and Hall, London, 1992.

[57] This type of viscometer is widely used in teaching laboratories. A discussion of the theory of the instrument can be found in many standard textbooks. See for example, *Scholarship Physics* by M. Nelkon, Heineman, London 1966, p. 68–69.

[58] J. Gleick, *Chaos, Making A New Science*, Viking Penguin Incorporated, New York 1987.

[59] S. K. Akhter, "Fineparticle Morphology and the Rheology of Suspensions and Powder Systems", M.Sc. Thesis, Laurentian University, 1982.

[60] P. V. Foukal, "The Variable Sun", *Scientific American 262* Number 2 (February 1990) 34–41.

[61] C. A. Long, "Intricate Sutures as Fractal Curves", *J. Morphology 185* (1985) 285–295.

[62] W. C. Hartwig, "Fractal Analysis of Suture Complexity", preprint provided by the author, Department of Anthropology, University of California, Berkeley, CA, 94720.

[63] A. W. Roberts, M. Ooms, S. J. Wiche, "Concepts of Boundary Friction, Adhesion and Wear in Bulk Solids Handling Operations", *Bulk Solids Handling* 10 (May 1990) 189–198.

[64] A. Zaltash, C. A. Myler, S. Dhodapkar, G. E. Klingsing, "Application of the Thermodynamic Approach to Pneumatic Transport at Various Pipe Orientations", *Powder Technology 59* (1989) 199–207.

[65] G. Kasper, S. Chesters, H. Y. Wen, M. Lundin, "Fractal Based Characterization of Surface Texture", *Applied Surface Science 40* (1989) 185–192.

[66] G. S. Upadhyaya, A. K. Jha, S. V. Prasad, "Surface Roughness Of Sintered 6061 Aluminum Alloy Based Particle Composites", *Powder Metallurgy Int. 21*, No. 4 (1989).

[67] J. Feder, *Fractals*, Plenum 1988.

[68] H. J. Schafer, H. J. Pfeifer, "Sizing Submicron Aerosol Particles by the Whisker Particle Collector Method", *Particle and Particle Systems Characterization 5* (1988) 174–178.

Chapter 9

Invisible Carpets, Swiss Cheese and a Slice of Bread

Seeing is believing!

The King's suit of invisible clothes
in the story by Hans Christian Andersen
was made of Sierpinski Silk
a piece of which is shown above.

The Sierpinski silk is ultimately invisible.

It has no area because it is formed from
an infinite number of infinitely thin threads
but
some Sierpinski silks have more threads than others.

Chapter 9 Invisible Carpets, Swiss Cheese and a Slice of Bread 387

Section 9.1 Fractalicious Bread and Cheese 389
Section 9.2 Exploring the Fractal Structure of Felts and Filters 403
Section 9.3 Characterizing the Porous Nature of Bone and Sandstone 408
Exercises 411
References 415

Chapter 9

Invisible Carpets, Swiss Cheese and a Slice of Bread

Section 9.1. Fractilicious Bread and Cheese

In *Figure 9.1* (a) a stylized replica of a slice of bread is shown. To create this replica a piece of bread was placed on a photocopying machine and the major holes present in the copy of the bread were traced with a pen and then blacked out. The image was made in this way to create a clear structure which could be used to illustrate the principles and concepts used in the fractal characterization of porous bodies such as bread. In a laboratory study using an image analyzer, the students at Laurentian University use their own slice of bread which is viewed directly through a television camera and characterized using the logic of the image analyzers.

The characterization of the structure of an object such as a slice of bread is important in the mass production of bread since customer acceptance depends upon the visual appearance and the "bite texture" of the bread. Both of these properties of bread are related to the structure and number of holes in the bread. In the past, one could characterize the bread structure by measuring the size and shape of the holes using shape factor ratios to characterize the holes. However, fractal geometry offers a new way of looking at the structure of bread as shown in Figure 9.1 (b). The concepts that will be developed in this section with respect to a study of the structure of bread have a wide application in applied science since the methods developed can be applied to characterize the structure of paper, filters, porous bodies such as ceramics, synthetic bone and powdered metal compacts. The concepts are also important in the study of movement of water through soil, the movement of oil through porous sandstone and the movement of reacting gases through catalysts.

Characterizing the structure of porous bodies is a branch of a subject known as *stereology* [1]. This name comes from the Greek word stereos meaning "solid". Stereology is a study of the techniques for the reconstruction of the three-dimensional structure of any body from the examination of two-dimensional sections through the system. In the case of the bread, we are trying to look at the overall structure of the loaf from looking at the slices of the bread. When studying parts of the body, stereologists sometimes use electronically generated images using devices such as computerized axial tomography. This technique is known in Canada as *CAT scanning*

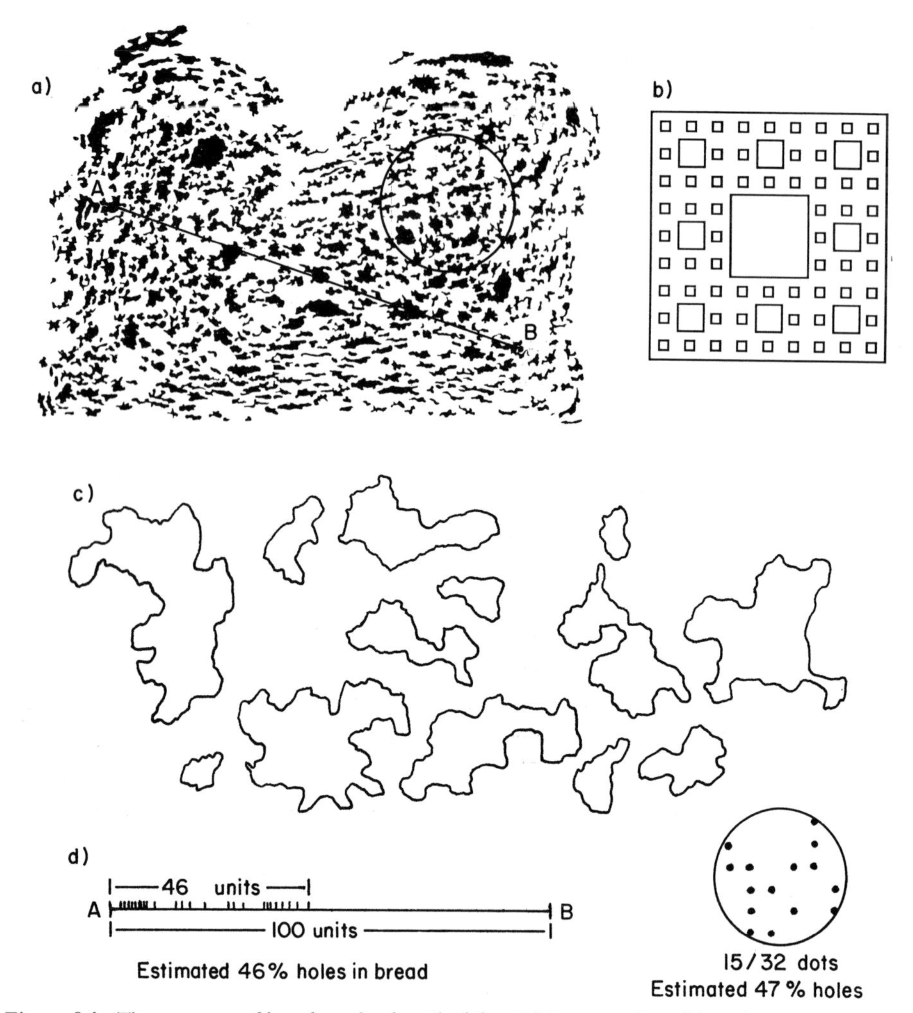

Figure 9.1. The structure of bread can be described from the perspective of fractal geometry using the Sierpinski carpet as a mathematical model. a) Replica of a full slice of bread. b) A Sierpinski carpet. c) Enlarged fractal holes from the body of the bread. d) Typical data for Rosiwal intercept by paper strip integration for estimating the 'holiness' of the bread.

[2]. The word *tomography* comes from the Greek word tomas, "a cut". A *Tome,* one volume of a series of books, was so called because it was a part cut from a large roll of writing when the large roll was too big to handle by itself. A *microtome* is a device for taking a thin slice of material. It comes from the Greek word micros meaning "small". Whole books have been written on stereology. The mathematical theorems of stereology form part of a subject known as *geometrical probability* [3].

In Figure 9.1(c), some of the holes of the slice of bread are enlarged so that the reader can make direct measurements of the fractal structure of these profiles. An important property of the bread is the percentage of the bread which has holes. This can be estimated using Chayes dot counting method discussed earlier, by placing a regular array of dots on a search circle such as that shown in Figure 9.1 (a). The percentage of the bread which has holes is calculated from the percentage of the search dots which fall within the holes of the bread. For this slice of bread and the search circle shown in the diagram, the percentage of bread flesh in the slice of bread is 46%.

When I first started to carry out measurements of this kind, my attention was directed to a method used by geologists to characterize the structure of ore specimens which is known as the *Rosiwal intercept method* [4]. The Rosiwal intercept method states that if one draws a line at random such as from A to B in Figure 9.1 (a) then the fraction of the line which lies within the flesh of the bread is the same as the percentage of bread flesh in the overall slice of bread. When adapting this method developed by the geologists to look at bread, we have to be a little careful because of the fact that the bread obviously has an outer rim which is low in holes and which we call the crust of the bread. The way in which we draw the line is also important because one can obviously see the lay of the dough as it was originally placed into the cooking pan. Vertical lines may have a slightly different hole content than horizontal lines; the readers can make measurements for themselves to check this possibility.

When I encountered the Rosiwal intercept method in the early 1960s, I personally found the method very surprising and had to carry out some measurements for myself before I could believe in the efficiency of the method. It is interesting to make measurements by the Chayes dot method and to compare it to the Rosiwal intercept method.

The paper strip integration technique used to measure the length of a polygon can be used to measure the total chord length lying within the holes of the image on a line such as A to B as illustrated in Figure 9.1 (a).

The word *integrate* has an interesting history. It is related to the word *integer* which means "a whole number"; a number not broken into parts. The word comes from in, which can mean "not", and the Latin word tangere meaning "to touch". Something which is integrated is a complete system which is not separated into parts. It has not been touched since it was created. A complete apple pie forms an integrated whole because it has not been touched with a knife. When we integrate a mathematical quantity we put it back together to make an untouched whole. A *section* is a part cut off from a whole; a word derived from the Latin word secare "to cut" – a section of a book was originally a part cut from the scroll which formed a whole book. When we integrate sections of the search line crossing the grids in the Rosiwal intercept method, we make a whole portion of the paper strip representing the total voids in the bread. The holes in the bread, and similar systems, are variously described as *voids* or *pores*. The word void means "empty space" from the Latin word vacare meaning "to make empty". A *vacant* lot is empty of buildings and a *vacation* was originally "empty days" when no work was required. The percentage of holes in a porous body is sometimes referred to as the *voidage* of the body. Other workers call it the *porosity* of the body. The word *pore* originally meant "a passage" from the Greek word poros – "a passage way". The pores of the skin are passage ways through the skin through which sweat evaporates to cool the body. In modern English, the pores of a piece of material is a general term for the holes in the substance and the term *blind pore* is used to describe the holes which do not connect with other holes in the body (the use of the term pore to describe the act of intense studies, as in the phrase "to pore over" is not related to the use of the term to describe holes. My dictionary is of little help in understanding this second meaning, it says "to pore – to gaze closely and attentively" – origin obscure). When one studies a slice of bread, it is difficult to know how many pores are blind and how many are interconnected. The ceramicist has the same problem.

Students find the fact that the Rosiwal intercept line density can give the area ratio of holes in the bread a very surprising result. It usually takes several sets of measurements to convince the students that the technique works. It is suggested that students who need convincing can carry out several dot counting estimates of area from different regions of the slice of bread and compare it with the Rosiwal intercept data. The variations in the estimate in the fractions of the hole in the slice of bread from various dot counting – Rosiwal intercept estimates will be Gaussian distributed. If a class carries out a group study, they could plot their data on Gaussian probability to estimate the average 'holiness' of the bread and the standard deviation of the variation in the hole content. Note that again, the variation in the 'holiness' of the bread from region to region is not an experimental error but is a real variation in the stochastically constituted variable – "the 'holiness' of the bread for a given inspection area".

Another theorem in the subject of stereology states that the average chord length on the Rosiwal intercept search is related to the average size of the holes. The distribution of chord lengths created as the search line moves across the holes in the bread are a Cantorian set that can be used to characterize the holes of the bread.

A body such as bread can contain many blind pores because of the way the pores are produced. Bread is turned into a porous body by means of what is known as a *blowing agent*. In home-made bread the blowing agent is yeast. A dictionary defines *yeast* as a substance consisting of certain minute fungi which causes alcoholic fermentation. Basically, the yeast cells attack the starch of the bread to produce alcohol and carbon dioxide. As the fermentation of the dough progresses, the bread rises in the pan prior to baking because of the production of the carbon dioxide. The baking drives off the traces of alcohol left by the fermentation process and the bubbles make the bread easier to eat. In mass production bakeries, the yeast is bypassed and carbon dioxide bubbles are forced into the dough to make it rise. Chemists mimic the production of bread when creating foamed plastic material by using chemical blowing agents. There are compounds such as ammonium carbonate which when heated produce only gases to create holes in the substance. Man-made foamed rubber and plastic made this way is often a very good insulator against cold because of the air trapped in the blind pores. Freon gases used to be used to make light weight foamed plastic for heat insulating applications.

The mathematical curve which serves as a model for the fractal structure of systems such as a slice of bread is the *Sierpinski carpet* shown in *Figure 9.2*(a) [5, 6]. At first sight, the two systems do not appear to be closely related because, as in the case of the boundary of the Koch island, ideal Sierpinski fractal systems are symmetrically structured whereas natural ones are not. The word *symmetry* is defined in a dictionary as "the state in which one part exactly corresponds to another in size shape and position". If one drew a line down the middle of the Sierpinski carpet, the two halves match each other exactly. The Sierpinski carpet is named after a Polish mathematician Waclaw *Sierpinski* (1882–1969) who taught mathematics in Warsaw. The Sierpinski carpet of Figure 9.2(a) (iii) is one of many different carpets that can be drawn using variations of the basic construction algorithm which is illustrated in Figure 9.2(a). To construct a Sierpinski carpet with a fractal dimension of 1.89, the carpet is first divided into nine squares and then the middle square is removed. A different Sierpinski carpet can be constructed by dividing the original carpet into 25 squares and removing $1/25^{\text{th}}$ from the middle of the square and so on. Many other divisions of the side of the square can be used to set up a construction algorithm for an infinite series of Sierpinski carpets [5].

The second stage of the fabrication of the Sierpinski carpet involves the dividing of each subsquare left after the removal of the central square with subsequent removal of the central portion as illustrated in Figure 9.2(a) (ii). The third stage of the construction algorithm is shown in Figure 9.2(a) (iii) and so on. Note that if one were to take small parts of part (iii) and enlarge them, they would look as if one was looking at the previous stage. The portion of part (iii) labelled A looks like part (i) and the portion marked B would look like part (ii). The series of carpets satisfy one requirement of the fractal system in that they "scale". That is, it is impossible to tell the order of magnification from the appearance of a subsection of a system. If one carries out the construction algorithm indefinitely, then the area of the carpet gradually disappears until one has a mathematical figure with no area composed of an infinite number of holes defined by an infinite number of infinitely thin threads! Mandelbrot, in his discussions of Sierpinski carpets, imagined that the carpet construction is carried out by geometrically inclined *termites*, which are insects similar to an ant that can eat into wood leaving a series of holes. Mandelbrot called the holes left by the geometrically trained termites *tremas* [5].

To understand how one can generate a useful fractal dimension for describing the disappearing carpet consider the Richardson plot of Figure 9.2(b). The ordinate of the graph is the fractional area of the carpet left when one can resolve a given hole size or trema. The abscissa is a measure of the size of holes which are being examined. The size of the hole is characterized by the square root of the area of the hole. The

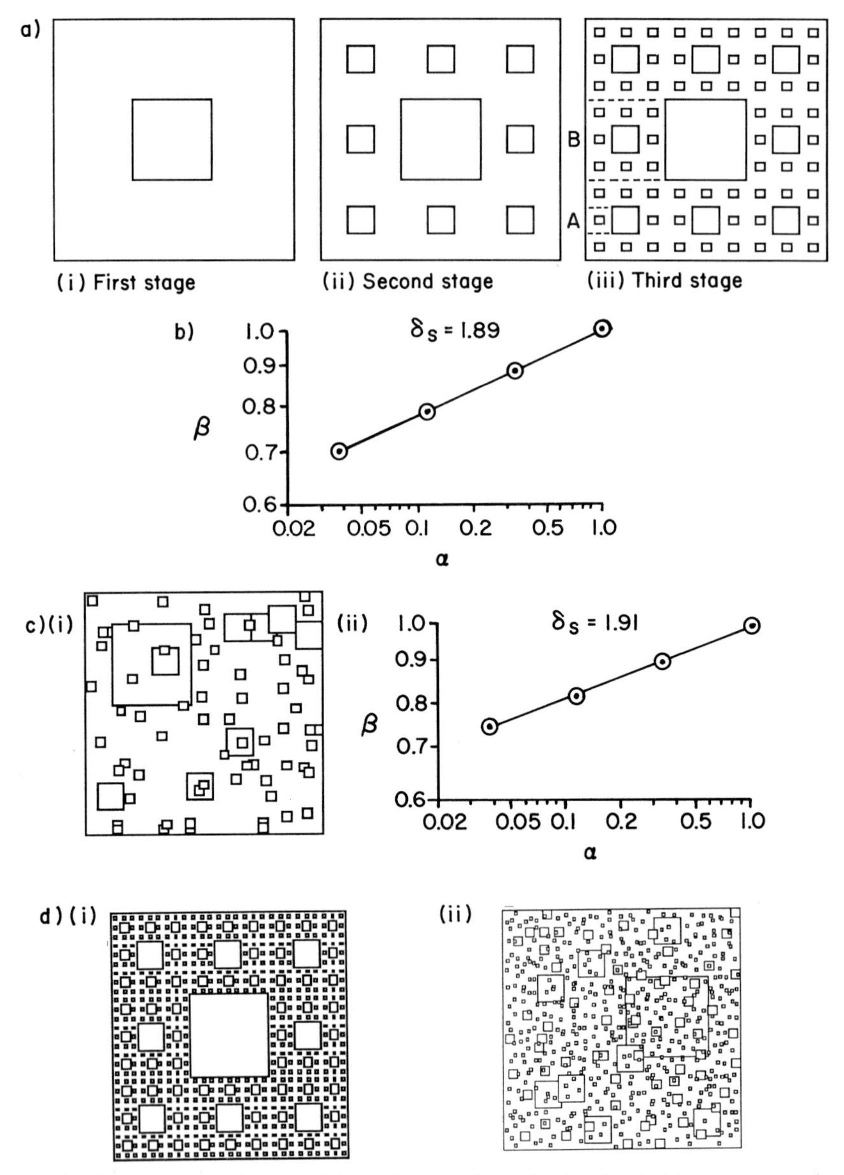

Figure 9.2. Typical construction algorithm and Richardson plot for Sierpinski carpet systems. a) Ideal carpet of fractal dimension 1.89. (i), (ii), (iii) Stages of construction. b) Richardson plot for calculation of the Sierpinski fractal of the carpet of (a). c) Statistically self-similar version of the Sierpinski carpet of part (a). i) Randomized Sierpinski Carpet of (a) (iii). ii) Richardson plot. d) Larger scale statistically self-similar Sierpinski carpet created on a computer.

first point on this graph is seen to be at coordinates l and l. This represents the carpet before it is attacked by the termites. The first stage of the carpet has one hole of size 1/3 which represents a fraction of the carpet area of 0.11 so that 0.89 of the original area remains. Then one looks at a carpet which contains both holes of size 1/3 and 1/9. The combined area of these holes leaves us with a carpet of area 0.78. We then look at the hole of the next smaller size and we are left with a carpet which has almost 0.7 in area of the original carpet. It can be shown that the slope of the line linking this "carpet disappearance" data subtracted from the number 2 generates a fractal dimension descriptive of the structure of the carpet. The Sierpinski carpet of Figure 9.2(a) has a fractal dimension of 1.89. In general, to distinguish this Sierpinski fractal from a structural fractal of the boundary, it is useful to use a subscript to designate the fractal dimension as shown in Figure 9.2(b). In many discussions, it is obvious which fractal dimension is being utilized to characterize a system and often the simple δ denoting fractal dimension without a subscript is used. When reporting experimental data the full subscript version of the descriptive fractal dimension should be defined at least once in the report. Construction by hand of stages beyond the third stage shown in Figure 9.2(a) becomes tedious. However, one can calculate the increasing loss of area as one carries the construction algorithm onto the Nth stage. In normalized units, the loss area at the first stage is 1/9, at the second stage it is 1/9 + 8 (1/81) and at the third stage is 1/9 + 8(1/81) + 64 (1/279) and so on. As a class project, students could calculate the series for Sierpinski carpets with the side length divided into 3, 5, 7 etc. The disappearing act of the carpets as calculated numerically can then be plotted on log–log graph paper so that the students can get a physical feel for the significance of the Sierpinski fractal. This study will also give the students some exercise in studying converging series in algebra.

The reader who has stayed with this discussion so far may have grasped the principle of the Sierpinski fractal but can be forgiven if they are doubtful about the utility of such a descriptive parameter for systems in the real world since systems such as a slice of bread do not look like ideal Sierpinski carpets. As in the case of fractal boundaries, the situation is that many porous bodies are statistically self-similar versions of ideal Sierpinski carpets. A randomized statistically self-similar version of Figure 9.2(a) is shown in part (c) of this figure. This statistically self-similar version was constructed by John Leblanc at Laurentian University. He cut out the holes of the carpet of Figure 9.2(a) (iii) and relocated them on a black sheet of paper selecting x,y coordinates for the cut-outs using a random-number table taking the bottom left hand corner of the square as the origin of the coordinate system (Mr. Leblanc cheated a little in that if the location of the square would have left part of it outside of the carpet he adjusted his random coordinates to fit the square just inside the carpet).

When creating the statistically self-similar version of the third-stage Sierpinski carpet of fractal dimension 1.89, one soon becomes aware of two important differences in the structure of the statisically self-similar carpet as compared to the ideal carpet. Basically, some of the relocated holes fall within the existing holes so that the self-similar carpet does not disappear as quickly as the ideal carpet for a given hole size. Secondly, as the construction of the carpet progresses, using smaller and smaller holes, the holes of the carpet in the statistically self-similar version become interconnected whereas in the ideal carpet no matter how many holes are created they are always isolated from one another. The ideal Sierpinski carpet, as a mathematical curve, models the structure of a woven type fabric with threads isolating every hole from one another as in the sieve cloth of Chapter 3.

The physical differences between the ideal and statistically self-similar version of the carpet are important for physical systems at high porosity. Because of the connectedness of the self-similar system, pathways exist through the pore structure of the randomized body whereas they do not exist in the idealized mathematical structure. The interconnectedness of the statistical self-similar version of the carpet obviously creates larger holes than those created by the construction algorithm for the ideal carpet. The interconnectedness of the pore structure is more obvious in the more detailed statistically self-similar Sierpinski carpet generated using a computer as shown in Figure 9.2(d). In Figure 9.2(c) (ii), the Richardson plot for the calculation of Sierpinski fractal of the self-similar version of Figure 9.2(c) (i) is shown.

Because of the loss of holes in the randomizing process, it can be seen that the measured fractal dimension of the carpet is slightly higher than the ideal curve for the reasons given above. This data was generated using an image analyzer to measure the size of the holes in the carpet. The reader should make careful note of the fact that in both of the graphs of Figure 9.2, the log scale on the ordinate is not the same as that on the abscissa. Because of this fact, the visual impression of the slope of the data line does not give a true impression of the magnitude of the Sierpinski fractal. If the two scales had the same magnitude, the slope of the line would have been much less obvious. In general data, the data line slope used to calculate the magnitude of the Sierpinski fractal is the mathematical slope of the line. It is not the visual one as judged from a graph with different logarithmic scale displays used in the study of Sierpinski fractals in this book.

To appreciate the way in which information on the size of holes present in a slice of bread are used to calculate the Sierpinski fractal of the bread consider the information displayed in *Figure 9.3*. Again the data used to construct a Richardson plot, from which the Sierpinski fractal is deduced, uses normalized data with the area the slice of bread without holes being the normalizing factor for the β or "residual area" parameter and the size of the largest hole to normalize the α or "resolved size" (square root of area) parameter. As a consequence of this choice of normalizing factors, the first point on the graph of β versus α is (1,1).

In a series of measurements which would be carried out by computer-aided image analysis, one would first of all measure the size distribution of the holes in the bread and list this information in the format shown in *Table 9.1*. The measurement of the size distribution of systems similar to the holes in the bread of Figure 9.1 is a very

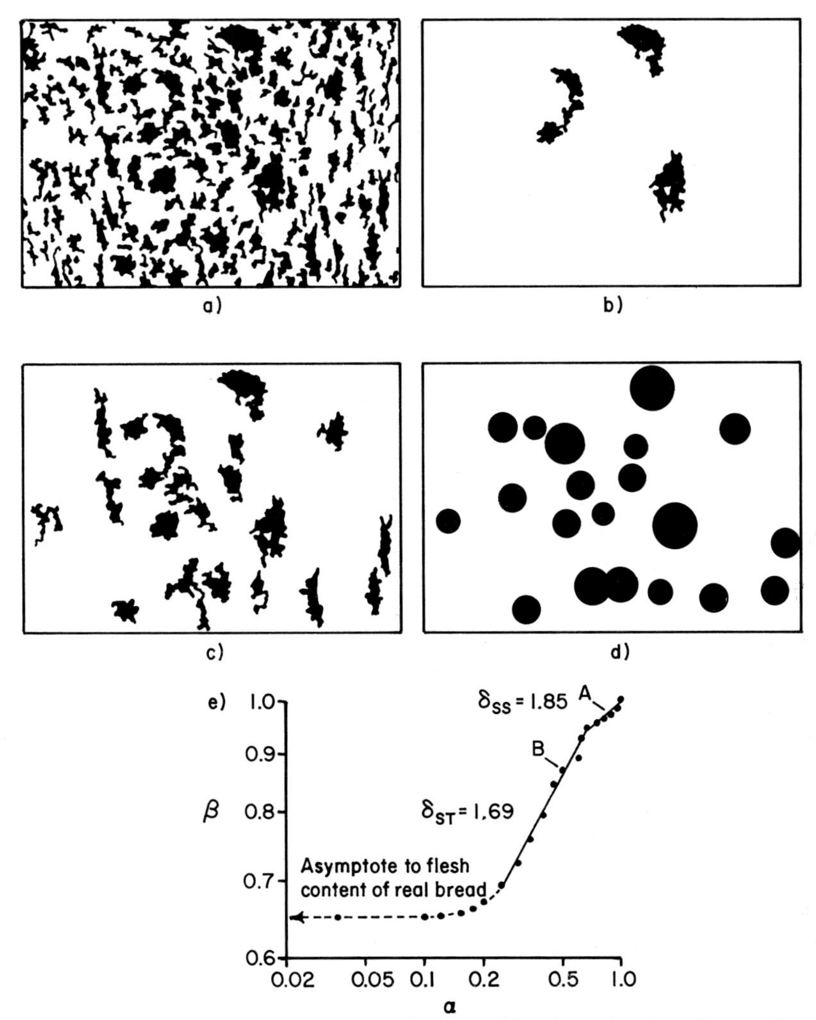

Figure 9.3. To characterize the fractal structure of a slice of bread one imagines that one inspects the structure of the bread at a series of increasing resolutions to deduce how quickly the bread disappears as the inspector sees smaller and smaller holes at higher and higher resolution. a) Magnified area of the slice of bread of Figure 9.1 (a). b) Appearance of the bread at a resolution showing only the largest holes. This data appears as point A on the graph of (e). c) Appearance of the bread at a higher resolution. This data appears as point B on the graph of (e). d) Appearance of the field of view when the holes of (c) are converted to circles. e) Richardson plot for the data obtained for all the holes of (a). α: The resolved size of the hole normalized by the size of the largest hole (the resolved size is defined as the square root of the area of the smallest hole we can see). β: The residual area of the carpet after all holes larger than or equal to the resolved size have been removed. This is then normalized by the total area of the carpet. δ_{SS}: The Sierpinski fractal of the carpet for the gross structure, that is for the largest holes. δ_{ST}: The Siepinski fractal of the texture of the carpet formed by the smallest holes.

Table 9.1. Hole size distribution data for the slice of bread shown in Figure 9.1.

a) Hole Number	Resolved Real Area (Pixels)	Resolved Size $\sqrt{A}$	Normalized Resolved Size	Residual Area (Pixels)	Normalized Residual Area
1	4706	68.60	1.0000	302494	1.0000
2	4567	67.57	0.9851	297927	0.9849
A 3	3703	60.85	0.8871	294224	0.9727
4	3104	55.71	0.8121	291120	0.9624
5	2717	52.12	0.7598	288403	0.9534
6	2120	46.04	0.6712	286283	0.9464
7	1987	44.57	0.6498	284296	0.9398
8	1949	44.14	0.6435	282347	0.9334
9	1836	42.84	0.6246	280511	0.9273
10	1791	42.32	0.6169	278720	0.9214
11	1782	42.21	0.6154	276938	0.9155
12	1770	42.07	0.6133	275168	0.9097
13	1767	42.03	0.6128	273401	0.9038
14	1751	41.84	0.6100	271650	0.8980
15	1740	41.71	0.6081	269910	0.8923
16	1368	36.98	0.5392	268542	0.8878
17	1280	35.77	0.5215	267262	0.8835
18	1226	35.01	0.5104	266036	0.8795
19	1189	34.48	0.5026	264847	0.8755
B 20	1185	34.42	0.5018	263662	0.8716

Note that the residual area is calculated based on a scanned image containing 307200 pixels.

important problem in fine particle science and many allied technologies. Even experienced technologists are sometimes unfamiliar with efficient methods of measuring the size distribution of an array of profiles such as that of Figure 9.1. Therefore an efficient strategy for carrying out this type of measurement forms the basis of Exercise 9.1

The logic sequence for transforming the size distribution data into a Sierpinski fractal is illustrated in Figure 9.3(b, c). We imagine that the computerized image analyzer is first set so that it can only see the large holes – the first size group of Table 9.1. From Table 9.1 we see that there are 3 holes of this size.

These large holes constitute a fraction 3% of the surface of the bread. This fact is plotted as the data point A at the top right hand corner of the Richardson plot. Although human observers see all of the holes all of the time, one has to realize that the computer-aided image analysis system can be set to see only the large holes and would see the bread as looking like Figure 9.3(b). We next imagine that the computer is set to see the next smaller hole size. To the computer, the bread would now look like Figure 9.3(c). The data point corresponding to this situation constitutes the data

Table 9.1. continued

b) Data Point Number		Resolved Real Area (Pixels)	Resolved Size $\sqrt{A}$	Normalized Resolved Size	Residual Area (Pixels)	Normalized Residual Area
	1	4706	68.60	1.0000	302494	1.0000
	2	4567	67.57	0.9851	297927	0.9849
A	3	3703	60.85	0.8871	294224	0.9727
	4	3104	55.71	0.8121	291120	0.9624
	5	2717	52.12	0.7598	288403	0.9534
	6	2120	46.04	0.6712	286283	0.9464
	7	1836	42.84	0.6246	280511	0.9273
	8	1740	41.71	0.6081	269910	0.8923
B	9	1185	34.42	0.5018	263662	0.8716
	10	975	31.22	0.4552	256107	0.8467
	11	755	27.47	0.4005	240776	0.7960
	12	580	24.08	0.3511	229519	0.7588
	13	434	20.83	0.3037	218998	0.7240
	14	297	17.23	0.2512	209423	0.6923
	15	189	13.74	0.2004	202299	0.6688
	16	144	12.00	0.1749	200418	0.6626
	17	108	10.39	0.1515	198653	0.6567
	18	68	8.24	0.1202	197653	0.6534
	19	48	6.92	0.1010	197263	0.6521
	20	6	2.44	0.0357	197154	0.6518

Note that for this table the Residual Area is based on the cumulative total of the areas of all holes larger than or equal to the stated size and is calculated based on a scanned image containing 307200 pixels.

point B on the Richardson plot. The process continues on to produce the whole of the data line as shown in Figure 9.3(e).

It is important to appreciate that the holes constituting the Sierpinski fractal do not themselves need to be fractally structured. The fact that the holes of Figure 9.1 are themselves fractally contoured is incidental to the calculation of the structural Sierpinski fractal of the slice of bread. To emphasize this fact, the large hole population of Table 9.1 as shown in the sketch of Figure 9.3(c) is converted to the equivalent circular holes distributed at random on the surface of the bread in Figure 9.3(d).

It is an empirical fact that all of the natural systems that we have examined at Laurentian University for characterization by their Sierpinski fractal have more than one data line as shown in Figure 9.3(e). We are not sure that this may be due to the fact that the holes have a certain size distribution which are then themselves manifest as a range of smaller sizes due to the intersection of the plane of inspection with the three-dimensional holes in the bread. If one imagines the many possible sections through a sphere, one can see that even a monosized set of bubbles would give a whole distribution in a section through the monosized randomly distributed spherical voids. The *culinary* experts amongst the readers (culinary is defined in a dictionary as pertaining to the kitchen or to cookery from the Latin word culina "a kitchen") can simulate the appearance of a section through a population of monosized randomly dispersed spheres by baking a cherry cake and cutting the cake into several slices. The many different sections of the cherries will form a random dispersion of circles of different sizes. The size of these dispersed cherry sections can then be calculated using the techniques of Exercise 9.1. The reader can check to see if the appearance of a sliced cherry cake has a Sierpinski fractal.

Many slices of different types of bread have been analyzed as laboratory projects by students at Laurentian University and every sample inspected gave rise to a two data line system such as that of Figure 9.3. The Sierpinski fractal of the coarse inspection section of the data line is one that seems to correspond to the visual impression of holes in the bread and could be related to customer acceptance of the structure of the bread. The second portion of the data line at high resolution appears to correspond to the texture of the bread that would be responsible for the sensation experienced when biting into the bread. For these reasons, we have chosen to call the Sierpinski fractal at coarse resolution of inspection the *structural Sierpinski fractal*. This fractal dimension is probably related to the breaking strength of the bread and other porous materials as well as the visual appearance of the material. We have called the second straight line data relationship of the Richardson plot for the hole structure of a porous body the *textural Sierpinski fractal*.

Because the study of the fractal structure of porous systems is such a wide topic, the reader is warned that these names are not yet in universal usage. As we gain a better appreciation of the physical reasons for the different data lines, it may be useful to adopt different names for different linear regions of the Richardson plot used to deduce Sierpinski fractals.

Technically, the scientist states that the holes of the bread and similar objects are supported by the matrix of food. The word *matrix* is one of those words which if the science student looks it up in a dictionary the definition can be more confusing than ignorance. Matrix is defined in my dictionary as "The womb from which a child is born".

The word matrix is closely related to the word mater meaning "mother" as in the phrase *Alma Mater* which people use to refer to the college where they were educated. The word alma means "good" or "gracious" and therefore the phrase Alma Mater refers to the idea that the college or school at which the person was educated played the same role in developing the student as a gracious mother.

In mathematics a matrix is an array of numbers which is used to generate other numbers by applying certain multiplication rules to pairs of numbers within the array of numbers. In other words, the matrix is the "mother" of all the numbers generated from the original array of numbers. In other sciences the word matrix is used to refer to the substance in which other objects are embedded or trapped. When one looks at the recovery of a valuable mineral from an ore, the worthless rock which forms the material in which the valuable mineral is trapped is referred to as the *rock matrix*. The act of crushing and grinding is said to *liberate* the valuable mineral from the matrix. The fragmented matrix left after the mineral is extracted is known as the *rock tailings* (see Chapter 8 for a discussion of the fractal structure of rock tailings).

Ron Lewis has pointed out that for a real piece of bread there has to be an ultimate region of the data where the line tails off to a constant value representing the part of the matrix of the bread providing the nutrition. One of my students pointed out that bread which had an ideal Sierpinski fractal would be an excellent dietary product since it would have an infinite number of holes and a vanishingly small amount of food and fiber. However, it would disappear at one bite. Indeed some "biscuits" provided for dieting humans are more holes than nutrition. The fractal structure of such puffed up biscuits should be interesting (see the suggestion by Mandelbrot that a human being is a vanishingly small amount of flesh between an arterial and venous network of infinite extent [5]).

It is interesting to note that when Mandelbrot was discussing the Sierpinski fractal curve, he discussed its similarity to the structure of two different types of Swiss cheese (hence the title of this chapter). Many other food products have a structure which can be characterized by means of a Sierpinski fractal. Consider for example the information displayed in *Figure 9.4*(a, b). Sections through two different varieties of aerated chocolate are shown. Customer acceptance of this type of chocolate bar depends upon the honeycomb material shown in the section of Figure 9.4 having a regular consistency. The pictures of Figure 9.4 are all black and white representations of originally colored photographs. The original photographs had been processed by a computer so that each size of hole is a different color [8]. For this reason, the relative size of holes can clearly be distinguished on the basis of the apparent grey color. The interested reader can enlarge the picture and carry out a size analysis of the holes to see if the holes can be described by a Sierpinski fractal.

Chocolate is basically a dispersion of solid fine particles of sugar, cocoa and milk solids within a continuous fat matrix. The size, shape and distribution of the ingredients of the chocolate affect the way in which the raw mix flows around the factory and also on the consumers evaluation of the quality of the chocolate. One of the most expensive stages in chocolate making is the grinding of the sugar to below 30 microns. If the sugar is coarser than this there is a gritty feeling between the teeth when biting into the chocolate. This is a sensation most of us have learned to associate with cheap chocolate and something which the chocolate manufacturer can avoid by grinding the sugar very fine. Not only does the dispersion of ingredients in the fat create a Sierpinski fractal but some of the individual fine particles have fractal shape as can be seen in Figure 9.4(c).

The food industry handles large amounts of dried milk solids and in Figure 9.4(d) a distribution of *casein* protein aggregates in a thin section of heat-treated milk powder is shown. Casein is the principle protein constituent of milk or cheese. Its

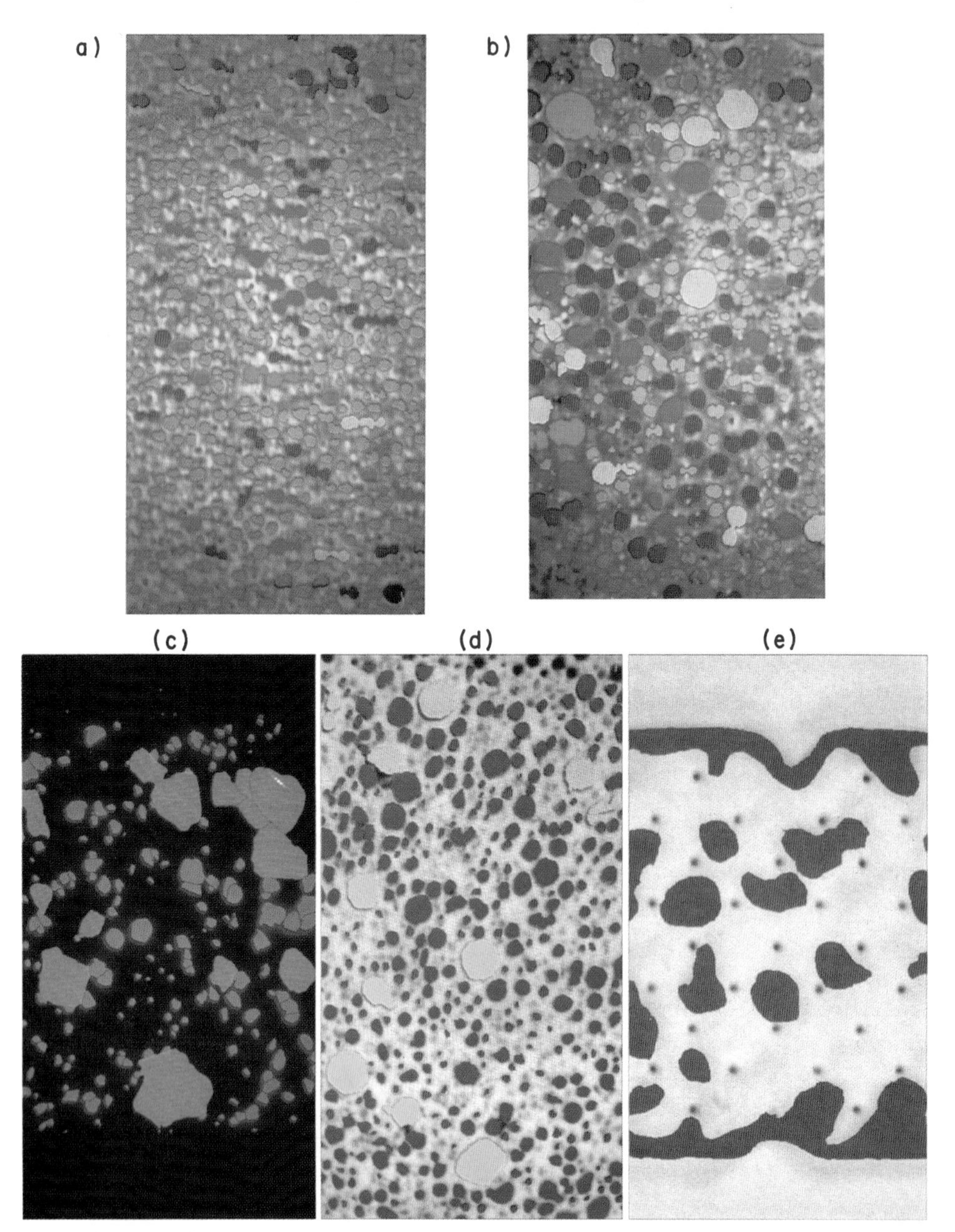

Figure 9.4. Many different food products look like Sierpinski carpets and can be characterized by means of a Sierpinski fractal. a, b) Sections through two different samples of aerated chocolate. c) Various ingredients dispersed in a fat matrix to create chocolate. d) A distribution of casein protein aggregates in a thin section of heat-treated milk powder. e) The browness patterns on a cooked biscuit look like a Sierpinski carpet and individual parts of the pattern have fractal boundaries [8]. Used by permission of Leica Cambridge Ltd., Clifton Road, Cambridge CB1 3QH, England.

name comes from the Latin word caseus for "cheese". Again, the difference in color of the different sized aggregates in the picture is an artifact produced by the fact that of Figure 9.4(d) is a false-color computer picture in which each size of aggregate of the figure is given a different color by the computer.

Figure 9.4(e) a double fractal system of interest to the food industry is shown. It is a computer-transformed image of the browness of a biscuit. The dark areas of the photograph are brown areas on the surface of the biscuit. If the biscuit is too brown it is rejected by the consumer as being burnt. On the other hand, if it is not sufficiently brown the consumer thinks that it is an uncooked batch of biscuits (remember that the European biscuit is the North American cookie). The browness boundaries are obviously fractal and it may be that the distribution of brown spots within the biscuit could constitute a Sierpinski fractal.

Section 9.2. Exploring the Fractal of Felts and Filters

When British sailors went on long voyages the water they carried for drinking sometimes became contaminated with bugs and slime. Over the years the sailors discovered that they could take a piece of bread and burn it black to create a carbon filter. By pouring the water through this filter they purified the water. This surprising anticipation of the purification of water with carbon filters illustrates how some of the strategies adopted by practical people anticipated modern technology to a surprising degree. The sailors creation of carbon filters by excessive toasting of bread parallels the other piece of empirical knowledge that a poultice made from moldy bread was very effective in curing wounds because the mold on the bread is a form of penicillin.

A carbonized piece of bread is described scientifically as a *depth filter* or *sponge-type filter*. The word filter has developed from the word *felt* which was the name given to a piece of nonwoven cloth. Felt was the earliest form of cloth developed before civilization had the technology to weave cloth using fibers spun into yarn. Modern paper making creates a felt from the cellulose fibers released from solid wood by mechanical or chemical means. The wood is turned into a suspension of fibers by a process described as *wood pulping*. The suspension of fibers is then poured over a mesh of wires which support the cellulose fiber network until it forms a system strong enough to exist by itself.

We can simulate the growth of felt in the paper-making industry by an extension of the experiment in which we sprinkled short fibers on parallel lines to carry out our Buffon's needle problem of Chapter 3. In *Figure 9.5*(a) the basic steps to be followed when adding a fiber to a felt being synthesized on a computer screen is shown. In general, the fibers will have a range of lengths. Therefore, one first

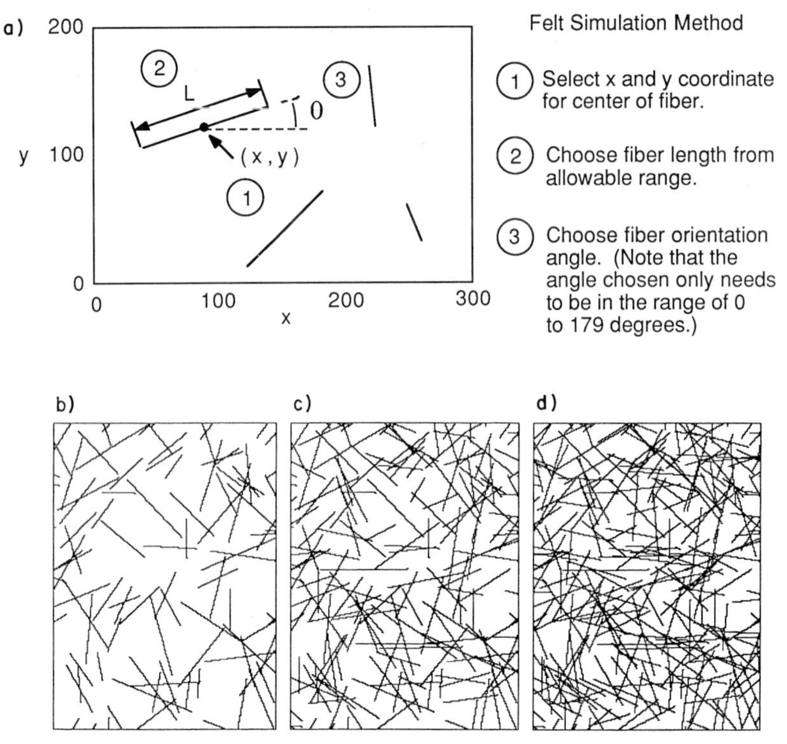
a)
200
100
0
y
0
100
200
300
x
L
(x,y)
0
1
2
3
Felt Simulation Method
1 Select x and y coordinate for center of fiber.
2 Choose fiber length from allowable range.
3 Choose fiber orientation angle. (Note that the angle chosen only needs to be in the range of 0 to 179 degrees.)
b)
c)
d)

Figure 9.5. Synthetic paper can be created on the screen of a computer by modifying and extending the technique used to model the Buffon's needle problem. a) The method used to simulate a fiber deposited on a surface. b) A felt of 100 fibers. c) The felt of (b) grown to 200 fibers. d) Resulting felt after 300 fibers have been deposited.

chooses a fiber of a given length by looking up the probability of a given length of fiber occurring in a population of fibers. This fiber is then placed on the screen by selecting coordinates for the center of the fiber, x and y, from random number tables. The direction of the placed fiber is chosen by choosing θ from a table of possible θ values chosen between 1 and 180° (there is no need to choose θ between 181° and 360° since after 180° the position of the fiber is a mirror image of values between 0 and 180°). In parts (b), (c) and (d) of Figure 9.5 the appearance of the felt as it grows is shown [9].

To study a completed simulated felt such as that shown in Figure 9.5(d), the Sierpinski fractal of the felt can be computed by focusing on the holes formed by the fibers. In *Figure 9.6*(a) the appearance of the felt of Figure 9.5 after 500 fibers have been deposited is depicted. The data for the holes of the filter over a wide range of sizes is summarized in the Richardson plot of Figure 9.6(b). It can be seen that the simulated felt has two characteristic Sierpinski fractals which again we designate as the structural and textural Sierpinski fractals.

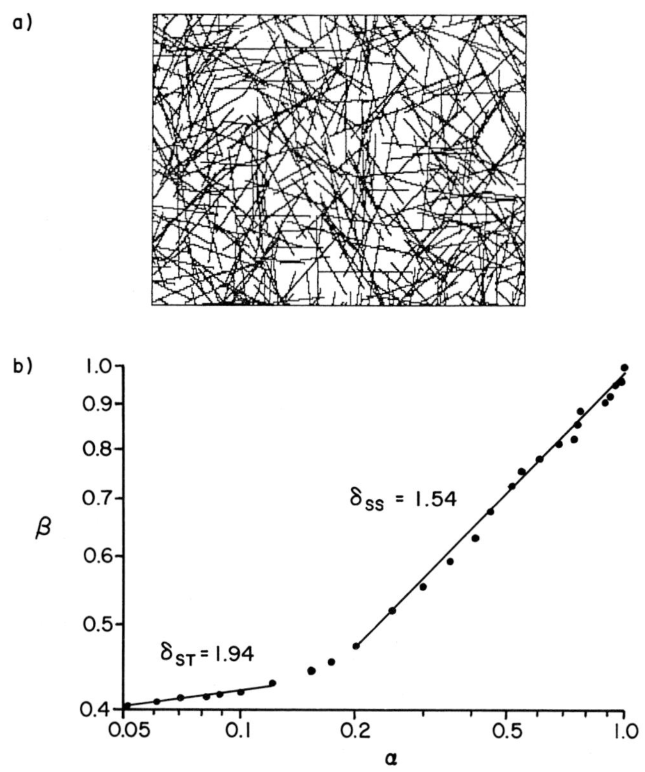

Figure 9.6. The Sierpinski fractal of a felt can be used to characterize its structure and probably its properties as a filter. a) Appearance of the simulated felt of Figure 9.5 after 500 fibers have been deposited. b) Sierpinski fractal of the felt of (a).

Figure 9.7. Filters used in scientific laboratories exhibit fractal structure [7]. Photo supplied by Gelman Science Inc.

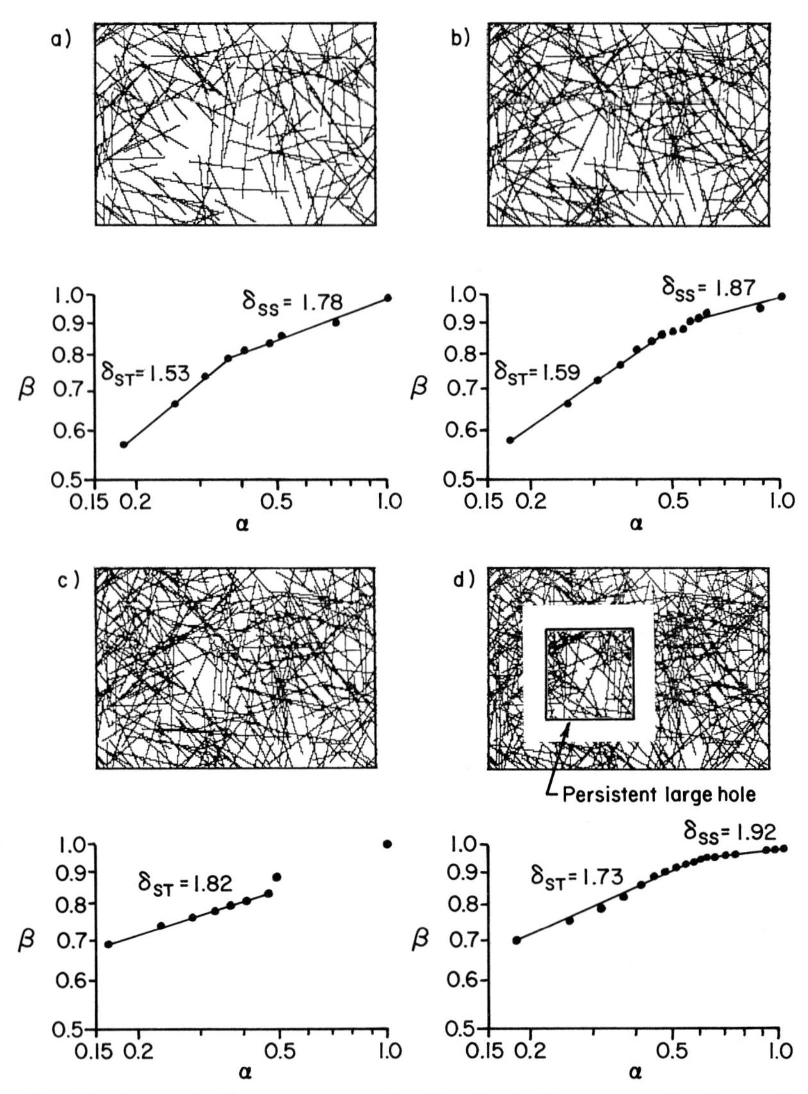

Figure 9.8. The changes in filter structure as the fiber density increases can be characterized by the Sierpinski fractal dimension of the felt.

Quite early in the development of technology in the early civilizations, it was found that one could clarify liquids by pouring the liquid through a piece of cloth. The felt used in this way was described as a *filter*. In *Figure 9.7* the appearance of fibrous filters used in scientific laboratories are shown. It can be seen that their structure is essentially the same as the felt of Figure 9.6. One of the important properties of a filter is the largest hole in the filter. This governs the size of the largest fine particle that can pass through the filter. This largest fine particle is an important

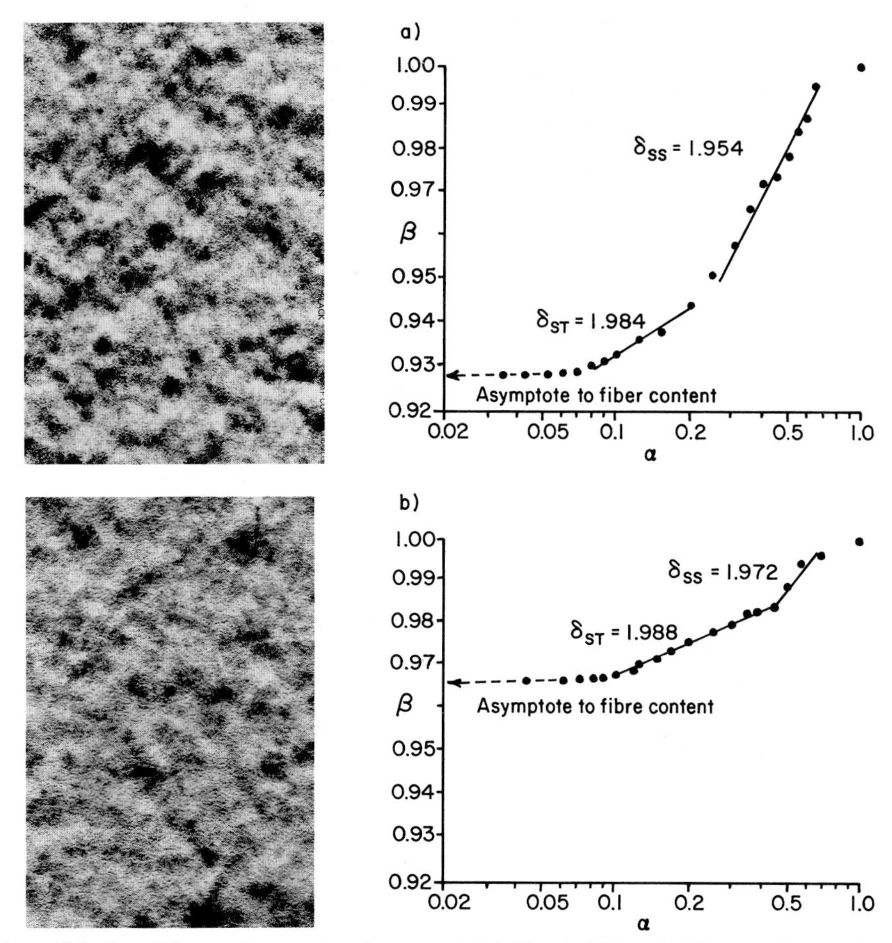

Figure 9.9. Two different photographs of paper and their Sierpinski fractals (photographs provided by the James River Corporation, 1915 Marathon Ave, P.O. Box 899, Neenah WI 54957–0899; used by permission).

quantity in the design of respirators for protecting workers and in the design of filters used to protect the clean rooms in which computers are manufactured and assembled.

In *Figure 9.8* the change in the filter structure of a fibrous filter as the density of fibers increased is shown. This series of synthetic filters illustrates an important problem which was discovered experimentally in the filter making industry. It can be seen that as the filter density builds up, a few of the larger holes created in the initial filter appear to persist even though the number of fibers constituting the filter increases dramatically. Fifty or sixty years ago people who made felts to be used in respirators knew about this problem. Therefore they used to expose their filters to a relatively coarse smoke before putting the filters into respirators. If one looks at the rate of the air flow through the various available holes of a filter such as that of Figure 9.8(b) the holes act as if they are pipes. Therefore the flow through a pipe of diameter D is controlled by Poiseuilles formula which was discussed in Chapter 6. This formula tells us that the flow through a hole is related to the 4th power of the size of the hole. This means that it is so much easier for the flow to move through the few large holes that in effect most of the holes in the filter have virtually no flow of gas through the filter. However, for the same reasons when one exposes the filter to a relatively coarse smoke, the coarser smoke fine particles are guided to the last few large holes in the filter blocking these holes so that when in use the gas to be filtered must flow through the smaller holes of the filter. The fractal structure of fibrous filters is proving to be an active area of applied research and the specialist reader will find more detailed information from this area of research in the literature [5, 7, 9].

Paper is essentially a felt of cellulose fibers and the Sierpinski fractal of paper may be an important way of characterizing the structure of paper. In *Figure 9.9*, the photographs of two specimens of paper along with their measured Sierpinski fractal dimensions are shown. The reader could find many similar photographs in textbooks on paper manufacture and could carry out Sierpinski fractal measurements on these pictures.

Section 9.3. Characterizing the Porous Nature of Bone and Sandstone

Some bones of the body are porous and constitute a structure which is strong and at the same time of low weight. When creating synthetic bone for use in the body it is better to create a porous replacement rather than a dense component to in order to help the body accept the replacement. In particular, if the bone replacement has the same pore structure as the original bone, the body will often accept the replacement and fresh bone created by the body will grow into the synthetic bone [10–14]. It may be that the best way to characterize the pore structure of desirable synthetic bone maybe to evaluate the Sierpinski fractal of the bone. To illustrate the use of fractal geometry in characterizing the structure of bone consider the two photographs shown in *Figure 9.10*. The first picture is of a section through healthy bone tissue whereas the second section is through the bone of a patient suffering from osteoporosis [15]. This is a disease which is most common in elderly people and represents a degeneration of the bone structure often associated with a lack of calcium in the diet and/or exercise, amongst other factors. The difference between the two bone sections is obvious but difficult to quantify.

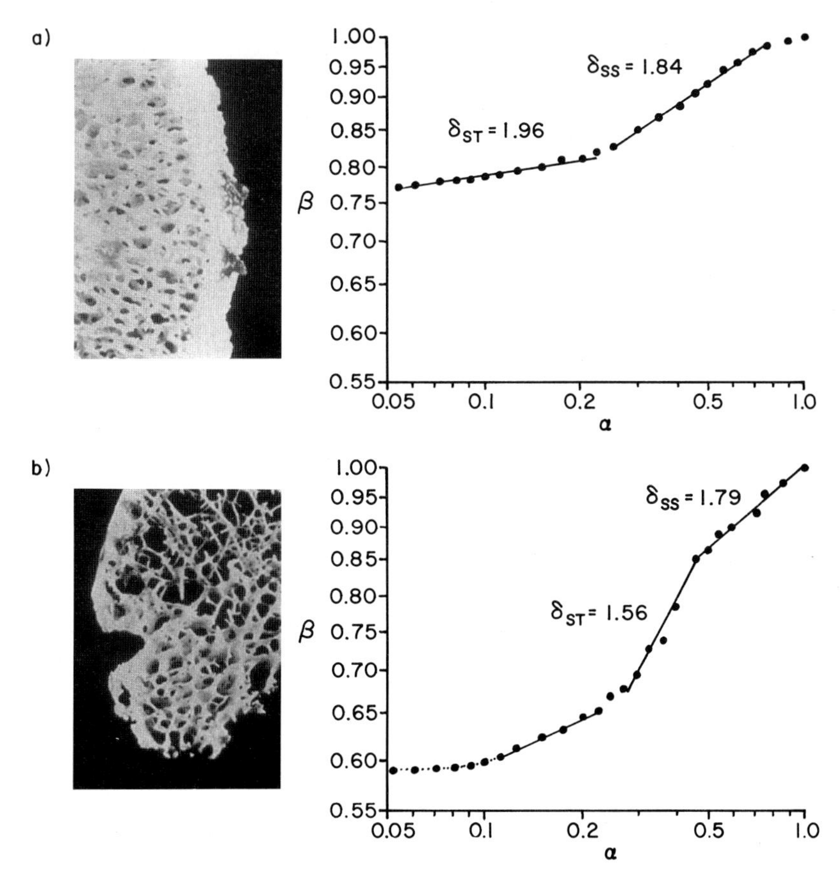

Figure 9.10. Healthy and diseased bone structures can be described by their Sierpinski fractals. a) Section through healthy bone and its Sierpinski fractal data plot. b) Section through bone with osteoporosis and its Sierpinski fractal data plots.

In the graphs of Figure 9.10 the size distribution of the holes present in the sections, presented as a Sierpinski fractal, is shown. The different regions of these curves may well be related to the overall properties and structure of a healthy and diseased bone. Another area of porous body research where the Sierpinski fractal could be a useful parameter for describing the properties of the system is in the structure of the sandstone constituting the mother rock of oil bearing rock.

In *Figure 9.11*(a) the section through a piece of sandstone taken from an oil reservoir is shown. The Sierpinski fractal determined from this rather sparse data is shown in Figure 9.11(b). Note that there is still an unsolved problem in such studies as to what constitutes an effective equivalent diameter for the pathway followed by the oil or gas moving through such reservoir rocks. Some scientists believe that the equivalent area pore is a useful measure, others maintain that because of the viscosity

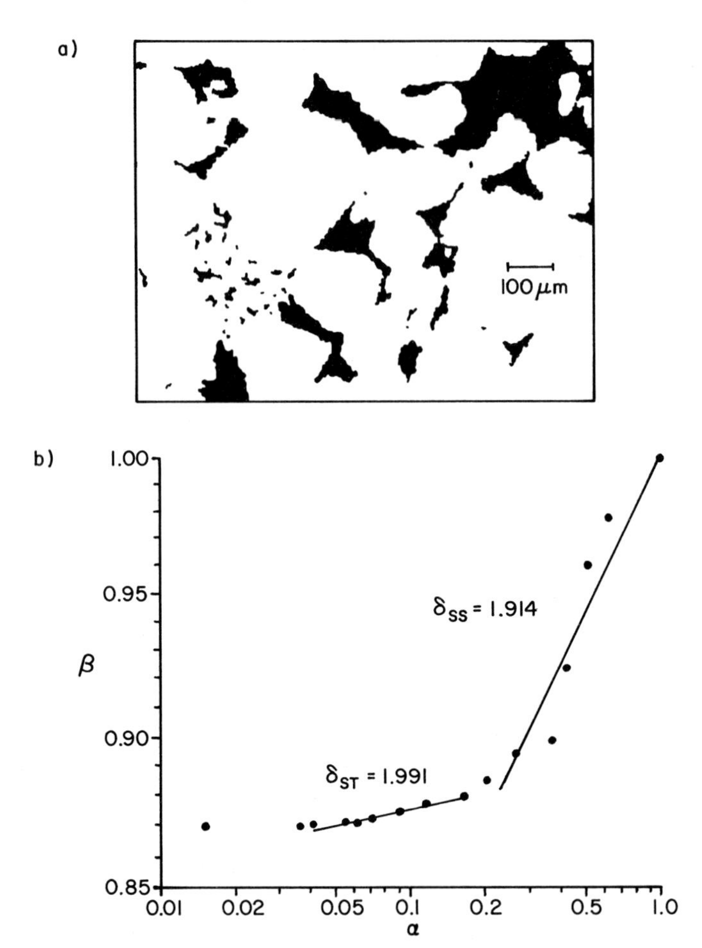

Figure 9.11. Oil-bearing sandstone can be described by means of a Sierpinski fractal dimension. a) Section through a sandstone. b) Sierpinski fractal data plot for the section of (a).

of the oil and the resistance to the flow created by the walls of the vessel that the smallest inscribed circle constitutes the useful measure of pore structure.

Many old buildings in Europe and other parts of the world are made from sandstone and other porous rocks [16]. The penetration of acid rain and other polluted water into such rocks is a major cause of scientific activity and the sealing of such rocks by the taking up of water-resistant chemicals is an area of research where Sierpinski fractals of a porous material may be useful in predicting the uptake of preservative chemicals and/or the movement of pollutants through the rock.

Exercises

Exercise 9.1. Stratified Count Logic for Efficient Characterization of a Set of Holes or Profiles

When faced with the task of characterizing an array of profiles such as those shown in *Figure E 9.1.1*(a) many workers start to characterize every fine particle encountered in the field of view as one starts from the inspection line A and moves across the field of view in the direction of the arrow as shown in Figure E 9.1.1. By the time they have reached the position shown by the dotted line B they have tallied the number of four different sized fine particles as shown by the tally given underneath the tally area. It can be seen that the analyst has encountered many more of the smallest fine particles than the larger profiles.

If the analyst continues such a characterization strategy the net effect will be to gain a great deal of information on the population on the smallest sizes per unit area of the field of view but sparse information on the population density of the largest profiles. By the time the simulated field of view of Figure E 9.1.1 (a) is completed the distribution of the various sizes encountered in the search will be as shown in the tally given in Figure E 9.1.1 (b). If the size distribution of the fine particles encountered up until the point C is converted into a weight distribution, the distribution function from the data of tally A to C is as shown as the solid line in *Figure E 9.1.2*(a). A useful way of deciding if one has really gained useful information on the size distribution function when carrying such a transformation is to plot the curves for + 1 and − 1 conversion curves in which one takes one from all the observed tallies in the size groups and calculates the distribution function and then one adds one to the size groups and repeats the calculation of the distribution function. This would generate the set of curves shown as the dashed lines in Figure E 9.1.2 (a).

The data of the tally up to the point C shows that there would be wild fluctuations in the distribution of the curve generated for + 1 and − 1 which is caused by the fact that we know so little about the population density of the larger fine particles which are rare events if one continues the first search strategy illustrated in Figure E 9.1.1. A far more efficient strategy for characterizing the size distribution of an array of profiles such as that of Figure E 9.1.1 is a procedure known as stratified count logic. The stratified logic count is not only more efficient but is more appropriate for a robotic analyzer as compared to a human being. For example, if the profiles of Figure E 9.1.1 were to be evaluated in total as the search scan moves from A through B to C then the microscope or television camera being used by the robot would have to be focused differently for each size of fine particle.

In the stratified logic procedure the robot will seek to characterize one size group at a time. The camera is initially set to see the largest fine particle known to be

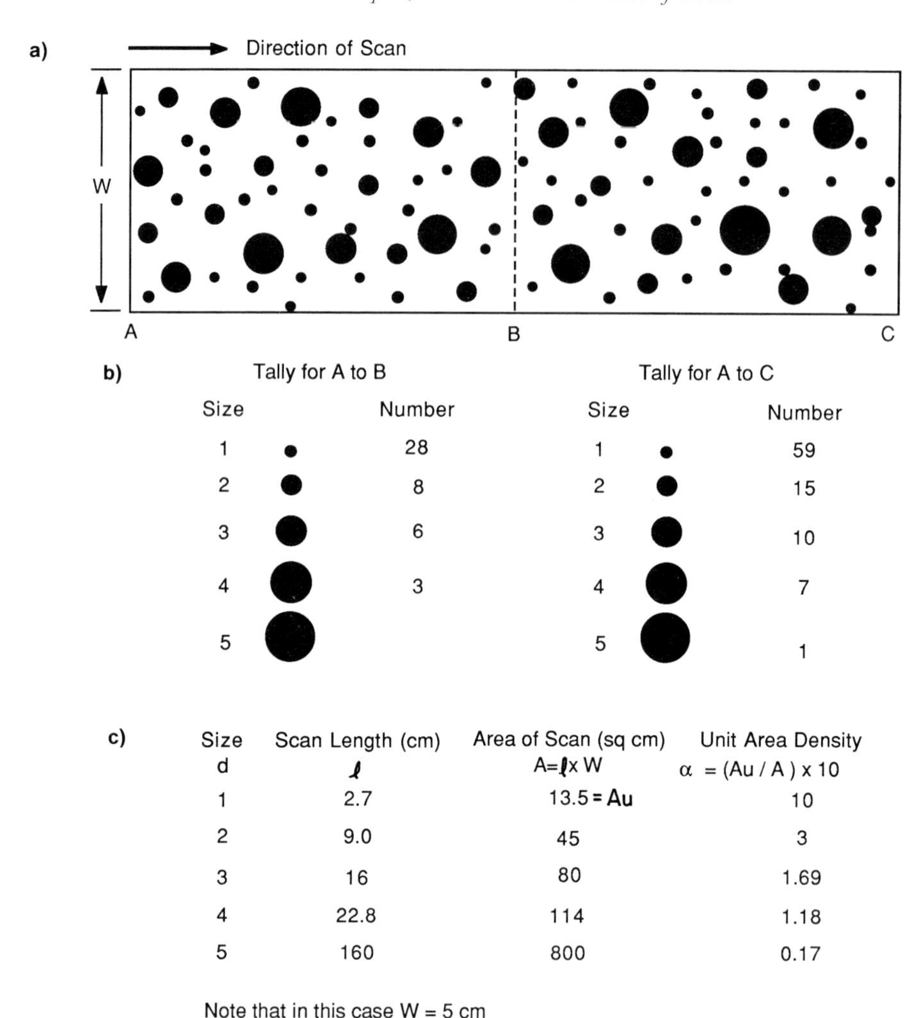

Figure E 9.1.1. In the stratified count logic size distribution procedure one seeks to estimate the population density of each size of fine particle present in a field of view. a) Simulated field of view. b) Data for a "size everything" scan. c) Data for stratified count logic exploration of the field of view. A: represents the unit area for 10 of the smallest features. l: The length of the scan of width W required to find 10 features of the stated size.

present in the slide or field of view. Then the field of view is inspected for only this largest fine particle and a linear search is inspected until ten of the largest profiles have been located. If in this search one discovers a larger profile, as for example the large profile encountered between B and C, then after it has inspected the field of view for the initial size for which it was focused it could then repeat the search for ten of the new largest profile. Assuming that this has been achieved one now looks

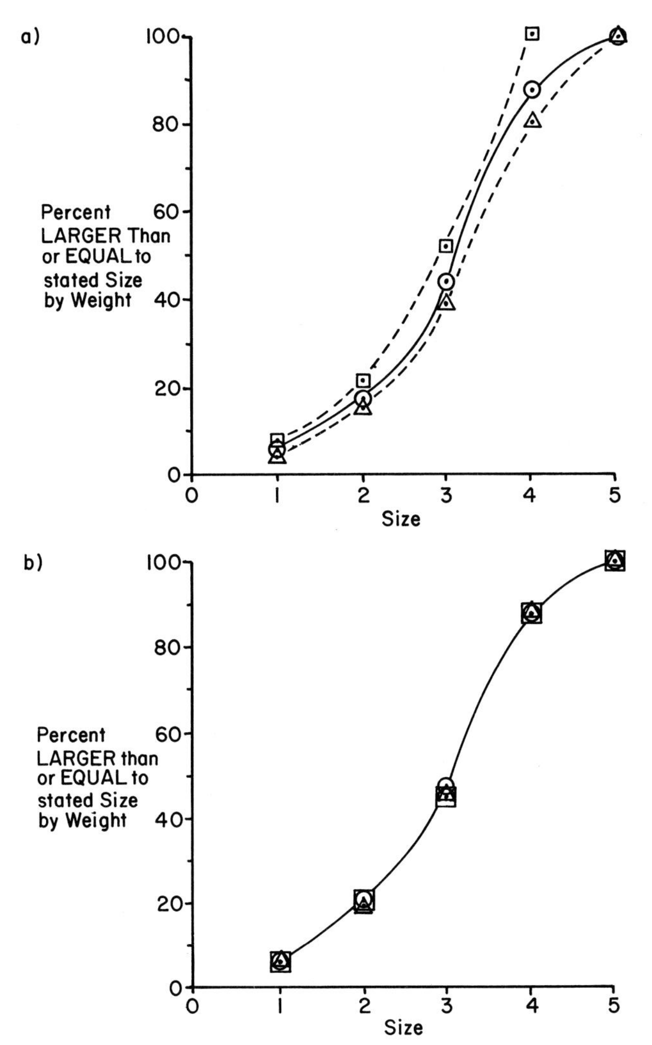

Figure E 9.1.2. Stratified count logic results in more confidence in the measured size distribution function for less data collection than "size everything" strategies. a) data for the "size everything" method. b) data for the "stratified count" method. Dashed lines represent the n − 1 and n + 1 curves for the data.

for the line scan length needed to discover ten of the next size of fine particle and so on. By adopting this strategy the camera and/or microscope is only adjusted at each switch to a new size group. The strategy results in information on the population density of each size of fine particle with approximately the same confidence levels.

The data is transformed into a size distribution function by first calculating the population density of each size per unit area. The calculation is illustrated in Figure E 9.1.1 (c). When this strategy is adopted the plus and minus one curves for the data are much more compatible with each other. The student is invited to use the data of Figure E 9.1.1 to calculate a Sierpinski fractal for the simulated field of view shown at the top of the figure.

Exercise 9.2.

The students should take a slice of bread and make a copy of it on a copying machine. One group of students should characterize the size of the holes in the slice of bread using a greatly enlarged copy of the bread, the holes can either be measured by dot counting or by visual comparison to a set of circles or by cutting out the holes and weighing them on a chemical balance. One group of students should use the count everything strategy of the previous exercise whereas the other students should attempt to carry out the stratified logic count. If there are enough students four groups can compare the variation in their estimated size distributions of holes by the two different techniques. The data can then be used to calculate the Sierpinski fractal of the piece of bread.

Exercise 9.3.

An important industrial technique is that of assessing the efficiency of pesticide spraying of leaves. The class can draw a paper leaf and lightly spray it with ink from a scent sprayer type bottle. The size distribution of the droplets can be measured and the efficiency of the coverage of the leaf estimated by calculating the Sierpinski fractal of the dispersion of dots on the simulated leaf. The students should then regard their leaf as an aerial photograph and calculate the number of trees in the area and the size distribution of the trees from the size of the canopies (the circles on the leaf). This is the technique used by foresters to assess the aerial photographs of sparse forests.

References

[1] E. E. Underwood, *Quantitative Stereology*, Addison Wesley, Reading, Mass. 1970.

[2] News story. "A Sideways Look at Scanners", New Scientist, December 6, 1979, Pgs. 782–785.

[3] M. J. Kendall, P. A. P. Moran, *Geometric Probability*, No. 10 of Griffith's Statistical Monographs in Courses, Charles Griffith, London, 1963.

[4] For a readily accessible discussion of the Rosiwal intercept method see G. Herdan, *Small Particle Statistics*, 2nd edition, Butterworths, London 1960.

[5] For a discussion of the Sierpinski Carpet see B. B. Mandelbrot an updated and expanded version of *Fractals Form Chance and Dimension* under the title *The Fractal Geometry of Nature*, W. Freeman, San Francisco 1983.

[6] For a discussion of the Sierpinski carpet see B. H. Kaye, *A Random Walk Through Fractal Dimensions*, VCH, Weinheim, 1989.

[7] See discussion of this idea in B. H. Kaye, *A Random Walk Through Fractal Dimensions*, VCH, Weinheim 1989, p. 405.

[8] The photographs of Figure 9.4 are taken from the commercial literature of the Cambridge Instrument Company which produces a sophisticated computer image analysis system for characterizing structures such as those in the figure.

[9] B. H. Kaye, "Describing Filtration Dynamics from the Perspective of Fractal Geometry", KONA 9 (1991), 218–236.

[10] D. Roy, "Bones From The Laboratory", *New Scientist* (October 21, 1976) 163–165.

[11] T. Kiely, "I Sing The Body Ceramic", *Technology Review* (November–December 1988) 42–45.

[12] News story, "Ceramics Make In Roads Into Bone Growth", *New Scientist* (March 30, 1972).

[13] P. Ducheyne, P. DeMeester, E. Aernoudt, "Isostatically Compacted Metal Fiber Porous Coatings For Bone In Growth", *Powder Metallurgy International 11* (1979) 115–119.

[14] R. M. Pilliar, J. B. Medley, "Porous Surface Structured Surgical Implants A New Design Concept", *Engineering Digest* (October 1978) 35–39.

[15] A. Purvis, "When Bones are Brittle", *New Scientist* (12 Nov. 1990) 57.

[16] W. B. Whalley, J. P. McGreevy, M. A. Summerfield, "Scanning Electron Microscopic Observations and Physical Attributes of Silcretes and the Implications for Sandstone Formation", *Scanning Electron Microscopy* (1982) 649–656.

Chapter 10

A New Wrinkle on Surface Fractals

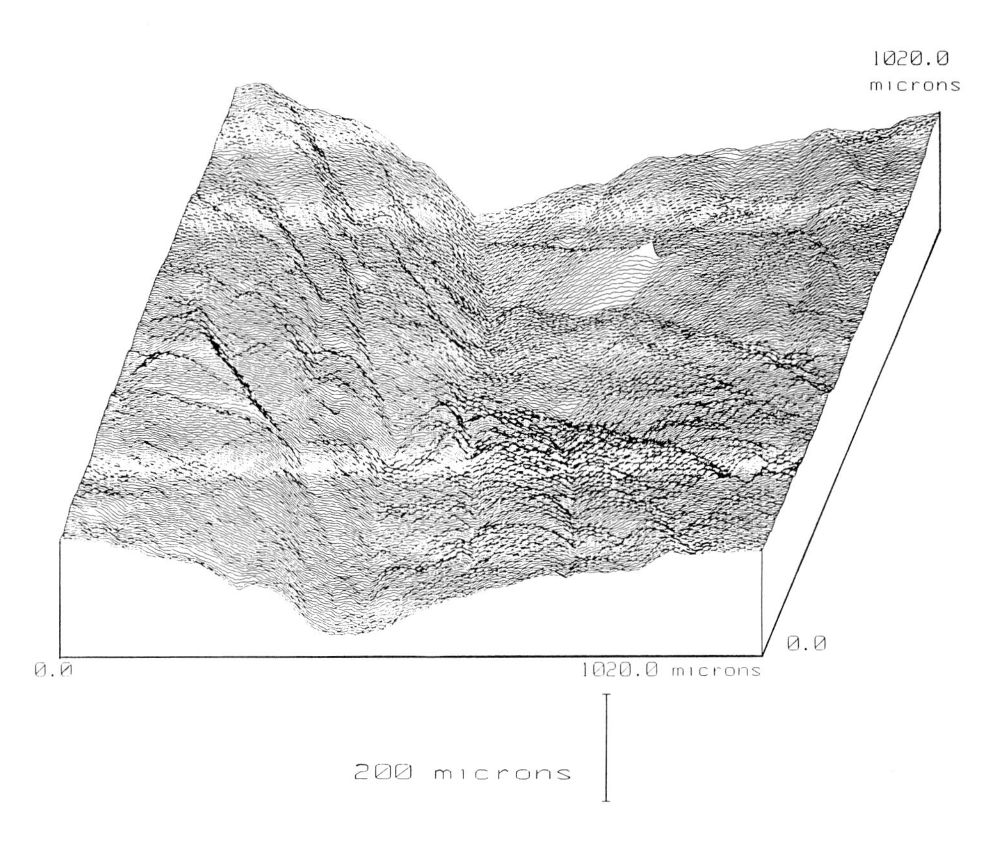

The canyons of the skin are fractal terrains

(Surface of skin after a dermatological treatment)
Surface contour graph used with permission of Micro Surface,
15 Rue des Saint Martin, B.P. 1313, 2500 Besançon Cedex, France

Chapter 10 A New Wrinkle on Surface Fractals 417

Section 10.1 Characterizing Canyons and the Effect of Sunshine 419
Section 10.2 Simulating Passoja's Method for Studying the Roughness of Metal Fractures 425
Section 10.3 How Many Islands are there in a Lake? 427
Exercises 434
References 436

Chapter 10

A New Wrinkle on Surface Fractals

Section 10.1. Characterizing Canyons and the Effect of Sunshine

After the publication of my book on fractal geometry, I was invited to visit various research laboratories where scientists were wondering if fractal geometry could be of help to them with their research projects. One manufacturer of a beauty cream wondered if fractal geometry could help to measure the smoothing of skin before and after using their cream. Another wondered if fractals could help to quantify the enhanced wrinkling (aging) of the skin when subjected to long exposure to sunshine (women of caucasian origin living in Australia often have skin as wrinkled as European caucasian women 5 years older). Scientists working with wood products asked me if fractal geometry could follow the progress of the improvement in the smoothness of the surface of a piece of wood being sanded by automated machinery. The answer to all three groups of scientists was yes.

The theory of the methodology they could use to study their problem had been worked out by a geographer studying the topography of canyons in a hilly area [1]. In fact if one looks at a low-magnification picture of a set of canyons without any other information, one does not know if one is looking at rocky countryside or wrinkled skin. We learned in Section 6.2 that geographers describe the "ups and downs" of a land area as its topography. In *Figure 10.1*(a) a section of a topographical map of a hilly area near Laurentian University is shown [2].

The manufacturers of the beauty cream use a word incorporating topos the same Greek root word meaning "a place", to make a word to describe creams and other products which are rubbed onto the skin. These products are described as *topical products* and *topical medication* and a lotion intended to protect the skin is known as a *topical cream*. On a topographical map, the geographer draws lines linking up all points on the ground that are at the same height above the level of the sea. The sea is a long way from the location represented on the map but "the level of the sea" is a reference point used by surveyors when drawing up their maps. These lines of equal height are called *contour lines*. Lines have been drawn in the topographical map of Figure 10.1(a) everytime the land level increases or decreases by 50 feet. The lines of Figure 10.1(a) are labelled 1300, 1350, 1400 etc. In the section of the map we have

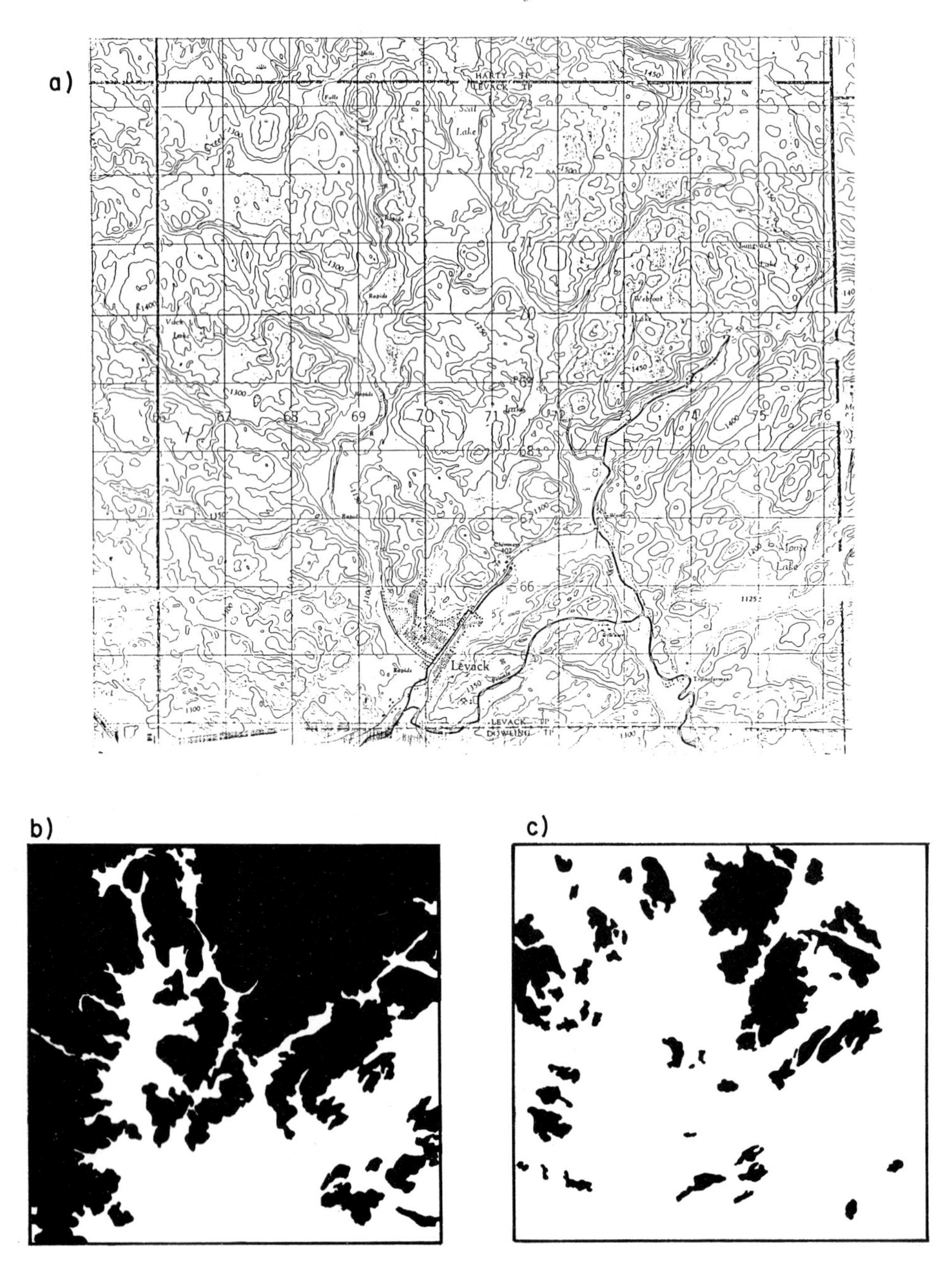

Figure 10.1. In a topographical map contour lines link up land at the same elevation above sea level. a) Topographical map of a region in Northern Ontario. b) Same area flooded to the contour line 1,250 feet above sea level. c) Same area flooded to the contour line 1,350 feet above sea level.

Table 10.1. Topographical map of the region shown in Figure 10.1(a) transformed in a matrix of elevation points above sea level.

	Distance East													
D		0	250	500	750	1000	1250	1500	1750	2000	2250	2500	2750	3000
i	0	1300	1300	1350	1300	1300	1450	1450	1300	1200	1200	1200	1200	1350
s	250	1300	1300	1350	1350	1400	1350	1400	1300	1250	1150	1150	1200	1400
t	500	1350	1300	1300	1350	1300	1300	1350	1350	1300	1250	1300	1200	1200
a	750	1350	1350	1300	1300	1300	1300	1350	1350	1250	1300	1300	1300	1250
n	1000	1200	1250	1250	1250	1300	1250	1300	1300	1300	1300	1300	1250	1250
c	1250	1300	1300	1250	1300	1250	1250	1250	1350	1350	1300	1300	1250	1250
e	1500	1350	1350	1300	1300	1300	1300	1250	1300	1300	1350	1350	1350	1250
	1750	1400	1350	1350	1300	1300	1300	1350	1300	1350	1350	1400	1350	1350
S	2000	1300	1400	1350	1250	1300	1250	1300	1300	1300	1350	1400	1300	1350
o	2250	1400	1350	1400	1350	1350	1300	1250	1250	1250	1350	1300	1300	1250
u	2500	1350	1350	1400	1400	1350	1300	1250	1350	1250	1300	1250	1250	1250
t	2750	1400	1400	1350	1350	1350	1350	1250	1350	1300	1200	1200	1250	1250
h	3000	1400	1350	1400	1400	1300	1350	1250	1350	1300	1350	1250	1150	1100

chosen to study there is a river running from the top to the bottom of the figure. We can gain a physical appreciation of the significance of the contour line by imagining the appearance of islands which would be created in a lake formed by building a dam on the river the top of which would be 1,250 ft. above sea level. These islands are shown in Figure 10.1(b). Another set of islands created by flooding to 1,350 feet are shown in Figure 10.1(c). It will be noted that the islands created by the flooding have fractal boundaries.

The geographer who worked out a way of characterizing the fractal dimension of the surface of land covered with canyons first transformed the topographical map into a set of elevation points. The set of elevation points for the area we have studied are shown in *Table 10.1.* If we attempted to measure the surface of every microscopically small grain of sand, soil and that of the blades of grass constituting the real surface of the land represented by the map, we could never finish measuring the area in an finite amount of time. Therefore, we conclude that the actual surface of the ground represented by the topographic map is infinity.

To gain an estimate of the surface of the rough ground, one can imagine that one replaces the actual surface with a series of triangles. The actual technique used here to estimate the surface with triangles is slightly different from that reported by Clarke but the logic procedures are essentially the same. The area of the triangles used to replace the actual surface can be calculated using Pythagoras' theorem from the elevation data. The first step is to calculate the area of the four facets spanning the space defined by the first four points of the elevation matrix. The area of a triangle forming a facet is calculated as described in *Figure 10.2*(a) and shown in Figure 10.2(b). The computer can now repeat this for the whole matrix of points.

a)

L = Resolution Length in m = the side length of the square cell on the base map at the given resolution. RA = Resolved Area = RL × RL. HD = Half the Diagonal of the square cell on the base map = $RL/\sqrt{2}$. a, b, c, d = the elevations at the four corners of the cell. e = the elevation at the center of the cell = (a + b + c + d)/4.

For the facet 1:

$$A = (RL^2 + (b - a)^2)^{1/2}$$
$$E = (HD^2 + (e - b)^2)^{1/2}$$
$$H = (HD^2 + (a - b)^2)^{1/2}$$

The area of the facet abe having sides of length A, E and H is given by:

$$FA1 = (s1 \times (s1 - A) \times (s1 - E) \times (s1 - H))^{1/2}$$

where: $s1 = \frac{1}{2}(A + E + H)$.
The calculation is carried out in a similar way for the three remaining facets and the surface area of the cell is: $SC_1 = FA1 + FA2 + FA3 + FA4$.
Calculation then proceeds to the next cell to give SC_2 and so on until the entire grid has been calculated at this resolution and the total surface area (TSA) estimate, that is the sum of all the cells at the given resolution, is calculated.

$$TSA_{RA} = \sum_{i=1}^{n} SC_i$$

The base map area (BMA) the map area contained by the grid of elevations, is then subtracted from the total to give the excess surface area (ESA) at the given resolution.

$$ESA_{RA} = TSA_{RA} - BMA$$

The entire series of calculations is then repeated from the next higher resolution. Finally the resolved map area and excess surface area values may then be normalized by dividing them by the base map area and the plotted as normalized excess surface area versus normalized resolved map area on log–log scales. The absolute value of the slope of the best fit line through the data points is then added to 2 to obtain the surface fractal dimension.

Figure 10.2. To calculate the fractal dimension of a rough surface from its topographical map the surface is plated with sets of various sized triangles. Plating the surface with triangles is the two-dimensional equivalent of replacing a rugged line with a set of linear chords. One does not need to do the plating physically. The process can be carried out mathematically using Pythagoras' theorem as illustrated above. a) Procedure for the calculation of the area of a plated triangle. b) Graphical summary of the facet area calculation.

The surface estimate of the area at the resolution represented by the separation of the points in the matrix, RL, is the sum of the area of all the triangles. This surface estimate is the highest resolution estimate of the surface for the matrix of points shown in the diagram. One now reduces the resolution of the inspection of the surface by plating the surface with larger triangles by using the elevations of every other point in the matrix. The resolution parameter is now 2 RL and the triangles with which the surface is plated can be used to calculate a coarser estimate of the surface than when using every point in the matrix. This is because the new larger

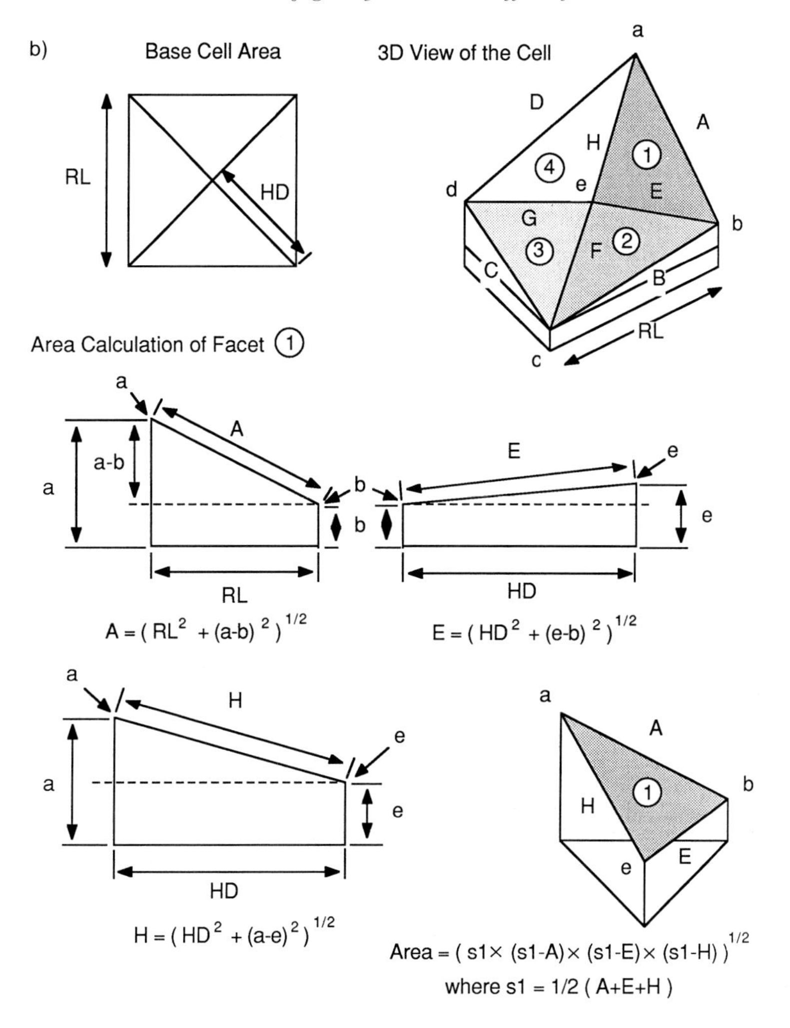

triangles will cut through some of the features which were present in the first estimate using the smaller triangles.

One now proceeds to make coarser and coarser inspections of the surface by using triangles with tips located at every third point in the height matrix then the fifth point etc. One now plots a graph of estimated surface area against the inspection resolution parameter to generate a Richardson plot from which the topographical fractal of the surface can be deduced. In *Figure 10.3* the data leading to the estimate of the topographical fractal dimension of a canyon area as calculated by Clarke is summarized [1]. When we were working with boundary fractals, we discovered that one had to be careful if the roughness of the boundary changed as one went around the perimeter of the profile. In the same way, one must be careful not to average out

a)

Resolved Side Length [m]	Resolved Map Area [m^2]	Excess Surface Area [m^2]
3840	14 745 600	42 988
1920	3 686 400	47 875
960	921 600	64 495
480	230 400	95 409
240	57 600	144 140
120	14 400	179 844
60	3 600	209 516
30	900	224 730

Normalized Resolved Map Area	Normalized Excess Surface Area
0.2246	0.000729
0.0625	0.000817
0.0156	0.001093
0.0039	0.001618
0.000976	0.002444
0.000244	0.003049
0.000061	0.003552
0.0000152	0.003810

These calculations were based on a 256 × 256 grid of elevations spaced at 30 m apart where the excess surface area represents the area remaining after the base map area is subtracted from the calculated surface area. Note that the base map area is given by the calculation:

$$(256 \times 30\ \text{m})^2 = 58\,982\,400\ \text{m}^2.$$

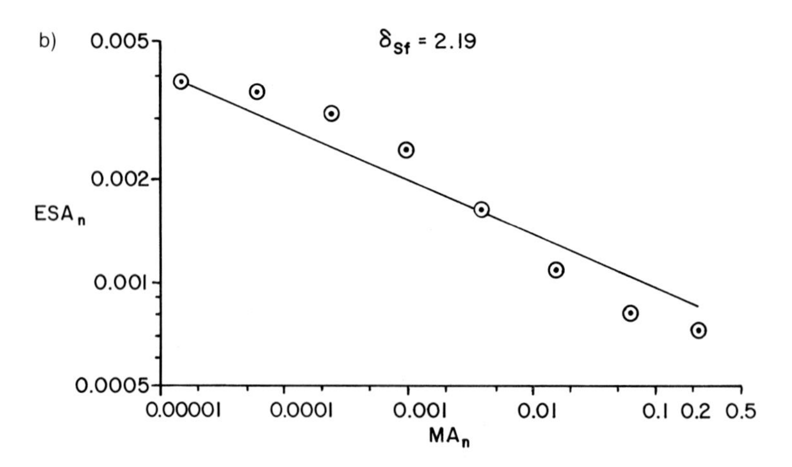

areas of very different roughness when one is looking at the topographical fractal of an area.

Ideally, if one measures the topographical fractal of an area such as that shown in Figure 10.1 (a), one should divide the area into four sections and compute the topographical fractal for each area separately and compare them. If they are very different, then one is amalgamating data from areas of significantly different roughnesses and one must change the experimental strategy. If the roughness of the surface is not changing as one covers the entire area under inspection then usually the topographical fractal dimension should be the same as the fractal dimension of the profile section of the surface plus one. If we take data such as that of the matrix of Table 10.1 and plot it as if it were a digitized set of points from a series of successes of profilometer traces, one would generate the curve representing a section through the area being studied. If we add 1 to the fractal dimension of this "profilometer trace", we can estimate the topographical fractal dimension of the surface. Note that a computer could make a quick check to see if the topography was changing from the top to the bottom of a matrix of points, such as those shown in Table 10.1, by doing a quick check on the linear fractal dimension of the roughness of the profiles represented by the first and last lines of the matrix of points.

Section 10.2. Simulating Passoja's Method for Studying the Roughness of Metal Fractures

In this section we will show how one can use the topographical map of Figure 10.1 (a) to simulate a study of fractured metals carried out by Passoja and co-workers [3, 4]. To characterize the roughness of the surface created when a metal bar was broken, Passoja and co-workers first embedded the fracture surface in an epoxy resin which was hardened by the appropriate treatment. The word *resin* is defined in a dictionary as "a sticky substance which oozes out of a tree". It comes directly from the Latin word that the Romans used to use to describe the sticky globs that came out of the bark of a tree, such as a pine tree, when the bark is damaged or if a twig is broken off from a branch. The dried out hardened resin is called *amber* and is used in the making of jewelry. A scientific dictionary expands the above

Figure 10.3. The surface area estimates of the surface of a map region studied by Clarke [1] at various inspection resolution can be used to deduce the topographical fractal dimension of the map area. a) Data caculated by Clarke for a map area containing a canyon. b) Richardson plot of the plating data of (a). MA_n: Two-dimensional map area representing the resolution of inspection normalized by the total map area of the region. ESA_n: Estimated surface area for the three-dimensional map at the stated resolution, normalized by the total map area of the region. δ_{Sf}: Surface fractal of the map region.

definition of resin with the information:

> *The term was originally applied only to vegetable products but now it is also used to describe any organic or silica organic material characterized by high molecular weight with a gummy or tacky consistency at certain temperatures.*

Modern science uses what is called *epoxy resins*. Their name comes from the fact that they contain a particular type of oxygen bond. Epoxy resins are used to create a hard transparent glassy substance, which looks like amber, by mixing the raw resin with a chemical liquid called a *hardener*. Commercially available packets of epoxy resin and chemical hardener package the two compounds in separate tubes. When the two substances are mixed together, they set within a specified period of time. To make the hardened resin visible when it is injected into a system such as a bed of glass beads, one can add carbon black or white pigments such as titanium dioxide to the resin.

Passoja and co-workers sliced the hardened resin surface they had created very thinly, using the techniques used by geologists to look at rocks. In this way, they created a smooth surface perpendicular to the length of the bar. The surface of the metal – resin composite was then polished gradually down towards the fracture until the first pieces of the rough surface of the fracture started to poke through the surface of the resin. If one looked at the fracture at this stage, the metal tips of the fracture formed tiny islands in the resin sea created by embedding the fracture in the resin.

As one continues to wear away the resin-metal surface, the islands of metal continue to grow as the embedded fracture is worn away. Passoja and co-workers showed that if one measured the perceived boundary of the island at each stage of the polishing of the embedded fractured surface against the area of the metal exposed in the surface that one could measure the fractal dimension of the rough surface by plotting a graph of the perimeter of the exposed islands against the area of the exposed islands on log – log graph paper. To carry out Passoja's studies on a real metal fracture requires expensive equipment and advanced laboratory skills. However, we can simulate Passoja's experiment using a series of floodings of a topographical area as we have shown in Figure 10.1.

If we imagine a very tiny ant surveyor exploring the metal fracture he would come up with a surveyor's map of the fracture in which contours of equal height would be joined to create a topographical map of the fracture. In other words, the human's view of the landscape of the topographical map of Figure 10.1 (a) would be mathematically indistinguishable from the tiny ant's view of a metal fracture. Embedding the metal fracture in the resin would be equivalent to flooding the area of the map of Figure 10.1 (a) until all of the hills were submerged. The polishing of the end of the metal is equivalent to letting the water drain out slowly to create a lake with islands. The repeated polishing of the ends would be equivalent to the changing structure of the islands as they emerge due to the falling water levels as the lake

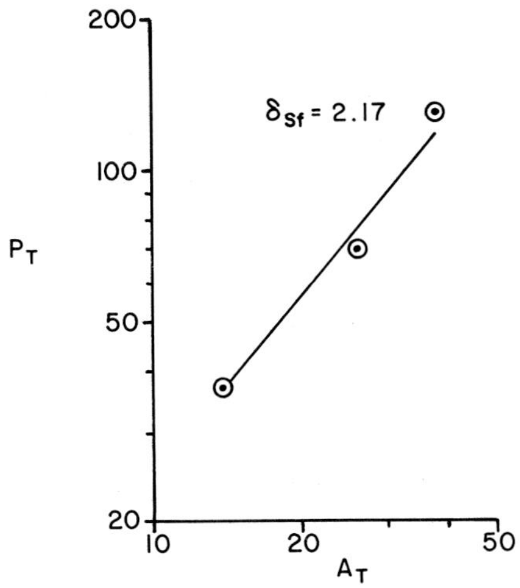

Figure 10.4. The topographic map of a rough land area can be used to simulate Passoja's method of measuring the topographical fractal of the surface of a metal fracture. A sequence of island archipelagos, such as those of Figure 10.1 (b), are created by flooding a larger region of the topographic map of Figure 10.1 (a) to a series of depths. The Passoja plot of total island perimeter versus total island area on log–log graph paper for the sequence of archipelagoes gives an estimate of the fractal dimension of the region. A_T: Total area of the islands in square kilometers. P_T: Total perimeter of the islands in kilometers. δ_{Sf}: Surface fractal of the map region.

drained away. We could regard Figure 10.1 (b, c) as two successive appearances of the polished embedded fracture. In *Figure 10.4* the area and perimeter of all of the islands for a series of simulated sections is shown on the plot of the total perimeter versus the total area of the islands for various flooding depths. The fractal dimension of the rough surface can be deduced from the data in the following manner.

Implicit in Passoja's treatment of the shorelines of the emergent islands is the idea that the fractal dimension of the islands themselves is directly related to the topographical fractal of the surface. The measured fractal dimension of an island of one of the various archipelagos of Figure 10.4 is shown in *Figure 10.5* (this is a structural fractal dimension not textural). It can be seen that the fractal dimension of this island is 1.23 which would yield an estimate of the topographical fractal 2.23 which compares with the value of 2.19 estimated by using the triangular plating technique of Clarke and 2.17 estimated from Passoja's method.

Section 10.3. How many Islands Are There in a Lake?

In Chapter 8, we briefly discussed a possible conversation between Alice in Wonderland and a fractal geometer. At that time Alice was introduced to the problem of

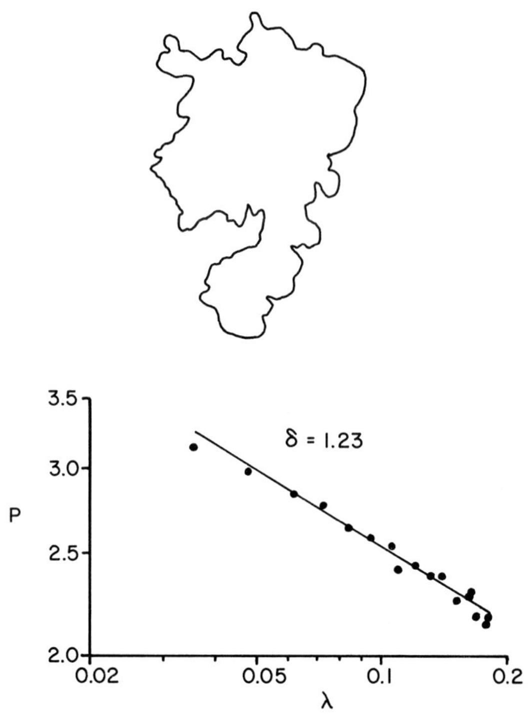

Figure 10.5. The islands of the artificial lakes created by flooding the land of Figure 10.1 (a) have fractal boundaries. The distribution of the fractal dimension of the island boundaries is probably related to topographical fractal dimension of the rough surface being subjected to flooding.

infinite coastlines. One can imagine that the conversation then proceeded to the question:

"How many islands are there in a lake?"

Again Alice would probably insist that the geometer tell her which lake was being discussed. Again the geometer would say that all lakes have the same number of islands – infinite. As so often happens in discussing problems of fractal geometry the paradoxical answer to the question, "how many islands in a lake?", comes from a failure to define what is meant by an island. For example, is a grain of sand lying half submerged on the beach an island? If so then when a ripple of water moves over the sand grains on the edge of a beach the number of islands varies with infinite uncertainty.

If we wish to describe the number of islands in a lake from the perspective of fractal geometry, it is necessary to carefully define what is meant by an island and to focus on some property of a set of islands that can be measured objectively and operationally. When discussing how to describe an island, Mandelbrot points out that (as we discussed in the previous section) if we attempt to measure the surface of each grain of sand, tree and grass on the island, the surface area of any island is infinite. Therefore the infinite number of small islands each have an infinitely large surface.

Mandelbrot points out that we cannot work with the actual, as distinct from perceived, perimeter of the islands since that itself is an infinite quantity. If one wants to work with a finite property of the islands, one has to work with what Mandelbrot has called the *map area* or *projected area* of the island [5]. Since the map area of an island, like the area of a Koch Island, is a finite quantity. The size distribution of the map area of the islands turns out to be a useful and interesting parameter for describing the number of islands in a given archipelago. In 1938, a Czech scientist Korcak suggested that the size distribution of the islands of the world can be described by what the mathematicians call a *hyperbolic function* [6]. It is unfortunate that the word hyperbolic and its close relative hyperbole, have two very different meanings in English. A dictionary defines the adjective *hyperbole* as "a figure of speech that produces a vivid impression by extravagant and obvious exaggeration" and *hyperbola* in mathematics as "a curve which is one of the conic sections."

Both words, hyperbole and hyperbola come from two Greek words hyper meaning "over and beyond" and ballein meaning "to throw". The use of the word hyperbole to describe an exaggerated statement captures the idea that the imagery used in a description has been stretched dramatically a lot further than it should be.

The mathematical use of the term hyperbola comes from the fact that one can regard the curve as being generated by an exaggerated cut through a conical body which matches the curving trajectory of a ball thrown upon the arc and forwards. When we come to Chapter 12 we will need to understand the word *parabola* which is also described in a mathematical dictionary as a *conic section* [8]. The way in which mathematicians have generated names for a set of curves using imagined sections through a cone is sketched in *Figure 10.6*. For the purposes of this discussion, we note that a hyperbolic mathematical function has the general structure:

$$f(x) = \lambda^{\alpha}$$

This can be rewritten in the form

$$\log f(x) = \alpha \log \lambda$$

Hyperbolic functions are discussed in detail in Chapter 11. Korcak suggested that the size distribution of the map areas of the islands of the world were describable by a hyperbolic function in which α was 0.5, when $f(x)$ is the cumulative frequency oversize distribution of the islands and λ, the square root of the map area of the island.

Mandelbrot tells us that he heard about Korcak's suggested empirical relationship in lectures given by Maurice Frechet. In his first book on fractal geometry, Mandelbrot mentions that it, (Korcak's empirical relation), was often mentioned in lectures

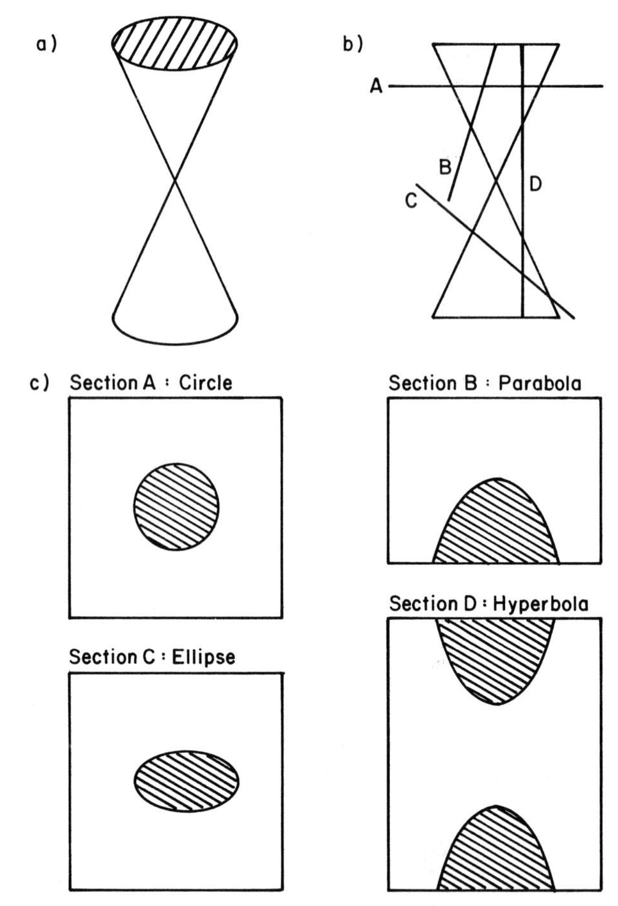

Figure 10.6. Greek mathematicians created names for a family of mathematical curves by studying the curves created by cutting through a double cone. a) Three-dimensional view of a cone. b) Two-dimensioal view of the cone showing sectioning planes. c) Resulting conic sections for the four planes shown in (b).

by Frechet and comments:

> "*It has long been left filed in the back of my mind and I am finally ready to take a first stab at it*" (*the theoretical derivation of the formula*).

Mandelbrot tells us that he was surprised by such a simple relationship and that it was "incredible that the slope of the line is 0.5."

Mandelbrot showed that in fact, the value of α could only be 0.5 if the islands of the world were circular and that theoretically, in general, α should be half the value of boundary fractals of the islands of the archipelago [5, 9].

To understand the implications of Korcak's relationship and Mandelbrot's interpretation of the structure of Korcak's empirical equation let us consider the size

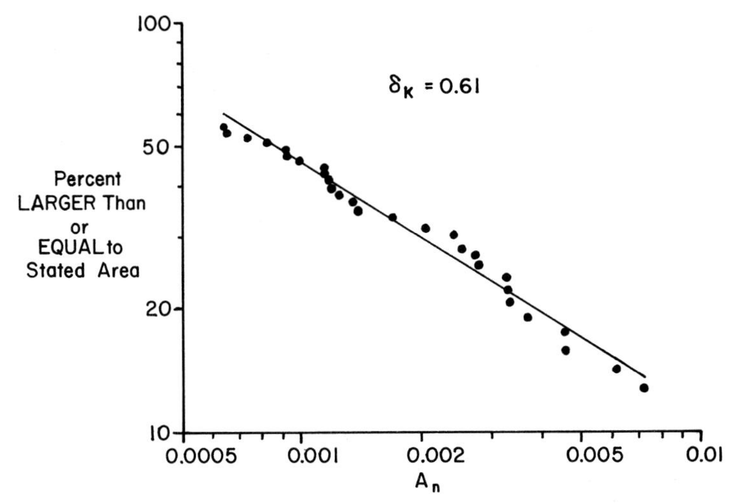

Figure 10.7. The size distribution of the synthetic islands created by flooding of the area of the topographic map of Figure 10.1 (a) can be described reasonably well by a hyperbolic function of the type suggested by Korcak. As predicted by Mandelbrot the slope of the Korcak relationship can be related to the fractal dimensions of the island boundaries. A_n: Island areas normalized by the total map area of the region.

distribution of the various islands of one of the archipelagoes created by flooding the map of Figure 10.1 (a). In *Figure 10.7* the size distribution data is plotted in the cumulative format on log–log graph paper. It can be seen that a straight data line can be drawn through the data points. From Mandelbrot's theory, the slope of this line should be one half of the average fractal dimension of the larger islands (it should be noted that the real boundaries of the islands are infinite and that in this type of discussion and measurement exploration we are making measurements on the perceived boundary at a given level of optical resolution and magnification).

The slope of the data line through the points is 0.61 which compares to the value of 1.23 for the profile of Figure 10.5 and as predicted from the Mandelbrot relationship is approximately one half of the fractal dimension of the boundaries.

Scientists started to interpret the α of hyperbolic functions such as the Korcak–Mandelbrot relationship in terms of fractal geometry. The α of the relationship becomes a fractal dimension in data space. This fractal dimension is no longer related to a physical structure such as the fractal dimension of the topography of an island. Rather, it is a fractal dimension in data space which describes the existence of a scaling function. The Korcak fractal dimension is therefore a measure of dispersion of the islands or the size distribution of the fragments from the impact crushing of a rock. Some scientists have criticized this new name for the scaling function, which existed in classical mechanics for log–log relationships long before the invention of fractal geometry, as a gimmick or a pointless shift in vocabulary. Whether one regards α as a fractal dimension or not is a matter of choice and perspective.

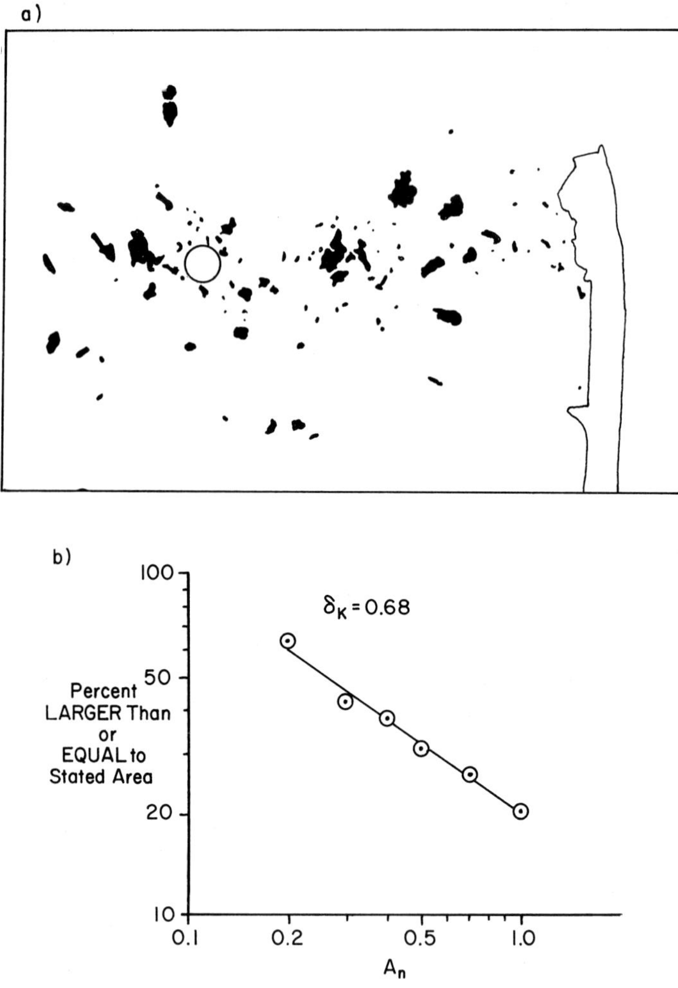

Figure 10.8. The fragments produced by the ballistic shattering of a piece of material can be described by means of the Korcak–Mandelbrot hyperbolic function (see page 352 of Reference 7).

In later sections of this book, we shall encounter several systems which are describable by a fractal dimension in data space. Extending the concepts of fractal geometry to data space is similar to the idea of describing the structure of information in systems which require nine parameters as constituting nine-dimensional data space. If we consider the energy distribution in recorded earthquakes, we will find that they can be described by a scaling function and thus with a fractal dimension. More conservative geologists may prefer to avoid fractal geometry by referring to the slope of the distribution function as a parameter rather than a fractal dimension. The choice is theirs as it is for specialists in other fields of endeavor.

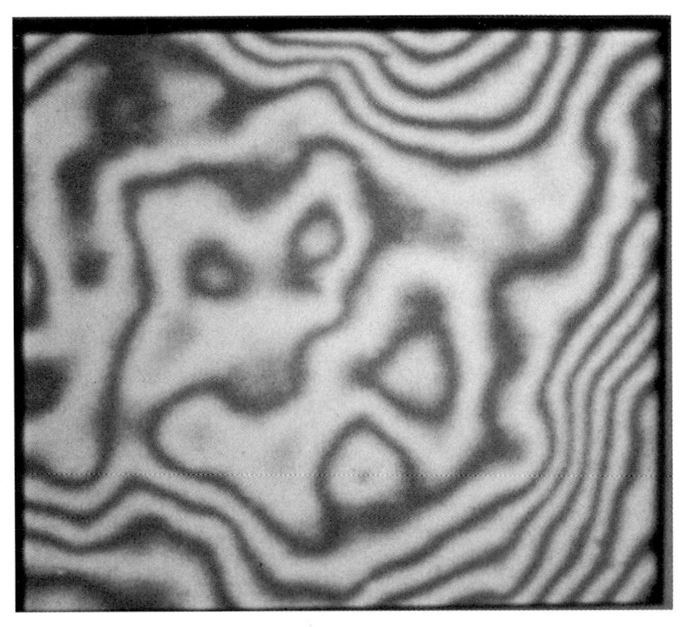

Figure 10.9. Newton's fringes (named after Newton who first studied such systems) map changes in air-gap thickness between two surfaces in contact. The fractal dimensions of the fringe boundaries may be a useful technique for describing the departure from ideal smoothness for the surface being inspected by the technique. Photo supplied by Zygo Corporation, Laurel Brook Road, P.O. Box AAB, Middlefield CT 06455-0448.

The Korcak–Mandelbrot relationship suggests a possible technique for measuring the topographical dimension of a surface which avoids the measuring of the structure of any of the boundaries of a system. If we were to take a fractured metal and embed it in epoxy resin and polish the resulting system following the procedure of Passoja then by measuring the size distribution of the islands we could deduce the topographical fractal dimension of the surface by multiplying the slope of the Korcak data line by 2 and adding one. This may seem to be a round about method to arrive at a parameter describing the surface but in practice many commercial computerized systems for analyzing images are already equipped to measure the size distribution function of the areas of the islands and this indirect method of evaluating the topographical fractal of a surface requires much less computer processing time and memory than the measurement of the surface using the data points derived from the topographic map using present processing technology.

In *Figure 10.8*(a) a sketch based on a high-speed photograph of the disintegration of a piece of material when it is the target of a high speed metal ball is shown. From the graph of Figure 10.8(b) it has been seen that the size distribution of the fragments in Figure 10.8(a) are describable by the Korcak-type relationship with a slope of 0.68 [7].

A widely used technique for investigating the smoothness of a surface, such as a piece of polished glass is to place a reference block of glass which is known to be very smooth on top of the surface to be evaluated. Next, using monochromatic light (*monochromatic* means of "one color", that is having light waves of one wavelength) from a laser, or other suitable source of light, one creates interference fringes modulated by the air gap between the standard glass block and the surface being investigated. These interference fringes are contours tracking increases in the air gap between the two surfaces. A set of such fringes is shown in *Figure 10.9*. Therefore, it is possible that the fractal dimension of the contour lines in the interference patterns can be used to deduce the topographical fractal dimensions of the surface being investigated. A full discussion of the concepts involved in this exploration of the fractal dimension of an almost smooth surface is beyond the scope of the book but the possibility of such a method is mentioned briefly here for the sake of completeness [10].

Exercises

Exercise 10.1.

The teacher can probably think of endless variations on the examples given in this chapter of the transformations of topographical maps into a data format for calculating topographical fractals of the surface of the land. The students should be encouraged to find a topographical map which can be used to create islands by flooding the map to a given contour line. The fractal dimension of the boundary of the islands should be measured and if possible compared to the value created by using the triangular plating technique or at least to the profilometer fractal boundary created by mapping the heights of the land in a linear manner.

The students can also select archipelago pictures and whilst some students measure the fractal dimension of the boundaries of the island, the others can measure the size distribution of the islands by the various techniques suggested in earlier chapters and the Korcak fractal can be plotted. The connection between the slope of the Korcak fractal and the fractal dimension estimated directly from the contours of the land may not be closely related and the students could discuss how in essence the boundaries of the islands in a real archipelago may have a different roughness from the topographical fractal of the island since the erosion of mountains and the sedimentation deposits in the river valleys could considerably reduce the fractal dimension of the surface structure. Sea erosion around the boundaries of the island could interfere with the expected data pattern. One might expect a better agreement between the islands created artificially by simulating the flooding of an area to a given contour line and the topographical fractal of the area subjected to simulating flooding since the erosive forces have been at work on the contour lines as well as on the general topography of the area. The molding forces are similar in nature whereas the difference between the weathering of the surface of the island and the boundary of a real island in the sea could be created by quite a different set of forces.

Even if complete agreement is not achieved in the different studies, the students will gain considerable experience in the use of log–log graph paper, size distribution measurements and boundary studies. Their attention will also be focused on the erosive forces that create land features of a geographical area. Students at Laurentian have generated interesting data by making measurements on the Orkney Islands, the Shetland Islands and the islands between Denmark and Norway.

Exercise 10.2. Studying the Size Distribution of Fragments Created by the Pulverization of Rock

If one takes a piece of relatively soft rock one can wrap it carefully in a piece of cloth and subject it to a single hard blow from a hammer. By wrapping the piece of rock in a piece of cloth one avoids the problem of flying fragments. In any case, the student should always wear safety glasses if they undertake this experiment. One could use a piece of brick in this experiment.

After the single blow has been applied to the piece of brick, the cloth can be unfolded and the larger fragments picked up and placed on a white card. The size distribution of the fragments can be assessed by weighing the fragments and the Korcak-type plot used to look at the size distribution of the fragments. In some cases, the fragments will have fractal boundaries in which case the students can carry out measurements on the projected boundaries. They could also place the larger fragments in a piece of resin which, after it is cured, can be cut in half using the appropriate equipment. Many schools will not be equipped to carry out this experiment and they could seek the cooperation of a local industrial operation that handles rocks or hard materials. This experiment should not be carried out unless it is closely supervised by a teacher.

References

[1] K. C. Clarke, "Computation of the Fractal Dimension of Topographic Surfaces Using The Triangular Prism Surface Area Method", *Computers and Geosciences 12* (1986) 713–722.

[2] Topographical map used for those studies is a 1 : 50 000 scale map of the Chelmsford area designated 41 1/11 W edition 2, ASC series A, 751 printed 1962.

[3] B. B. Mandelbrot, D. E. Passoja, A. J. Paullay, "Fractal Character of Fracture Surfaces of Metals", *Nature 308* (1984) 721–722.

[4] B. B. Mandelbrot, D. E. Passoja, A. J. Paullay, "Fractal Character of Fracture Surfaces of Metals", in B. B. Mandelbrot and D. E. Passoja (Eds:) *Extended abstracts of the Meeting on Fractal Aspects of Materials: "Metal and Catalyst Surfaces, Powders and Aggregates*, Materials Research Society, Pittsburgh, 1984, p. 7–9.

[5] See discussion in which B. B. Mandelbrot first described his theories of fractal geometry in his book entitled *Fractals Form Chance and Dimension*, W. Freeman, San Francisco 1977.

[6] Korcak's work is not readily accessible in English. However Mandelbrot has given a good review of his work in English in B. B. Mandelbrot, *Fractals Form Chance and Dimension*, W. Freeman, San Francisco 1977.

[7] See Chapter 9 of B. H. Kaye, *A Random Walk Through Fractal Dimensions*, VCH, Weinheim 1989.

[8] See discussion in I. Asimov, *Biographical Encyclopedia of Science and Technology*, Doubleday and Company Inc., Garden City, New York 1972.

[9] For an application of the Korcak–Mandelbrot relationship to island archipelagos see Chapter 9 in B. H. Kaye, *A Random Walk Through Fractal Dimensions*, VCH, Weinheim 1989.

[10] The Newton's fringe pictures of Figure 10.9 are taken from the commercial literature of Zygo Corporation, Laurel Brook Road, P.O. Box 448, Middlefield, CT 06455-0448, U.S.A.

Chapter 11

Zipf's Law
and the
Surprising Patterns of Word Occurrences

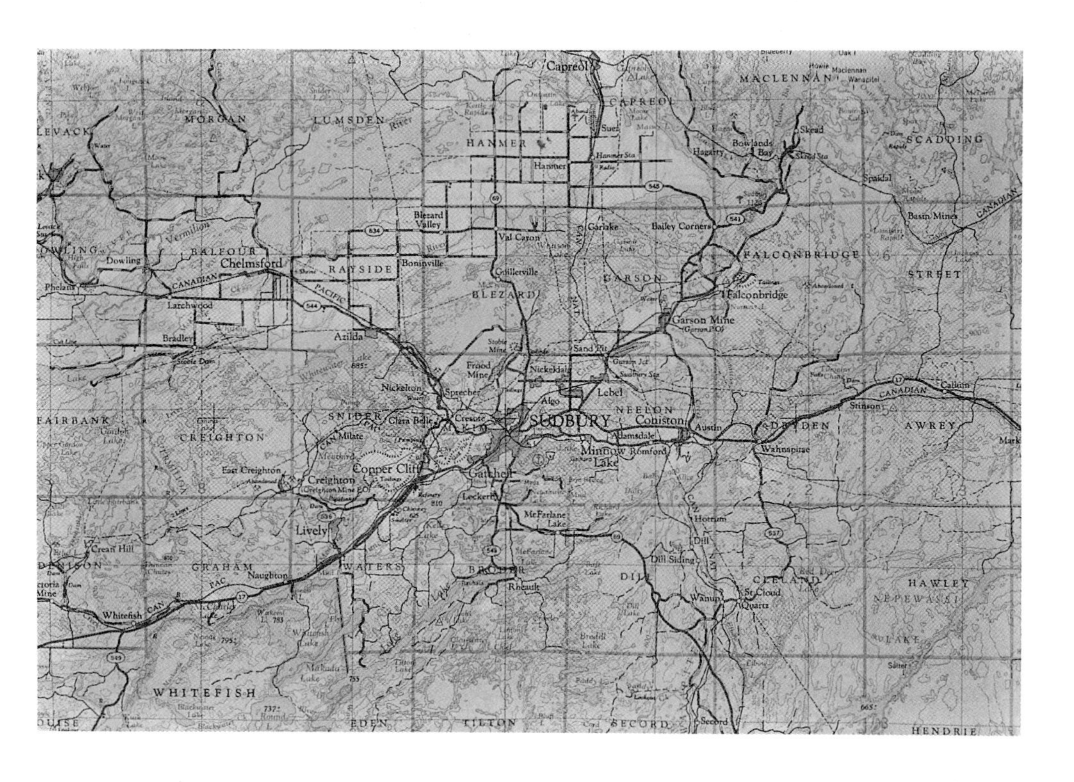

As you Zipf across Canada you will find that the size of cities fits a surprising law

Chapter 11 Zipf's Law and the Surprising Patterns of Word Occurrences 437

Section 11.1 Hyperbolic Word Frequencies 439
Section 11.2 The Size Distribution of Population Centers 443
Section 11.3 Beware of Procrustean Thinking 445
Section 11.4 Levy Flights – A Generalized Theory of Brownian Motion .. 450
References 461

Chapter 11

Zipf's Law and the Surprising Patterns of Word Occurrences

Section 11.1. Hyperbolic Word Frequencies

If one were to take a piece of written text and count the frequency of the words in the piece of text, would one expect a pattern to be discovered in the frequencies in which the different words were used? Mandelbrot discusses the possibility that a pattern existed in such word frequencies and states that:

> "*One might expect any (discovered) relationship to vary widely according to the language being used and the speaker and/or writer.*"

It appears that the first person to study the frequency of word usage in a written text was George Kingsley *Zipf* (1902–1950). Zipf started his scholarly career as a philologist. *Philology* is defined in the dictionary as "the science of language which concerns itself with the sounds of speech, the history of sound changes and the origin of words". The term comes from two Greek root words philos meaning "loving" and logos, one meaning of which is "words".

Zipf lectured at Harvard University for 20 years. Towards the end of his career he preferred to call himself a "statistical human ecologist". The word *ecology* comes from two Greek root words oikis meaning "home or house" and logos, which can mean "to study". The dictionary definition of ecology is "the study of the relations of organisms – animate and inanimate – to their environment both physical and biological."

In Chapter 12, when we study the fluctuations in the populations of animals in a given environment, we will be looking at the *ecology of populations*. Perhaps, a simple definition of *human ecology* is that it is the field of study concerned with the exploration of the interaction of the human animal with its environment and co-inhabiters. Zipf tackled such a study using the mathematical discipline of statistics. Zipf collected all of his ideas together and published them at his own expense in a book "Human Behavior and the Principle of Least Effort" [1] (the hint in the title is that human beings, like students, will try to find the path of least resistance to their ultimate goal. Why choose a difficult course if one can take a simpler study with higher marks?).

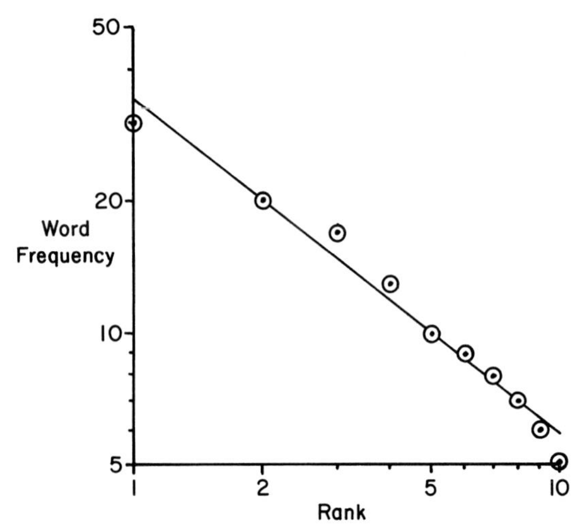

Figure 11.1. Zipf's law of word frequency states that $F = (1/R)^{\alpha}$ where F = frequency, R = rank and α = a constant.

In his study of word frequencies, Zipf showed that if one plots the frequency of a word against the rank of that word on log–log graph paper one obtains a straight line relationship as illustrated by the word frequencies of *Figure 11.1* [2]. This relationship is now known as *Zipf's law* of word frequencies. Using the language we have developed earlier in this book, we see that the frequency with which a word occurs in a written text is a scaling property and that the relationship linking the information is a hyperbolic function similar in structure to the one that we discovered in Chapter 10 for the size distribution of islands (Korcak's relationship). Studies have shown that the same basic relationship describes word frequencies used by different writers and for word frequencies in different languages [3].

In *Figure 11.2*(a), a word frequency study based on a piece of written French is compared to word frequency studies for two different authors writing in English (Figure 11.2(b)). It may be that the slope of the line is characteristic of the vocabulary used by a specific author (see Exercise 11.1) [4–6].

Mandelbrot tells us that Zipf's law of word frequencies played an important role in the thought chain that led to his development of fractal geometry [7]. He tells us that his lifetime involvement in studying scaling phenomena was triggered in 1951 by a casual interest in Zipf's law, saying "this empirical regularity concerning word frequencies had come to my attention through a book review".

Mandelbrot, when a student, in a search for "light reading on the Paris subway", had retrieved the book review from a "pure mathematicians" waste basket. Mandelbrot reports that when he read the book review, the relationships seemed too simple to be true. For a time Mandelbrot was heavily involved in studying word frequencies in the hope that such a study would lead to more efficient methods of sending

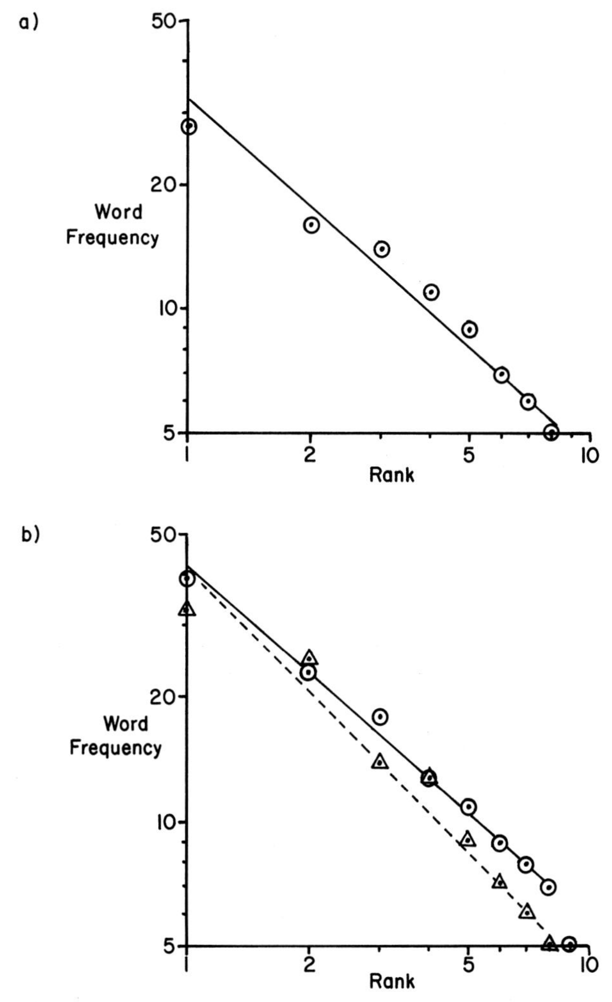

Figure 11.2. Zipf's law appears to describe the frequency of word occurrence in different languages and as used by different scholars writing in the same language. a) Word frequency versus rank distribution for a portion of written French [4]. b) Word frequencies of two different selections of written English [5, 6].

messages over communication systems but in his own words "The birth of mathematical linguistics was helped by his work but the study of word frequencies was a self-terminating enterprise."

Mandelbrot says that through his study of Zipf's law "I became sensitive to analytical empirical regularities in diverse fields beginning with economics. Though astonishingly numerous, these regularities were viewed as of little consequence to the established field of study".

Although Zipf's law served as a stimulus to his later thinking, Mandelbrot has this to say about Zipf's book:

> *"This is one of those books in which flashes of genius projected in many directions is nearly overwhelmed by a gangue* of wild notions and extravagance. On the one hand, it deals with the shape of sexual organs and justifies the 'Anschluss' of Austria into Germany because it improved the fit of a mathematical formula. On the other hand, it is filled with figures and tables that hammer away ceaselessly at the empirical law that in social science statistics, the best combination of mathematical convenience and empirical fit is often given by a scaling probability distribution" (scaling probability distribution is Mandelbrot's phrase for a hyperbolic relationship, such as Zipf's law and Korcak's relationship).*

* *Gangue* is a term from mineral processing and describes the worthless material in which a valuable ore is embedded. Rock tailings released when freeing nickel ore from its rock would be described by the mining engineer as gangue. When Mandelbrot refers to the Anschluss of Nazi Germany or Austria he is referring to the invasion by Nazis of Austria in 1939 and its political annexation to Germany.

Mandelbrot has some interesting comments on how *social scientists* (people such as economists, sociologists and geographers) viewed the hyperbolic laws that Zipf insisted could be used to describe many phenomena of interest to the sociologist etc.

> *"Natural scientists (physicists, chemists etc.) recognize in Zipf's law, the counterparts of the scaling laws which physics and astronomy accept with no extraordinary emotion. When evidence points out their validity therefore physicists find it hard to imagine the fierceness of the opposition when Zipf, and Peratoe before him, followed the same procedure with the same outcome in the social sciences. The most diverse attempts continue to be made to discredit in advance all evidence based on the use of doubly logarithmic graphs (scaling functions). But I think that this method would have remained uncontroversial were it not for the nature of the conclusion to which it leads. Unfortunately, a straight doubly logarithmic graph indicates a distribution that flies in the face of the Gaussian dogma which long ruled uncontested (in the social sciences). The failure of applied statisticians and social scientists to listen to Zipf helps account for the striking backwardness of their fields. Zipf brought encyclopedic fervor to collecting examples of hyperbolic laws in social sciences and unyielding stamina to defending his finding and analytical findings by others".*

The review of Zipf's book that Mandelbrot read on the Paris subway was written by the Mathematician J.L. Walsh. Commenting on this review Mandelbrot states:

> *"By only mentioning what was good, this review influenced greatly my early scientific work. Its indirect influence continues. Therefore I owe a great deal to Zipf through Walsh. Although Zipf's influence (on the main body of science) is likely to remain marginal, one sees in him in clearest fashion the extraordinary difficulties that surround any interdisciplinary approach to scientific problems".*

Recently, West and Shlesinger have discussed a number of phenomena which can be described by hyperbolic functions. They show that, very often, the small number of events in the tail end of a distribution, the main portion of which can be describable by functions such as the Gaussian probability function, become hyperbolically distributed and are describable by relationships such as Zipf's law [8]. A full discussion of this structure of the tails of populations is beyond the scope of this study but interested readers will find a good introduction to the topic both in the publication by West and Shlesinger and in an earlier publication by Montroll and Shlesinger [9].

Section 11.2. The Size Distribution of Population Centers

After my original study of Mandelbrot's work in this area, my second encounter with Zipf's law occurred when a student came to me for help in drawing a line on graphs being prepared for a geography assignment. The project involved exploring the relationship between the population of a city and its rank in a listing of the population of the various cities of Canada. The table which the student had been given is shown in *Table 11.1*. The laboratory script which had been given to the student stated that the rank size rule is an empirical rather than a theoretical observation based on actual city population sizes and their ranking. The students were told that if they plotted the population of the city against its rank on log–log graph paper, they would obtain a straight line relationship, a fact which was first noted by Felix *Auebarch* in 1930 but which had been developed and popularized by George K. Zipf. The laboratory script went on to say that:

> "*Application of the rank–size rule has shown some intriguing empirical regularities in many parts of the world especially for the larger industrialized countries. The use of Zipf's relationship provides a means of reflecting differences amongst various urban hierarchies and serves as a technique for deriving estimates for the population of the city and assists in making generalizations concerning population distributions*".

The major difficulty which the student was having was concerned with the drawing of the best straight line through the data as displayed in the graph reproduced here as *Figure 11.3*. I suggested to the student that in fact the data could best be described by drawing two data lines which met at the data point representing Edmonton as shown in Figure 11.3. I said that this so called *bimodal* (i.e. two-data-lines interpretation of the data) would probably indicate that some different forces were at work to create the major cities of the country as compared to the smaller population centers. I also pointed out that Montréal and Toronto were about the same size and that perhaps the population pattern for Québec might be different

Table 11.1. Populations of the 50 largest cities in Canada.

Rank	City	Population (1971)	Rank	City	Population (1971)
1	Montréal	2743208	26	Kingston	85877
2	Toronto	2628043	27	Sherbrooke	84570
3	Vancouver	1082352	28	Sault St. Marie	81270
4	Ottawa-Hull	602510	29	Brantford	80284
5	Winnepeg	540262	30	Sarnia	78444
6	Hamilton	498523	31	Moncton	71416
7	Edmonton	495702	32	Peterborough	63531
8	Québec	480502	33	Guelph	62659
9	Calgary	403319	34	Shawinigan	57246
10	St. Catherines	303429	35	North Bay	49187
11	London	286011	36	Prince George	49100
12	Windsor	258643	37	Cornwall	47116
13	Kitchener	226846	38	St. Jean	47044
14	Halifax	222637	39	Drummondville	46524
15	Victoria	195800	40	Kamloops	43790
16	Sudbury	155424	41	Timmins	41473
17	Regina	140734	42	Lethbridge	41217
18	Chicoutimi	133703	43	St. Hyacinthe	39693
19	St. John's	131814	44	Granby	39307
20	Saskatoon	126449	45	Naniamo	38760
21	Oshawa	120318	46	Barrie	38176
22	Thunder Bay	112093	47	Fredricton	37684
23	Saint John	106744	48	Valleyfield	37430
24	Trois Rivières	97930	49	Kelowna	36956
25	Sydney	91162	50	St. Jerome	35335

from the rest of Canada because Montréal was essentially a French speaking city and that early settlement in Québec followed different cultural patterns than in Ontario.

The student seemed quite impressed with my analysis and turned the graph in with the two lines shown in the solid format in Figure 11.3. The student assistant grading the geography project was unimpressed by the analysis presented and gave me, via the student, a failing grade. He then corrected the student's efforts by drawing in the dotted line shown. Incidently, this dotted line shows that Canada should have at least one city with a population of 7 million which is hard to imagine when the total population of Canada was only of the order of 26 million at the time that the population data was assembled.

If one looks at the data used to generate Figure 11.3, we see that one difficulty in applying Zipf's Law to population centers is defining exactly what constitutes a city. For example, the data for Sudbury in the table is a population figure for the greater Sudbury area. Sudbury the city proper, has a population of one hundred thousand.

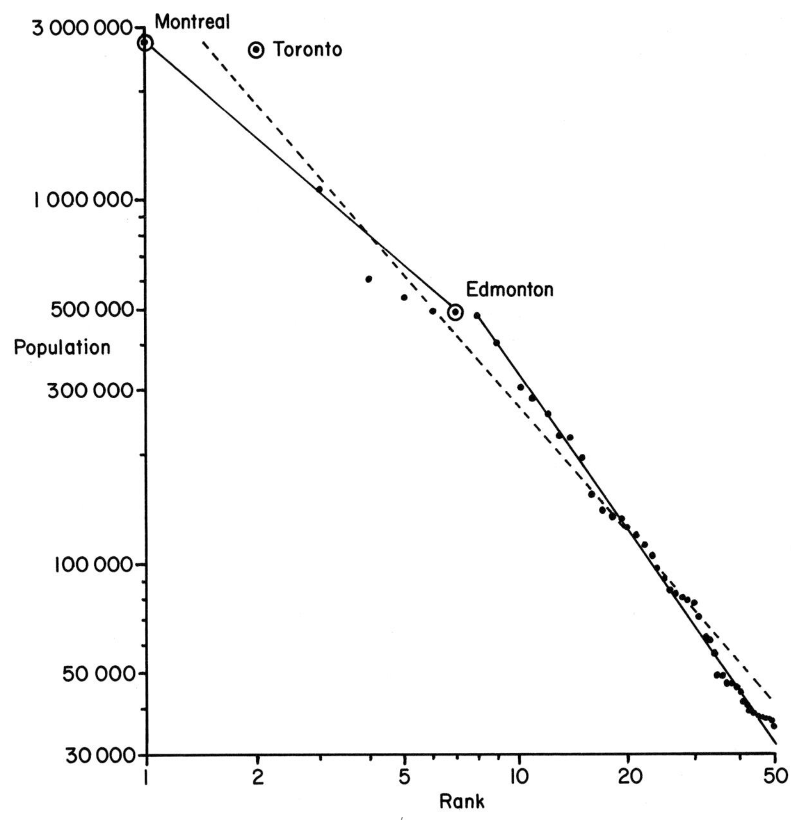

Figure 11.3. Can one draw one or two data lines through the logarithmic plot of city size versus rank for the 50 largest Canadian population centers?

Many large cities have contiguous suburban areas which are divided from the main city for political reasons and not for any natural geographic consideration. If one were to use a listing of populations of urban areas from the back of an atlas, one would have to be very careful about how the listing treated contiguous suburban areas. A question arises whether Zipf's law applies to the overall urban area as a unit rather than the political subdivisions.

Section 11.3. Beware of Procrustean Thinking

To console the student when reviewing the grade assigned by the marking assistant for the geography class, I pointed out that student assistants are not there to think about the assignment and that the person who set the assignment and the marking scheme was guilty of procrustean thinking. *Procrustean* is defined in a dictionary as "the achievement of uniformity by violent means". In Greek the word literally means "the *stretcher*". In a procrustean approach to a theory, the scientist stretches and pulverizes any data until it fits the preconceived ideas. The idea of achieving uniformity by violent means goes back to a story told by the Greeks about an individual called Procrustes who ran a hospitality center outside of Athens. Travellers could stay free at this hospitality center but they had to fit the bed that Procrustes provided. If people were too tall, part of their heads were chopped off. If they were too short, they were stretched until they were the same length as the bed.

To demonstrate how even experienced scientists can be seduced into procrustean thinking because of a lack of awareness of the insensitivity of logarithmic and cumulative plot scales, we will briefly discuss a size distribution function frequently used by a fine-particle scientist to describe systems as different as pulverized coal and the droplet sizes produced by the breakup of a jet of water. This distribution function is known as the *Rosin–Rammler* distribution function [10, 11]. Although some of the readers may find this distribution function rather complicated, a discussion of its usage is included in this chapter because many fine-particle specialists use it in their studies and are often unaware of how insensitive their data plots are when processing experimental information.

The basic structure of the Rosin–Rammler function is shown in *Table 11.2*. The original work was carried out by the two German scientists who were interested in the size distribution of crushed coal. To estimate this size distribution, they placed a sample of the powder coal at the top of the set of sieves stacked one on top of the other (a system known as a *nest of sieves*). The engineers who developed such techniques for characterizing powders wove different sized wire cloths and the set of sieves had apertures that decreased down the nest in the way shown in Table 11.2. The mesh number is the number of wires per inch in the woven wire surface. The engineer would shake the powdered coal through the set of sieves and after a given sieving time would assume that the fractionation process was complete [12]. The weight of powdered coal remaining on the surface of each sieve was determined to create a set of numbers as shown in Table 11.2. This is then converted into a "percentage oversize" distribution giving the weight of coal greater than or equal to a given size. This weight of coal greater than or equal to a given size is the variable P in the Rosin–Rammler equation. The other terms in the Rosin–Rammler distribution are e, the magic number that we met in Section 1.5, and d the size of the fine particles. b and n are two constants.

The Rosin–Rammler equation is usually transformed into a different format by taking logarithms of each side of the equation twice as shown in Table 11.2. As a result of this double transformation, it is suggested that if one were to plot the logarithm of a 100/P against d on log–log graph paper, one would have a line of slope n. When P equals 36.8% there would be a value of a that is a characteristic

Table 11.2. The Rosin–Rammler function has been widely used to describe size distribution of pulverized coal and other finely divided material.

Mesh Number	Aperture Size [μm]	Sieve Residue W [gm]	Cumulative Weight ΣW [gm]	Percent Oversize % ≥
270	53	30	455	100.0
230	63	16	425	93.4
200	75	42	409	89.9
170	90	65	367	80.7
140	106	22	302	66.4
120	125	82	280	61.5
100	150	75	198	43.5
80	180	65	123	27.0
70	212	52	58	12.7
60	250	6	6	1.3

The Rosin–Rammler function: $P = 100\,e^{-bd^n}$

where: d: is the size of the fine particle. e: is the exponential function and e ≈ 2.718. P: the percentage of material greater than or equal to the stated size. b and n: constants.

The equation may also be written: $e^{-bd^n} = \frac{100}{P}$ and taking the log we get: $bd^n = \log\left(\frac{100}{P}\right)$ taking the log a second time:

$$n\log(bd) = \log\left(\log\left(\frac{100}{P}\right)\right)$$

$$n(\log b + \log d) = \log\left(\log\left(\frac{100}{P}\right)\right)$$

$$a + n\log d = \log\left(\log\left(\frac{100}{P}\right)\right)$$

where $a = n\log b$ is a constant since b is a constant.

parameter describing a given powder. To facilitate the exploration of the possibility that a powder could be described by the Rosin–Rammler distribution, scientists have developed a special graph paper of the form shown in *Figure 11.4*(a) [11]. One can plot the percentage of the material greater than a given size directly on this graph paper. The data from Table 11.2 has been plotted on this special Rosin–Rammler graph paper to generate the data line of Figure 11.4(b). This data line demonstrates the insensitivity of the Rosin–Rammler relationship since the data of Table 11.2 is not an actual set of data from a sieving experiment but is a simulation of the weight residue on a nest of sieves generated using a random number table. Even when the weight of powder on the various members of the nest of sieves are varying completely at random, the data appears to fit the Rosin–Rammler distribution function.

In *Figure 11.5*(a) a random number table is reproduced. The lines drawn around the columns show how one can simulate the weight residues on a nest of sieves and

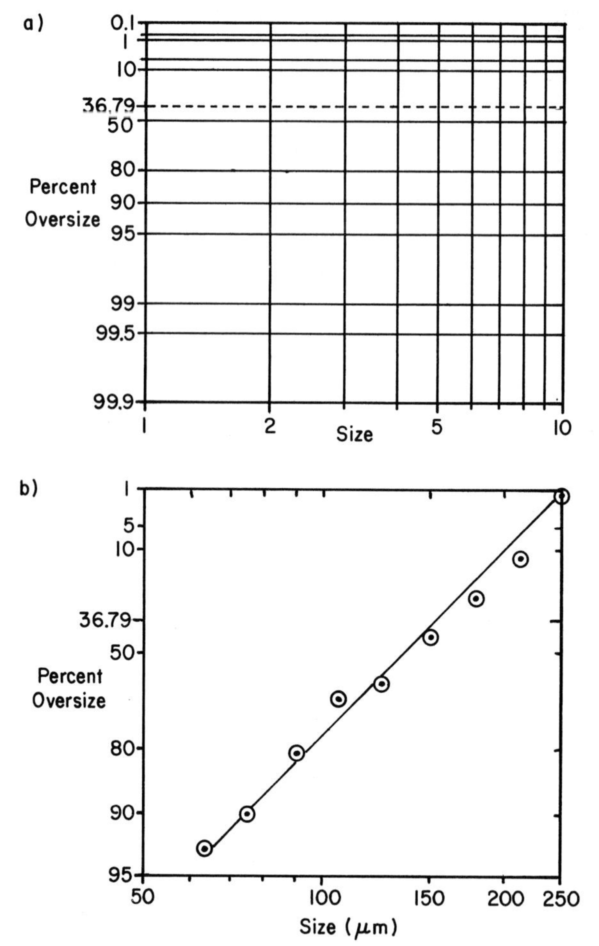

Figure 11.4. Typical plot of size distribution data on Rosin–Rammler graph paper. a) Simplified version of Rosin–Rammler graph paper. b) Data from Table 11.2 plotted on Rosin–Rammler graph paper.

how one can convert such data to simulate the size distribution of a powdered material. The simulated blocks of data from Figure 11.5(a) are plotted on the Rosin–Rammler paper in Figure 11.5(b–e). As it can be seen again, most of the data simulation studies, according to the judgement of an inexperienced scientist, are describable using the Rosin–Rammler distribution. The data of Figure 11.5 indicate how cautious one should be when using complicated log–log function to transform data because of the fact that we usually lack experience when attempting to interpret scatter of data points on such complicated graph paper.

Beginning in the late 1970s, scientists interested in powders began to characterize suspended fine particles in a liquid by studying the diffraction pattern created when laser light was passed through a suspension or cloud of the fine particles [13]. A full

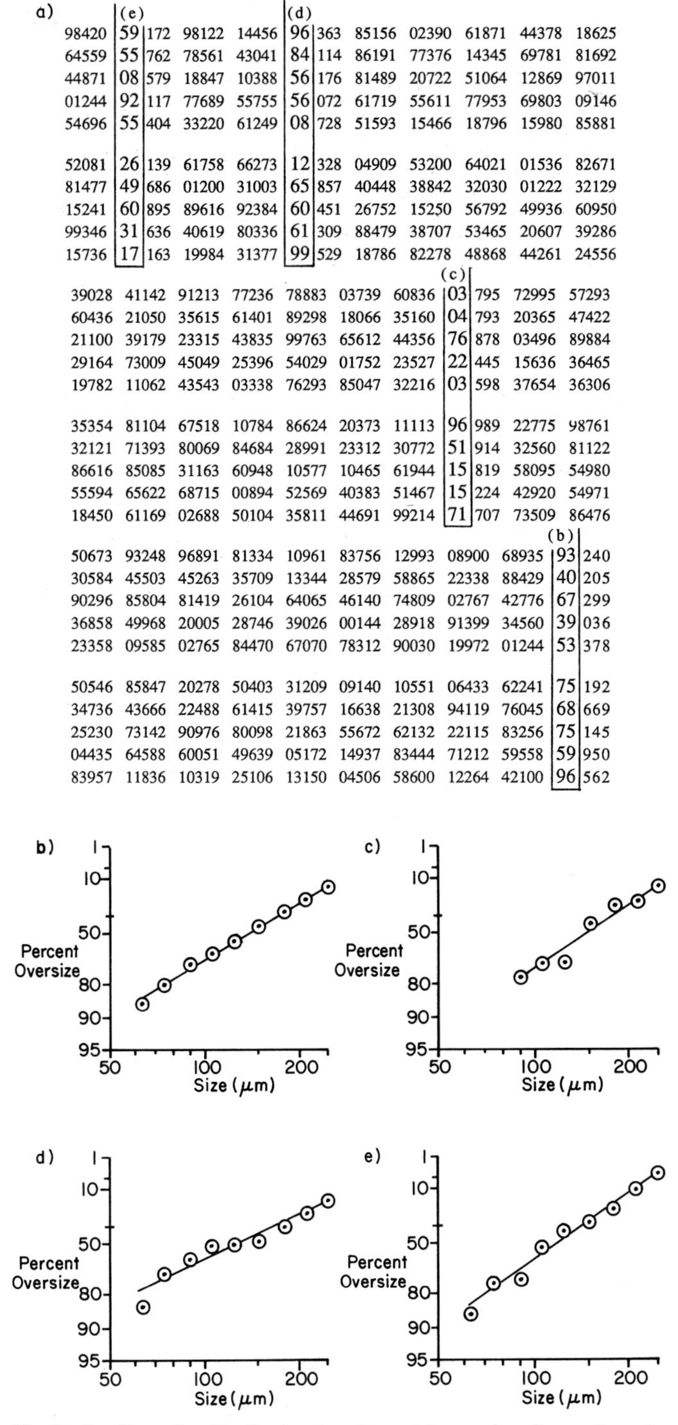

Figure 11.5. The Rosin–Rammler distribution is so insensitive to data fluctuations that even randomly fluctuating sieve residue data appears to fit the distribution function. a) Simulated data for the powder retained by a nest of sieves. b–e) Rosin–Rammler plots for the simulated nest sieves data from Part (a).

discussion of these methods is beyond the scope of this book but the interesting thing to note is, that because of the difficulties in transforming the diffraction pattern into a size distribution, the instruction manuals of many of these instruments contain a statement (in small print) that to transform the data of the diffraction pattern into a size distribution function, it is assumed that the size of the fine particles in suspension being inspected by laser light are describable by the Rosin–Rammler distribution. Not surprisingly much of the data generated by such instruments therefore are describable by the Rosin–Rammler distribution.

The evaluation of the validity of data from any particular diffractometer is difficult because the manufacturers do not share with their customers the details of the computer software programs built into their instruments. Their computer programs are described as the *software* of the computer systems. I feel that if a manufacturer is using the Rosin–Rammler distribution to interpret the data from a diffractometer the instrument should bear a sticker saying:

"Procrustes is alive and well and busy in your instrument"

It should be noted that one of the books describing methods for characterizing fine particles makes the statement: "Agreement with the Rosin–Rammler distribution is often deceptive since even large discrepancies can be made to appear small by the act of taking logarithms twice" [14].

Section 11.4. Levy Flights – A Generalized Theory of Brownian Motion

In Section 1.8 we discussed the irregular motion of small fine particles such as pollen and paint pigments when viewed through the microscope, a motion known as Brownian motion. In that discussion it was pointed out that Brownian motion arises from the uneven molecular bombardment of small fine particles in suspension and that Brownian motion consists of a set of random steps which never become a continuous function no matter how intensely one scrutinizes the zig-zag movement of the pigment fine particle undergoing motion

In our earlier discussion of Brownian motion, we showed how the average displacement after n random steps all of the same size L was L $\sqrt{n}$. However in that discussion, we did not discuss the variations in the value of the displacement vector after a given number of steps.

From a consideration of the fact that the longer average displacements in a random walk require a favorable combination of the random steps one could anticipate (following the reasoning set out in Chapter 4) that the distribution function of the displacement vectors from a series of random walks of the same number of steps

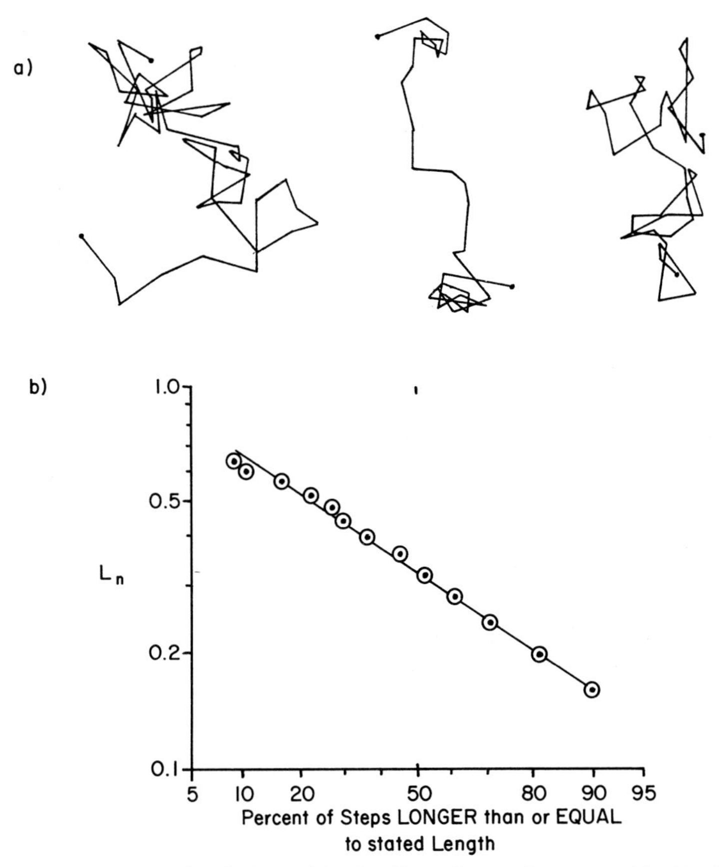

Figure 11.6. The step size distribution of the first Brownian motion reported by Perrin, appear to be describable by the log–normal distribution function. L_n: Length of step normalized by the largest step encountered.

would be log–normally distributed. In *Figure 11.6* the distribution of the steps of the Brownian motion observed by Perrin in his pioneering study and reported widely in other textbooks, are shown [15–18]. It can be seen that the step sizes appear to be describable by a log–normal probability function. To see if computer modeling would generate a similar result for simulated random walks, an experiment was carried out to generate 100 random walks of 400 equal steps, with each step having an equal probability of moving in four directions mutually at right angles (north, south, east and west!).

Simple diffusion theory would indicate that the median value of the set of data should be 20 steps whereas, the data of *Figure 11.7*(a) shows that the experimentally determined median value is 18. This can be considered to be good agreement for the relatively small amount of data involved. However, it can also be seen from the

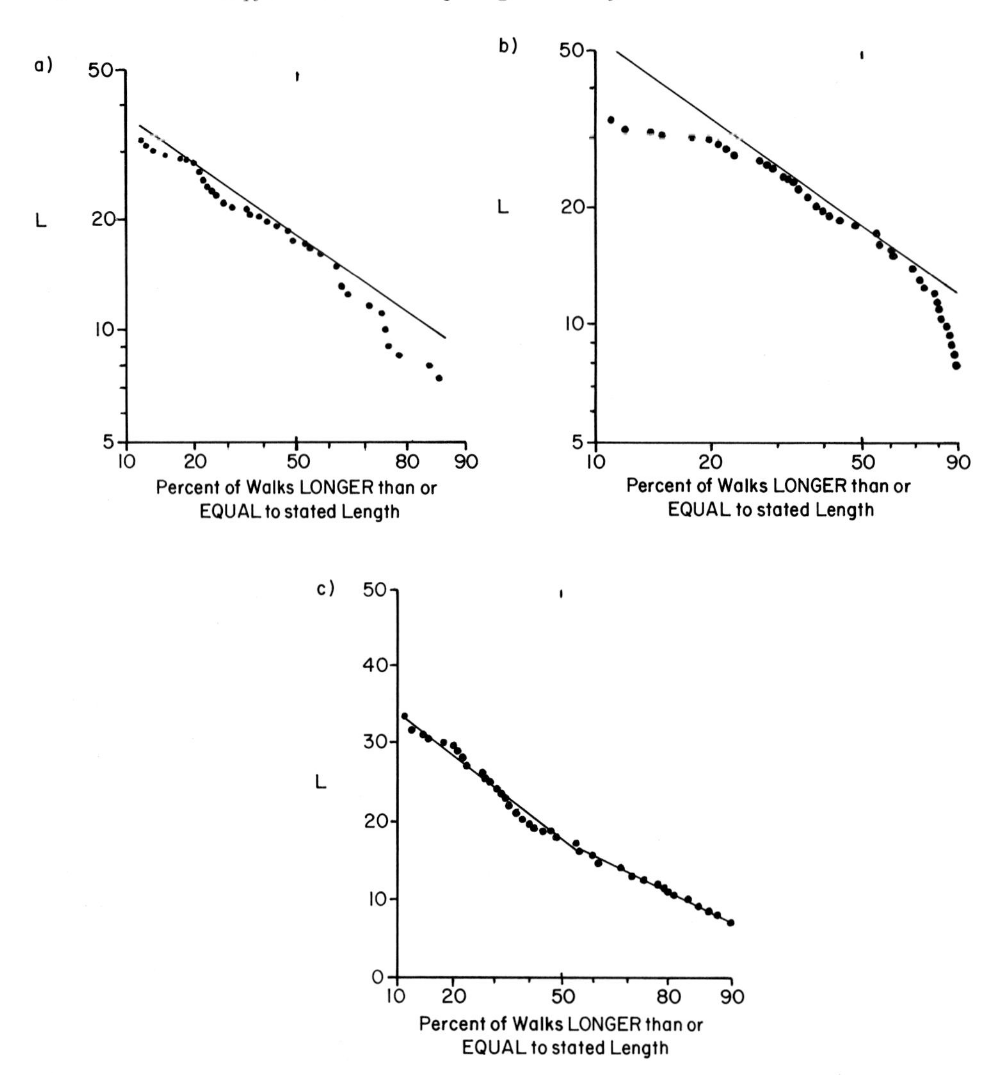

Figure 11.7. The distribution of displacement vectors in a series of 100, 400 equistep random walks is not describable by the three classic probability function discussed in this text. a) Distribution plotted on log–probability graph paper. b) Distribution plotted on log–log graph paper. c) Distribution plotted on Gaussian probability graph paper. L: Length of the displacement vector in arbitrary units. $\% \geq$: Percentage of walks longer than or equal to the stated length.

distribution of displacement vectors as summarized in Figure 11.7(a) that the distribution of final displacement vectors for the random walks cannot be described by the log-probability function since the data does not plot as a straight line relationship on log–probability graph paper. When J. Gratton and I carried out this experiment, we were surprised by this result since in view of the success of the distribution function in describing the data of Figure 11.6, we had anticipated that the same type of distribution would have been obtained by the mathematical modeling.

The distribution data from the random walk experiment was also plotted on log–log graph paper and Gaussian probability graph paper as shown in Figure 11.7(b, c). It can be seen that neither of these graphical displays resulted in straight line relationships so we can assume that none of the major types of probability function fit the data.

If one draws a tangent to the data curve on the log–probability graph paper for the simulated random walks, then the physical interpretation of the data curve is that there are fewer longer and shorter walks than one would have anticipated from a log–normal distribution. The fact that the real data of Perrin fits the log–normal distribution data would indicate that in the real Brownian motion there were other factors that increased the probability of large displacement vectors after a given period of random walking. This could have been some mutual repulsion of the fine particle by other fine particles in the suspension.

Indeed many suspensions incorporate dispersing agents that help the individual fine particles to repel each other. Whatever the reason, real Brownian motion may differ from the simple postulated model for the primary movement of equal probability in all directions at all times. We plotted the data of Figure 11.7 on log–log graph paper because we had just read a scientific paper by Juri Muller [19, 20] who had indicated that many distributions that scientists had studied using the log–normal probability distribution function actually where scaling functions describable by the log–log relationships of the type described by Zipf when one had more information on the rare events in the tails of the distributions.

We had hoped that if the distribution function did not fit the log–normal distribution then perhaps, because of the much larger data set for Figure 11.7 as compared to that of Figure 11.6, the data might manifest a straight line relationship on log–log graph paper. It can be seen however, that the data curved continuously on this type of graph paper. The final data plot of Figure 11.7 shows that the distribution functions of the random walk did not fit the Gaussian probability distribution function. If the distribution of displacement vectors had been a scaling function (straight line on log–log) then we would have been able to show that the general random walk was a special case of a more general random diffusion theory developed by Paul Levy known as *Levy Flights*. Paul *Levy* was a French mathematician (1886–1971) who taught Mandelbrot at the University in Paris. Mandelbrot tells us that Levy was probably the most influential teacher that he encountered during his days as a student. It is rather interesting that although in recent years there have been several studies of Brownian motion using large computers there have been no reported studies of a distribution function of the displacement vectors [21–23].

In modern fine-particle science, the study of Brownian motion has become very important because a method of measuring the size of small biological fragments and the study of latex spheres used in large quantities in the paint industry involves the study of the Doppler shift in laser light reflected by the randomly diffusing fine particles being studied [24].

The Doppler effect is named after an Austrian scientist Christian Johan *Doppler* (1803–1853). Doppler was interested in the well known fact that if a sound is emitted by an object moving towards an observer it sounds differently from sound heard when its source is moving away from the listener. Anyone who has heard the siren of an ambulance approaching them and leaving them has heard this difference in the quality of the sound as the ambulance passes the listener. The change in frequency caused by motion is now known as the *Doppler shift*.

Sound waves are created by a series of compressed air packets moving through the air created by the vibrating object generating the sound. If we imagined a tuning fork on a trolley moving towards us then the time interval between the compressions made by the oscillation of the tuning fork would be just less than those if the tuning fork was stationary. Because the time interval between the compressions was less, the ear would interpret the information as meaning that the note was of a higher frequency. The opposite effect is created when the source sound is receding from the listener. As the sound source recedes each time a compression is emitted by the moving of the arm of the tuning fork the arrival of successive compressions is delayed because of the movement away of the source of sound. This is interpreted by the ear as meaning that the receding source of sound has a lower frequency than the approach sound.

When Doppler had worked out his theory as to the reason for the sound change as the source of the sound moves past the listener, he carried out an experiment. The experiment was carried out in 1842 long before there was precise instruments for measuring the frequencies of sound. To test his theory Doppler arranged for a locomotive to pull a flat car back and forth at different speeds. On the flat car were trumpeters sounding this note or that. On the ground, musicians with a sense of absolute pitch recorded the sound of the note as the train approached them and then again as it receded. Doppler's theories correctly predicted the shift in the sounds noted by the stationary musicians played by the moving musicians depending on the speed of the locomotive. The Doppler effect also takes place when sound is reflected off a moving object, and it also occurs with light waves. The modern police equipment for measuring the speed of cars utilizes the Doppler effect. A laser light of known frequency is directed toward the moving object and the signal reflected off of the moving object has its frequency shifted by an amount depending upon the moving speed of the object. With modern electronics, the shift in the frequency can be interpreted immediately and displayed by a quick calculation at a measured speed of the movement of the car or other object.

The interpretation of the measured Doppler shifts in laser light from randomly moving fineparticles depends upon the assumed distribution of velocities of the fine-particles undergoing Brownian motion which in turn assumes some type of basic kinetics in the suspension. If the basic kinetics of Brownian motion are different from those involved in assuming equal random motion at any one time in any one direction then there could be systematic bias in the interpretation of the Brownian motion in terms of the sizes of the fine particles undergoing Brownian motion. Certainly the data of Figure 11.7 would indicate that some mechanism is enhancing the longer leaps in a real suspension as compared to those that would be predicted from the simple kinetic model. We are currently trying to add a modifying factor into the modeled random walk so that we can modify the type of distribution data obtained in Figure 11.7 to obtain a better fit to the log–normal type data line of Figure 11.6. Such modifying factors may be related to known forces existing within colloidal suspensions.

To understand the way in which Levy generalized Brownian motion to create a more comprehensive theory of random diffusion, it is useful to look at the data of *Table 11.3* from a different perspective. Instead of regarding them as a set of experimental data, they can be regarded as constituting a probability table describing the probability of a step of a given size in a Markovian chain that constitutes a diffusion

Table 11.3. Frequencies of occurrence for the magnitude of the resultant displacement vectors for the 100 walks of 400 steps of Figure 11.7.

Length of Walk (Total Displacement) (Steps)	Cumulative Frequency $\Sigma \geq$	Percent Greater than or Equal to Stated Length $\% \geq$
1	400	100
2	396	99
3	392	98
4	388	97
5	384	96
6	376	94
7	364	91
8	356	89
9	344	86
10	336	84
12	312	78
14	272	68
16	224	56
18	192	48
20	152	38
22	136	34
24	124	31
26	54	27
30	36	18
35	16	8
40	6	3
45	4	1

Table 11.4. Series of step lengths for a Levy flight generated by using the frequency of occurrence data of Table 11.3 as the probability function for the step length.

Step Number	Random Number	Selected Length	Random Direction
1	72	14	2
2	54	16	3
3	54	16	1
4	52	16	1
5	93	6	4
6	33	22	3
7	61	14	4
8	6	35	2
9	44	18	2
10	5	35	3
11	11	30	3
12	45	18	3
13	31	24	3
14	7	35	1
15	68	14	1
16	75	12	2
17	22	26	3
18	17	30	3
19	94	6	3
20	71	12	4
21	40	18	3
22	29	24	3
23	64	14	2
24	73	12	3
25	85	9	1

random walk. One can now allocate random numbers to the table according to the frequency with which a given step size occurs in the system as indicated in Table 11.3. One can now synthesize a random walk without having to undergo the myriad of small steps taken when each step is a unit step with the probability of four directions being equal.

We now choose a step size from the probability of displacement vectors of a given magnitude of 400 steps and use them to synthesize a random walk for simulated diffusion. Two digits are chosen from the random number table and the appropriate step size chosen. We then randomize the direction by allocating numbers to the possible direction with the subsequent choice of the direction from a random number table. In *Table 11.4* a series of steps using the probability distribution from the data of Table 11.3 are shown, and in *Figure 11.8* we have very quickly simulated a longer random walk using the data of Table 11.3 as a probability distribution set for possible steps in space.

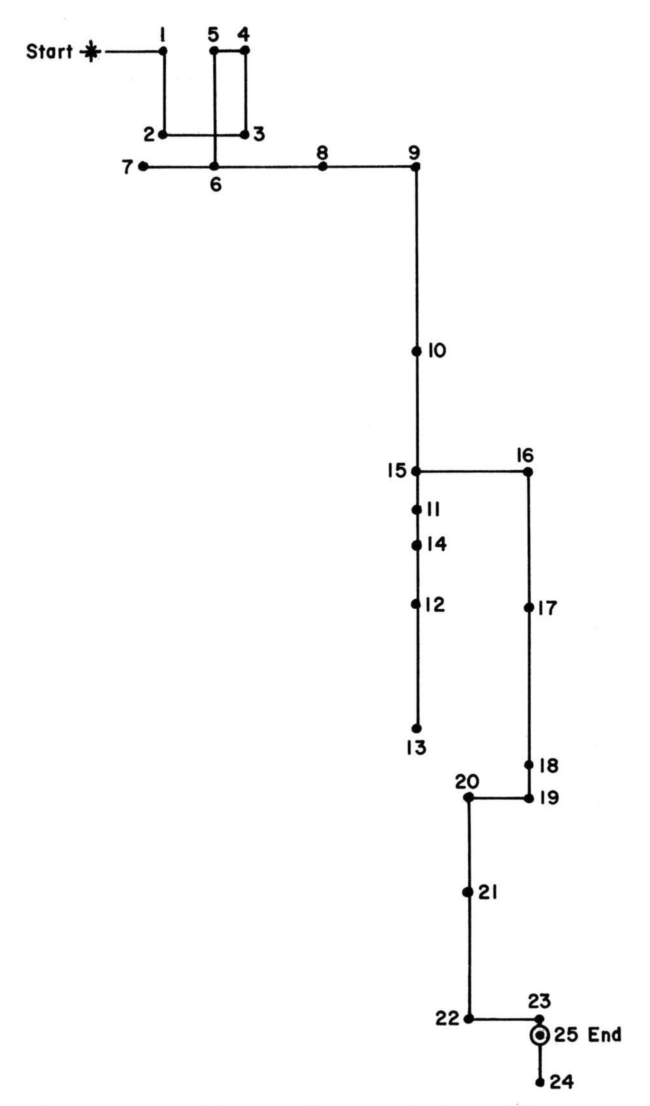

Figure 11.8. Simulated Levy flight created by using the data of Table 11.4.

If we were to look at the tangent drawn on log–log graph paper of the data of Figure 11.7, this straight line can be regarded as defining a probability distribution of step sizes in space with the values being those listed in *Table 11.5*. The straight line on the log–log graph paper as we have learned defines a hyperbolic distribution function. Therefore we can see that this distribution function can be written in the following form, when P_λ is the probability of steps equal to or greater than λ.

$$P_\lambda = K(1/\lambda)^f$$

Table 11.5. Step length probabilities extracted from the line drawn through the log–log graph of Figure 11.7(c).

Step Length	Percent Greater than or Equal to Stated Length
11	100
12	90
13	80
14	70
15	60
18	50
21	40
25	30
33	20
40	15
53	10
57	9
62	8
68	7
76	6
86	5
100	4

This type of distribution of probable steps in space is known as a Levy flight in honor of Paul Levy who first studied such generalized diffusion equations. It differs from the Brownian motion in that rare event long steps are possible. Mandelbrot has extended the concepts of Paul Levy by interpreting the slope of the line such as that drawn tangent to the curve of Figure 11.7 as a fractal dimension in data space. For the line drawn tangent to the Brownian motion the value of f in the equation is a fractal dimension of 0.68. This fractal dimension has the physical interpretation that it describes the way in which the diffusing objects cover the available space. In two-dimensional space, Levy flights of different fractal dimension cover the available space in a different way as illustrated by the data shown in *Figure 11.9*.

In our earlier discussion of Brownian motion, we discussed the fact that one can model diffusion by looking at the behavior of randomly staggering drunks away from a friendly lamppost. If we consider staggering drunks to be undergoing Levy flights, we can now regard the longer leaps in space for a low-dimension Levy flight as being undertaken by the blind meanderings of a drunk who is allowed to catch buses and travel random distances on the bus before staggering around the location at which he is ejected from the bus before he boards another bus for a random length further journey. Levy flights are becoming very important in applied science because they are being used to model problems as different as the spread of a forest fire, the mixing of powders in a chaotically tumbling container and the spread of cancer in the body.

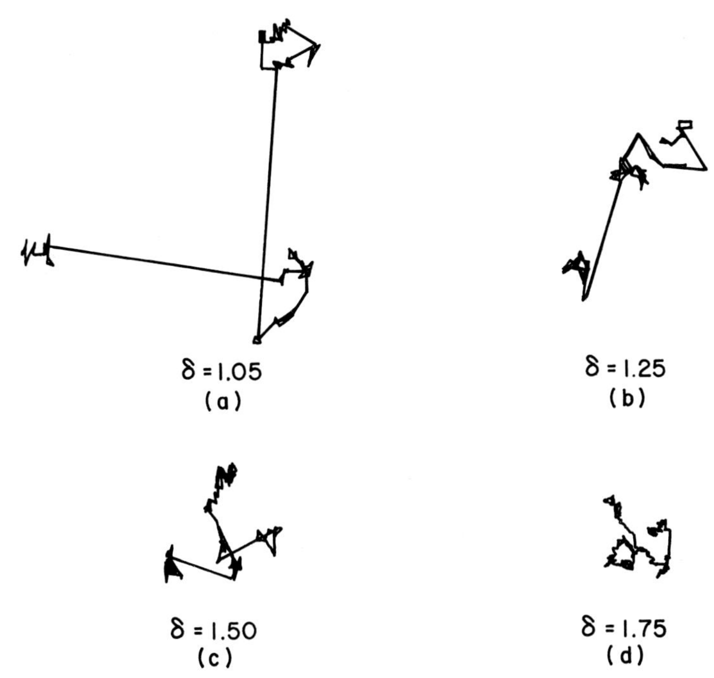

Figure 11.9. Simulated Levy flights of various fractal dimensions the value of which is shown below each.

One of the main difficulties faced by doctors treating cancer patients is what is known as *metastasis*. Metastasis is the technical term used to describe the spread of the cancer around the body. If a cancer is growing at a certain location in the body, the spread of the cancer cells can be modeled by a localized random diffusion of cancerous cells. When a cancerous cell enters the bloodstream and travels to a distant location where it attaches itself to the surface of an organ and initiates a secondary cancerous growth, one can see that the journey in the bloodstream to the new location is like the long leap in a low-dimensional Levy flight. In the same way, if there is a disease spreading amongst animals on a farm and a bird carries the virus on its feet to a different location, the spread of the disease begins to look like a Levy flight. In the same way leaping sparks in a forest fire can be built into a Levy flight model which can be used to look at the spread of a forest fire.

In an earlier discussion of the use of Levy flight modeling, I suggested that sometimes when the distribution of possible steps is not strictly hyperbolic but a combination of high-probability small steps and low-probability large steps that the appropriate terminology for such diffusing random walks is pseudo Levy flights [25]. In computer modeling, one can set up any probability set for modeling a distribution. In *Table 11.6* an arbitrary set of probability steps are shown. A sequence of selection from this table along with randomization in space to model a dispersion

Table 11.6. An arbitrarily chosen set of step length probabilities to be used to simulate a Levy flight.

Step Length	Percent Greater than or Equal to Stated Length
1	100
2	80
3	74
4	58
5	46
6	34
7	24
8	18
9	15
10	12
12	9
15	7
17	5
20	3
25	2
30	1

Step Number	Random Number	Selected Length	Random Direction
1	38	5	1
2	74	3	4
3	16	8	2
4	63	3	4
5	48	4	4
6	1	30	4
7	24	7	4
8	44	5	2
9	4	17	3
10	96	1	3
11	1	30	3
12	18	8	2
13	20	7	4
14	54	4	4
15	35	5	3
16	14	9	1
17	14	9	2
18	28	6	2
19	55	4	4
20	66	3	2
21	14	9	4
22	13	9	3
23	4	17	3
24	86	1	1
25	77	2	3

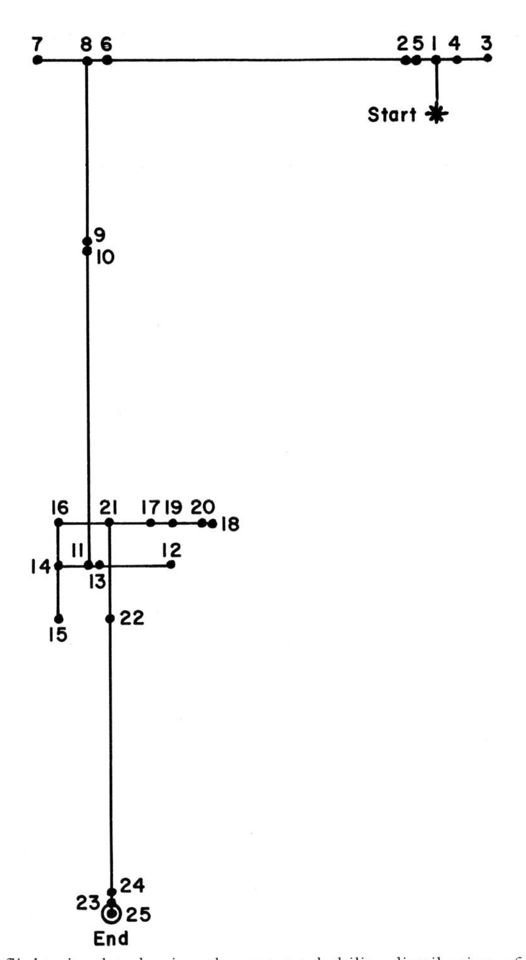

Figure 11.10. A Levy flight simulated using the step probability distribution of Table 11.6.

process as illustrated in *Figure 11.10* is shown. Readers can simulate many different types of diffusion mechanisms by setting up their own set of probability steps, and by comparing different dispersion tracks they can compare them with real situations that determine which models best fit real data.

References

[1] G. K. Zipf, *Human Behavior and the Principle of Least Effort*, Addison Wesley, Cambridge, MA 1949 (Hefner reprint).

[2] Words for this English language word frequency study were taken from the article "Faldo", *Sports Illustrated*, April 8, 1991.

[3] J. R. Pierce, *Symbols, Signals and Noise, The Nature and Process of Communication*, Harper Modern Science Series, New York, 1961.

[4] Words for the French frequency distribution study were taken from the article "Les mamans prematurees", by Pauline Cyr, *Chatelain 29 No. 3* (March 1988) 47.

[5] Words for the first English frequency distribution study were taken from "Police Brutality", by Alex Prudhomme, *Time* (March 25, 1991).

[6] Words for the second English frequency distribution were taken from "Patagonia Puma; The Lord of Land's End", by William L. Franklin, *National Geographic 179* (January 1991) No. 1.

[7] See Chapter 38 of B. B. Mandelbrot, *The Fractal Geometry of Nature*, W. Freeman, San Francisco 1983. See also Page 3 and 4 in this book.

[8] B. J. West, M. Shlesinger, "The Noise in Natural Phenomena", *American Scientist 78* (January–February 1990) 40–45.

[9] E. W. Montroll, M. F. Shlesinger, "On 1/f Noise and Other Distributions With Long Tails", *Proc. NAS 79* (1982) 337.

[10] For a discussion of the Rosin–Rammler distribution see Green and Maloney (Ed:) *Perry's Chemical Engineers Handbook*, 6th edition, McGraw Hill, New York 1984.

[11] For a discussion of the structure of the Rosin–Rammler graph paper, see the paper by Kaye and Clark, entitled "Dangers of Curve Fitting in the Deduction of Size Distribution from Diffraction Data", *Proc. Powder and Bulk Solids Conf.*, Rosemont, May 6–9 1991, Cahners Exposition Group, Cahners Plaza, 1350 E. Touhy Avenue, P.O. Box 5060, Des Plaines, Il, 60019-9593.

[12] For a discussion of the sieve analysis technique, see B. H. Kaye, *Direct Characterization of Fine-Particles*, Wiley, New York 1981.

[13] For discussion of the theory and instrument performance of various laser diffractometers for determining the size of fine particles see B. B. Weiner, "Particle and Droplet Sizing Using Fraunhoffer Diffraction", Chapter 5, in H. G. Barth (Ed:) *Modern Methods of Particle Size Analysis*, Wiley, 1984.

[14] T. Allen, *Particle Size Analysis*, 4th ed., Chapman and Hall, London 1991.

[15] J. Perrin, "La discontinuité de la matiere", *Revue du Mois 1* (1906) 323–344.

[16] J. Perrin, "Mouvement Brownien et Realite Moleculaire", *Annales de Chimie et de Physique VIII, 18* (1909), 5–114. Trans. F. Soddy, as "Brownian Movement and Molecular Reality", Taylor & Francis, London 1910.

[17] J. Perrin, "Les Atomes", Alcan., Paris 1913, "Atoms", the English translation by D. L. Hammick, London, Constable.

[18] See discussion of Perrins work in B. B. Mandelbrot, *Fractals Form Chance and Dimension*, W. Freeman, San Francisco 1977.

[19] J. Muller, J. P. Hansen, A. T. Skjeltorp, J. McCauley, "Multifractal Phenomena in Porous Rocks", *Mater. Res. Soc. Symp. Proc. 176* (1990) 719–722.

[20] J. Muller, "Fractals for Managers; From Telephone Lines to Oil Recovery", Institutt for Energiteknikk (no further details available).

[21] S. Wolfram, "Computer Software in Science and Mathematics", *Scientific American* (September 1984) 193–195.

[22] R. Hearsh, R. J. Greigo, "Brownian Motion and Potential Theory", *Scientific American 220* No. 3 (March 1969) 66–77.

[23] B. H. Lavenda, "Brownian Motion", *Scientific American 252, No. 2* (February 1985) 70–85.

[24] V. B. Elings, D. F. Nicoli, "A Recent Advance in Submicron Particle Sizing By Dynamic Light Scattering", *American Laboratory* (June 1984) 34–39.

[25] See discussion of Levy flights in B. H. Kaye, *A Random Walk Through Fractal Dimensions*, VCH, Weinheim 1989.

Chapter 12

Climbing Fig Trees
to
Discover Fascinating Numbers

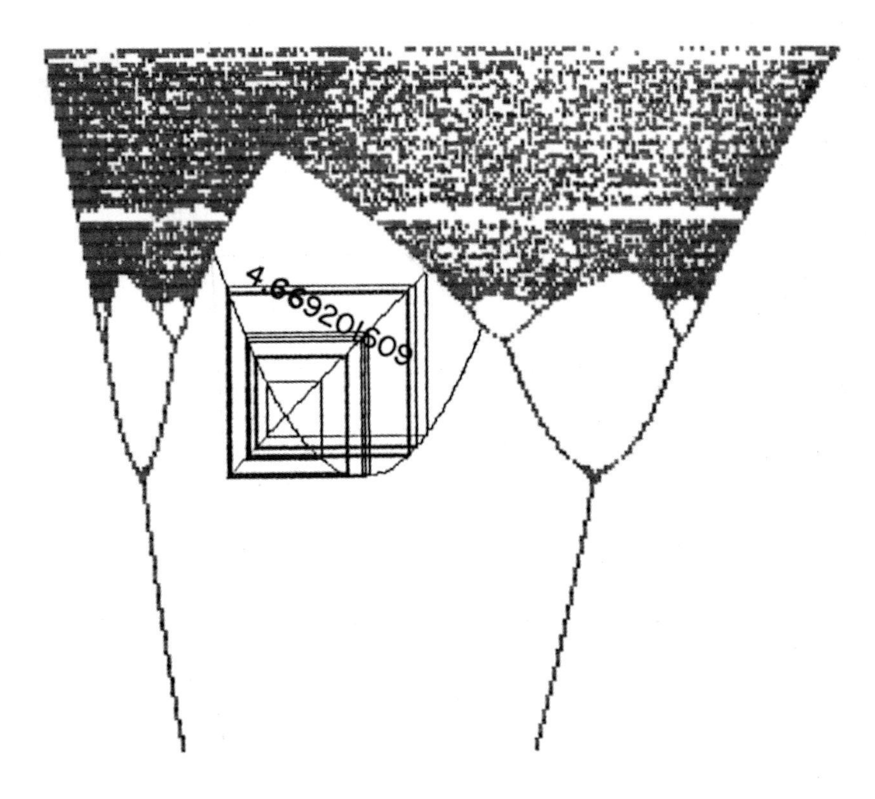

You will be surprised at what you can catch in the cobwebs at the top of attractive fig trees

Chapter 12 Climbing Fig Trees to Discover Fascinating Numbers ... 463

Section 12.1 Magic Numbers for Communicating with Space Aliens? 465
Section 12.2 Population Ecology – Malthus Modified 466
Section 12.3 Climbing Attractor Fig Trees by Means of Parabolic Cobwebs 477
Exercises 482
References 482

Chapter 12

Climbing Fig Trees to Discover Fascinating Numbers

Section 12.1. Magic Numbers for Communicating with Space Aliens?

Fantasy television shows dealing with space travel usually feature space ships equipped with universal translators so that no matter how strange an alien group is encountered during the voyage earthling astronauts are always able to communicate with the aliens through their computer. In reality, if the human race ever makes contact with an intelligent life form in outer space, communication with aliens is likely to be very difficult [1, 2]. Those who have studied the possibility of communicating with space aliens have suggested that mathematics and mathematical relationships may be the first type of intelligent message that can be exchanged with the space aliens. It has been suggested that the first message received from outer space may be a series of pulses such as the following sequence of zeros and ones (in this sequence a 1 indicates a positive pulse of energy and 0 no energy).

0111000101111010111110111111110110111111

Assuming that 000 is a decimal point we could interpret this sequence as being equivalent to the number 3.1415926

The aliens transmitting the message would assume that intelligent human beings would be able to recognize that this was a sequence of pulses representing the universal constant, π, approximately equal to 3.1415926. This sequence of numbers would indicate that the aliens were seeking to know if humans were intelligent enough to have discovered the universal constant governing the ratio of the perimeter to the diameter of a circle. In his book "Does God Play Dice", Ian Stewart suggests that the scientific community prior to 1976 would have been baffled to have received the following set of pulses [3].

1111000111111011111101111111110110010111111001111111110

4.669201609

Again assuming that 000 meant a decimal point and 00 meant zero, we would assume the number being transmitted was 4.669201609. Prior to 1976, astronomers would have assumed that such a series of pulses would indicate an intelligent transmitter but the deduced number would have been meaningless to the scientific community of the time. In this chapter, we will discover how this number, which is now known as the *Feigenbaum number*, has come to be another magic number in science comparable in importance and significance to π and the exponential number e, that we discovered in Chapter 1 [4]. The reader with some knowledge of German will have already recognized that Dr. Feigenbaum's name means "fig tree" in German and hence the title of this chapter. However, before we discuss the significance and structure of the fig tree number it is necessary to make a detour and study a subject known as population dynamics.

Section 12.2. Population Ecology – Malthus Modified

Population biologists are very interested in the fluctuations of the populations of a given animal species in a particular self-contained area. We can consider the population of cats and mice in a large warehouse. Traditionally, one plots the rise and fall of the populations in time graphs such as that of *Figure 12.1*(a). Our time series graphs tells us that at the beginning of our study, the number of mice in the warehouse increased rapidly in the absence of any cats. Presumably at the time t_1 of our graph, the growing population of mice attracted the attention of cats on the outside of the warehouse who invaded the populated area to avail themselves of the food supply. Prior to time t_1, a population ecologist looking at the mice could predict that the rising population of mice threatened to take over all of the available space in the warehouse.

The introduction of the cats into the problem is typical of the kind of disturbance that comes to predictions of infinitely increasing populations and can be regarded as a feed-back mechanism. From the cat's point of view, the feed-back in this case would be literal; however, in scientific problems generally the term feed-back is used to describe the modification of a system by a signal generated by the system. Therefore, the rising heat of a room activates a thermostat which causes a *feedback signal* to turn off the heat-generating mechanism until the temperature of the room falls back below the position of the thermostat which then activates the heating mechanism again. The population of cats is controlled by a feed-back mechanism in which the fall in the population of mice causes some of the cats to leave the area thus leaving more food for the remaining cats.

Malthus who was an English economist (1766–1834) was the first to study the rise and fall of human populations. Specialists who follow human populations and their movement inside a country are called *demographers* a word created from the Greek word demos meaning "people". This same root word has given us the term *democ-*

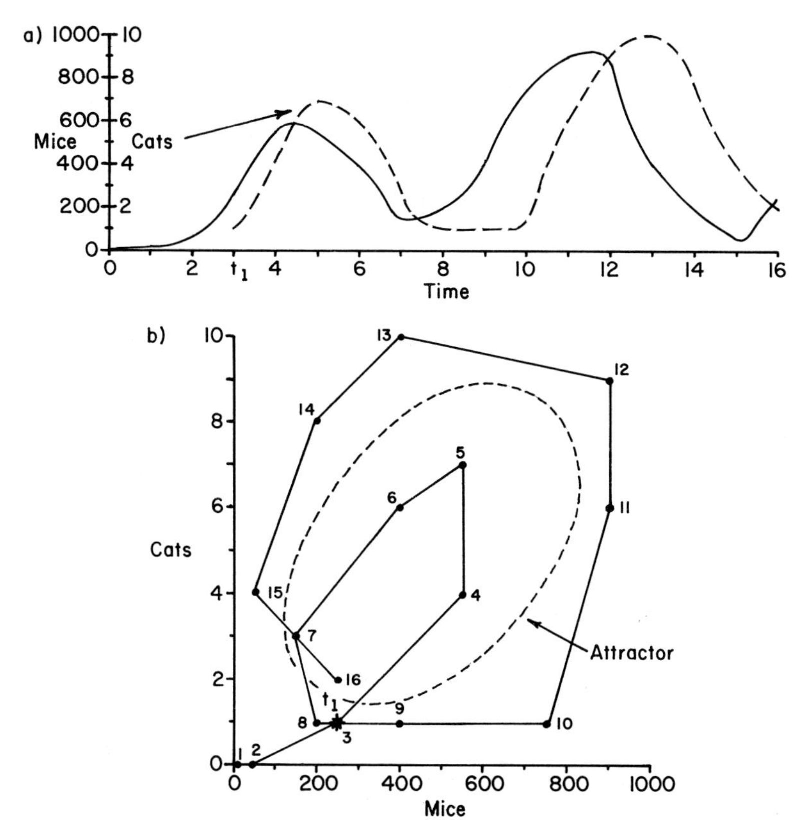

Figure 12.1. The variations of a predator–prey population in a specified location can be represented either as a time series or a phase diagram. a) Record of cat and mouse populations presented as a time series. b) Predator–prey ecology cycle, for the cat/mouse populations.

racy for a form of government in which people control politicians by means of their votes (the Greek word krateein means "to rule"). Malthus studied the demographics of the population of Europe and came up with a gloomy forecast for the future of the Earth. He outlined his views in a book published in 1798 entitled "An Essay on the Principle of Population". According to Malthus, populations tend to increase faster than the supply of material for the needs of the people and this depresses the standard of living of the community. Malthus put forward the idea that the only way to avoid continual pressure from ever increasing population was a policy of

> "*Moral restraint of the passion between the sexes to keep the population within manageable proportion.*"

This does not seem to be very practical advise in a real world! Malthus had anticipated the basic problem of modern society in which the continuous pressure of population expansion causes environmental crisis. In our cat/mice time series of Figure 12.1, the increasing population of mice attracts the cats which bring on the population crash. In modern society, scientists are predicting that the tremendous increase in population will create a disaster scene which may either be a new disease such as AIDS or another virulent disease such as a new flu bug which will sweep through a population weakened by malnutrition.

In recent years, population biologists and human demographers have started to present information on the rise and fall of populations in a given location using a non-time series graphical technique. The technique that they use is similar to the phase space technique we used in Chapter 1 to summarize the dynamics of a pendulum in a different way than the traditional time series. The new way of presenting the information on the population fluctuations is shown in Figure 12.1 (b). This type of graph is called a predator–prey ecology cycle [3]. Note, the word *predator* which means the one who catches and eats the other animal which is described *the prey*. Both words come from the word the Roman soldiers used to use to describe the goods that they seized when they conquered another person. In this case, the soldier was the predator and the conquered people the prey.

To plot a point on the phase space representing population fluctuations in the predator/prey group, one measures the population of cats and the mice at a given time such as t_1 and then uses this information to create a point on the graph such as the asterisk shown in Figure 12.1 (b). As the populations vary, provided no other factor comes into the situation (such as the introduction of guard dogs by the owners of the warehouse which could cause further catastrophe for the cats) the two populations would vary in such a way that the data point of the phase space diagram would wind their way up and around the basic dotted line shown in the diagram.

In mathematical terms, this most probable shape of the phase space of the predator/prey relationship is called the "strange attractor" of the population of cats and mice. It is the most probable path to which the populations will tend to be attracted. Note again, there is nothing absolutely predictive about the strange attractor. The attractor is a description of a most probable situation. The real variations form an infinite number of possible paths around the strange attractor curve.

Another type of data summary used by population ecologists is known as the iterative mapping of the population being studied in successive time periods [3]. To illustrate what is meant by this term, consider for example an alternative method of describing the mouse population growth dynamics in the absence of the cats. Let us assume that at the end of each year there are five times as many mice as at the beginning of the year. Thus per annum, we write the relationship:

$$R_{n+1} = R_n \times 5$$

Now, to calculate how fast the mouse population increases, we write the following table for the first seven years.

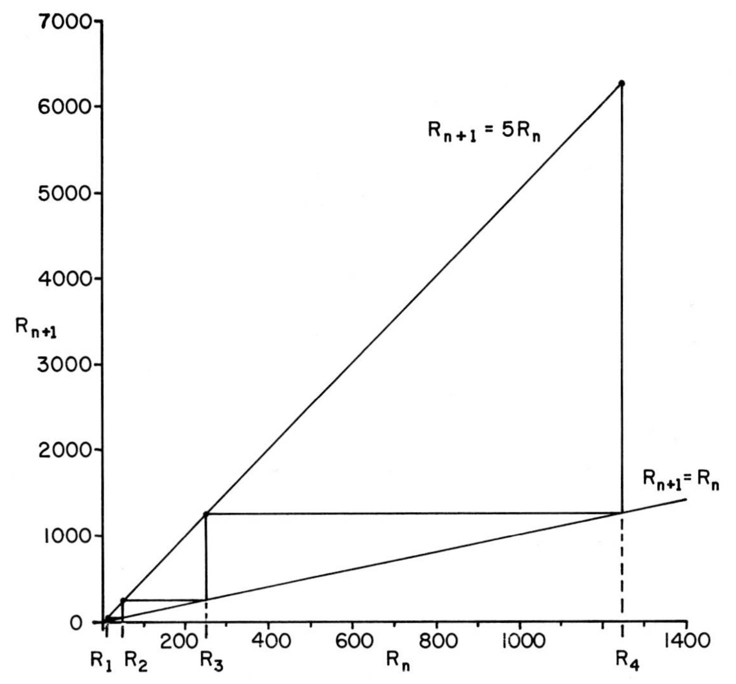

Figure 12.2. Iterative mapping of the growth of the mouse population in the absence of any cats using the equation $R_{n+1} = R_n$.

Year	**Population**
Year 1	10
Year 2	50
Year 3	250
Year 4	1250
Year 5	6250
Year 6	31250
Year 7	166250

When drawing an iterative map to show such a rapid growth in the mouse population, the population biologist uses the system shown in *Figure 12.2.*

Until recently, the physicist did not normally draw a phase space description of the performance of the physical system. However, the utility of such a diagram can be appreciated from the information summarized in *Figure 12.3*(b). If we take the time series of listed temperatures given in Figure 12.3(a), we can construct a phase space description of the thermostats performance by using sequential pairs of the time series to plot a graph in which the abscissa is temperature at the nth time interval and the ordinate is the temperature at the $n + 1$ time interval. The first point in this diagram is T_1 on the abscissa against T_2 on the ordinate the next point is T_2 on the

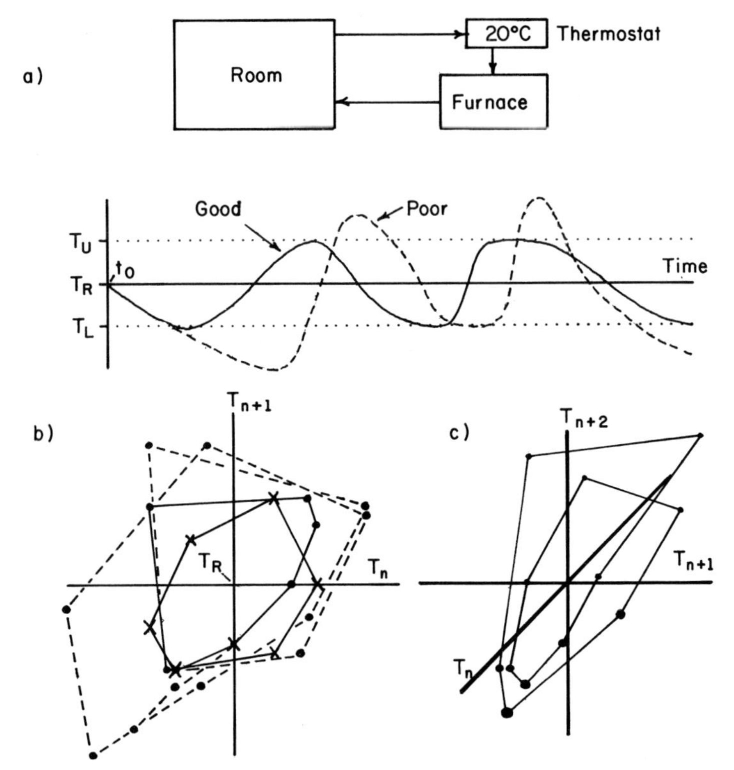

Figure 12.3. A phase space description of the performance of a physical system, such as the temperature of a room heated by a furnace the operation of which is controlled by a thermostat, can be a useful summary of the system over a long period of time. a) Traditional time series description of the performance of a thermostat. The attractor is the desired room temperature. b) Phase space description of the thermostats performance. c) Three-dimensional phase space diagram of thermostat performance.

abscissa and T_3 on the ordinate and so on. In this way, the first complete cycle of the time series from $T = T_0$ to $T = T_x$ will trace out the curve containing the data points marked by the symbol X.

For this system, the data points describing the actual performance of the thermostat are said to be trying to fall into the point attractor which represents the desired room temperature (to the student new to the subject such vocabulary can be very confusing since the "attractor" in chaos theory jargon does not do any physical attracting but is a concept used to describe point locations in phase space). The shape and the area of the performance curve in the phase space description of the thermostat performance of Figure 12.3(b) is a measure of the efficiency of the thermostat control of the room. If one were to imagine an older thermostat which was sticking and not performing as well as a new thermostat then the performance curve moves away from the point attractor as shown in the figure.

The use of pairs of data from the time series to construct a phase space description of a system is a widely used technique of summarizing data used by scientists studying deterministic chaos phenomena. The utility of this type of diagram can be appreciated from Figure 12.3(b) since it shows how a very extensive time series with temperature fluctuations can be summarized as a series of paths in the phase diagram. The phase space diagram is a very concise and vivid record of the performance of the thermostat. A years data could easily be compressed into such a diagram. The actual paths would also have infinite variations about the fixed attractor point of the diagram. Sometimes scientists find it useful to create a three-dimensional phase diagram using triplets from the time series as indicated briefly in Figure 12.3(c). We will use more phase space time series graphs in Chapter 13 when we study catastrophic tumbling of rocks.

In the early 1970s a mathematician, R. May, began to experiment with a modified growth equation to describe population variations when the growth in the population produced population stress which modified the actual growth curve. He started with a very simple equation which is usually written in the following form where x_{n+1} is the population one time interval later than the value of the population at the nth time interval [3, 4].

$$x_{n+1} = k\,x_n(1 - x_n)$$

The quantity k is some modifying constant governing the population growth. In our first discussion of the growth of the mouse population, k was taken to be 5. The term in the bracket obviously grows smaller as x grows bigger. This equation embodies the basic idea that the feedback of the population becomes stronger the higher the population of the animal or insect being considered. This equation can also be written in the alternative form:

$$x_{n+1} = k\,x_n - k\,x_n^2$$

Because it contains a term involving x_n^2, it is described by mathematicians as a *second order equation*. Before we begin to discuss the use of this equation to predict populations, it is useful to review some of the terms used to describe systems which are said to be in equilibrium. The meaning of some useful and basic terms connected with the concept of equilibrium are illustrated in *Figure 12.4*. The meaning of the word *equal* comes from the Latin word meaning "level". At first sight, this is a strange origin for the word but it comes from the technology of weighing objects. To determine the weight of an unknown object, such as the apple shown in the balance pan of Figure 12.4, we add weights to the other balance pan until the beam of the balance (the long piece of material supported on the pivot and from which two pans are suspended) is level. When the beam of the balance is level, we say that the weight of the object in one pan is the same weight as the object in the other pan. When we describe them as being equal, we are referring back to the fact that if we

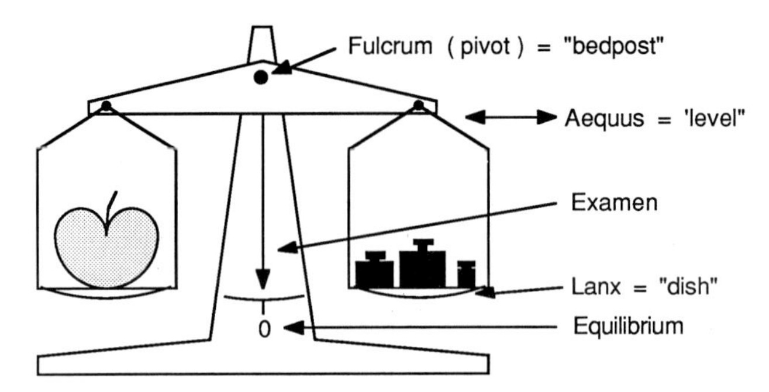

Figure 12.4. To maintain equilibrium when starting population dynamics one needs a clear understanding of equality in mathematics. LIBRA: Latin for a set of scales used to weigh an object. Balance – BIS (two) + LANX (a dish). Equation – a statement of the equality of two quantities e.g.: $y = x - 4$. Equality – two things that are the same. Root of an equation – the value of the unknown quantity, x, which satisfies the equation for $y = 0$ e.g.: $0 = x - 4$. Equal sign (=) – means that the values on either side of it are equal. Not Equal sign ($\neq$) – means that the values on either side of it are not equal. Equivalent – equally strong from the Latin VALENTIS for "worth or value". Linear equation – one which contains only unknown quantities of one dimension e.g.: $y = 5x$. Nonlinear equation – contains unknowns of higher dimension. Thus $y = kx(1 - x) = kx - kx^2$ is a second order nonlinear equation $y = x^3 - 1$ is a third order nonlinear equation. Deliberate – something done consciously after weighing all the evidence or consequences from DE a prefix meaning "to do something".

carried out an actual weighing, the use of the same weight in the other balance pan would make the beam of the balance level.

The Roman word for "scale" was libra. When the beam was balanced the system is said to be in *equilibrium*. It is interesting to note that the Romans started to use technological terms from the act of weighing to describe activities related to the control of society. The Goddess of justice is always shown blindfolded holding a set of scales in which to "weigh" the intangible evidence. As one loads the balance pan with weights, the movement of the beam is monitored by watching the movement of a metal finger attached to the beam. In Latin this finger was called an examen from the Latin word for "the tongue". The person carrying out the act of weighing would follow the movement of the examen and so an intense study of a process came to be described as an *examination* of the performance of the object. When students take a test, that test in which their ability is "weighed" is called an exam or an examination. My favorite word involved in the description of the act of weighing is the word for the pivot – fulcrum. This is Latin for "a bed post". One can imagine a Roman making his first balance by taking hold of a bed post and balancing a piece of wood with two pans on it!

In the foregoing discussion, we have used the word *balance* interchangeably with equilibrium and the basic meaning of the two terms in every day speech is very close. When we write down a statement of what is placed in each of the two pans to create an equal relationship the statement is regarded as an *equation*. The root of an equation is a value of the unknown quantity which satisfies the equation.

How would May's nonlinear second-order equation predict fluctuations in the population of our mice in the warehouse. Note that our term $(1 - x)$ in the bracket indicates to the biologist that something reduces the population of the mice but it does not tell us whether it is cats, dogs or mouse traps that is achieving the change. Population biologists have developed a graphic method of following the fluctuations predicted by May's equation which is described as a parabolic *cobweb diagram*. The meaning of this term will become apparent later in this discussion.

Mathematicians will recognize that the expanded form of the Malthus modified equation, which we will describe as the *May equation*, is the equation of a mathematical curve known as a parabola (the parabola was encountered as a conic section in Figure 10.6). It is useful to describe population variations as taking place between the lower limit of 0 and an upper limit of 1. The maximum population ever recorded is divided into all of the other populations to scale them on a scale of 0 to 1 (as we learned earlier, this is known as normalizing the data). To create a cobweb diagram, the parabola representing May's equation is plotted on the piece of graph paper by calculating the value of x_{n+1} for each value of x_n. Such an iterative map is shown as part of *Figure 12.5*(a). In this diagram the parabola for the value of $k = +2$ is shown. The next stage in drawing the cobweb graph is to draw the line representing $y = x$ on this graph. The table of value for the calculation of the curve of the parabola is as follows:

Points on the Parabola $x_{n+1} = 2x_n(1 - x_n)$

$\mathbf{x_n}$	$\mathbf{x_{n+1}}$
0.0	0.00
0.1	0.18
0.2	0.32
0.3	0.42
0.4	0.48
0.5	0.50
0.6	0.48
0.7	0.42
0.8	0.32
0.9	0.18
1.0	0.00

Let us now consider what happens as we move from an initial population of $x = 0.2$. If we place the value 0.2 in the May equation this gives us the value 0.32 for the population for the next time interval x_{n+1}. This is the value (a) on the parabola of Figure 12.5(a). If we look at the horizontal line ab this moves across the graph to the point b which is the value of x_{n+1}. This value is the one placed in the

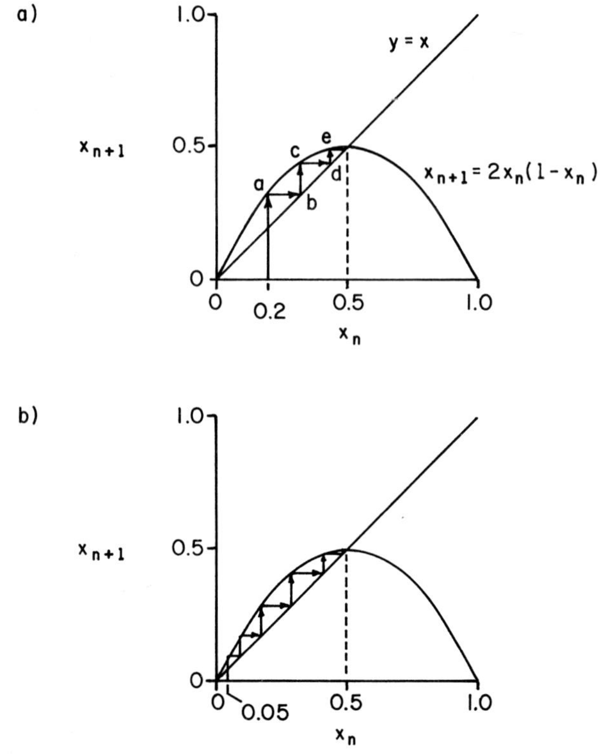

Figure 12.5. Iterative mapping of population trends helps the population ecologist visualize the fluctuations in a population of a given animal in a self contained system such as the mice in the warehouse. a) Iterative map for $k = +2$ with a starting value of $x_1 = 0.2$. b) Cobweb created for a starting value of $x_1 = 0.05$.

May equation to give us the value of c for the next value of the population of animals. Likewise moving from c to d gives us the next value to be inserted in the May equation.

The curve abcde and so on is known as the *trajectory of the calculation.* A dictionary definition of *trajectory* is the curve described by a body or a number under the action of a given force or transformation from the Latin word trans "across" and jacere "to throw". This latter root word has given us the word *javelin* for a spear that is thrown and the term *adjective* for a word thrown onto another to modify its meaning. To describe the trajectory of how the population changes as it climbs the parabola as a cobweb trajectory because of its vague similarity to the web woven by a spider, involves the use of the word cobweb as an adjective thrown at the trajectory curve. If the reader feels that the word cobweb is a rather fancy term for the staircase curve of Figure 12.5(a) the suitability of the name will become more clear as we proceed with the explorations of the predictions of the May equation for other populations.

If one looks at the cobweb trajectory for the same parabola but with a different starting population, one notices that the curve "homes in" on the same ultimate equilibrium population of 0.5 as shown in Figure 12.5(b). This equilibrium population for the May equation when the constant is 2 is known as the phase space attractor of the population. In the words of the mathematician, the fluctuating populations take some time to fall into the attractor of that population situation. In the case of the data for Figure 12.5(b), the fluctuations have died down (the population has fallen into the phase space attractor) by the sixth iteration of the calculation.

In *Figure 12.6* some situations involving equilibrium states which are of general interest throughout this book are illustrated. A ball resting on the top of the hill is

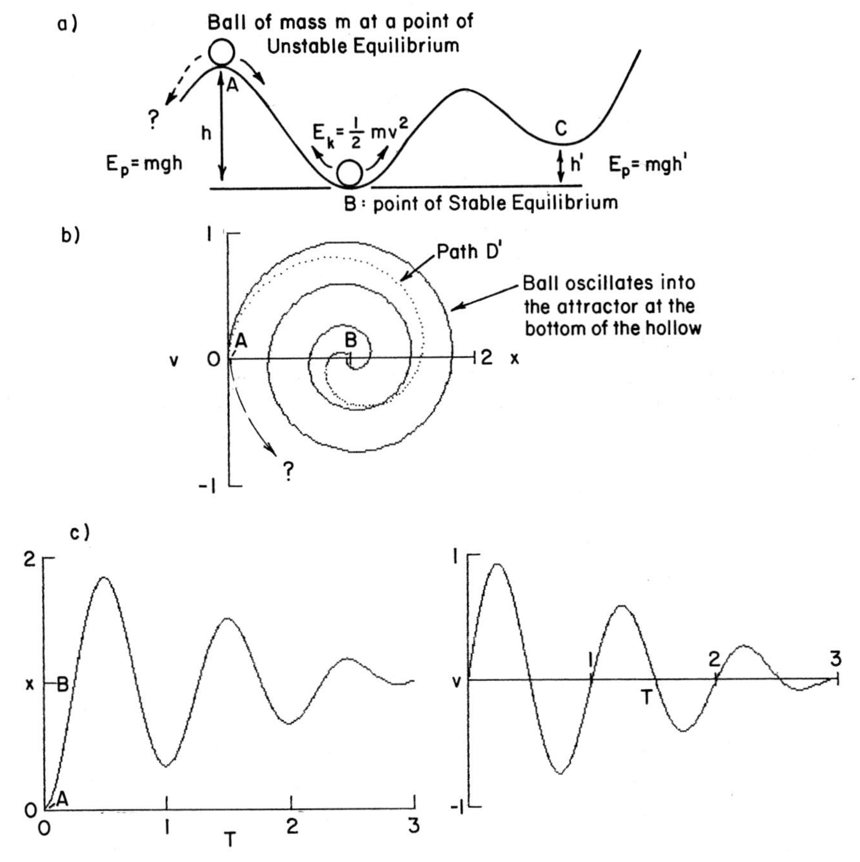

Figure 12.6. Phase space diagrams are useful for describing systems which can be in states of stable and unstable equilibrium. a) A ball in a position of unstable equilibrium at the top of a hill. b) Phase space diagram of the ball oscillating in the hollow. The path D′ represents the path followed if the ball experiences additional friction. c) Time series for the sliding ball showing position versus time and velocity versus time.

described as being in a state of *unstable equilibrium*. The slightest disturbance of its position will lead to the ball sliding down the hill (we say sliding because if the ball were rolling some of the energy it had at the top of the hill would become invested in the rotary motion of the ball which complicates the situation). If the ball moves to the left, its future is unpredictable because we do not know the terrain it will encounter. This is shown in the velocity–position curve by the rapid movement of the ball down to the question mark. Note, the fact that the ball is moving to the left is represented by the negative value of the velocity. If on the other hand, the ball slides down to the right when it reaches the bottom of the concave hollow, it will start to oscillate.

When the ball is at the top of the hill, it is said to have *potential energy*, a phrase which means "energy by virtue of its position". The word potential comes from a Latin word posse meaning "power". Someone who is said to have the possibility of becoming something is said to have the internal power to achieve that position (the Posse of the western movie, which sets out to capture the villain, is so called because it is given the power of arrest when it catches up with the fugitive). As the ball slips up and down in the bottom hollow it loses energy and finally comes to rest at the bottom of the hollow. It is then said to be in a position of *stable equilibrium* since if one moves it either way for short distances it falls back to its stable position. When the ball slips down to its first encounter with the bottom of the hollow it has energy because it is moving, a type of energy which is called *kinetic energy*. Assuming that the ball is on the Earth's surface, the potential energy it has lost by moving down to the bottom of the hollow is mgh. The kinetic energy that it has retained at the bottom of the hollow is $1/2mv^2$. If there were no frictional losses the following relationship would hold and the phase diagram would be a circle as shown in Figure 12.6(b) (i).

$$mgh = 1/2\,mv^2 \quad \text{and} \quad v = \sqrt{2gh}$$

In a real world $1/2\,mv^2 < mgh$ and the energy is gradually lost as the ball oscillates back and forth. This is depicted in the phase diagram of Figure 12.6(b) (ii) by the way in which a trajectory of the moving ball spirals into the attractor of the position on the velocity equal 0 line. If friction increases in the second experiment, the sliding ball would follow the path D′ into the attractor. It is left to the student to visualize what happens if the friction is low and the ball manages to slide up the secondary hump and down into a second concave. Depending on just how much energy it has left when it reaches the point C, where it will have potential energy mgh^2 plus some kinetic energy, it can fall back into the hollow at B or can oscillate until it achieves a stable equilibrium at the point C. The reader is invited to draw the phase space diagrams for both possibilities.

In Figure 12.6(c) the traditional time series representation of the velocity of the sliding ball as it slips down into the hollow and oscillates before it comes to rest in the position of stable equilibrium is shown (note the term stable comes from a Latin word which means "to stand". The stable where the horses are kept was originally a place where they could stand and rest overnight and stable equilibrium implies that the thing can stand in its position and is not easily disturbed from that position).

Section 12.3. Climbing Attractor Fig Trees by Means of Parabolic Cobwebs

When May started to change the value of k from 2, he was quite surprised to find that at k = 3.2 the cobweb trajectory on the parabola started to go round in squares being alternately attracted to one of two attractors. The parabolic cobweb for the value of 3.2 was shown in *Figure 12.7*(a). The time series and phase diagram constructed from the time series are also shown in this diagram. A further surprise lay waiting for May as he tried a value of k such as 3.5. The fact that the population did not settle down to a single attractor value intrigued May but he did not continue to study the problems since he assumed that any pattern of this kind would soon be lost in noise if he tried to apply it to populations of living creatures in the real world. The total pattern of events which May had started to unravel had to await the interest of the mathematician Feigenbaum.

Feigenbaum started to carry out calculations varying k in small increments. At k = 3.5, the population attractor had four stable values as shown in Figure 12.7(b). At k = 3.56, the number of stable values of the population attractor is eight as shown in Figure 12.7(c). If one looks at the values of k for the population attractor to be 2, 4 and 8, it is apparent that the changes in k leading to a doubling of the number of stable population attractors in dropping drastically.

At the time that Feigenbaum carried out his original explorations of the problem, computers were far slower and more cumbersome than today and he found himself wanting to speed the process by which the value at which the attractor doubled again as he increased k. To improve his guess in these attempts to predict the next value at which the population attractor would double, he set up a ratio. In effect his formula compared the magnitude of the gap between the value of k at which one doubling of the number of attractors occurred and the next gap to the next doubling. By comparing the ratio of these gaps for some of the sequences that he had already worked out Feigenbaum guessed at the next gap. He was amazed to find that as the number of doublings increased, the value of the ratio given above tended to the value 4.699. This means that the increase in k leading to a doubling of the increase in the number of population attractors decreased rapidly. If the ratio held from the point where the 2 value of k differs by 0.007 the next doublings would occur at the values 0.00125 0.002250 0.00045. In fact, by the time that the value of k reaches the order of 3.58 there is an infinite number of possible values for the population

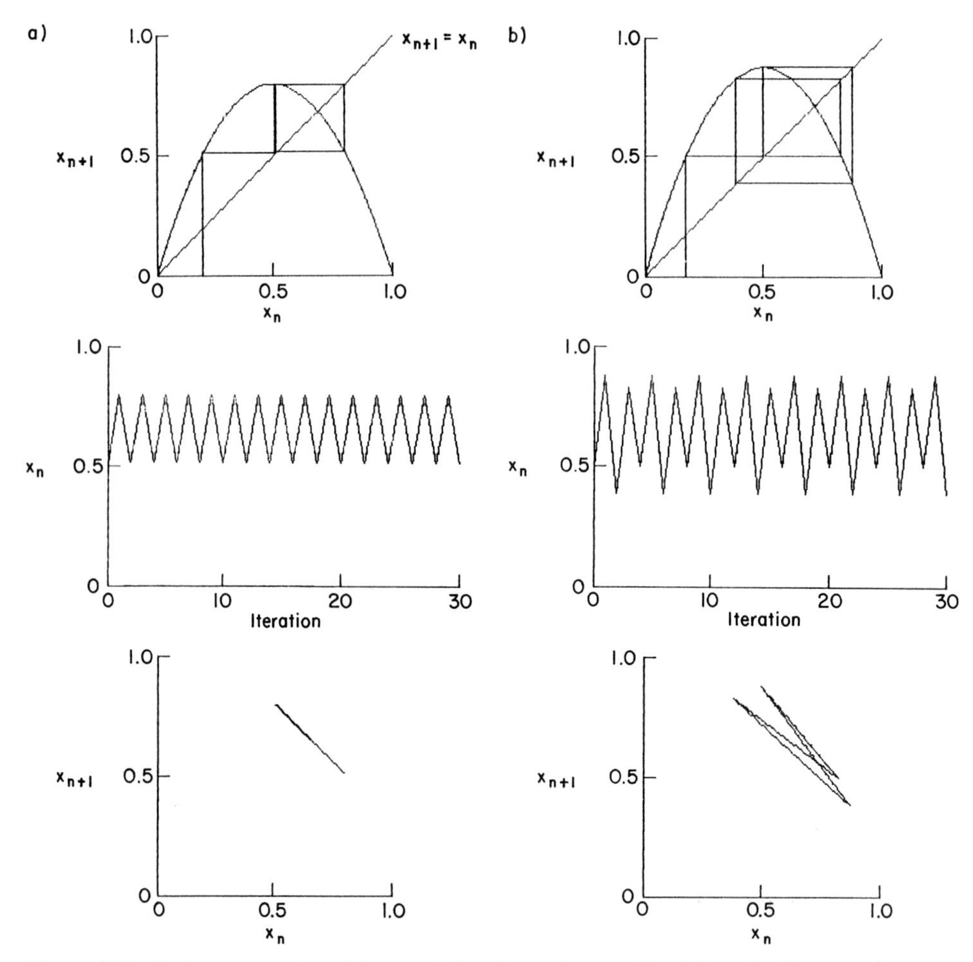

Figure 12.7. Further exploration of the population fluctuations predicted from the May equation, as k increased from 3.2, led to patterned chaos. a) Cobweb for k = 3.2. b) Cobweb for k = 3.5. c) Cobweb for k = 3.56.

attractor. A slight movement above that value results in chaotic variation in the population.

The number 4.669, which appears to be a universal number describing the behavior of systems which are describable by the May equation is known as the Feigenbaum number. More surprises awaited Feigenbaum as he pursued the value of k in May's equation beyond the value 3.58. Above this magnitude he found values of k which were not chaotic. If one investigates the fluctuations in population for a value of k equal to 3.835, the population attractor settles down to three values 0.1520744, 0.4945148, and 0.9586346 (if one attempts to investigate the behavior of the values of k for oneself, one should bear in mind the warning given to all explorers of the

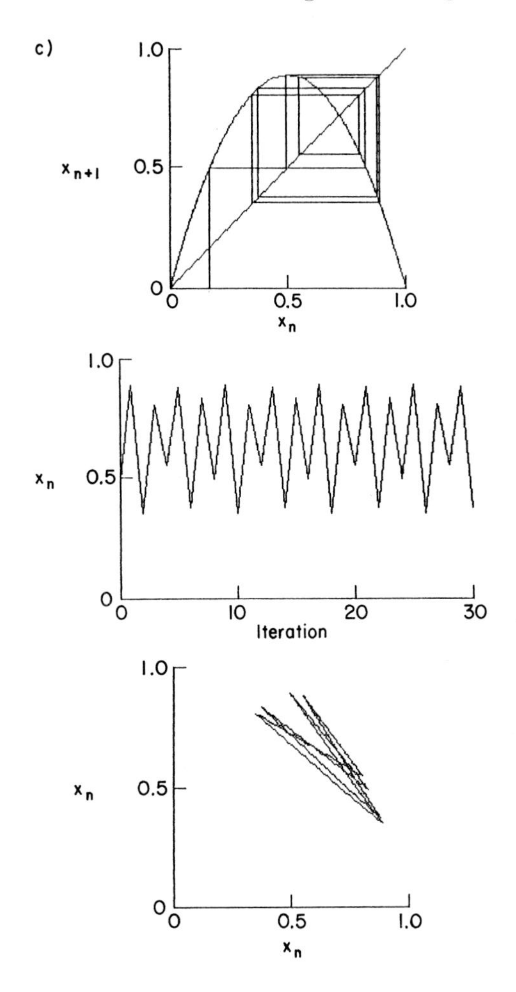

May equation by Stewart. He pointed out that small differences in the internal performance of a specific computer logic operation can shift these values slightly both in the value of k and in the population dynamics corresponding to the value of k). The values given here for k and the population attractors corresponding to the insertion of the value of k in the May equation are those given by Stewart [3]. For 3.739 there are five stable population attractors. These are 0.8411372, 0.4996253, 0.9347495, 0.2280524 and 0.6583204.

The startling pattern of events discovered by Feigenbaum are usually shown in a diagram such as those shown in *Figure 12.8.* When I was first investigating this pattern, I was constantly frustrated because the books showing such diagrams

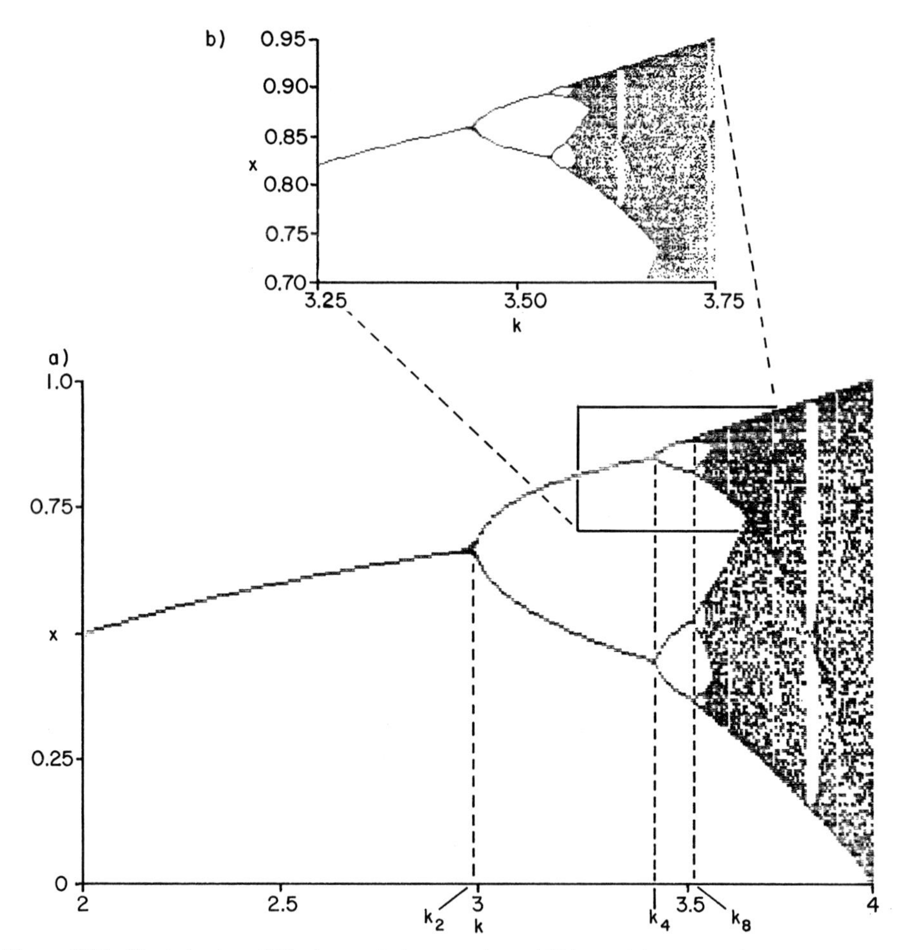

Figure 12.8. The behavior of May's population equation. a) Values of the roots of the May equation as k is varied from 2 to 4 show bands of chaos in the iteration of the equation. b) An enlarged region of (a) demonstrates self-similarity within the region.

committed a cardinal sin of data presentation, one which physics students are told is unforgivable. It is that of presenting a graph without labels on the axes. The diagrams of Figure 12.8 did not carry any labels in their original presentation.

In *Figure 12.9* the ratio of the length of the branches in the number tree which would obey the Feigenbaum ratio are shown. It can be seen that within four branchings of the number, the magnitude of the separation between the values of k would be miniscule if shown on a regular linear graph. To show the true complexity of the number tree of the population attractor would require a greatly magnified scale which changes with the value of k. Mathematicians use a term bifurcation to describe branching curves. This comes from the Latin word bi which means "twice

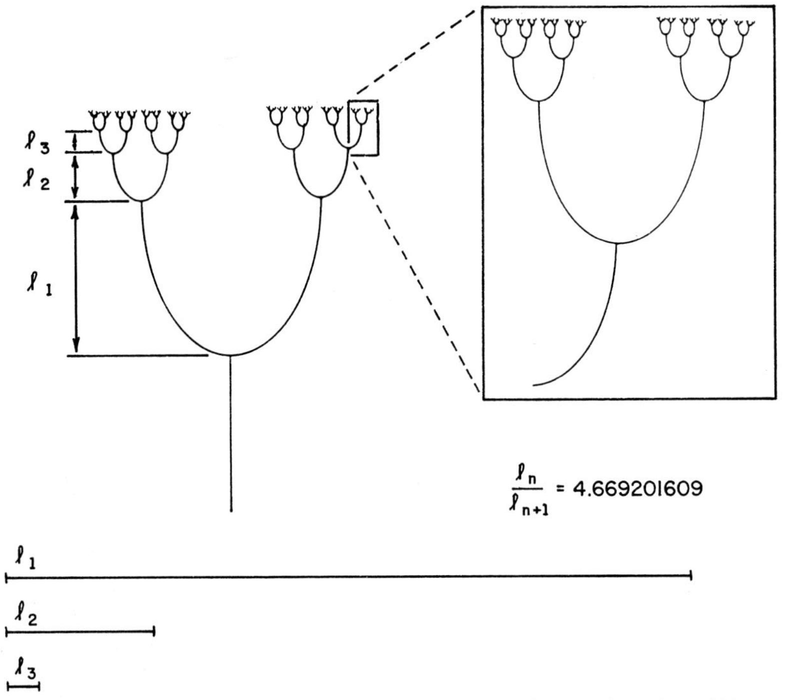

Figure 12.9. The way in which the Feigenbaum number varies when tracking down bifurcation population attractors can be appreciated by considering the structure of an idealized numerical "fig tree" of the type suggested by Stewart.

or two" and furka meaning "a fork". Stewart has suggested that one can understand the structure of the patterns of Figure 12.8 by considering the structure of an idealized bifurcation pattern which he suggested should be called "a fig tree" from the fact that, as discussed earlier, Feigenbaum in German means "fig tree". A stylized numerical fig tree is shown in Figure 12.9. Note however, the bifurcation lengths are not shown to scale since they would very rapidly shrink to invisibility.

The use of such a stylized tree however, illustrates the fact that the fig tree of numbers is a fractal system which is statistically self-similar. If we were to take a region of the tree such as that shown by the box of Figure 12.9 and enlarge it, it would look exactly like the tree from which it was broken, as suggested by the sketch of Figure 12.9. Stewart also points out that the way the branches and twigs of the number tree open out is symmetrically describable by another universal number 2.50290750957.

Returning to the overall display of the population attractors as calculated from the May equation summarized in Figure 12.8, the white bands in the pattern represent the surprising regions in which the chaotic values of the population attractors settle down to a small number of values. As can be seen clearly for the quiet band shown in the lower diagram, the trajectory of the population attractors have regions which are replicas of the overall pattern. The enlarged region of the diagram draws attention to a number tree which is the same as the overall number tree. The pattern of population attractor trajectories is fractal in a double sense in that it contains miniature structures self-similar to the large overall structure. Stewart, in his discussion of the Feigenbaum number, points out that a "green tentacled mathematician" of some remote galaxy could consider the Feigenbaum number just as universal as the value of π and hence the suggestion at the beginning of this chapter that a message from outer space may consist of the number 4.669201609. Other functions can also generate "attractor numerical trees" as the modifying constant such as k is changed.

Exercises

Exercise 12.1.

The mapping of the webs for various values of May's equation is an interesting class exercise. In my experience, the students are very amazed to find out that below three all of the solutions home in on one value. The more adventurous students can soon map this program on the computer. Biology students interested in demography can find summaries of interesting data experiments in Reference 3.

References

[1] C. Sagan, F. Drake, "The Search for Extraterrestrial Intelligence", *Scientific American*, May 1975.
[2] See page 195 of Reference 3.
[3] I. Stewart, *Does God Play Dice – the Mathematics of Chaos*, Blackwell Limited, Oxford 1989.
[4] The history of the evolution of the Feigenbaum number is discussed in Reference 3 and in J. Gleick, *Chaos: Making a New Science*, Viking Press, New York 1987.

Chapter 13

Coincidences, Clusters and Catastrophes

A "Non-PURRcolating" cluster of cats

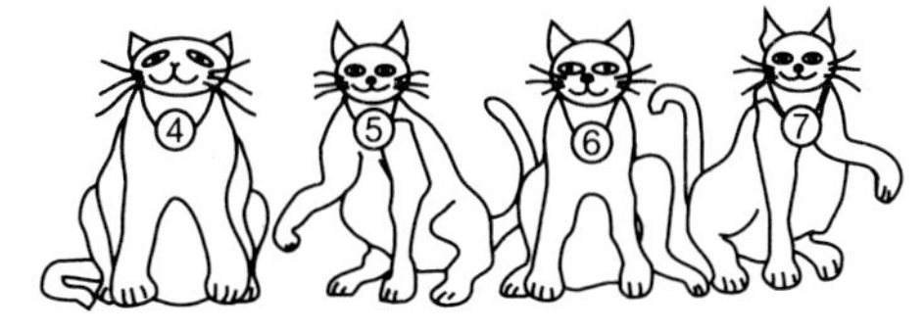

Set of cats with "PURRcolating" paws

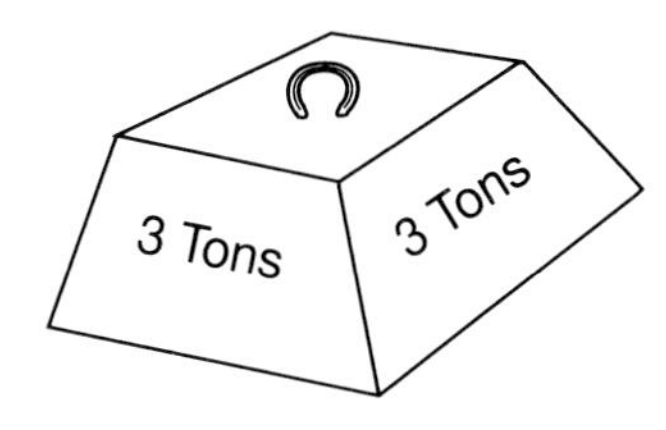

CATastrophes come in threes don´t they?

Chapter 13 Coincidences, Clusters and Catastrophes 483

Section 13.1 Strange Coincidences and Significant Clusters 485
Section 13.2 Simulating Significant and Nonsignificant Patterns of Accidents and Diseases 487
Section 13.3 The Importance of Understanding Coincidences and Clustering in Fine Particle Science 499
Section 13.4 The Catastrophic Behavior of Dripping Taps and Tumbling Rocks 515
Section 13.5 Avalanches and Earthquakes 528
Exercises 530
References 532

Chapter 13

Coincidences, Clusters and Catastrophes

Section 13.1. Strange Coincidences and Significant Clusters

It is not a coincidence that we are going to discuss catastrophes in Chapter 13. Popular superstition in the western world links the number 13 with disaster. In North America many hotels do not have a thirteenth floor; their number system leaps from the 12th to the 14th floor. The rooms on a given floor often pass over 13 with the sequence 711, 712, 714 etc. for rooms on the seventh floor. If a hotel management tried to ignore possible superstition in its visitors they would soon find that some hotel visitors would decline being placed in room 1313 on the 13th floor.

A *coincidence* literally means things that happen in the same place or same time but in popular speech the term coincidence is used to describe an unexpected or surprising simultaneous "happening". The term surprising coincidence is reserved in everyday speech for something which appears to have a very small probability of occurrence. For example, when we moved to live in a suburb of Chicago, my wife had an unexpected visit from a local police officer who asked "Does Brian Kaye live here?" Nervously my wife answered yes, whereupon she was advised "You will have to keep better control of him after school; he has been caught smoking in empty buildings."

Fortunately I was available to establish the fact that the Brian Kaye that lived at the house the police were visiting did not smoke and was old enough to organize his own time. The confusion had arisen from the fact that a small boy who lived in the house immediately behind ours was also called Brian Kaye. Once this strange coincidence had been established, the police were able to visit the other home to convey their message. The odds against having someone with the same name living adjacent to one's house obviously varies. For example in Wales, it is not necessarily a strange coincidence if people with the name Jones live beside each other. The number of surnames in use in Wales is relatively small and hence the cultural pattern of calling people by a nickname to establish different identities.

When we hear of coincidences such as the fact that the neighborhood "wild child" had the same name as the writer of this book, one is amused. However, if coincidence is involved in things such as the occurrence of cancer then people start to be uneasy at the presence of coincidences and start to look for mysterious causes which are bringing about the coincidences. For example, a few years ago several clusters of cancer cases occurred in a small town near Sudbury and there was a flurry of correspondence in the local newspaper in which some residents blamed a local industrial establishment for the occurrence of the cancers.

To study the cluster of cancer cases in this small town an epidemiologist was consulted. *Epidemiology* is defined in a medical dictionary as "a branch of medical science that deals with the incidence, distribution and control of disease in a population" The word was coined from two Greek root words epi meaning "upon" and demos "the people". An *epidemic* is defined as a disease that attacks a great number in one place at one time (in Chapter 11 we discussed briefly how Levy flight theory can be used to model the spread of an epidemic). In its strictest sense epidemiology is not concerned with the cause of a disease only in studying the pattern of a disease amongst population [1]. Indeed in many cases, the causes of a disease or a sickness that creates the pattern is not known.

Sometimes, the first warning society receives that a new disease is beginning to attack it is a pattern of a spreading disease of unknown origin. The main task faced by a epidemiologist, from a mathematical perspective, is to be able to decide whether a clustering set of events indicate that special causes are starting to be operative as distinct from chance clustering of events by random chance. The problem the epidemiologists face when attempting to present their conclusions to the general public is that the public often have very little appreciation of how events can cluster, either in time or in space, by the random interaction of multiple causes.

In Section 13.2, we will look at how we can simulate the occurrence of accidents in an industrial plant to gain synthetic experience with regard to the type of clustering of accidents that can arise from pure chance. We will demonstrate how the average citizen would look at the same pattern of the clustering of accidents and reach a different decision with regard to the significance of the clusters as compared to an assessor from an insurance company who must view such data with cold precise detachment. In the same section, we will show how a random number table can be used to study the clustering of disease events in communities and how the epidemiologist tackles the problem of distinguishing between random chance and possible new causes [2, 3].

In the physical sciences, the clustering of objects from the random interaction of many causes is important in such areas of study as the design of paint films, composite materials and pharmaceutical mixtures of medication with the powders used to deliver different forms of a medication. In Section 13.3 patterns of cluster formation in such physical systems will be studied by using Monte Carlo routines based on various recipes for converting random number tables into simulated physical structures.

In everyday speech, the term *catastrophe* is used to describe either large-scale accidents or the sudden failure of a system. A very large earthquake is called a catastrophe whereas a small one is called an earth tremor. Folklore has always recognized that catastrophes are major disasters triggered by minor events. It is said that it is the last straw that breaks the camels back. In other words, the camel can sustain a certain load and everything is functional until suddenly the system becomes overloaded by one more straw which causes a catastrophic failure of the camel's back.

In Section 13.4 we will look at the surprising patterns of catastrophic events such as earthquakes and avalanches. In recent years, deterministic chaos has been extended to study such systems and the ideas and techniques used to study catastrophic failures of various systems will be explored briefly by discussing a new method for measuring the shape of rocks from their catastrophic tumbling behavior. The reader will discover how to model and describe earthquakes by looking at the avalanching of powder heaps.

Section 13.2. Simulating Significant and Nonsignificant Patterns of Accidents and Diseases

It is part of the folklore of superstition that accidents or bad news comes in threes [4]. Part of the reason for belief in this pattern of events is an unwillingness to remember events that don't cluster. Consider for example, the pattern of accidents over a three year period shown in *Figure 13.1*. If one looks at the data for year 1, one could see that the events in May and June appear to be a sequence of three within 4 weeks. When one comes to September and October one would remember the clustering on the 18th and 19th of September and the 11th of October and would forget that there was another one on the 28th. Again, in year 2, the events of July would reinforce the "bad news comes in three" theory.

The reader can demonstrate to themselves that the data of Figure 13.1 is compatible with a Poisson fluctuation with an expected population of 1.5 accidents per month. The superstition that bad things happens in threes can be attributed to selective memory which is unfamiliar with the type of clustering one gets with a stochastic variable. An accident in an industrial situation is very rarely caused by a single event. Usually, the unpleasant event arises from the interaction of several variables such as tiredness, pressure to achieve production, ill health in the operator, emotional anger in the operator and the misplacing of equipment by co-workers. Accidents in an industrial area are the concern of many people but two main groups of interested bystanders are the insurance companies, which must assess the levels of premiums to be paid by an organization to cover losses from accidents, and the other is the health and safety committee of the factory.

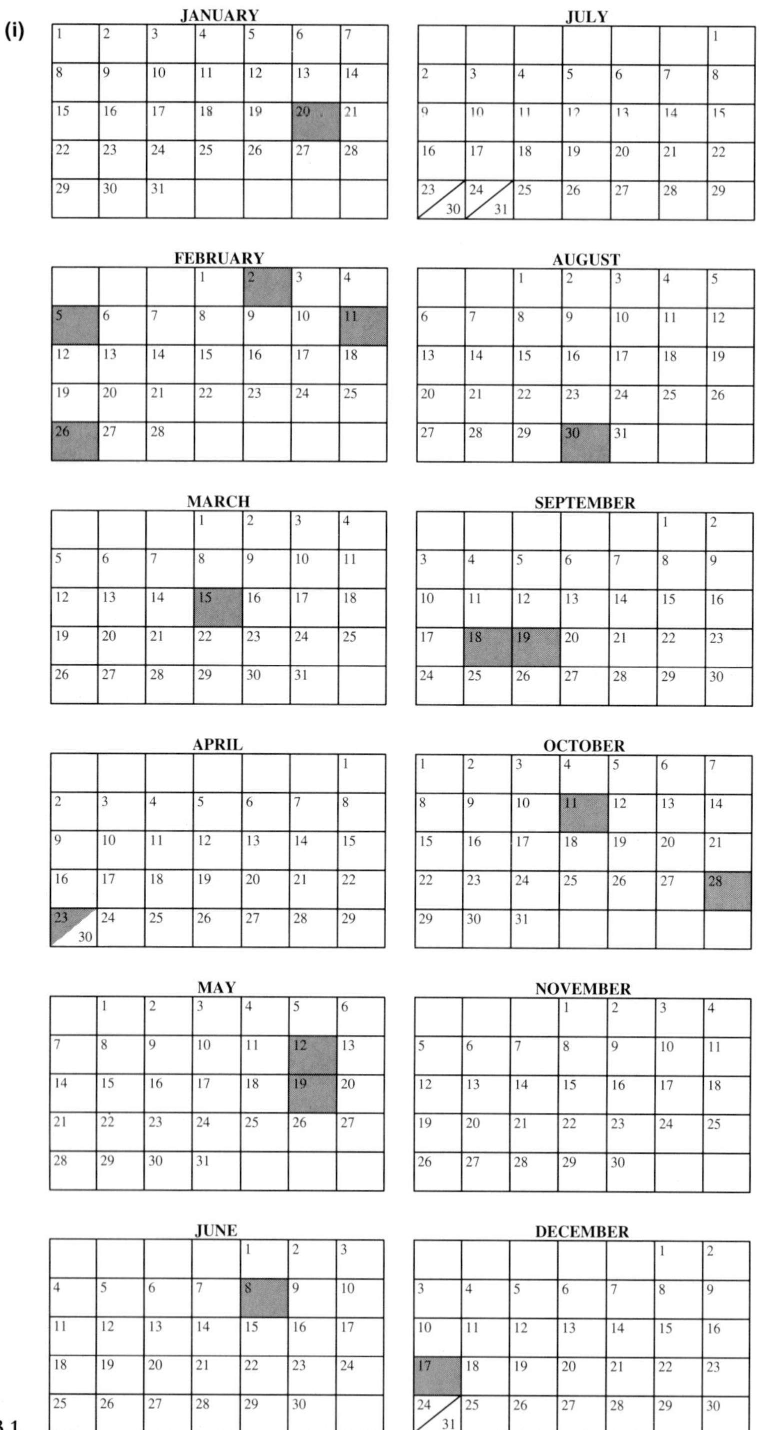
(i)
JANUARY
JULY
FEBRUARY
AUGUST
MARCH
SEPTEMBER
APRIL
OCTOBER
MAY
NOVEMBER
JUNE
DECEMBER

Fig. 13.1.

(ii)

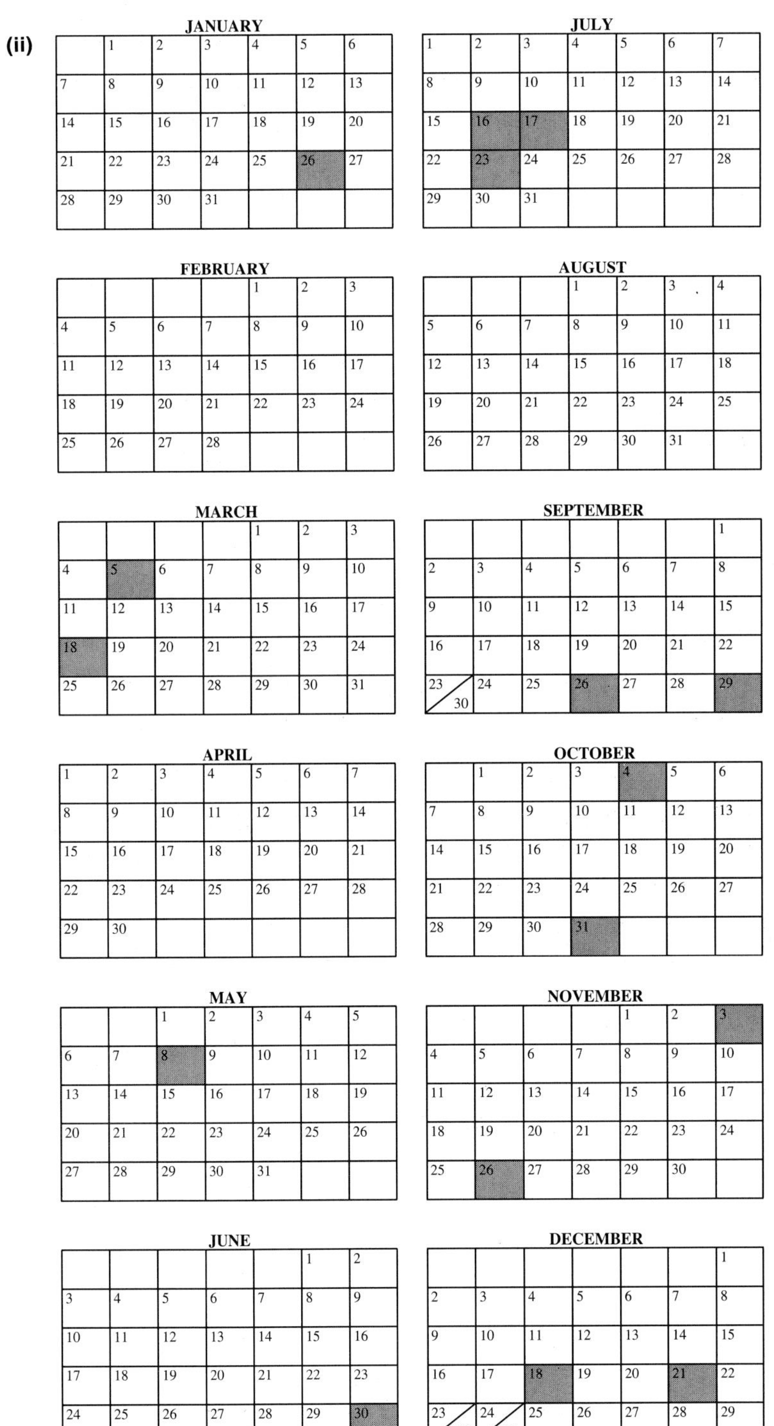
JANUARY
JULY
FEBRUARY
AUGUST
MARCH
SEPTEMBER
APRIL
OCTOBER
MAY
NOVEMBER
JUNE
DECEMBER

Fig. 13.1.

(iii)

JANUARY

		1	2	3	4	5
6	7	8	9	10	11	12
13	14	15	16	17	18	19
20	21	22	23	24	25	26
27	28	29	30	31		

FEBRUARY

					1	2
3	4	5	6	7	8	9
10	11	12	13	14	15	16
17	18	19	20	21	22	23
24	25	26	27	28		

MARCH

					1	2
3	4	5	6	7	8	9
10	11	12	13	14	15	16
17	18	19	20	21	22	23
24 / 31	25	26	27	28	29	30

APRIL

	1	2	3	4	5	6
7	8	9	10	11	12	13
14	15	16	17	18	19	20
21	22	23	24	25	26	27
28	29	30				

MAY

			1	2	3	4
5	6	7	8	9	10	11
12	13	14	15	16	17	18
19	20	21	22	23	24	25
26	27	28	29	30	31	

JUNE

						1
2	3	4	5	6	7	8
9	10	11	12	13	14	15
16	17	18	19	20	21	22
23 / 30	24	25	26	27	28	29

JULY

	1	2	3	4	5	6
7	8	9	10	11	12	13
14	15	16	17	18	19	20
21	22	23	24	25	26	27
28	29	30	31			

AUGUST

				1	2	3
4	5	6	7	8	9	10
11	12	13	14	15	16	17
18	19	20	21	22	23	24
25	26	27	28	29	30	31

SEPTEMBER

1	2	3	4	5	6	7
8	9	10	11	12	13	14
15	16	17	18	19	20	21
22	23	24	25	26	27	28
29	30					

OCTOBER

		1	2	3	4	5
6	7	8	9	10	11	12
13	14	15	16	17	18	19
20	21	22	23	24	25	26
27	28	29	30	31		

NOVEMBER

					1	2
3	4	5	6	7	8	9
10	11	12	13	14	15	16
17	18	19	20	21	22	23
24	25	26	27	28	29	30

DECEMBER

1	2	3	4	5	6	7
8	9	10	11	12	13	14
15	16	17	18	19	20	21
22	23	24	25	26	27	28
29	30	31				

Fig. 13.1.

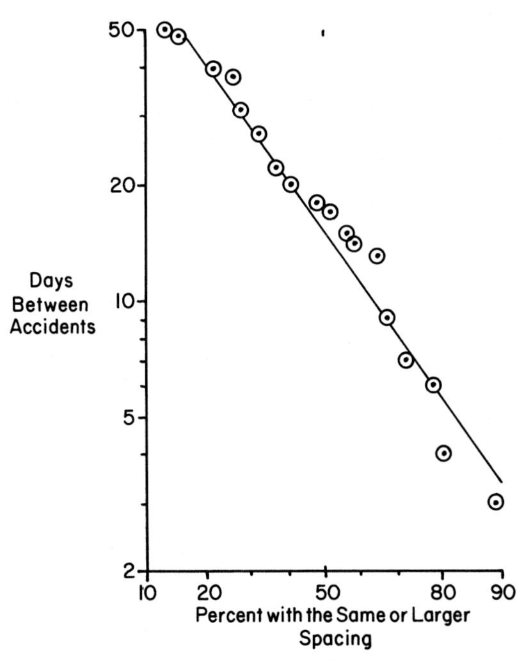

Figure 13.2. The accident pattern of Figure 13.1 is actually a stochastic fluctuation in an expected rate of 1 accident every 20 days or a Poisson expectation of approximately 1.5 accidents per month. Days between accidents plotted on log–Gaussian graph paper manifests a straight line relationship.

The insurance assessor would look at the pattern of events in the three years as shown in Figure 13.1 and would have to decide what was the expected accident rate for the factory. It is assumed that the factory was working on an everyday schedule; that is working 365 days a year. In fact, the insurance assessor would be able to show that the data for the three years was compatible with an expectation of 1 accident every 20 working days as demonstrated by the log–probability plot of the days between accidents as shown in *Figure 13.2.*

The head of the health and safety committee would find it very difficult to take such an objective view of the accident rates and I am quite sure that in any factory where there was a series of events involving accidents starting on January 20th through to the 5th accident on February 26th there would be an emergency meeting of the health and safety committee to discuss the reason for the high accident rate.

One could imagine that the workers would not believe that the fluctuations were just due to chance and would insist that there be a tightening up of safety procedures. The fact that there was then only 2 accidents in March and April would convince

◄

Figure 13.1. Is the pattern of accidents shown above compatible with the hypothesis that the accident rate at the factory is 1 accident for every 20 days of operation? Do accidents happen in threes?

most of the members of the committee that they had solved a problem which was a real problem. The reader should discuss the pattern of accidents in Figure 13.1 with another person and ask them about the clustering of events and whether they thought that things had gone sloppy and had been tightened up. There would probably be comparable concern and investigations in year 3 following the 5 accidents in March and April.

In fact the data of Figure 13.1 is simulated data using a random number table to generate a pattern of events when the accident rate expected was 1 per 20 working days (and hence the amazingly accurate perception of our fictitious insurance assessor). To generate the pattern of events in Figure 13.1 we used two random-number tables. To simulate a rate of 1 in 20, we assumed that a sequence of digits in the random number table were working days and that 9s were potential working days with accidents. However, only half of the 9s could become accident rates if the accident rate was 1 in 20. Therefore, when we encountered a 9 in one random-number table, we looked at the sequence of digits in another random-number table and accepted the 9 as an accident day, if we had an even number in the subsidiary table. In fact, the crisis meeting after the cluster of January and February of year 1 was unnecessary and things didn't improve after the meeting, they just fluctuated according to Poisson expectations.

When one studies the pattern of accidents at a given factory, there will always be a tendency of events to cluster due to random chance. This makes it very difficult to decide when an accident rate has dropped sufficiently for the insurance company to consider reducing the insurance premiums for that factory. Let us consider for example that in our hypothetical factory, a new director of safety was appointed who went through all of the factory devising less stressful procedures for the workers and better safety equipment. He would probably have to wait for a considerable period before the insurance company could be persuaded that a reduced insurance premium was warranted. Let us assume for example that the accident rate was indeed reduced to a rate of 1 in 33 days (approximately 1 a month). We can now simulate a series of events at the new occurrence level by using our random-number table in the following way. If a 9 was encountered in the sequential set of numbers representing the days then one would go to a subsidiary table, ignore the digit 0, and turn the 9 into an accident day if we found the digit 1, 2 or 3 in the subsidiary table (this corresponds to a chance of 1 in 3 that the day should become an accident day). A three-year simulated performance under the new safety director would have the appearance shown in *Figure 13.3*(a).

In Figure 13.3(b), the two log–probability plots of days between accidents before and after the new inspector was appointed are shown. Although the data is sparse,

Figure 13.3. Simulated accident pattern when the expectation is 1 accident in 33 days. a) Calendar day pattern. b) Days between accidents plotted on log–Gaussian paper.

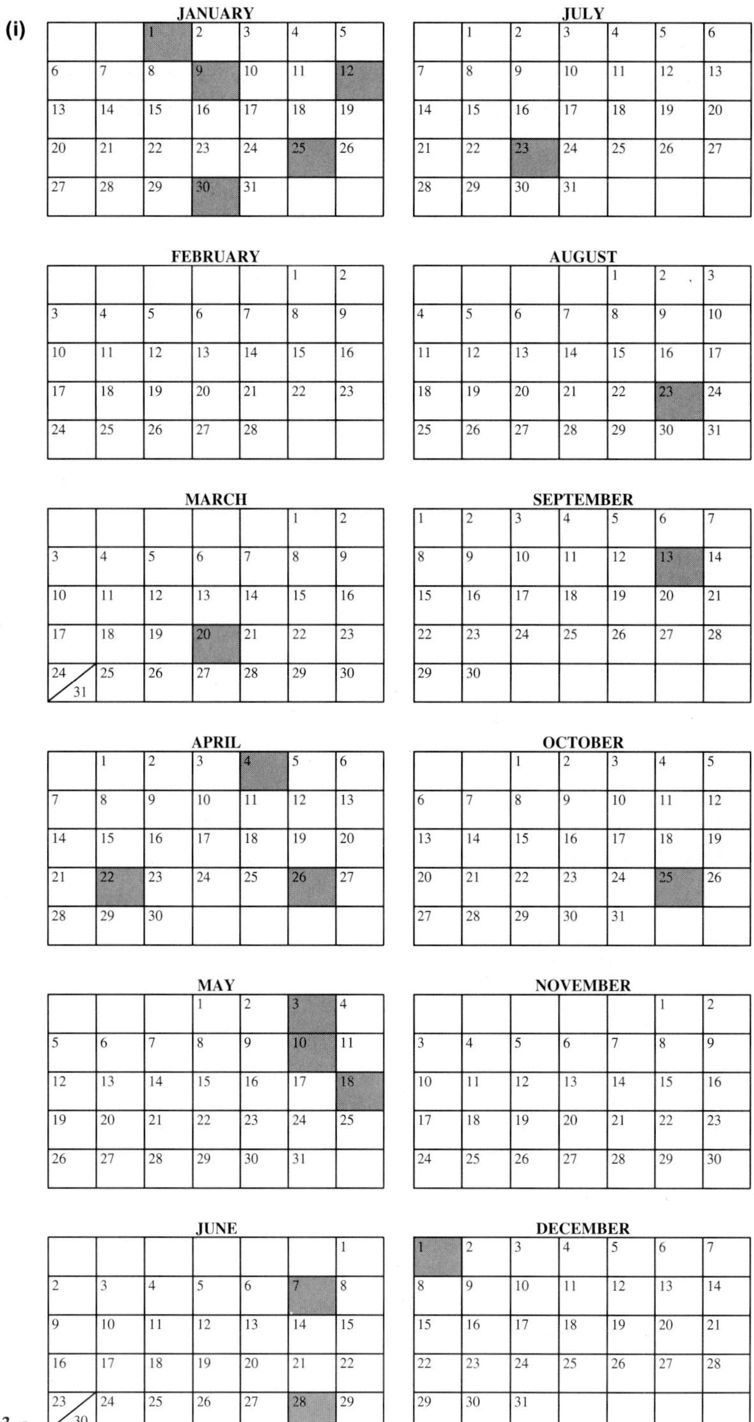
(i)
JANUARY
JULY
FEBRUARY
AUGUST
MARCH
SEPTEMBER
APRIL
OCTOBER
MAY
NOVEMBER
JUNE
DECEMBER

Fig. 13.3. a

(ii)

JANUARY

			1	2	3	4
5	6	7	8	9	10	11
12	13	14	15	16	17	18
19	20	21	22	23	24	25
26	27	28	29	30	31	

FEBRUARY

						1
2	3	4	5	6	7	8
9	10	11	12	13	14	15
16	17	18	19	20	21	22
23	24	25	26	**27**	28	

MARCH

						1
2	3	4	5	6	7	8
9	10	11	12	13	14	15
16	17	18	19	20	21	22
23/30	24/31	25	26	27	28	29

APRIL

		1	2	3	4	5
6	7	8	9	10	11	12
13	14	15	16	17	18	19
20	21	22	23	24	25	26
27	28	29	30			

MAY

				1	2	3
4	5	6	**7**	8	9	10
11	12	13	14	15	16	17
18	19	20	21	22	23	24
25	26	27	28	29	30	31

JUNE

1	2	**3**	4	5	6	7
8	9	10	11	12	13	14
15	16	17	18	19	20	21
22	23	24	25	26	27	28
29	30					

JULY

		1	2	3	4	5
6	7	8	9	10	11	12
13	14	15	16	17	18	19
20	21	22	23	24	25	26
27	28	29	30	31		

AUGUST

					1	2
3	4	5	6	7	8	9
10	11	12	**13**	14	15	16
17	18	19	20	21	22	23
24/31	25	26	27	28	29	30

SEPTEMBER

	1	2	3	4	5	6
7	8	9	10	11	12	13
14	15	16	**17**	18	19	20
21	22	23	24	25	26	27
28	29	30				

OCTOBER

			1	2	3	4
5	6	7	**8**	9	10	11
12	13	14	15	16	17	18
19	20	21	22	23	24	25
26	27	28	29	30	31	

NOVEMBER

						1
2	3	4	5	6	7	8
9	10	11	12	13	14	15
16	17	18	19	20	21	22
23/30	24	25	26	27	28	29

DECEMBER

	1	2	**3**	4	5	6
7	8	9	10	11	12	13
14	15	16	17	18	19	20
21	22	**23**	**24**	25	26	27
28	29	30	31			

Fig. 13.3.a

(iii)

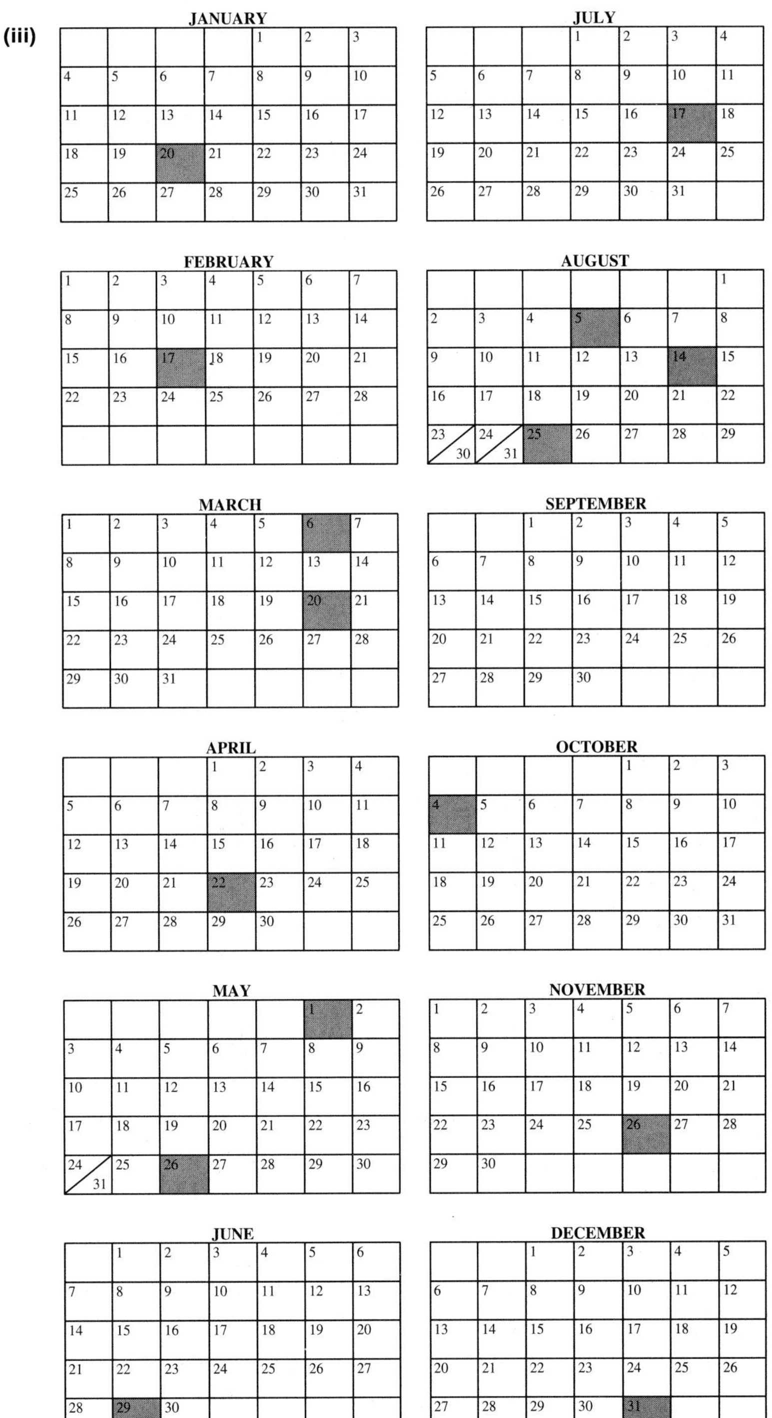
JANUARY
JULY
FEBRUARY
AUGUST
MARCH
SEPTEMBER
APRIL
OCTOBER
MAY
NOVEMBER
JUNE
DECEMBER

Fig. 13.3.a

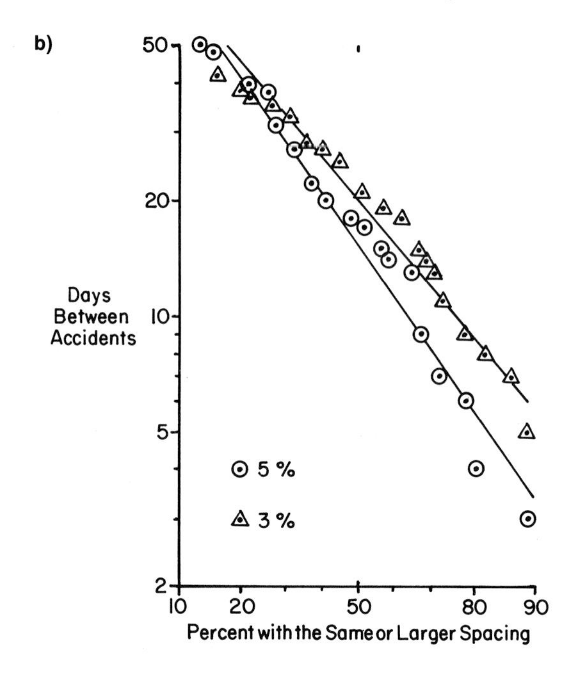

Fig. 13.3.b

there would be an indication to the insurance assessor that indeed the accident rate had changed. However, the data of Figure 13.3 shows how difficult it is to take decisions when sparse events are spread over a long period of time. The statistician would like to wait until there is ten years of data available before taking a decision, however sometimes one has to take a decision in shorter time because a customer may not want to wait so long. The statisticians working for the insurance companies make their estimates as to the probability of the occurrence of accidents using techniques similar to those used in the Monte Carlo routines used to generate the data of Figures 13.1 to 13.3. What we have done in this section is to use Monte Carlo routines to gain synthetic experience of a difficult type of problem where relatively rare events have important consequences.

In recent years, there has been considerable controversy as to whether or not there is an increased incidence of leukemia in communities near nuclear power stations [1–3]. Again, the problem is complicated by the fact that one is studying rare events. Resolving the conflict as to whether or not leukemia is caused by living near a nuclear reactor is essentially a problem of deciding what is a valid stochastic fluctuation in a low expected population. To illustrate the difficult type of problem facing epidemiologists studying leukemia cures, let us imagine for example that a rare disease that we will call netronella (a fictitious disease) is occurring with a probability of 1 in 1,000 in the general population. Let us assume that we select for study 20 communities of 2,500 people each, 3 of which are located near chemical pesticide plants which the local inhabitants claim is causing an increase in the disease. Let us

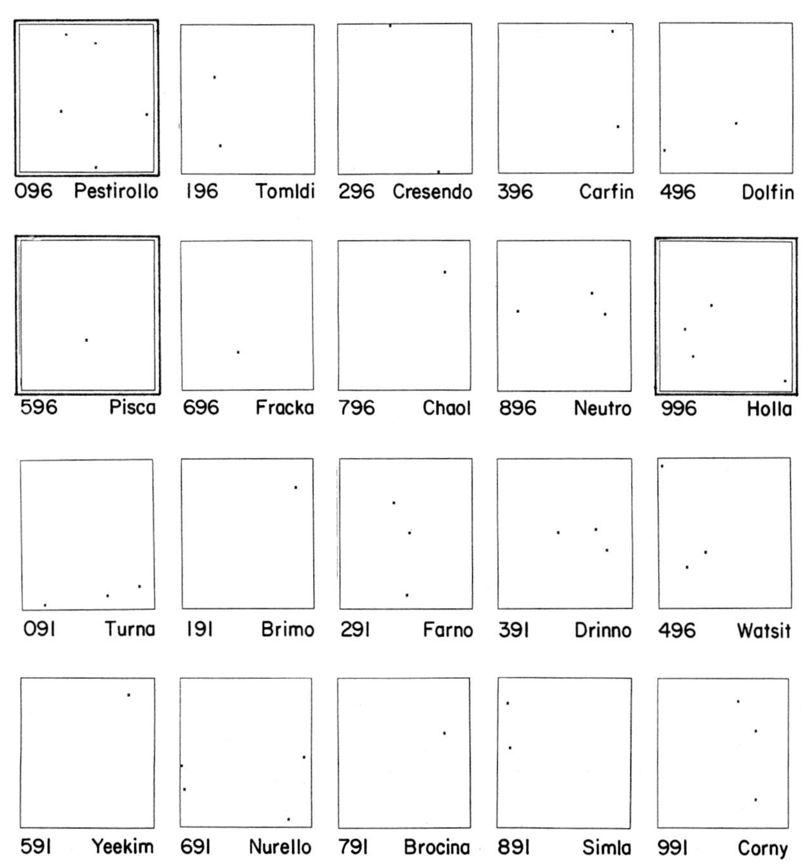

Figure 13.4. Each square represents a community of 2,500. The black pixels indicate the location of a netronella patient. Does the above pattern of events support the hypothesis that pesticide manufacturers' plants in the communities of Pisca, Pestirollo and Holla are causing more cases of netronella than the expected 1 in 1,000.

assume that the 20 squares shown in *Figure 13.4* are maps of the 20 communities with cases of netronella marked by the small black squares inside the squares (for obvious reasons the towns have been given fictitious names). We are told that if the towns had a normal rate for occurrence of the disease netronella there would be on average 2.5 cases per city. Let us also assume that the cities with the pesticide plants are Pisca, Holla and Pestirollo. These have been marked with double boundary squares in the picture of Figure 13.4.

I can tell you that through discussing similar patterns of events with citizens lacking a background in the size of fluctuations in cases of disease which can be expected by chance, and from class discussions of the patterns of Figure 13.4, that many individuals support the claim that something funny must be happening in Pestirollo and Holla in spite of the fact that Pisca only had one case. The fact that the cases in Pestirollo is double the expected number appears to be significant to the average citizen. Nonmathematically trained people are often extremely suspicious of the epidemiologist who assures them that the events in Pestirollo can occur by natural chance. In fact all of the squares of Figure 13.4 were created by turning a 2,500-digit block of a random-number table into a community of 2,500 and simulating the occurrence of netronella cases by letting one digit in 1,000 become a patient. Thus the digit 9 appears in a block of 2,500 random digits with a frequency of 250.

To create the patterns of Figure 13.4, we only permit a 9 to become a medical case in 1 case out of 100 to simulate an occurrence of a disease of 1 in 1,000 inhabitants. The three number sequence at the bottom of each square shows that the 9 was allowed to become a case of netronella if the digit above it and below it were the digits shown in the three digit number. The number 096 at the bottom of the square representing Pestirollo indicates that a 9 became a disease case location if the number above it was a 0 and the number below it was a 6. Convincing the public that a pattern of events is purely by chance, and not because of some other hidden cause, becomes even more difficult if one were to be looking at cases occurring by chance in a community of 40,000. For example, social activists shown a map of netronella cases in a city of 40,000 as shown in *Figure 13.5* are tempted to draw the squares labelled α and β and ask why one area was disease free whereas the other had 9 cases.

Sometimes, scientists warn each other of *Murphy's second law*. *Murphy's first law* states that if something can go wrong it will. Murphy's second law states that the probability of a discovery is directly proportional to its desirability.

To create Figure 13.5, we put 16 squares like those of Figure 13.4 next to each other and made a reduced photocopy of the array. Again, the array of Figure 13.5

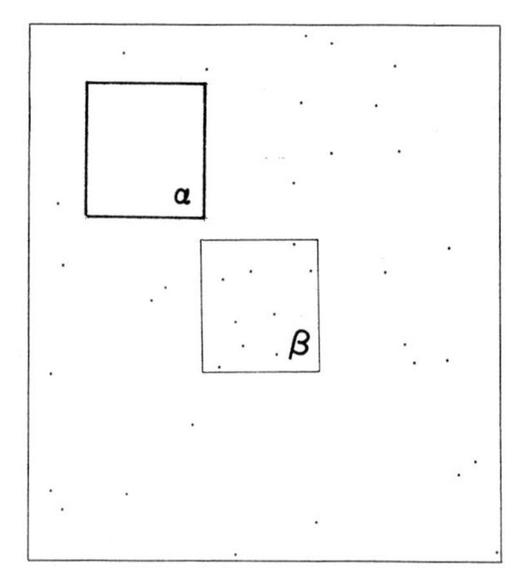

Figure 13.5. Desiring to establish a pattern of events can influence the way we draw boundaries on a field of events. The square drawn around nine events becomes a desirable boundary if we wish to prove there are areas of increased incidence of disease.

is a pattern of disease which can arise purely by chance in the city of 40,000. The inexperienced find it very hard to believe that the differences between squares α and β is a chance fluctuation in an expected disease rate of 1 in 1,000 in a city of 40,000. As communities become more aware of the possibility that there may be real disturbances to the health of small communities from factories and chemical processes, it becomes very important that the community as a whole becomes aware of the surprising patterns of disease that can occur just by chance. When discussing probable events with anxious citizens, all the formulae in the world will be less convincing than the type of Monte Carlo simulations shown in Figures 13.1 through to Figure 13.5. Students can be given sheets of random numbers and can simulate the kind of fluctuations that can occur by chance both in accident rates and in disease patterns. In this way they can gain a synthetic experience that removes the surprise from the observed patterns when they explore the real world of industrial hazards and commercial effects of industrial activity.

Section 13.3. The Importance of Understanding Coincidences and Clustering in Fine Particle Science

It is difficult to give an exact definition of fine particle science. One useful definition is that it is the study of materials in the finely divided state when the properties of the materials are created by an interaction of surface forces, such as surface area, and mass forces such as weight and structural strength [5, 6]. Face powder is a fine particle system whereas a pile of cannon balls is not. An important branch of fine particle science is the determination of the sizes of the fine particles in a powder, mist or spray. A very important group of methods for sizing fine particles is what is known as *stream counters* [7]. One important version of a stream counter is the *resistazone stream counters.*

The basic system of a resistazone stream counter is shown in *Figure 13.6*(a). Fine particles suspended in an electrolyte are caused to flow from the general container through a hole in the cylindrical tube placed in the middle of the bath of the electrolyte. Electrodes are placed on each side of the hole in the wall of the container. The electrical signal between the electrodes is interrupted if a fine particle enters the hole as the liquid flows from one container to the other. This causes a pulse to be registered by the electronic monitoring system of the counter. The size of this resistance pulse is related to the size of the fine particle in the orifice as shown in Figure 13.6(b) (i) and (ii). By causing a stream of fine particles to go through the orifice and measuring the succession of pulses, the sizes of the fine particles in the suspension can be measured.

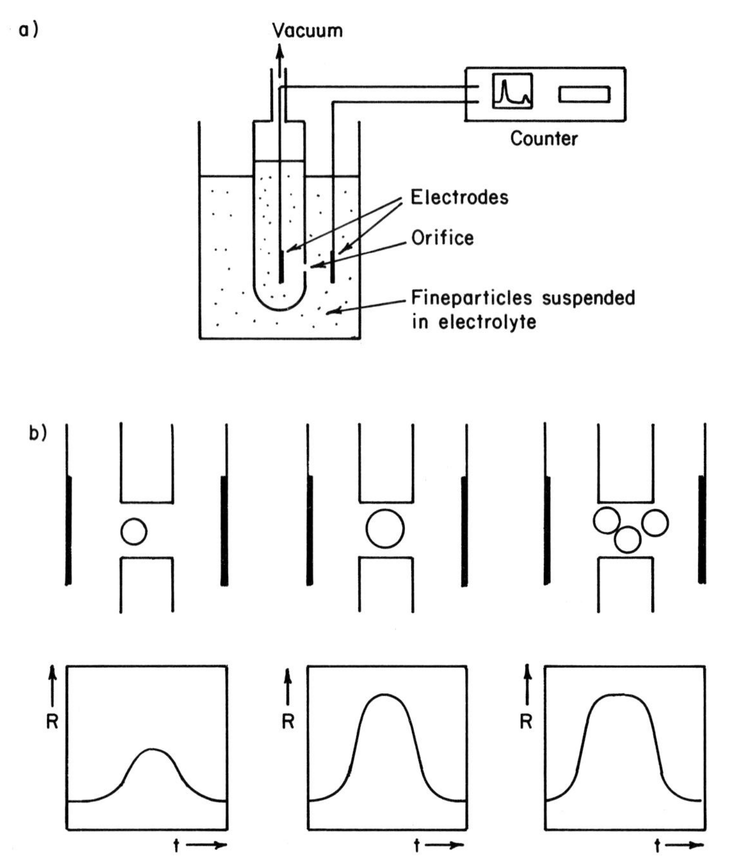

Figure 13.6. In a resistazone stream counter the size of a fine particle is deduced from the electrical resistance change it creates in an electrolyte filling the hole between two electrodes. The method is vulnerable to the coincidence errors described as primary count loss and secondary count gain. a) Schematic of a resistazone stream counter. b) Counting and sizing fine particles. i) and ii) the effect of size. iii) The effect of multiple occupancy of the orifice.

This technique is widely used in hospitals to measure the size of healthy and diseased blood cells. When carrying out fine particle sizing by this method, one must use a very dilute suspension to avoid errors known as *coincidence errors* as shown in Figure 13.6(b) (iii). If, for example, three small fine particles are present in the orifice between the electrodes, the electrode registers a higher pulse than it would have done for either of the single smaller fine particles. Therefore, the electronics register the presence of a large fine particle which does not exist.

On the other hand, the count of the smaller fine particles is less by 3 due to the misinterpretation of the signal caused by the coincidence of the three fine particles in the measuring zone. For this situation, in the language of fine particle science, the machine is said to generate coincidence errors consisting of a *primary count loss* of 3 and a *secondary count gain* which is the registering of false larger fine particles due to the coincidence of smaller fine particles in the measurement zone. To avoid these types of coincidence errors, one has to carry out measurements at a series of suspension dilutions until the measurements registered are independent of further dilution. At this suspension concentration it can be presumed that coincidence errors are no longer significant.

It is hard to visualize the frequency and extent of coincidence errors in resistazone counters and perhaps it is easier to appreciate the problems of coincidence errors when one is studying the problems of measuring the size of fine particles on a microscope slide. At the time this book was written, there was a great deal of research activity aimed at using robots to characterize the size and shape of fine particles deposited on a glass slide [8]. It is very difficult for a robot to decide whether two fine particles on a glass slide have been placed next to each other by random chance or have actually arrived together as a fused agglomerate. This is a very important problem in monitoring air pollution or looking at the efficiency of spray generators in pharmaceutical science and pesticides.

A major question relevant to the designs of robotic machines to characterize dust is: "How many dust fine particles can be allowed onto the slide before the robot is likely to make errors by treating two fine particles as one?". Note the robot will make the same kind of mistakes as the electronics in our stream counter. It will count two or three small fine particles touching each other as being a false large fine particle and in the process will fail to count the constituent primary fine particles so that again, as in the case of the resistazone counter, there will be secondary count gains and primary count losses.

Pioneer workers in the field of dust assessment were aware of the problem of creating agglomerates by touching on the deposition surface and recommended that the density of coverage on the microscope slide to be inspected for dust deposition should not exceed 5% by area coverage [9]. The problem with this recommendation is that human beings are not very good at judging area coverage of a glass slide by deposited fine particles. This fact is illustrated by the system shown in *Figure 13.7*. This system is a set of 5% covered simulated dust slides created by converting a specified digit in a random-number table into a pixel representing a dust fine particle according to the following algorithm. When a specified digit, such as 9, was encountered in the table then if a subsidiary table gave an even number, the digit was converted into a pixel representing a dust fine particle. Since any given digit had a 10% chance of occurring and if we rejected half of them at random, we simulate a 5% coverage.

Readers should show the simulated arrays of Figure 13.7 to different people and ask them to estimate the percentage area covered. They will be surprised at the results. The views of Figure 13.7 have been used to train operators to estimate the percentage of the area of a microscope slide covered by dust. If one inspects the various fields of view of Figure 13.7, it is obvious that even at 5% coverage

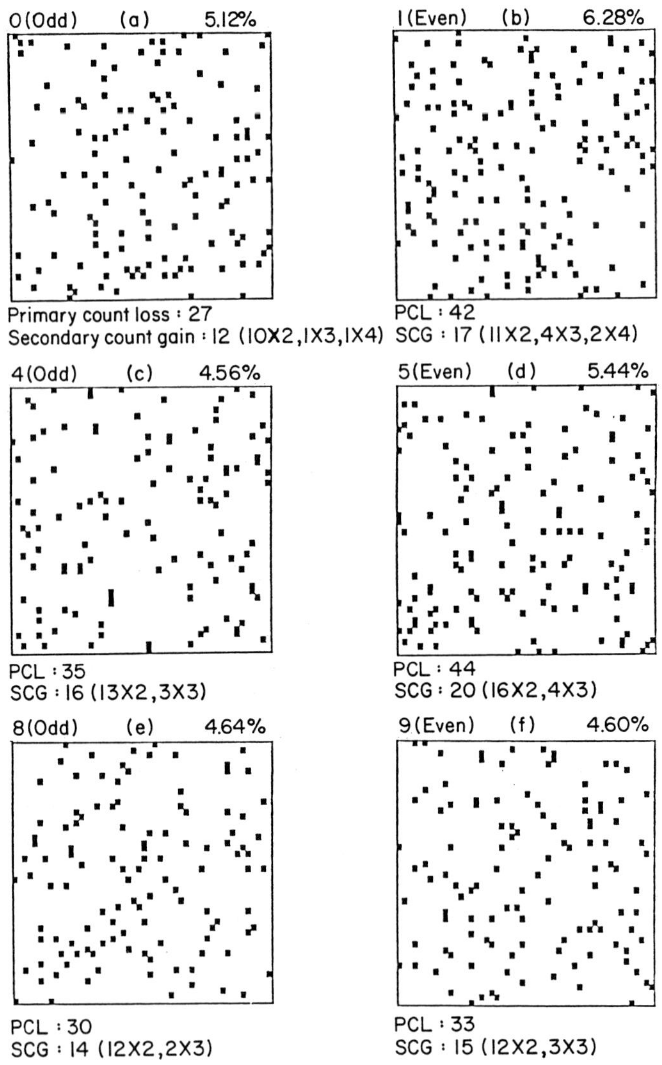

Figure 13.7. Human operators find it difficult to judge the area of a microscope slide covered with deposited fine particles.

coincidence errors are occurring. In *Figure 13.8*, the size distribution of the clusters present in the combined fields of view of Figure 13.7 is shown.

The technical term for agglomerates created by touching on the deposition surface is *juxtaposition agglomerate* from the Latin word juxta meaning "near or close to". The problem of distinguishing between real agglomerates and juxtaposition systems is a major problem in assessing dust levels to which workers in an industrial environment are exposed. Consider for example, the array of fine particles shown in *Fig-*

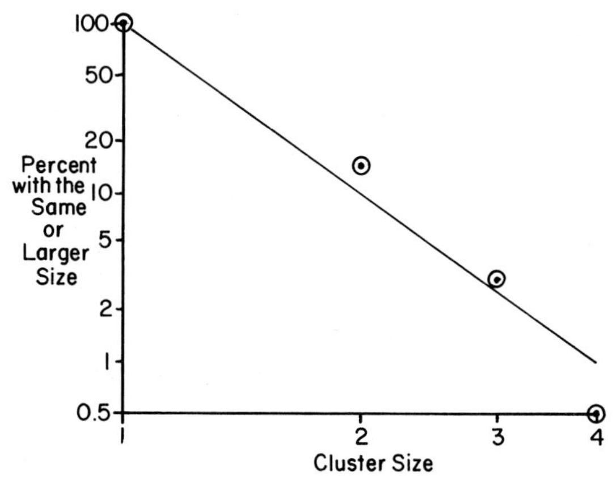

Figure 13.8. Juxtaposition cluster size distribution created by the random deposition of a monosized aerosol on a microscope slide at five percentage coverage of the field of view.

ure 13.9(a). This photograph is part of an excellent study of the size of aluminum oxide fine particles in which the investigators, Schafer and Pfeifer were using the fine particles to create visible streamlines in a wind tunnel [10]. When I looked at their research publication, I was surprised to see the size distribution that they reported from the Figure 13.9(a) as shown in Figure 13.9(b). Their reported size distribution showed no fine particles bigger than 1 micron whereas a quick glance at the systems of Figure 13.9(a) show some agglomerates larger than 5 microns. The explanation of their surprising data soon became apparent upon reading the text of their publication. They said that the systems of Figure 13.9(a) were "obvious agglomerates" and they resolved the agglomerates into the visible subunits before making their measurements. In other words, they treated the agglomerates as a bunch of grapes and measured the size of the individual grapes not the dimensions of the bunch.

If one looks at the field of view shown in Figure 13.9(c), taken from a scientific publication by Ensor and Mullins, one can see the type of clusters that have built upon the fibers of the filter [11]. These types of clusters are known variously as *dendritic agglomerates* or *capture trees.* In Greek the word dendron means "a tree" and so the two names are equivalent in meaning. In the case of the study carried out by Ensor and Mullins, they knew that they were studying a monosized *aerosol* (aerosol is the scientific term for any cloud of dust or spray droplets) and therefore they knew that the capture trees of their filter represented juxtaposition clusters in three-dimensional space.

Kaye and co-workers recently suggested that the fractal dimension of the capture trees such as those built up in a filter from monosized aerosol could become a technique for deducing the efficiency of capture forces operating between the aerosol fine particle on the fiber in the filter being used to capture the aerosol [12]. The fact that capture trees are built up from monosize aerosol as in the case of Ensor or

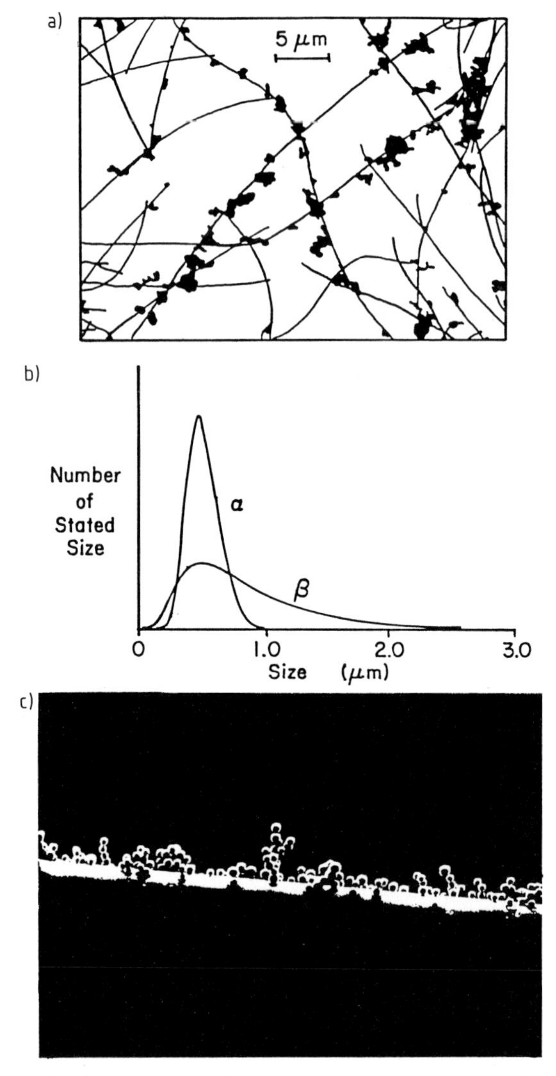

Figure 13.9. Differentiating between real and juxtaposition agglomerates in a dangerous dust assessment study is a major unresolved scientific problem. a) Alumina fine particles captured on the whisker collector [10]. b) Size distribution of the alumina fine particles of (a) by two different methods [10]. c) Dendritic capture trees formed by monosized spheres depositing on a single fiber [11].

Mullins is not a sufficient reason to interpret all the agglomerates of Figure 13.9(a) as being juxtaposition agglomerates. As in the case of the one-dimensional space of the stream counter, the only sure way to make sure that the agglomerates of Figure 13.9(a) are real is to carry out studies with decreasing deposition densities until the measured size distribution of the agglomerates is independent of the deposition density.

In the case of the aerosols used by Schafer and Pfeifer, and those by Ensor and Mullins, whether or not the agglomerates of their pictures are real or not is a somewhat minor academic problem. However, inhalation of aluminum oxide dust has been implicated as a positive factor in an industrial disease called *Shaver's disease* (named after the doctor who first described the disease amongst workers handling aluminum oxide). If we were to try to estimate the health hazard posed by the aluminum oxide fine particles of Figure 13.9, there is a world of difference between a size distribution reported from Figure 13.9 treating agglomerates as permanent entities and the size distribution shown in Figure 13.9(b). The number and surface area of dust fine particles entering the lung, if the size distribution was that of Figure 13.9(b), would represent a very different health hazard from the dust of Figure 13.9(a) if the various fine particles in the field of view represented permanently agglomerated subsidiary units present in the inhaled powder.

Assessing the exact respiratory hazard from such a field of view is an extremely contentious and debatable exercise and could result in prolonged debate over scientific evidence presented at a legal investigation into the hazards faced by the workers. The problems faced by industrial safety experts with regard to the monitoring of dust is extremely labor intensive with current technology and as part of an investigation into the possibility of using robots to monitor dust levels from photographs of filters the data shown in *Figure 13.10* was generated [8]. In this study, different levels of dust deposition were simulated to see if one could discover the permitted dust deposition limit if robots were not to be faced with the task of deciding between real and juxtaposition agglomerates.

The simulated fields of view of Figure 13.10(a) represent simulated deposition levels of 10, 9, 8 etc. percentage coverage. It can be seen that by the time we reach the percentage coverage of 2%, we did not have any obvious agglomerates. The term obvious is used here because there are four unit pixels touching the boundary of the simulated field of view and they could be parts of agglomerates from the surrounding areas of the slides sticking into the field of view. The occurrence of boundary-intercepted fine particles is a major problem in determining the accuracy of data reported by industrial scientists who study deposition levels and definite rules have been established as to whether one can count intercepted agglomerates or not in one's size distribution estimates [13, 9].

Again, the reader can cover up the information giving away the details of coverage in Figure 13.10 and discover why most people find it difficult to estimate the coverage of the viewed areas. Most people are also surprised at the level of clustering in the fields of view. The investigation from which the data of Figure 13.10(a) is taken established that in fact, one should really work with deposition levels as low as 0.1% area coverage if one wishes to avoid the formation of juxtaposition errors [8, 14].

In Figure 13.10(b, c) two more sets of data drawn from this study of the possibility of avoiding coincidence errors on microscope slides are shown. These fields of view

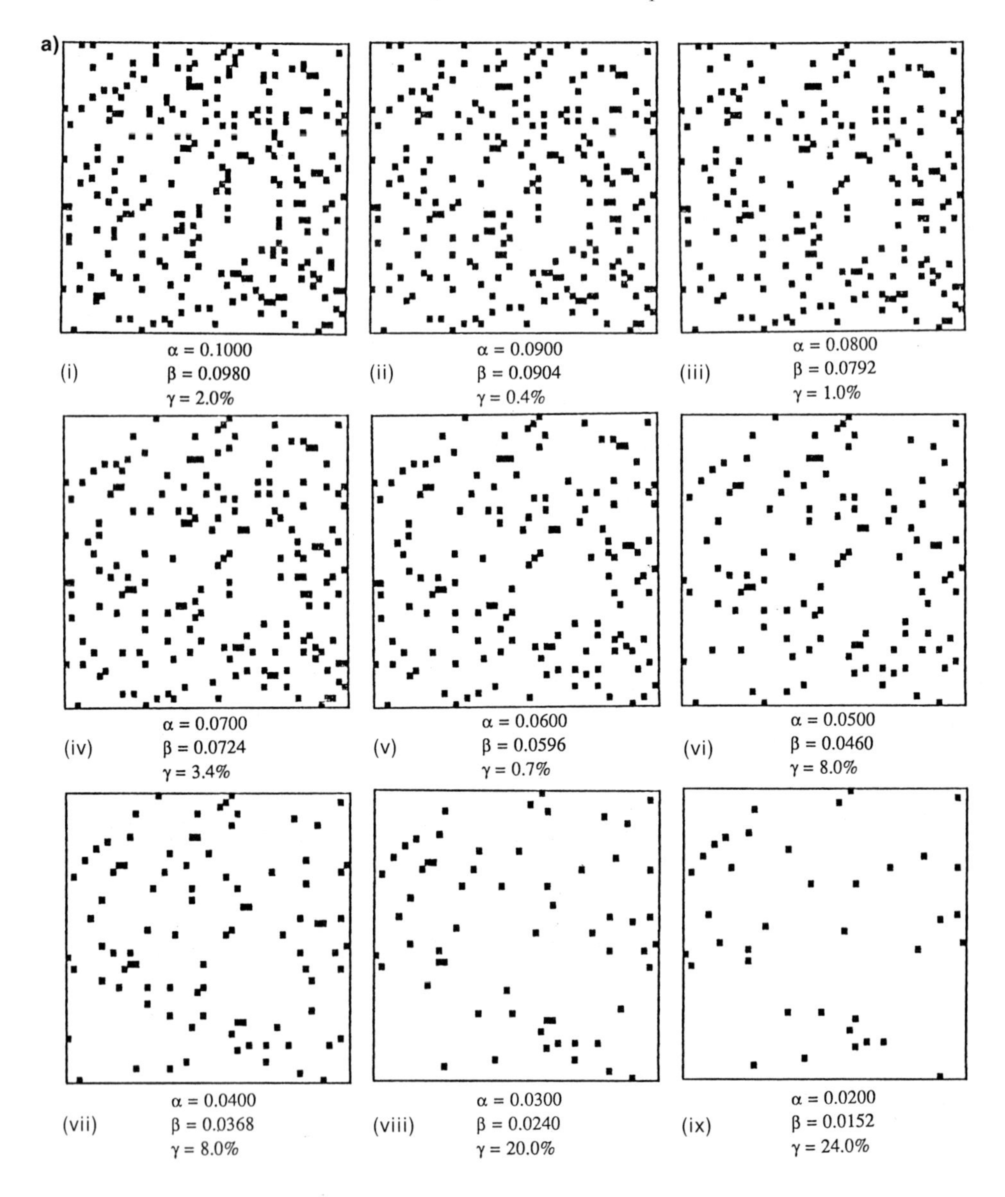

Figure 13.10. The persistence of juxtaposition clustering in two-dimensional space, as the fractional area covered by a deposited monosized aerosol is reduced, is surprising to the untrained observer. a) Sequence of simulated fields of view of decreasing percentage coverage; α = Theoretical value of simulated coverage. β = Actual value of simulated coverage. b) Simulated fields of view at an anticipated coverage level of 1.0%. c) Simulated fields of view at an anticipated coverage level of 0.1%.

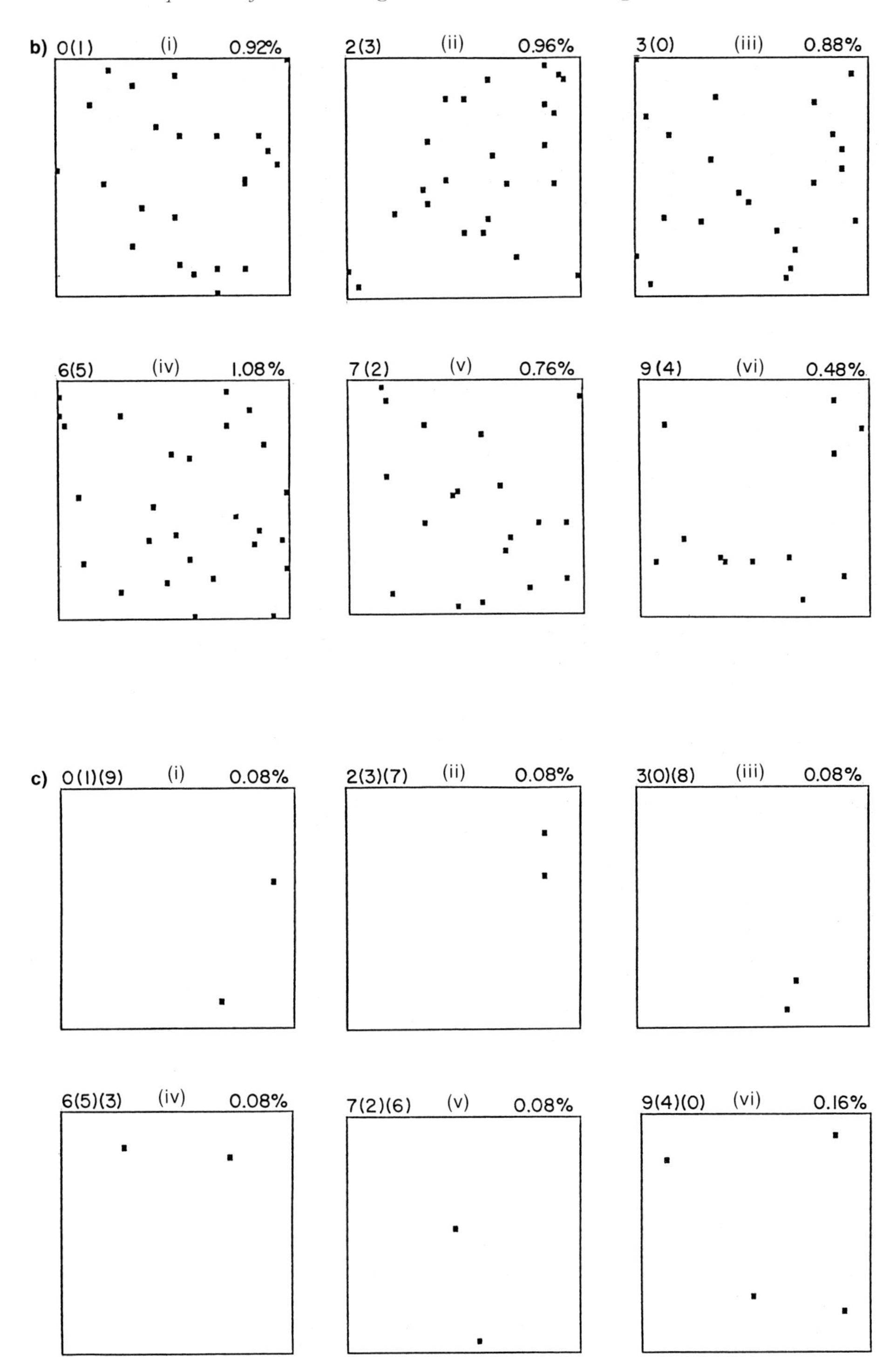

b) 0(1) (i) 0.92%
2(3) (ii) 0.96%
3(0) (iii) 0.88%
6(5) (iv) 1.08%
7(2) (v) 0.76%
9(4) (vi) 0.48%
c) 0(1)(9) (i) 0.08%
2(3)(7) (ii) 0.08%
3(0)(8) (iii) 0.08%
6(5)(3) (iv) 0.08%
7(2)(6) (v) 0.08%
9(4)(0) (vi) 0.16%

illustrate how variable the coverage of a slide is just by random chance. The field of view labelled (vi) in set (b) is from a simulation where the expected coverage was 1%. One can see the simulated coverage was actually 0.48%. When the study into avoiding juxtaposition clusters was completed it created a major dilemma for those who would like to create robots able to measure dust levels. If we were to try to avoid coincidence errors by using very low deposition rates, the act of collecting and photographing with the electron microscope the low density fields of view to be studied would become totally uneconomic. As a consequence, studies have begun to see if one could use electrostatic and/or dynamic manipulation of the dust fine particles to make them sit down in rows so that the robot would know where to look for them. It would also know that because of the way the dust fine particles had been collected there would be no coincidence errors because the deposited dust fine particles would repel each other due to electrostatic charge.

In *Figure 13.11* two filter systems are shown. The design of both of these filters involves the properties of random clustering of events in space. The filter of Figure 13.11 (a) is known as a *surface filter*. The commercially available version of this type of filter is made by the Nuclepore Corporation [15].

The type of filter shown in Figure 13.11 (a) has revolutionized the ability to filter fine particles and then photograph them with an electron microscope. The filters are manufactured by bombarding a thin film of the plastic polycarbonate with neutrons. *Neutrons* are very tiny components of all atoms. They have little mass but very high energy. The mass of a neutron is 1.67×10^{-27} kilograms. Because the neutron has no electric charge, it can pass through material such as plastic without being diverted by the cloud of electrons around the atoms of the plastic. As they whiz through the plastic, they weaken the bonds in the plastic along their line of flight. In the manufacturing process, a 15 micron thick plastic film is bombarded with a collimated beam of neutrons from the disintegration in a nuclear reactor of uranium atoms (the term *collimated* comes from a Latin word which means "travelling together in a straight line". The collimated neutrons are selected neutrons which pass through a hole in the shielding material of the reactor and only those travelling in a certain direction emerge from the hole).

After the plastic has been treated with the neutrons, it is placed in a mildly caustic solution (the word kaustikos in Greek means "to burn" and *caustic soda* is a com-

Figure 13.11. Creating the physical properties of different types of filters involves factors dependent on coincidences and clustering phenomena. a) Coincident holes can spoil the structure of Nuclepore® track etch filters. b) One method of manufacturing sponge filters involves the clustering of soluble fine particles to create percolating pathways through the filter. c) Nuclepore® TM track etch filters are useful for studying very small objects collected on the surface of the filter (scanning electron micrograph pictures taken from the trade literature of the Costar Corporation manufacturer of Neclepore® track etch filters [15] showing *Bacilus subtilis* on a 0.4 µm polycarbonate membrane filter (upper part), and on a 0.45 µm tortuous path membrane filter (lower part). d) Chrysotile asbestos on the surface of a Nuclepore® filter. e) Comparative structure of surface and depth (sponge) filters.

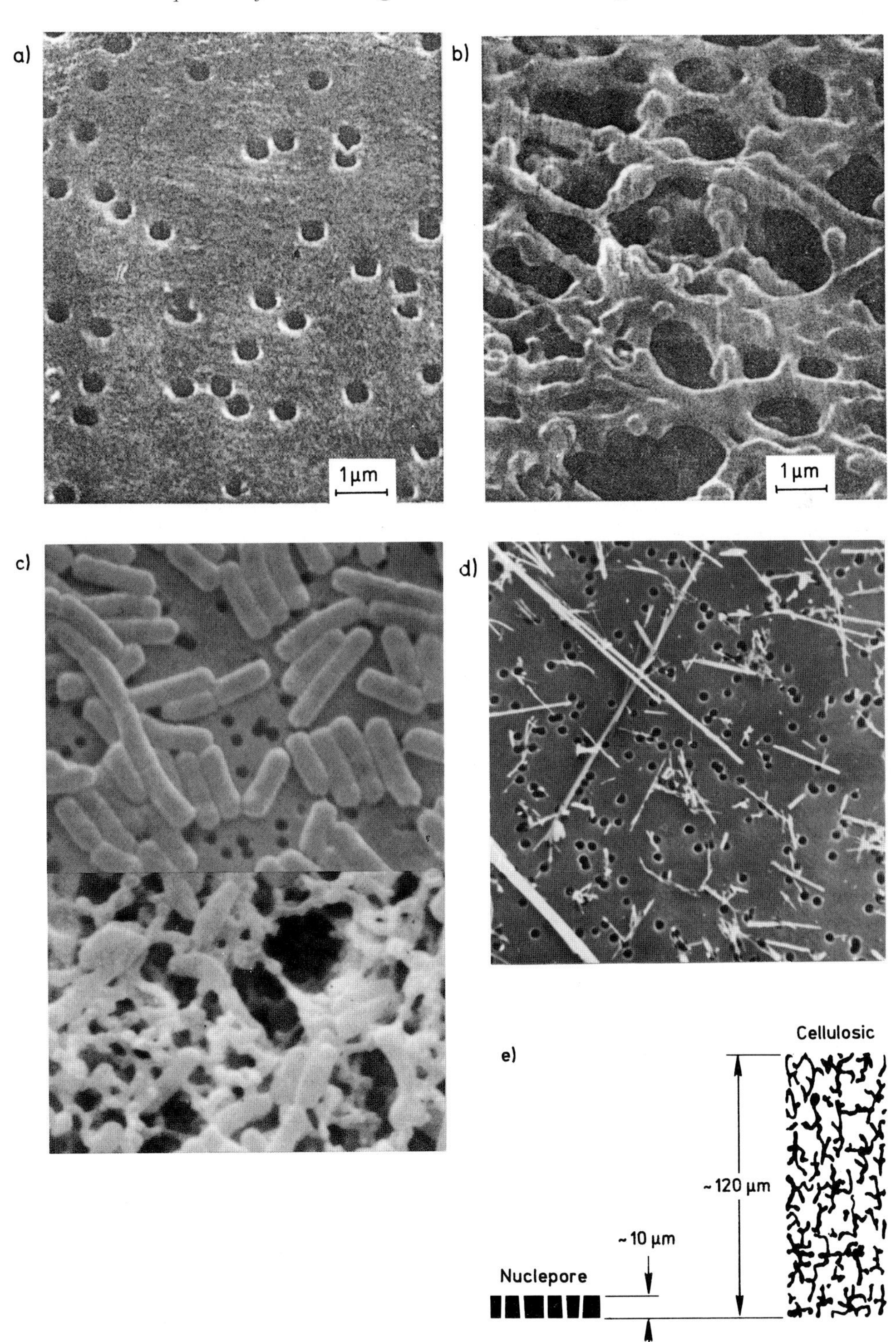
a)
1 µm
b)
1 µm
c)
d)
e)
Cellulosic
~120 µm
~10 µm
Nuclepore

pound of sodium, sodium hydroxide, which when in solution will burn holes in cloth and flesh). The caustic solution can only attack the plastic where it has been weakened by the passage of the neutron. As the caustic attack on the plastic continues, cylindrical holes develop which make the track of the neutron visible. The size of the holes made in the carbonate by the etching process depends upon the length of time that the material spent in the caustic solution [15–17].

In the case of the filter shown in Figure 13.11 (a), the holes were etched until they were 0.4 microns in diameter. As can be seen from the diagram, the set of holes created by this process is very uniform. The wavelength of yellow light is 0.5 microns and the holes of the filter shown in Figure 13.11 (a) are about the same size as the wavelength of blue light. The flux of neutrons through the carbonate is a completely random process with the probability of impact being such that all x, y coordinates for a given neutron-hit location are equally probable. The limitation on the accuracy of the filter made in this way is the probability of two adjacent holes growing into each other to create an unexpectedly large hole.

In the filter of Figure 13.11 (a) there are four pairs of holes that look as if they are about to fuse into one hole. Another pair looks as if they have already fused to make one hole double the size of the others. Making a good filter by this process is a stochastic process and the manufacturers try to have as many holes as possible to increase the flow of liquid or air through the filter at the same time attempting to keep the fusing of adjacent holes to a minimum. The reader can measure the percentage area of the filter of Figure 13.11 (a) by Chayes dot method to see what fraction of the filter consists of available holes. One can simulate the manufacturing process by making sets of locations of neutron hits at random in space by converting the digits of a random-number table using the appropriate algorithm and then drawing circles of different sizes on those holes to see how many holes fuse together at a given level of neutron flux and etched hole size.

If one looks through scientific magazines, one often sees photographs of tiny objects which have obviously been captured on surface filters of the type shown in Figure 13.11 (a). The big advantage of this type of filter for scientific study is that the material being filtered stays on the surface of the filter. In Figure 13.11 (c) two photographs of the bacteria "bacillus subtilis" on a 0.4 micron hole polycarbonate filter and on a depth-type filter of the type shown in Figure 13.11 (b) are shown. It can be seen that the Nuclepore holes form a background to the rod-like bacteria on the surface of the filter (the first type of *bacteria* studied in biology looked like the ones of Figure 13.11 and hence their name, because in Greek bakterion means "a little rod").

The bacteria are not so obvious on the sponge-type depth filter. In Figure 13.11 (d) a picture of chrysotile asbestos fibers on the surface of a Nuclepore filter are shown. Readers can see for themselves how the Nuclepore holes cluster at random in a very similar way to the simulated fields of view shown in Figure 13.7 and 13.10.

One of the first techniques used for manufacturing filters of the type shown in Figure 13.11 (b) was to fill a plastic with a powder such as pulverized salt (sodium chloride). The mixture was stirred up and poured into a thin film and then the salt was dissolved out. This type of sponge filter has tortuous paths through it which widen and narrow. The minimum diameter of this fluctuating path determines the size of filtered material that can go all the way through the filter.

The difference in structure of the two types of filters shown in Figure 13.11 (a) and (b) is illustrated in Figure 13.11 (e). The size of fine particles which cannot pass through a sponge filter cannot be deduced from the holes visible on the surface of the filter. Such filters have to be characterized using calibration tests. Using the

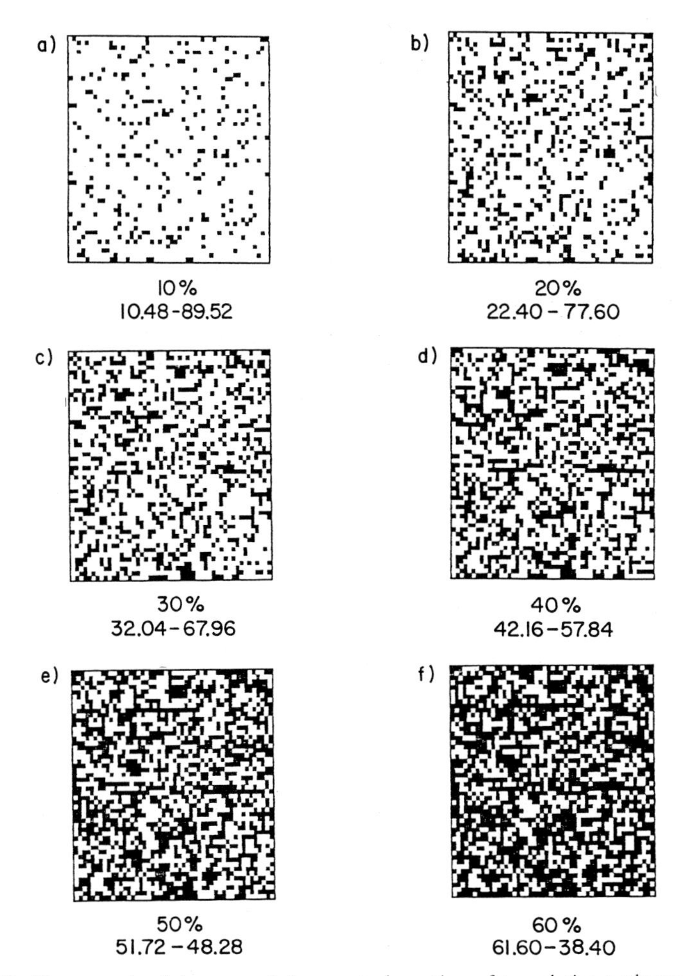

Figure 13.12. The growth of clusters and the eventual creation of percolating pathways in a composite material and/or filter can be modeled by using random-number table conversion Monte Carlo routines.

appropriate test method, the filter of Figure 13.11(b) was found to have a stopping power, described in terms of the fine particle that could be stopped by the filter, of 0.45 microns. It can be seen from the diagram that surface holes in this type of filter are as large as 4 or 5 microns. The stopping power rating must mean that further down in the filter the available pathways narrow down to much smaller diameter tubes.

In the study of fine particle systems, the term *percolating* has acquired two different meanings both of which are involved in the study of porous bodies. The term percolating comes from two Latin words per meaning "through" and colare meaning "flow". One meaning of the word percolation is "the physical flow of fluid or other objects through a porous body". When we talk about water percolating through sandstone, we are describing the physical movement through a percolating material. In the same way, if we discuss the filtering of a fluid with a depth filter, we are interested in the percolation of the liquid through the holes. Physicists recently adopted the word percolation to describe the existence of continuous paths through a material. Before we turn our salt-loaded plastic into a sponge filter, the continuous path through the salt crystals is dispersed in the plastic is discussed as the percolation of the salt crystals. In this case there is no movement of the crystals just continuity of pathways. The two different meanings of the word can become quite confusing and the reader is advised to be very careful in interpreting the word percolation when it occurs in the scientific literature. In particular, the physicist often talks about percolation theory of clusters when the clusters have not yet built up into continuous paths.

Whole books have been written on the percolating pathways (or the lack of them) the material in References 18–20. Researchers in many different disciplines of applied science are interested in the build up of clusters when powder is dispersed in a material, and in the eventual creation of percolating pathways in composite bodies. For example, paint technologists trying to optimize the scattering power of paint pigments would like to be able to stop the build up of clusters in the paint film (for a discussion of the optimization of paint reflectants in the design of paint for spacecraft, see the discussion given in Reference 18).

Pharmaceutical scientists very often mix a drug such as aspirin in an inert powder, such as starch, which is known as the *excipient powder*. When a tablet is placed in water it must break up within a specified time and for this to happen the starch grains of the composite tablet must form a percolating pathway. Pharmaceutical technologists are applying the latest theories of fractal geometry and chaos theory to the creation of percolating pathways of excipients in medication tablets and we can expect to see many studies in this area [21].

The reader can simulate the build up of clusters in a composite material as the percentage of material dispersed in a supporting matrix is increased. By varying the amount of material assumed to be dispersed in the support matrix, one can discover at which point percolating pathways start to exist in the mixture. In *Figure 13.12* simulated dispersions of black powder in white powder at various volume concentrations are shown. Scientists who have been studying the growth of clusters in such systems have worked out theories for how the clustering occurs when clustering is defined as groups of unit pixels which are joined to each other orthogonally (diagonal touching does not constitute the formation of a cluster in strict percolation theory). Scientists who have studied percolation theory have shown that if one looks at the size of orthogonally joined clusters then the number of clusters having n units in the cluster is describable by the simple relationship where F is the frequency of clusters having n or more subunits and $\tau = 2.054945$.

$$F(n) = (1/n)^{\tau}$$

As Schroeder points out, this means that the ratio of the number of clusters of two different sizes is independent of cluster size and depends only on the size ratio of the clusters [22]. Readers can test this surprising result for themselves by looking at the size distribution of the clusters in any of the simulated systems of Figure 13.12. In *Figure 13.13* the size distribution of the orthogonally structured juxtaposition clusters formed by chance at a fractional percentage cover of 20% is shown. The theoretical line for the cluster size distribution predicted from percolation theory is shown by a dotted line on this diagram. Using the vocabulary that we developed in Chapter 11, the cluster size distribution of juxtaposition agglomerates has a fractal dimension in data space of 2.05 [23].

Other specialists interested in percolating pathways are those attempting to make biodegradable plastics (*biodegradable* means "broken down" by the processes in the environment including physical and biological mechanisms). One of the ways of creating biodegradable plastic is to incorporate starch granules into a matrix of plastic. For this type of composite material to break down in the environment, the starch granules should have percolating pathways through the plastic. Some interesting diagrams very similar to those of Figure 13.12 are depicted in an article on biodegradable plastic showing how the available pathways in a composite material increase as the loading of the plastic with starch increases [24].

In the mining industry, new techniques are being developed for processing low-grade ores (ores containing only a small amount of valuable mineral) in which bacteria are allowed to grow on crushed heaps of the material. The compounds formed by the bacteria are leached out by percolating fluid moving down through the heaps of powder. In this case the interest in percolation concerns both the existence of percolating paths within a powdered heap of ore and also the rate of physical percolation of the fluid containing the bacteria through the heap and the removal of the valuable chemical from the leaching fluid after it is digested by the bacteria.

The accessibility of the ore to the bacteria is governed by the structure of the percolating pathways within the fragments and the physical movement of the fluid is controlled by the different widths of the pathway through the heap being treated.

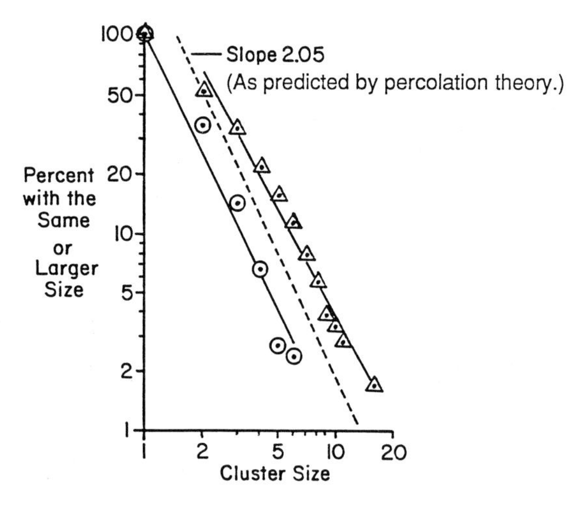

Figure 13.13. Size distribution of the orthogonal juxtaposition clusters formed by chance at 20% field of view coverage plotted on log–log graph paper.

All of these problems can be simulated on a computer using techniques similar to those discussed briefly in this section but studies of these particular projects are large-scale Masters and/or Ph.D. projects and will not be discussed further in this text [25, 26]. Another specialist interested in the existence of percolating pathways in a randomly structured mixture is the technologist concerned with making modern low-weight compact dry-cell power sources [27].

One of the surprising results of percolation theory is that there is a very sharp division between the volume concentration of stochastic loadings of monosized particles in a matrix with and without percolating pathways. Theory predicts that, for orthogonal juxtaposition clusters, percolating paths exist when the volume loading is greater than 59.28% [18, 19, 28]. The reader can test the physical reality of these limits by looking at Figure 13.12(d) and (f). Figure 13.12(d) was intended to be a 40% coverage created by turning four digits of a random-number table into black pixels. When the simulation was carried out, one actually created a field of view covered with black pixels to 38.40%. However, this can also be regarded as a 61.60% "covered white" matrix. The reader can use a pencil to track through the matrix and show that at the 61.60% level percolating pathways exist through the matrix.

The simulated field of view of Figure 13.12(f) was intended to be a 60% covered field of view created by turning six digits in the random-number table into black pixels. However, the actual coverage achieved for the random-number table conversion because of stochastic fluctuations was 57.84. Again, the reader can discover that no pathways exist through the black pixels of Figure 13.12(f). The possibility of a pathway through a random array of pixels depends to some extent on the size of the array. For the smaller arrays, one always has a chance of not having pathways when one should have them and vice versa. The reader should create fields of view such as those of Figure 13.12(d) and (f) at different total size of the array to investigate the existence–nonexistence of percolating pathways.

Section 13.4. The Catastrophic Behavior of Dripping Taps and Tumbling Rocks

As we discussed briefly in the introduction to this chapter, the term catastrophe is reserved in everyday speech for spectacular accidents. A dictionary defines a *catastrophe* as a final event, an unfortunate conclusion, a clamity. The word comes from Greek root words kata meaning "down" and strophe "turning". The word catastrophe is very similar in meaning to the modern everyday phrase used when we say that "things took a turn for the worst". In science, the term catastrophe is used to describe the behavior of a system when the consequences of the interaction of forces changes suddenly to produce a radical shift in the behavior of the system [29].

Middleton, in a glossary of terms used in the new subject of "deterministic chaos" gives the following definition of catastrophe:

> "*The term catastrophe theory was introduced by E. C. Zeeman to describe a theory originally proposed by the French mathematician, Rene Thom. Catastrophes are changes arising as a sudden response of a system to a smooth change in external conditions*" [29, 30].

To understand the technical meaning of the term catastrophe, let us consider a simple deterministic chaos system which has received a great deal of attention in

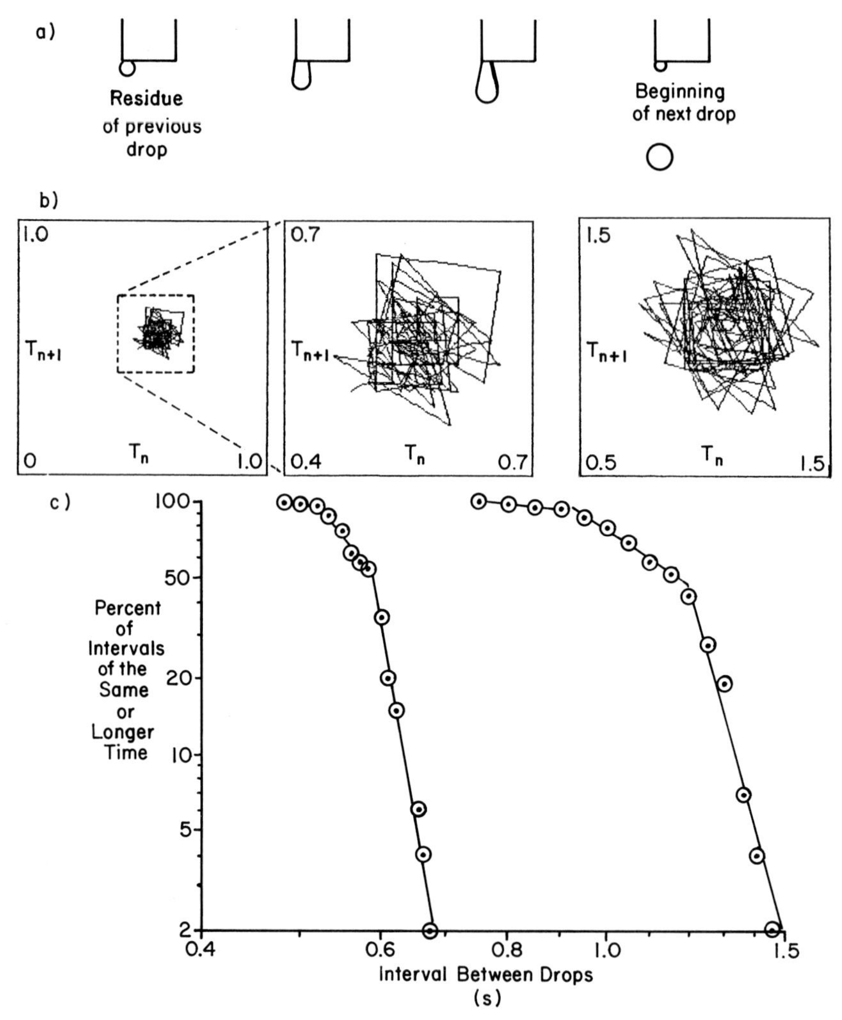

Figure 13.14. The stochastic variation in the time series generated by recording the behavior of a dripping tap can be used to generate strange attractor diagrams and descriptive fractals in data space. a) The growth of a water drop. b) Strange attractors for a tap dripping at two different rates. c) Scaling function description of time series data for dripping tap.

articles and books on deterministic chaos, a dripping faucet or tap [31–33]. Drips form at the end of the tap because the valve which is supposed to cut off the water supply to the tap when it is turned off does not completely close. The small amount of liquid which drains through the faulty valve collects at the mouth of the tap to form the drip. This very small leakage which is "the smooth change in external conditions" of our definition of a catastrophe causes a drop to start to grow at the edge of the mouth of the faucet or tap as illustrated in *Figure 13.14*(a). The shape of the drop depends upon the competition between the forces of gravity and the surface tension forces holding the drop to the tap.

To understand what is meant by surface tension, consider what happens when a drop of water is placed on a surface which has been waxed. It appears to pull itself into a spherical ball as if there was a surface force pulling the water into itself. This force is actually due to the molecular attraction of the molecules in the drop for each other but it is called the *surface tension* of the liquid because when it was first studied people thought there was a kind of tension in the surface pulling the drop together. If one adds a small amount of detergent to a water droplet on a waxed surface, it suddenly spreads out and the liquid is said to wet the surface. In fact the molecules of the *detergent* have one end which likes to be attached to water molecules and the other end which likes to be attached to wax so that the detergent forms a kind of bridge between the water and the surface. Scientists say the surface tension of the water has been altered by the addition of the detergent.

The drop at the end of the leaking tap gradually grows in size because of the continuous supply of water through the leaky valve. Suddenly it breaks away. The breaking away of the drop is the catastrophe which is the abrupt change arising as a sudden response to the cumulative effect of the smooth supply of small amounts of water added to the drop by the flow through the leaking valve in the tap. The size of the drop when it breaks away is a consequence of the interaction of several variables – environmental vibration, wind drafts, consistency of supply through the leaking valve, the roughness or smoothness of the edge of the tap, the purity of the liquid dripping from the tap and local heat changes.

The dripping tap is typical of a complex deterministic system in that if we knew all the exact magnitudes of the variables interacting to produce the drop we would be able to predict the size of the drop and the time at which it would move away from the tap using the laws of physics. In practice, we can only know the average rate of dripping and the observed time between successive drops of a real system in a stochastic variable. Expressed another way, the deterministic chaos expert says that the time series of the intervals between drops is an unpredictable set of events which might as well be chaotic even though the average behavior of the system can be described in some detail.

Recently, scientists have developed two ways of describing the behavior of the dripping tap. In the first method, the data from a time series between drops is used to construct what is known as a two-dimensional graph known as a discrete time map of the type shown in Figure 13.14(b). In Figure 13.14(b), pairs of successive time intervals in the time series between drops is used to generate the strange attractor in phase space. The lines joining successive points on the strange attractor have no physical significance but are added to the phase space diagram to create a visual image of the structure of the strange attractor.

In other situations, it is useful to display the strange attractor as a cloud of points in space. This type of data presentation will be discussed later in this section. It should be noted that if the time period between the drops leaving the tap was regular and predictable, the strange attractor would be a point attractor in space equidistant from both axes. In some cases, one can follow systems which move from deterministic chaos behavior to regular behavior with the strange attractor shrinking in space down to a point attractor (see discussion of the study of erosion of a piece of brick tumbled in a rotating cylinder discussed later). In Figure 13.14(b), the strange attractors for two rates of dripping from a tap are shown.

One can also study the time intervals between drips by drawing a graph of the cumulative occurrence of time intervals greater than or equal to a given value. When this form of the data is plotted on log–log graph paper as in Figure 13.14(c) one can see that it generates two straight lines. For 50% of the time intervals there exists a function which is a scaling function with a fractal dimension in data space of $F = 15.41$. The reason there are two lines is that the time sequences between drops leaving the end of the tap are not independent events. The exact way in which one drop breaks away determines the residual amount of liquid clinging to the end of the tap. Sometimes there will be a short time interval between drops caused by the fact that the last drop did not break away cleanly. One can obviously carry out many different experiments using roughened tap surfaces and different leaky taps etc. One could put a piece of vibrating machinery near the tap, or one could generate local drafts from an oscillating fan to vary the drip rate and the catastrophic behavior of the drop under various conditions could be summarized on the graphs.

Early on in my own studies of catastrophe theory, it seemed to me that there was a possibility of adapting the dripping tap experiment to develop a new method for studying the tumbling behavior of small rock fragments and of items such as gravel rocks taken from the beds of rivers.

The study of the different shapes of rocks generated by the fragmentation process and the effect of the fragment shapes on the subsequent use of these fragments to build stable, well-drained rock beds for such things as railway line support beds, dam embankments and foundations for buildings is important. The structure and properties of such packed beds of rock depend upon the shape and size distribution of the fragments. For example, in the Sudbury area, one of the local industries fragments cooled slag from the nickel production process to produce a kind of synthetic gravel which is used to make the road bed of a railway system. A set of the type of fragments produced by this process is shown in *Figure 13.15*(b).

Some people make the mistake of using "river rounded" gravel to make pathways in a garden or other public area only to find that the friction between the pieces of rock is too low because of their smooth rounded structure. Visitors to the public areas then find themselves deep in round rocks instead of being able to walk comfortably on tightly packed angular fragments.

To understand how one can characterize the structure of rocks from a study of the tumbling behavior, consider what happens to an angular piece of rock placed inside a rotating foam-lined cylinder. In the equipment shown in Figure 13.15(a), as the

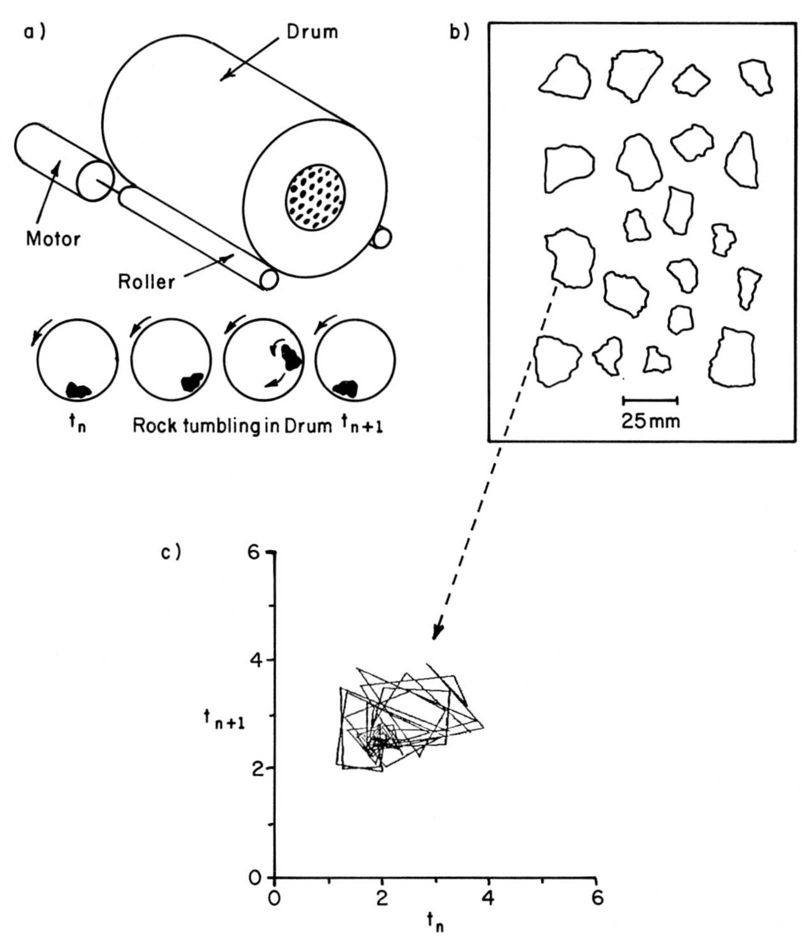

Figure 13.15. Various shaped rock fragments can generate strange attractors when their tumbling behavior is studied in a rotating cylinder. a) Equipment used to study catastrophic tumbling. b) Slag fine-particles studied. c) Typical strange attractor for the slag fine particle indicated in (b).

container rotates slowly, the rock is lifted up until it reaches a point at which it suddenly falls down and tumbles along the surface of the revolving cylinder. As the rock moves up the wall whether or not it falls is determined by the balance between the frictional forces preventing it from falling, the pull of gravity on the rock and the effect of other external influences such as vibration, air drafts etc. The interaction of these forces are affected by the position that the rock assumes after its last tumble so that successive tumbles are not entirely independent events.

The cylinder must only rotate slowly otherwise centrifugal force will hold the rock against the wall of the rotating cylinder and prevent it from tumbling down the surface. The speed of rotation obviously affects the point at which the rock will

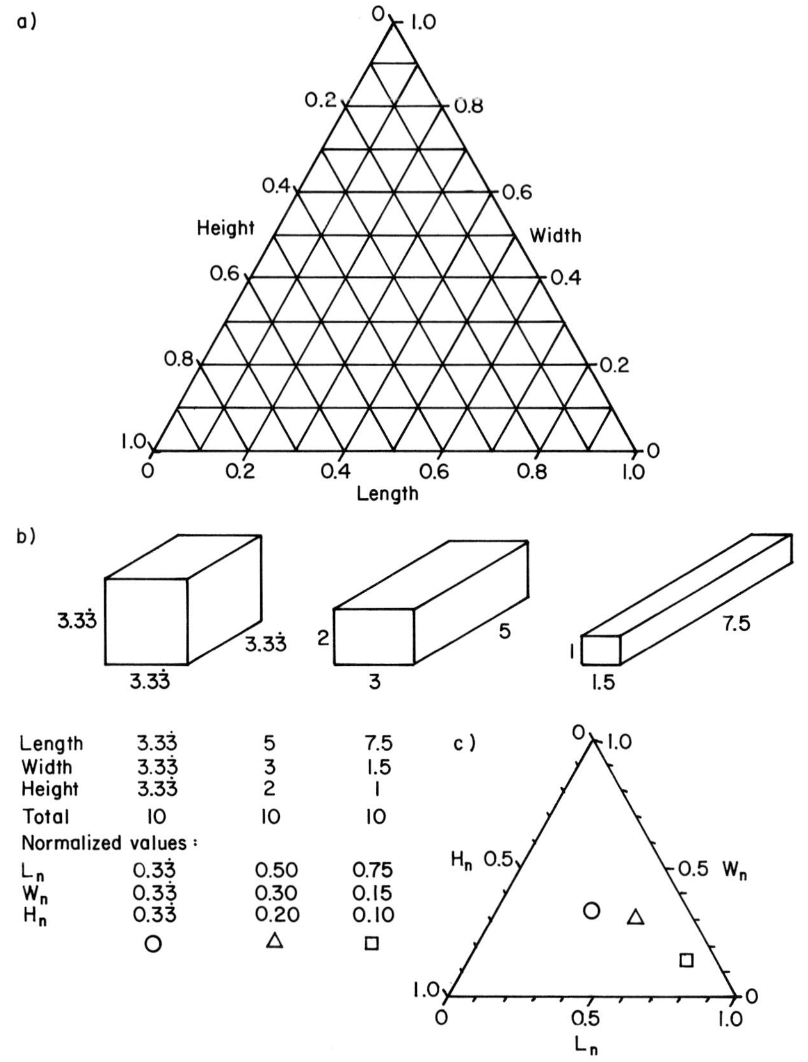

Figure 13.16. Geologists have developed a specialized triangular mesh diagram for summarizing information on the three-dimensional structure of rock fragments. a) Special graph paper. b) Representation of the technique used to normalize the data for the length, width and height of a piece of rock. c) Illustration of the technique for locating a data point on the graph descriptive of the structure of a rock piece.

tumble so that when observing the tumbling one must specify both the radius of the cylinder, the texture of the surface and the speed of rotation. The point at which the rock tumbles and the extent to which it tumbles down the surface will again be a function of the surface and also of the three-dimensional shape of the rock which affects the way in which it tumbles. Therefore if one specifies carefully the conditions

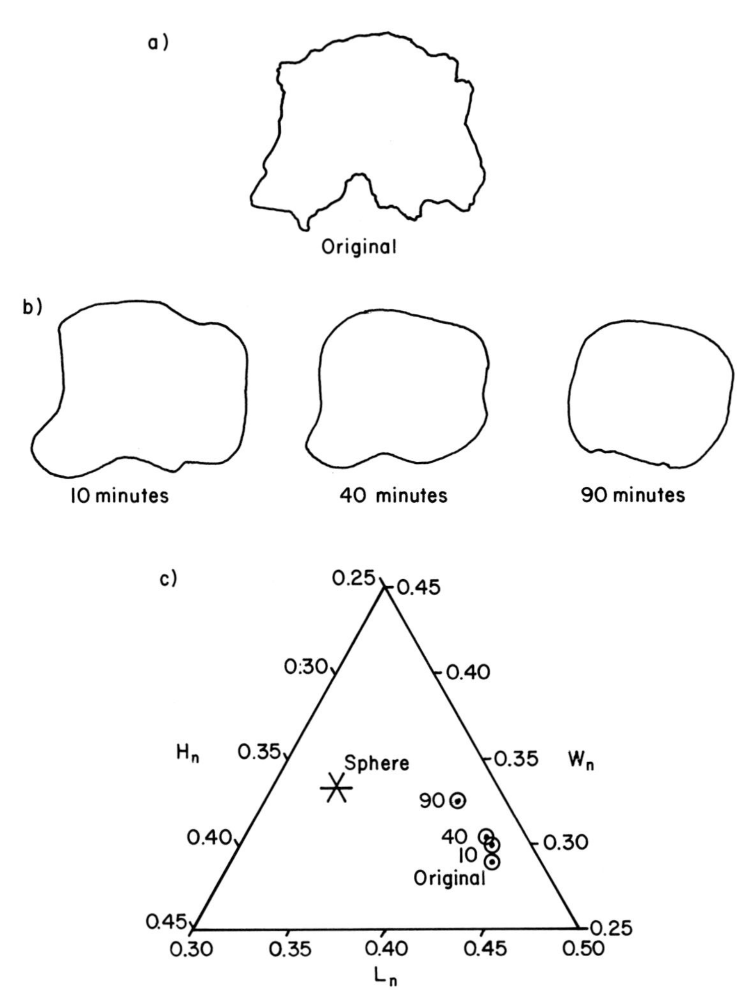

Figure 13.17. Shape changes of the eroding piece of brick studied in the experiment of J. Gratton [36]. a) Original piece of soft brick. b) A display of the shape changes of the piece of brick as it erodes when tumbled in a small ball mill for the time shown under the picture of the fragment. c) Display of the changes in the structure of the piece of brick as it erodes shown on the triangular mesh graph of Figure 13.16.

under which the tumbling rock is observed then the time series between catastrophic tumbles will contain information on the three-dimensional shape of the rock.

In Figure 13.15(c) a typical strange attractor for tumbling a slag fine particle is shown. To demonstrate the way in which one can use such tumbling studies to gain information on the structure of a piece of rock consider the data summarized in *Figures 13.16, 13.17*, and *13.18*. The studies from which these data are taken was carried out by a Laurentian University student for a senior project and involved the simulation of the type of erosion manifest in a tumbling piece of rock being moved

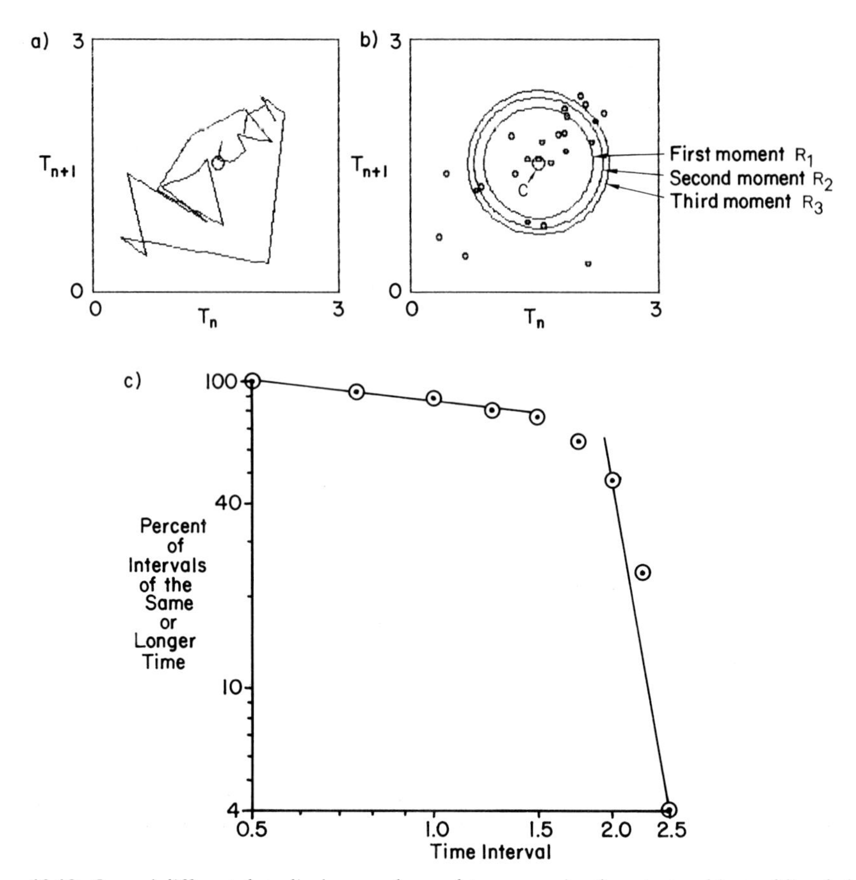

Figure 13.18. Several different data displays can be used to summarize the catastrophic tumbling behavior of the piece of brick. a) Usual format of the strange attractor of a tumbling rock. b) Data cloud transformation of the strange attractor data. c) Time series distribution function. R_n: The nth order moment of the data points. C: Centroid of the data points.

along the bed of a river by the flow of the water. In order to speed up the simulation, a relatively low strength piece of brick was used in the study. The original piece of brick used is shown in Figure 13.17(a). This piece of rock was placed in a small ballmill 11 cm in diameter filled with a slurry of sand and tumbled at a rate of 34 rpm.

Before summarizing the changes in the structure of the piece of brick during the tumbling experiments it is necessary to develop two new graphical display concepts. The first of these is the use of triangular mesh graph paper to summarize the three-dimensional shape of rock fragments. When using this special graph paper, geologists have developed a technique for representing three measurements on the dimensions of the pieces of rock. They measure the longest length of the fine-particle, then the width of the fine particle is described as the maximum dimension measured at right angles to the direction of the length. The height of the piece of rock is then defined as the thickness of the fragment at right angles to the direction at which the width has been measured. These three measurements for four Euclidean cuboid fine particles are summarized in Figure 13.16(b). Before plotting the data on the graph paper shown in Figure 13.16(a), the measurements are normalized by adding the length, width and height of the individual pieces and dividing the values for each piece of rock by this cumulative total for each rock as shown in Figure 13.16(b).

In Figure 13.16(c), the three Euclidian shapes of Figure 13.16(b) are plotted on triangular mesh graph paper. It can be seen that the center of the graph paper represents equidimensioned pieces of rock. The weakness of this type of data presentation is that one cannot represent the surface texture/structure of the pieces of rock. The center point represents spheres, cubes and other equidimensioned fragments (note however that Davies has suggested that one could use a series of triangular mesh graphs stacked one above the other to plot fragments of the same basic shape but changing in texture in the series of stacked graph paper [34, 35]).

To carry out the investigation of the erosion rate of the piece of brick it was tumbled for a specific time and then taken out of the ballmill. The length, width and height of the fragments were measured and then the fragment was tumbled in a foam lined cylinder 28 cm in diameter at 4 rpm to generate the strange attractor. To describe the structure of the strange attractor in quantitative terms, the strategy outlined in Figure 13.18(b) was adopted. The points defined in the structure of the strange attractor were treated as if they were unit mass points and the centroid of the cloud of data was calculated. This centroid is labelled C in the diagram of Figure 13.18(a). As expected it is equidistant from both axes since it represents the period of tumbling that one would have if there was no variation in the individual catastrophic tumbles of the rock in the cylinder. Technically it is the period of the equivalent complex oscillator.

One can then calculate the linear average of points from the centroid, the second and third order moments, and then generate the circles of radii R_1 and R_2 and R_3 as illustrated in the diagram. It is clear that the smaller the variation in the scatter of the points of the strange attractor about its centroid the smaller the value of R_1. Furthermore, the values of R_2 and R_3 show the presence of larger values of the time series produced by relatively rare events. One can use the ratio of these various moments of the data points about the centroid to specify descriptive parameters which contain information on the three-dimensional behavior of the tumbling piece of rock.

Another way of treating the time series data is to plot a cumulative occurrence of the time between tumbles in the same way that we plotted time intervals between successive drips leaving the faucet in Figure 13.14(c). This is shown in Figure 13.18(c).

In *Figure 13.19* the strange attractors generated at each stage of the erosion of the brick, along with the data cloud form of the strange attractor data are summarized.

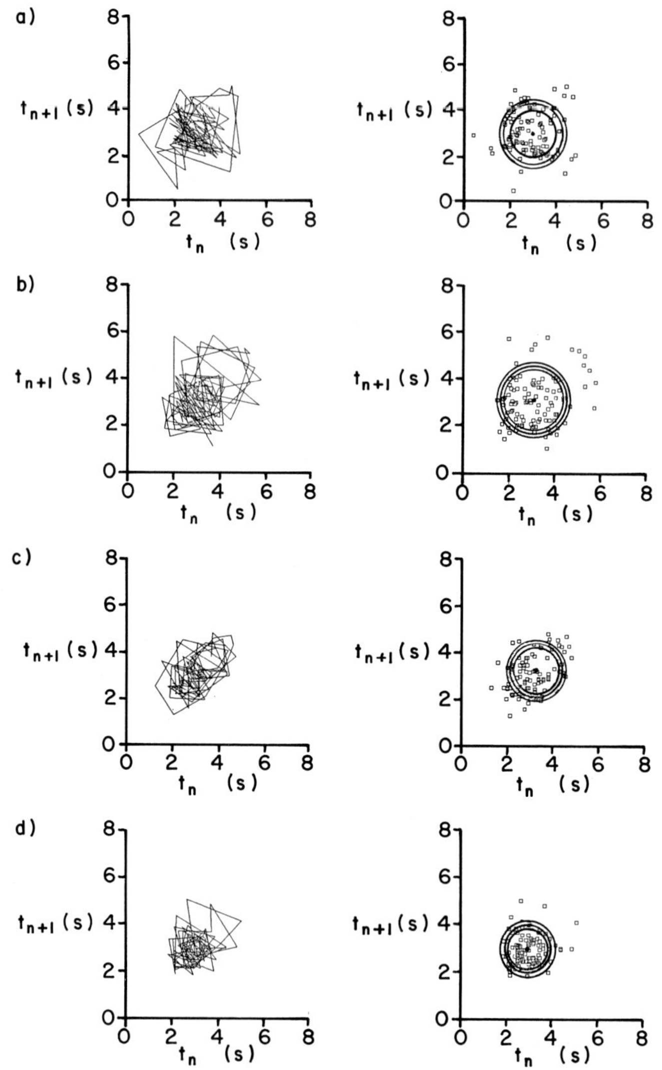

Figure 13.19. As the piece of brick eroded the strange attractor tended to shrink into itself and the centroid coordinates of the cloud diminished. a) Attractor for the original uneroded fragment. b) Attractor for the fragment after 10 minutes of erosion. c) Attractor for the fragment after 40 minutes of erosion. d) Attractor for the fragment after 90 minutes of erosion.

It can be seen that as the rock eroded, the centroid representing the average time of tumble decreased and the strange attractor became more compact as demonstrated both by the visual structure of the strange attractors [36].

To investigate the relationship between the structure of an agglomerate and its strange attractor generated by catastrophic tumbling, Y. Liu carried out a series of tumbling studies with a set of fabricated agglomerates of equal mass [37, 38]. To

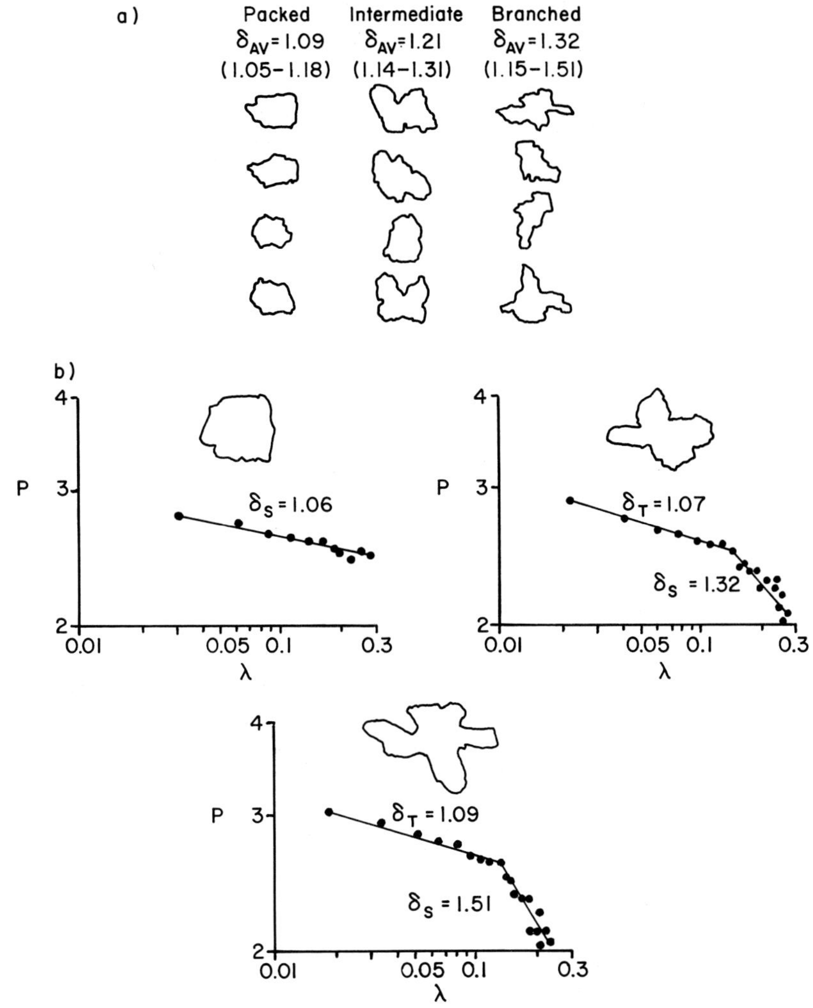

Figure 13.20. To investigate the relationship between the structure of an agglomerate and the strange attractor generated by catastrophic tumbling Y. Liu carried out a series of tumbling studies of a set of fabricated agglomerates of equal mass and various fractal dimension and morphology [37]. a) Physical appearance of the series of agglomerates made by gluing pieces of slag together. The constituent slag fragments were weighed before adding them to the agglomerate and the total weight of the assembled agglomerate kept within a defined limit. The fractal dimension of the agglomerates shown is the average value for 12 orientations in space and the range of structural boundary fractal dimension manifest in the rotation of a specific agglomerate is specified in the brackets located below the stated average fractal dimension. b) Typical Richardson plot for the three agglomerates as viewed in a position of maximum stability on a flat surface.

make these agglomerates, several pieces of slag were glued together. The total weight of each agglomerate was made approximately the same but their spatial configuration and structure was changed as can be seen from the profiles of the three agglomerates given in *Figure 13.20*(a). The fractal structure of such agglomerates varies in space. This is demonstrated by the series of projections of agglomerates summarized in Figure 13.20 (a). To create these images the agglomerate was mounted on a thin wire and rotated in front of a television camera. The reader will see that rotating the agglomerate 90° around one axis and again at 90° around an axis at right angles to the first axis will create all possible projections of the agglomerate into two-dimensional space from rotation in three-dimensional space.

For each projection of the agglomerate into two-dimensional space the fractal dimension of the profile was measured and a typical Richardson plot for each profile is presented in Figure 13.20 (b). The experimental measurements made by Liu shows that the structural fractal dimension for the fabricated agglomerates varied in that most of them had essentially the same texture fractal in all orientations (for a full description of the measurements see Reference 37).

The tumbling behavior of these agglomerates was studied using the foam-lined cylinder used by Gratton in the earlier studies of the erosion of the piece of brick. In *Figure 13.21* (a) the structure of the strange attractors of the tumbling agglomerates at a rotational speed of two revolutions per minute are shown. It can be seen that as the fractal dimension of the structures increased the centroids of the strange attractor increased and the data point cloud became more widely dispersed. The change in the tumbling behavior of the agglomerates as the speed of rotation increased is summarized in Figure 13.21 (b). It can be seen that if one plots the centroid of the strange attractor against the average fractal dimension of the agglomerate linear relationships are obtained at the three speeds used in the experiment. The slope of the linear relationship decreased with increasing speed of rotation.

The suggestion from the data of Figure 13.21 is that if one plots the slope of the line of Figure 13.21 (b) at any specific speed against the speed one obtains a graph of the form shown in Figure 13.21 (c). This indicates that by the time the cylinder is rotating at 6 rev/min the difference between the tumbling behavior of the rocks of different fractal dimension would disappear. The physical interpretation of this fact is that if the agglomerates were being poured onto a heap at a low delivery rate the fractal dimension of the agglomerates would be important in determining the structure of a heap. However if the agglomerates were falling fairly rapidly onto the heap its structure would be independent of the morphology of the arriving agglomerates. This would indicate that in static heaps of a powder the structure of the individual members of the heap is more important than when the powder is avalanching down the side of the heap and that at high speeds of avalanching the shape and texture of the profiles is only a minor factor to the dynamics of the avalanching behavior of the heap. In the next section we will look at powder avalanches as another interesting branch of deterministic chaos.

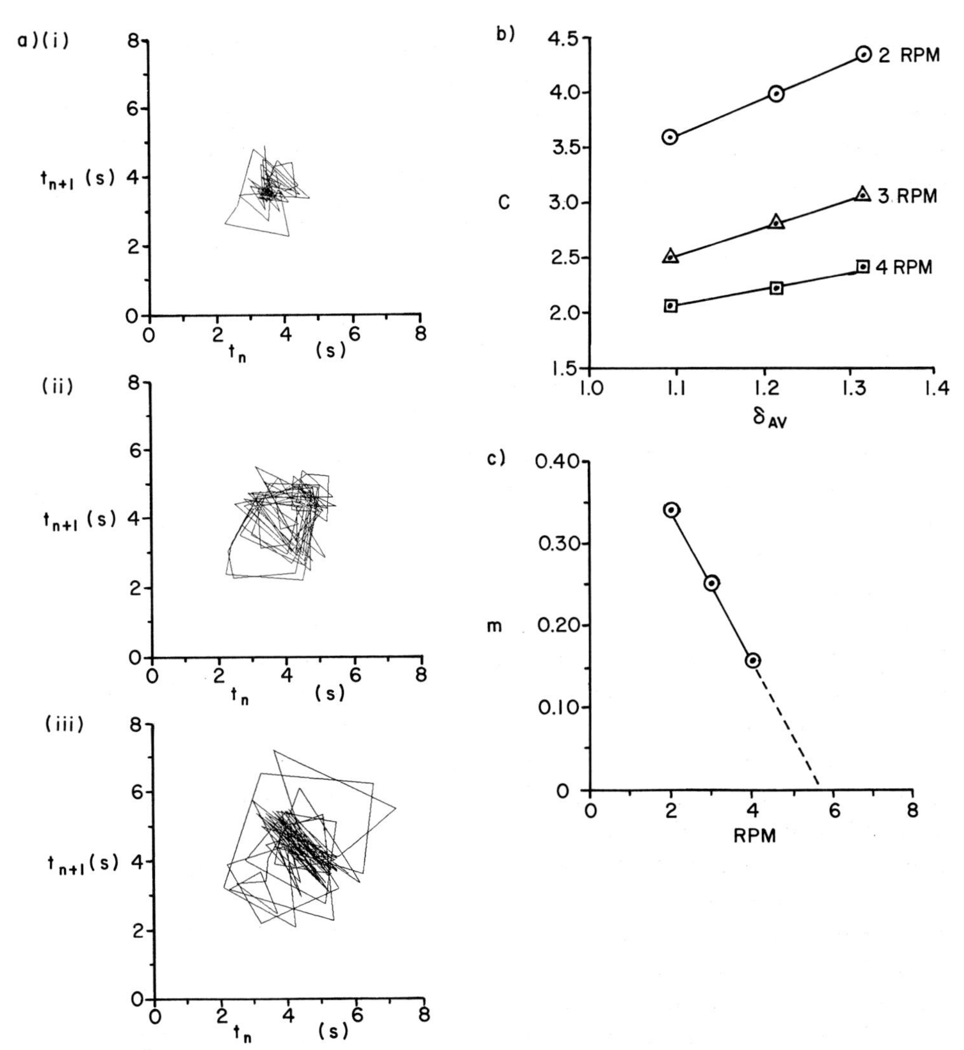

Figure 13.21. The structure of the strange attractor of the tumbling agglomerates can be linked to the fractal dimension of the agglomerate at low speeds of tumbling. a) Tumbling behavior of the various slag agglomerates at a rotation speed of 2 rpm. b) Changes in the centroid of the strange attractor at various speeds of rotation. C: Centroid of the strange attractor in seconds. δ_{AV}: Average fractal dimension of an agglomerate. c) Studies of the strange attractors of agglomerates tumbling at various speeds indicates that at high speeds of revolution the shape of the structures is no longer an important factor governing the tumbling behavior of the agglomerates. m: Slope of the centroid versus fractal dimension line.

Section 13.5. Avalanches and Earthquakes

Avalanches have always been a disastrous fact of life for people living in mountainous regions of the world. Popular stories of avalanches have acquainted the public with the idea that something as harmless as a shout can deliver enough sonic energy to an unstable mass of snow to initiate the avalanching behavior of the snow on the sides of the mountains. From the view point of deterministic chaos the sonic energy of the yell is the tiny trigger that initiates the catastrophic change in behavior and is similar to the straw that breaks the camels back.

The scientific study of avalanching is an important branch of science but a detailed discussion of the various factors involved is beyond the scope of this book. Occasionally, in the world of play or building, there will be tragedies where children or workmen are digging holes in the sand and what appears to be a stable wall of sand suddenly collapses when triggered by a very minor event. In our discussion of avalanching we will restrict our discussion to a more industrial type problem which involves the behavior of powder leaving a storage vessel.

As a powder flows onto a heap the heap builds up for a while and then collapses and if one is pouring a powder of two or more different sizes in a mixture of powders the resultant intermittent avalanching behavior is a major cause of segregation of the ingredients of the mixture [39, 40]. Although powder technologists knew that avalanching behavior was important for their technology there was very little study of avalanching behavior prior to 1990. Many of the studies that were carried out were on the behavior of static heaps, hoping that such information would prove to be useful in the ultimate study of flowing powder systems. However, as the interesting studies of Liu on agglomerates (discussed briefly in the previous section) hint there is probably very little direct relation between the parameters describing the structure of static powder heaps and the dynamic behavior of such systems.

One of the reasons why there were so few studies of avalanching behavior in the powder technology literature prior to 1991 was the absence of descriptive technology for studying such systems. However, in the early 1990s several scientists focused their attention on what were technically known as *critically self organized systems*. P. Bak. and colleagues started to look at things like earthquakes, which are major energy releases from strained rock strata released by a small trigger.

In one sense, critically self-organized systems is a branch of catastrophe theory and, as described in an interesting review article "The Avalanching Behavior of a Powder Heap", has many features in common with the behavior of earthquakes [41, 42, 44]. In particular, it has been shown that if one plots a graph of the frequency of earthquakes of energy more than or equal to the specified energy level the data line is a self similar data line on log–log graph paper. Another scientist G. Held [43] simulated this type of catastrophic phenomena by studying the behavior of heaps of very clean uniformly sized sand in which the sand being added to the heap arrived

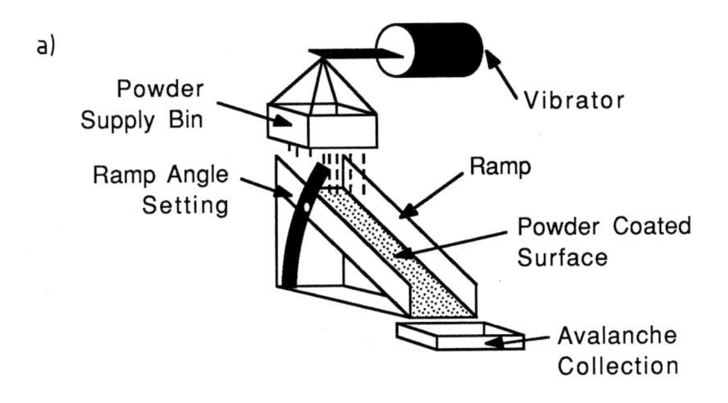

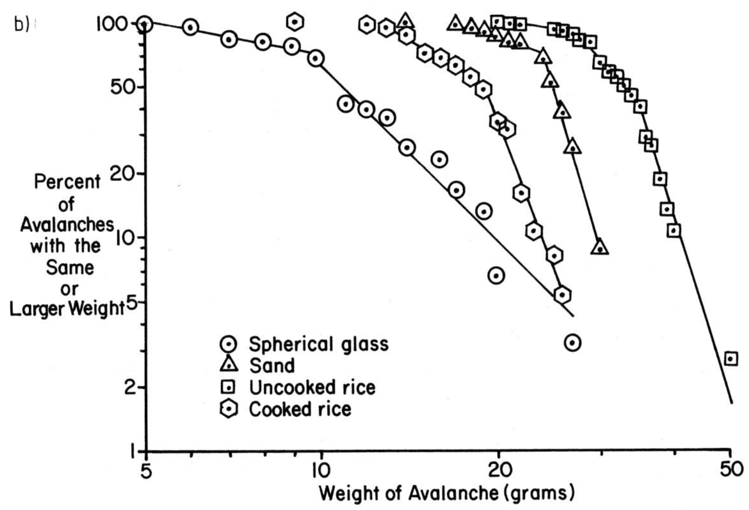

Figure 13.22. Equipment used to study the avalanching behavior of various powders and a graph of typical results. a) The ramp used to study avalanching of powders. b) Graph of number of avalanches

one grain at a time. He was able to show that the avalanching behavior of the heap of sand was very similar to that of an earthquake.

The publication of the studies by Bak and Held and co-workers soon stimulated many other studies [45–47]. A type of experiment that readers can develop easily for themselves to study the catastrophic behavior of poured sand is illustrated by the equipment and data reported by J. Gratton [36]. The equipment that she used is shown in *Figure 13.22*(a). It consisted of a sloping surface to which a layer of the powder being studied had been glued to create a surface which, to the arriving powder, looked like a powder heap. The slope of this treated surface could be changed by raising or lowering one end of the surface.

The initial angle in a series of experiments fixed at a value very close to that at which one could anticipate avalanching as determined by some preliminary experiments. A shaker which contained a supply of powder which was continually agitated by a vibrator, the amplitude and frequency of which could be controlled, was then placed over the top of the slope. At the end of the sloping plane a container was placed to catch an avalanche when it occurred. This experiment was not intended to be a sophisticated wide-ranging research investigation but was built to demonstrate the avalanching phenomena of real powder heaps with relatively low cost equipment available to investigators in high school and undergraduate laboratories.

The powder was shaken onto the sloping surface until an avalanche occurred, at which stage the container at the bottom of the slope was removed and replaced by another one without interrupting the flow of powder onto the surface. The weight of the powder representing an individual avalanche was recorded. After a series of avalanches the frequency of the weight of avalanches equal to or greater than a specific value was plotted on the log–log graph paper as shown in Figure 13.22(b).

The data of this graph represents four separate experiments. In the first experiment clean sand of average size 600 microns was studied. It can be seen that, over a considerable range of avalanches, the distribution function is a self-similar one with a fractal dimension of $F = 9.28$. The tail end of the distribution is not describable by this function and this is probably due to the relative insensitivity of the equipment to smaller avalanches.

When experiments were carried out with glass powders and food powders of distinctive shape the data shown in the graph was obtained. The bimodal distribution of the graph, having two different fractal dimensions, indicates that the avalanching phenomena on the slope is not one in which individual avalanches are independent of earlier events. It appears that smaller avalanches are governed by a different set of interactions than the larger ones. Attempts to understand the physical significance of such investigations are underway although the fascinating set of data presented in this chapter demonstrates that real powders obviously differ in their behavior from the ideal system of monosized clean smooth sand. Now that techniques and descriptive technology for studying powder avalanching have been developed, we can expect to see many research investigations in this area.

Exercises

Exercise 13.1.

A useful project for a geology class would be to take a brick and fragment it with a single blow (before doing this, the material should be covered with several layers of fabric so there are no flying fragments). The weight of the different fragments produced by the blow would be weighed and the size distribution of the fragments plotted on log–log graph paper to see if the size distribution fitted the Korcak distribution described earlier. The class could then take many measurements on the dimensions of the fragment to plot the shape distribution of fragments on the triangular mesh graph paper described in this chapter. The fragments could then be tumbled in a cylinder to determine the strange attractors. Next, several of the larger fragments could be studied as they eroded in a ballmill system of the type described in this chapter. The change in the three-dimensional structure could be plotted on the triangular mesh graph paper and the change in the tumbling behavior mapped as a series of strange attractors.

There are many variations in this experiment. Similar fragments could be tumbled using different ballmills, different slurries etc. One could also repeat the experiment with different bricks of different strengths. One could also section the brick and determine the Sierpinski fractal of the brick to see if there is any correlation between the Sierpinski of the brick and the erosion behavior of the fragments.

Exercise 13.2.

The students could easily find gravel around their home environment or the school yard. It would be interesting to map the tumbling behavior of different pieces of gravel discovered in their environment. One could also get a selection of pebbles from a stream and look at the type of pebbles found in different regions of a stream particularly if the stream is located in a hilly area. If one walks along a beach very often the gravel of the beach has been segregated by the wave action of the coastline. One could look at the tumbling behavior of rocks taken from different parts of the beach.

Exercise 13.3.

The use of equipment such as that shown in Figure 13.22 is endless. We are surrounded by commercial powders such as ground coffee, instant coffee, sugar, salt etc. Many class investigations of the avalanching behavior of such powders can be carried out. It is instructive to build a humidity control environment in which one would be able to study the modifications of avalanching behavior at high and low humidity. The more ambitious class may like to build a temperature control environment and study the avalanching behavior of powdered snow or crushed ice. The temperature control of such environment is not necessarily difficult but involves more expensive investment in the equipment than is normal in a traditional high school. However, a group of students might like to carry on this kind of study as a science fair project. They could study how modifying the addition of powder to powder heaps in different ways could modify avalanching behavior.

References

[1] For a review of how epidemiologists try to track down the cause of diseases and how specialists argue over rare events of diseases see the articles F. Pearce, "A Cartography of Cancer", *New Scientist* (September 8, 1983) 682–683. D. Taylor, D. Wilkie, "Drawing The Line With Leukemia", *New Scientist* (July 21, 1988) 53–56.

[2] News story *New Scientist* (June 19, 1986) 28, "Cancer Fears for Pastry Cooks". This short review indicates that truck drivers show an excess bladder cancer from possible exposure to diesel fumes.

[3] *New Scientist* (June 5, 1986) news story "Is Your Card Marked For Cancer". This news story covers the possibility that some people are predisposed genetically to the effects of certain chemicals or working conditions.

[4] E. Fairstein, "Do Accidents Occur in Threes?", (Letter to the Editor), *American Scientist 72* (1984) 232.

[5] A broad ranging textbook on fine particle science is J. K. Beddow, *Particulate Science and Technology*, Chemical Publishing Company Incorporated, New York 1980. Fine particle science is often referred to as powder science.

[6] A comprehensive technical level book on powder science is M. E. Fayed, L. Otten, *Handbook of Powder Science and Technology*, Van Nostrand Reinhold, New York 1984.

[7] See discussion in Chapter 6 of B. H. Kaye, *Direct Characterization of Fine Particles*, Wiley, New York 1981.

[8] B. H. Kaye, "Monte Carlo Studies of the Effect of Spatial Coincidence Errors on the Accuracy of the Size Characterization of Respirable Dust", *Particle* & *Particle Systems Characterization 9* (1992) 83–93.

[9] British Standards 3406, 1961, "Methods for the Determination of Particle Size of Powders", information available from the British Standards Institute, 2 Park Street, London W1.

[10] H. J. Schafer, Pfeifer, "Sizing of Submicron Aerosol Particles by the Whisker Particle Collector Method", *Particle and Particle Systems Characterization 4* (1988) 174–178.

[11] D. S. Ensor, M. E. Mullins, "The Fractal Nature of Dendrites Formed by the Collection of Particles on Fibers", *Particle Characterization 2* (1985) 77–78.

[12] R. Trottier, B. H. Kaye, I. Stenhouse, "Possible Links Between the Fractal Structure of Dust Capture Tree Deposits in a Fibrous Filter and Loading Effects", Proc. 5. Annu. Conf. Aerosol Soc., Loughborough University of Technology, Loughborough, England, 25–27 March 1991.

[13] See discussion of microscope methods for sizing fine particles, in B. H. Kaye, *Direct Characterization of Fine Particles*, Wiley, New York 1981.

[14] B. H. Kaye, "Operational Protocols for Efficient Characterization of Arrays of Deposited Fine Particles for Robotic Image Analysis Systems", published as Chapter 23 in *Particle Size Distribution II: Assessment and Characterization*, the proceedings of a symposium sponsored by the American Chemical Society, Boston MA, April 22–27, 1990.

[15] See the trade literature of the Nuclepore Corporation, 7035 Commerce Circle, Pleasanton, California, USA, 94566.

[16] M. C. Porter, "A Novel Membrane Filter for the Laboratory", *American Laboratory* (November 1974) 63–76.

[17] M. C. Porter, H. J. Schneider, "Nuclepore Membranes for Air and Liquid Filtration", *Filtration Engineering*, January–February 1973.

[18] See discussion of percolation phenomena in B. H. Kaye, *A Random Walk Through Fractal Dimensions*", VCH, Weinheim 1989.

[19] D. Stauffer, *Introduction To Percolation Theory*, Taylor and Francis, London 1985.

[20] A. L. Efros, *Physics and Geometry of Disorder – Percolation Theory*, translated from the Russian by V. I. Kisin, published by Mir Publishers, Moscow 1986.

[21] This aspect of pharmaceutical technology is discussed at length in B. H. Kaye, *Mixing Powders*, to be published by Chapman and Hall, London 1993.

[22] M. Schroeder, *Fractals, Chaos and Power Laws: Minutes from an Infinite Paradise*, W. H. Freeman, New York 1991.

[23] B. H. Kaye, "Computer Aided Image Analysis Procedures for Characterizing the Stochastic Structure of Chaotically Assembled Pigmented Coatings", *Particle* & *Particle Systems Characterization 9* (1992) 157–170.

[24] "Biodegradable Plastics – New Technologies for Waste Management", by T. Studt, *R & D Magazine* (March 1990) 51–56.

[25] K. H. Debus, "Mining With Microbes", *Technology Review* (August–September 1990) 50–57.

[26] B. H. Kaye, "Fine Particle Characterization Aspects of Predicting the Efficiency of Microbiological Mining Techniques", *Powder Technology 50* (1987) 177–191.

[27] See news story "Paper Gives New Meaning To Flat Batteries", *New Scientist* (December 10, 1987).

[28] R. Orbach, "Dynamics of Fractal Networks", *Science 231* (1986) 814–819.

[29] E. C. Zeeman, "Catastrophe Theory", *Scientific American 234 No. 4* (April 1976) 65–83.

[30] Dr. Middleton's glossary is given in Volume 9 of the notes for a short course first held in Toronto in 1991 by the Geological Association of Canada, Geological Association of Canada, Department of Earth Sciences, Memorial University of Newfoundland, St. John's, Newfoundland, Canada, A1B 3X5, entitled "Nonlinear Dynamics, Chaos and Fractals".

[31] See discussion of the dripping faucet in J. Gleick, *Chaos, Making A New Science*, Viking Penguin Incorporated, New York 1987.

[32] R. Shaw, *The Dripping Faucet As A Model Chaotic System*, The Science Frontier Express Series, Aerial Press Incorporated, P.O. Box 1360, Santa Cruz, California, USA, 95061, 1984.

[33] K. Dreyer, F. R. Hickey, "The Route To Chaos In A Dripping Water Faucet", *Am. J. Phys. 59 (7)* (July 1991) 619–627.

[34] R. Davies, "A Simple Feature Based Representation of Particle Shape", *Powder Technol. 12* (1975) 111–124.

[35] This method of displaying shape data is reviewed in B. H. Kaye, *Direct Characterization of* Fineparticles Wiley, New York 1981, 347–349.

[36] J. Gratton, senior project – Department of Physics and Astronomy, Laurentian University 1992. "A Study of the Dynamics of Rock Fragments and Powder Flow from the Perspective of Deterministic Chaos".

[37] Y. Liu, *M.Sc. Thesis* 1992, Laurentian University. "The Development of Intelligent Machines for Characterizing the Shape of Fine Particles."

[38] B. H. Kaye, G. G. Clark, Y. Liu, J. Gratton, "Characterizing the Three-Dimensional Structure of Rock Fragments From Their Catastrophic Tumbling Behavior", poster presentation at the Conference "25 Years of Particle Size Analysis", Loughborough University, Loughborough, England, September 17–19, 1991. A more extensive presentation of this data is B. H. Kaye, Y. Liu, J. Gratton, G. Clark, "Dynamic Shape Factors From Catastrophic Tumbling Studies and the Catstrophic Collapse of Powder Heaps", Proc. 5. Europ. Symp. Particle Size (Partec), Nürnberger Messe GmbH, Nürnberg 1992.

[39] N. Pilpel, "Cool Powders Run Fast", *New Scientist* (October 29, 1991) 313–315.

[40] The current literature on powder rheology relevant to avalanching behavior is reviewed in B. H. Kaye, *Mixing Powders*, to be published by Chapman and Hall London 1993. See also B. H. Kaye, "Fractal Dimensions in Data Space; New Descriptors for Fineparticle Systems" Submitted to Part. *Part. Syst. Charact.* April 1993.

[41] P. Bak, K. Chen, "Self Organized Criticality", *Scientific American* (January 1991) 46–53.

[42] P. Bak, C. Tang, K. Weisenfeld, "Self Organized Criticality", *Physical Review A 38, No. 1* (July 1, 1988) 364–374.

[43] G. A. Held, D. H. Solina, D. T. Keane, W. J. Haig, P. M. Horn, G. Grinstein, "Experimental Study of Critical Mass Fluctuations in an Evolving Sandpile", *Physical Review Letters 65* (1990) 1120–1123.

[44] P. Bak, K. Chen, *Physica D 38* (1989) 5–12, "The Physics of Fractals", this is also to be found in *Fractals in Physics – Essays in Honour of Benoit B. Mandelbrot*, edited by A. Aharony, J. Feder, Elsevier, Amsterdam 1990.

[45] News story in *Technology Review* (February–March 1990) 21–22, by Steve Nadis, "Sandbox Scholars". This discusses and reviews how Leo Kadanoff and Sydney Nagel are looking at the avalanching behavior of sand.

[46] C. Zimmer, "Sandman", *Discover* (May 1991) 58–59, a popular review of the work on sandpiles carried out by G. Held and colleagues.

[47] P. Plessis, B. H. Kaye, "Powder Sampling From Mixing Chambers", *Proc. 1991 Powder and Bulk Solids Conf.*, May 6–9, 1991, Rosemont, Illinois.

Chapter 14

Mathematical Watersheds and Rooting Around in Drainage Basins

Solutions to algebraic equations can flow from exotic routes

Chapter 14 Mathematical Watersheds and Rooting Around in Drainage Basins 535

Section 14.1 The Concepts of Fluxions 537
Section 14.2 Using Newton's Method for Discovering the Roots of Equations 543
References 552

Chapter 14

Mathematical Watersheds and Rooting Around in Drainage Basins

Section 14.1. The Concepts of Fluxions

In this chapter, we are going to learn how to track the fascinating journey of pixels in two-dimensional space as we program their movements to discover solutions to certain mathematical equations. In books describing the fantastic data patterns which constitute graphical summaries of attempts to solve equations by the process of guessing at a possible solution to the equation, followed by iterative calculations aimed at moving towards better and better estimates of the real solutions, the pattern of *Figure 14.1* is often shown in glorious technicolor [1].

This particular pattern is a summary of the behavior of pixels starting off from various locations in two-dimensional space, specified as initial estimates of solutions, as they seek out the locations of the solutions to the equation shown, $z^4 - 1 = 0$. The mathematical procedure used to move the pixels from their starting point towards their ultimate locations, representing a numerical solution of the equation, is known as *Newton's method of approximation.*

To be able to understand the mathematical techniques used in Newton's method for solving equations, one must have knowledge of the basic concepts of *differential calculus*. The reader with a good background in differential calculus can skip the rest of this section and begin to track pixel journeys as mapped out in Figure 14.1 by following the travel instructions set out in Section 14.2. Those readers who have never wrestled with the concepts of differential calculus should follow the rest of this section to acquire a basic survival kit from the techniques and vocabulary of differential calculus before they undertake pixel journeys exploring the patterns of Figure 14.1.

If typical undergraduates in science are asked which subject they disliked (or feared) most they would have a hard time deciding between physics and what is known as calculus. In my opinion both of them rate highly on the student hate scale because they are usually taught in todays classes with excessive use of symbolic logic which is often remote from the experience of the student. One of my colleagues, who teaches philosophy, tells me that students of logic have the same fear of symbols and

Figure 14.1. The fractal boundaries between the "watersheds" of the "drainage basins" in two-dimensional space, of pixels starting off from various starting points and seeking to migrate to the roots of the equation $z^4 - 1 = 0$, following the guidelines of Newton's methods for solving such equations by iterating x, constitute what is known as a Julia set [1].

that his classes have a high drop out rate the moment he starts to use abstract symbolism on the blackboard to express logical arguments and conclusions.

I have argued elsewhere that a course in *Blissymbolics* (a symbol-based writing communication system for use with handicapped children who cannot read the alphabet or phonic-based writing, due to damage in the speech area of the brain) can do much to overcome the student's fear of symbolic logic. It is my view that Blissymbols, which were the invention of Dr. Charles Bliss, should be introduced early in the educational system, not only as a fun way to have international penpals, but as a disguised method of introducing symbolic logic into the school curriculum. It is no accident that Chinese students, who use a symbolic logic written language, (and from which Blissymbols were derived) take to calculus like ducks to water. Their reading experience has prepared them for manipulating the mysterious symbolism of unpronounceable mathematics [2, 3].

Students are often introduced to differential calculus (the full name of the subject) by being taught a series of magic spells for doing tricks with conglomerates of symbols without being taught the history of the development of the subject, or how the development of the idea that one could work with what has come to be known as "the differential of a function" was one of the exciting turning points in the history of the development of modern science. Students are taught that Newton, when he was hit on the head by an apple, made the great intellectual leap to conceive the universal law of gravity but are rarely taught of the amazing intellectual creativity involved in the development by Newton of the mathematical techniques we now know as the differentiation and integration of mathematical functions. It is not crystal clear whether or not Newton or the famous German mathematician Leibniz was the first to work with what we now know as differential calculus.

Gottfried Wilhelm *Leibniz* (1646–1716) was born in Saxony and died in Hannover, Germany. He was a brilliant philosopher and mathematician. Some scholars regarded him as being the last human being who knew everything. By the age of 8 he knew Latin and by the age of 14 he was fluent in Greek. He obtained his first degree in law at the age of 19 and worked not only as a scholar but as a diplomat-secret agent for the Electors of Hannover. He invented a similar system of mathematics to the differential calculus of Isaac Newton, the contemporary British scientist (1642–1727).

Leibniz called his system "The study of fluxions". The basic idea in both mathematical systems was the study of the way in which a variable being studied changed with time – Newton emphasized the study of the differences in a variable at the beginning and end of a very small time interval and created the name the "calculus of differences" which later became the phrase "*differential calculus*" whereas Leibniz using Latin looked at the way in which things flowed with time and hence his name *fluxions*. Leibniz published his ideas on the study of the rate of change of variables in 1684 but several eminent British scientists accused him of having stolen the ideas from Newton. Amongst scholars the stealing of ideas is given the technical term *plagiarism*. This word is defined in the dictionary as:

> *"The act of taking the thoughts or writings of another and claiming to be the originator of the ideas".*

The word comes from the Latin word plagiarius a "kidnapper" with the basic idea built into the word coming from the word plaga meaning "a net". Leibniz vigorously denied the charge of plagiarism and, as Isaac Asimov has pointed out, it is probable that the advancement of science had brought scientists to the point where they needed to invent the subject we now know as calculus before they could make any further significant advances in the development of their study of the mechanics of moving bodies and therefore the ideas would occur independently to several scholars about the same time.

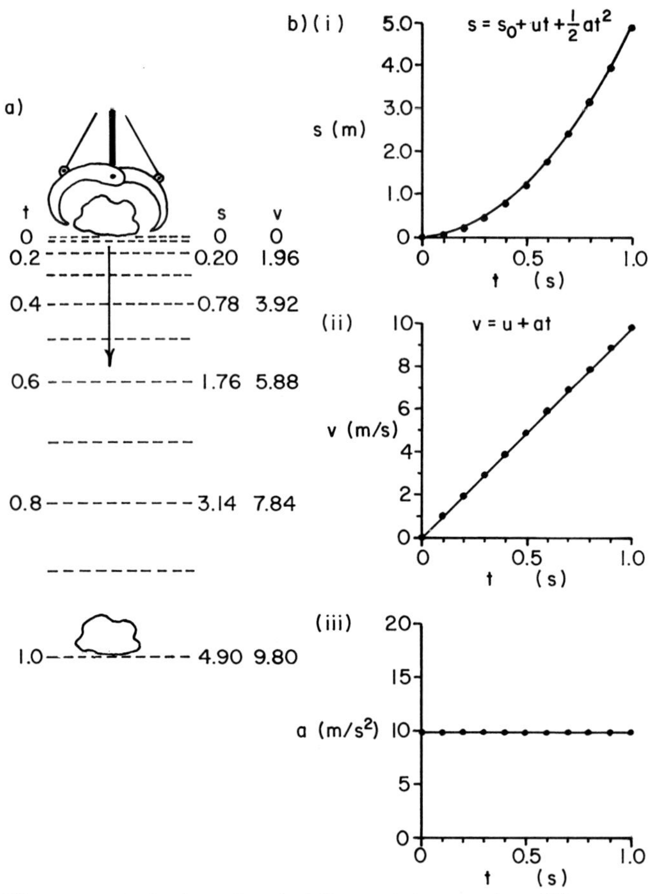

Figure 14.2. When we study the dynamics of a falling rock there are three different ways of summarizing the important aspects of the trajectory of the rock. s = distance. v = velocity. a = acceleration. u = initial velocity = 0. t = time. k = a constant. a) Sketch of a rock falling from a resting state. b) (i) Graph of position versus time. (ii) Graph of velocity versus time. (iii) Graph of acceleration versus time.

The basic concept involved in the development of the ideas of differential calculus can be appreciated from the sketches of *Figure 14.2.* If one were to study the dynamics of a rock dropped from some height as illustrated by the diagram, then one could record the dynamics of the rock in three important ways. These are the graphs of displacement – time, velocity – time and acceleration – time as shown in the sketches. In fact, we know that the rock falls with constant acceleration as determined by the attraction of the earth for the rock. Therefore, the simplest of the three dynamic relationships which can be used to describe the rock is the acceleration – time graph shown as Figure 14.2(b) (iii).

For the sake of our elementary discussion of the concepts of calculus, graphs are shown in which the velocity of the rock at the moment it began its fall was zero. The

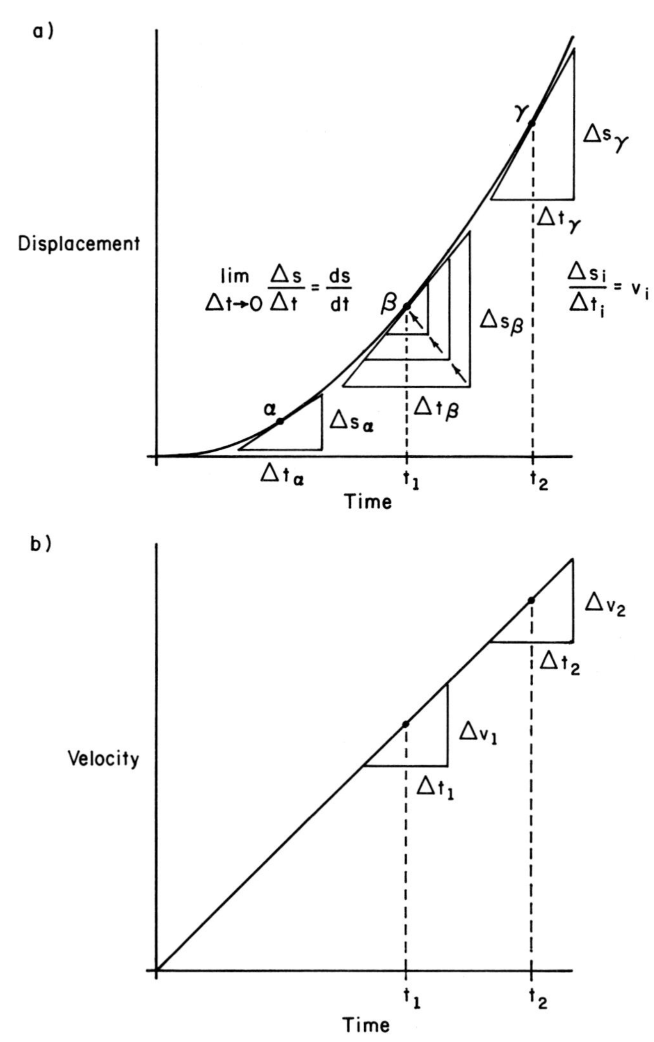

Figure 14.3. The genius of Newton was to link the three curves of Figure 14.2 by looking at how the curvature of the first two curves changed with time. He linked the rates of change in one variable to the magnitude of another. In so doing he created the subject now known as differential calculus or as it is better known "calculus". a) Differential calculus is a method of studying the rate of change of a function. b) The differential study of the velocity–time relationship of the falling stone.

mathematical formula describing the structure of the three different curves are given in the sketch. The brilliance of Newton and Leibniz was to develop a method of relating the three curves of Figure 14.2(b) to each other by looking at the rate of changes of each of the data summary curves. In *Figure 14.3* the distance–time curve for the rock falling under the influence of gravity is shown again. If one looks at a specific point of the curve such as that at time t_1 then if at that time we draw a tangent to the curve then that tangent is the instantaneous rate of change of the distance being travelled by the rock in a very tiny time interval.

To be able to express their ideas in mathematical symbols Newton and Leibniz invented new symbols to capture and represent their ideas. It is a strange twist of fate that today we use the symbolism of Leibniz rather than that of Newton to express symbolically the ideas of calculus. To represent a small difference in time the symbol Δ was chosen (presumably because it was the Greek form of the first letter of the word difference). To describe the slope of the tangent at time t_1 on the distance–time graph the scientists said that we have to consider the small change in distance, Δs, which occurred in the small time difference Δt. They then said that to draw a tangent on a graph, such as that drawn to β in Figure 14.3(a), one had to use relatively large values of Δs and Δt but that one could imagine that the same value was obtained as when one considered smaller amounts of distance and time as suggested in the figure.

Eventually, when carrying out this shrinking of the small time and distance intervals, one imagines that one reaches infinitely small amounts which are now written by the symbols ds and dt. Mathematicians make the statement that, in the limit, as Δt tends to zero, the value of the tangent is given by the differential relationship ds/dt. Symbolically this is written:

$$\lim_{\Delta t \to 0} \frac{\Delta s}{\Delta t} = \frac{ds}{dt}$$

It is obvious that such a function is the instantaneous velocity of the falling rock at the time t_1 and so scientists made a statement that the differential version of the distance–time graph is the velocity–time graph. This is the second of our dynamic graphical summaries of the falling rock movement shown in Figure 14.3(b). This is shown by the tangents drawn at α, β, γ of Figure 14.3(a). The act of calculating the tangents of the various points such as α, β, γ described by Newton as the calculation of differences, a phrase which eventually was turned around to differences calculations and eventually was shortened to differential calculus. Calculating the instantaneous value of a function is called taking the *differential* of that function and the quantity was called the *differential function.* In today's scientific jargon, the velocity–time curve is described as the differential function of the distance–time curve. The jargon obscures the simple idea that the velocity–time curve is developed from the distance–time curve by looking at the way in which the distance–time curve changes at each instant of time.

If one now carries out the same type of mathematical operation on the velocity–time curve, we find that at times such as t_1 and t_2 the tangent drawn to the curve representing the range of change in velocity with time is the same at all the points of the curve. We say that the differential function of the velocity–time curve is a constant and equal to the acceleration. This is because dv/dt is the rate of change of velocity which is by definition the acceleration of the object. We can now take the three relationships from Figure 14.2 and write them in a new way as shown below.

$$s = ut + 1/2\,at^2$$

$$\frac{ds}{dt} = v = u + at$$

$$\frac{d^2s}{dt^2} = \frac{dv}{dt} = a$$

These are described as the *differential equations* of motion describing the motion of the rock. When writing down the expression for the acceleration we note that we have taken small amounts of a quantity which itself is expressed in terms of small amounts of distance. Mathematicians have agreed that when one carries out a study taking small quantities of a small quantity this operation shall be written using the following symbolism:

$$\frac{d}{dt}\left[\frac{ds}{dt}\right] = \frac{d^2s}{dt^2}$$

The differential equations of motion are often written in the following format:

$$\frac{d^2s}{dt^2} = a$$

$$\frac{ds}{dt} = u + at$$

The first equation is described as being a second order differential equation and the second equation is written as a first order differential equation. Another way that scientists write the differential form of the function f(x) is to use the following notation:

$$\frac{d}{dt}(f(t)) = f'(t)$$

Now that we have developed a basic knowledge of the concepts of the terms and vocabulary of differential calculus we can explore the utility of Newton's methods for finding the solution of difficult equations.

Section 14.2. Using Newton's Method for Discovering the Roots of Equations

For historical reasons, the numbers which are the solution to an equation are known as the *roots of the equation*. We know that for the equation $z^2 - 1 = 0$ there are two numbers which can be substituted in the equation to give the value zero. These are + 1 and − 1. Thus + 1 and − 1 are said to be the roots of the equation. Both of these solutions for this particular equation exist in one-dimensional space which we can describe as line space. To illustrate how one can use what is known as Newton's methods for finding the roots of an equation let us assume that we did not know that the roots of Equation 14.1 are + 1 and − 1.

$$z^2 - 1 = 0 \tag{14.1}$$

Newton's method for locating the roots of an equation written symbolically as $f(z) = 0$ consists of iterating the relationship shown in Equation 14.2, where z_1 is a better estimate of a root deduced from a beginning guess of z_0.

$$z_1 = z_0 - \frac{f(z)}{f'(z)} \tag{14.2}$$

For $z^2 - 1 = 0$, since the differential function of $z^2 - 1$ is $2z$ we obtain Equation 14.3.

$$z_1 = z_0 - \frac{z_0^2 - 1}{(2z_0)} \tag{14.3}$$

Before carrying out the calculation and subsequent iteration of the calculation it is useful to simplify the relationship 14.3 as follows:

$$\begin{aligned} z_1 &= z_0 - \frac{z_0^2 - 1}{(2z_0)} \\ &= \frac{2z_0^2 - z_0^2 + 1}{2z_0} \\ &= \frac{z_0^2 + 1}{2z_0} \end{aligned}$$

Details of this calculation for a complex number $z = x + yi$ are shown in *Table 4.1*. Iteration of the calculation from a first guess of $z_0 = 1.5 + 0.5i$ gives us the sequence:

Real Part	**Imaginary Part**
1.05	0.15
0.99167	8.33333×10^{-3}
0.99999	-7.00230×10^{-5}
0.99999	4.10410×10^{-11}
1.00000	-8.80914×10^{-20}
1.00000	0.00000

As can be seen from this sequence, the iteration homes in on the root 1.000 + 0i very quickly. If we carry out a series of calculations from various guesses we soon

Table 14.1. Details of the calculations required to iterate the equation $z^2 - 1 = 0$ by Newton's method for a complex number $z = x + yi$.

For the equation

$$z^2 - 1 = 0$$

to solve for the roots of this equation using Newton's Method we need

$$f(z_{n+1}) = z_n - \frac{f(z_n)}{f'(z_n)}$$

where $f(z_n) = z_n^2 - 1$ and $f'(z_n) = 2z_n$

$$\text{so } z_{n+1} = z_n - \frac{(z_n^2 - 1)}{2z_n} \qquad z_{n+1} = \frac{z_n^2 + 1}{2z_n} \qquad \text{Eq. 1}$$

The division operation for complex numbers is as follows: for two complex numbers $z_1 = a + bi$ and $z_2 = c + di$

$$\frac{z_1}{z_2} = \frac{a + bi}{c + di} = \frac{ac + bd}{c^2 + d^2} + \frac{bc - ad}{c^2 + d^2}i \qquad \text{Eq. 2}$$

if we now let $z = x + yi$ then

$$z^2 = (x + yi)(x + yi) = (x^2 + xyi + xyi + (yi)^2) = (x^2 - y^2 + 2xyi)$$

(Remember that $i^2 = -1$.)
So our Equation 1 becomes

$$z_{n+1} = \frac{z_n^2 + 1}{2z_n} = \frac{(x^2 - y^2 + 1) + 2xyi}{2x + 2yi}$$

and therefore by Equation 2

$$a = (x^2 - y^2 + 1), \quad b = 2xy, \quad c = 2x, \quad d = 2y.$$

Now we can calculate the final version of the equation which can be used to iterate the values using the complex number $z = x + yi$.

$$ac = (x^2 - y^2 + 1)(2x) \qquad bd = (2xy)(2y)$$
$$bc = (2xy)(2x) \qquad ad = (x^2 - y^2 + 1)(2y)$$

When we carry out the calculations we find that

$$ac = 2x^3 - 2xy^2 + 2x \qquad bd = 4xy^2$$
$$bc = 4x^2y \qquad ad = 2x^2y - 2y^3 + 2y$$

then

$$ac + bd = 2x^3 + 2xy^2 + 2x$$
$$bc - ad = 2x^2y + 2y^3 - 2y$$
$$c^2 + d^2 = 4x^2 + 4y^2$$

Now from Equation 2 we find

$$z_{n+1} = \frac{2x^3 + 2xy^2 + 2x}{4x^2 + 4y^2} + \frac{2x^2y + 2y^3 - 2y}{4x^2 + 4y^2}i$$

This can be now be used to calculate the iterations of Equation 1 for a complex number such as $z = x + yi$.

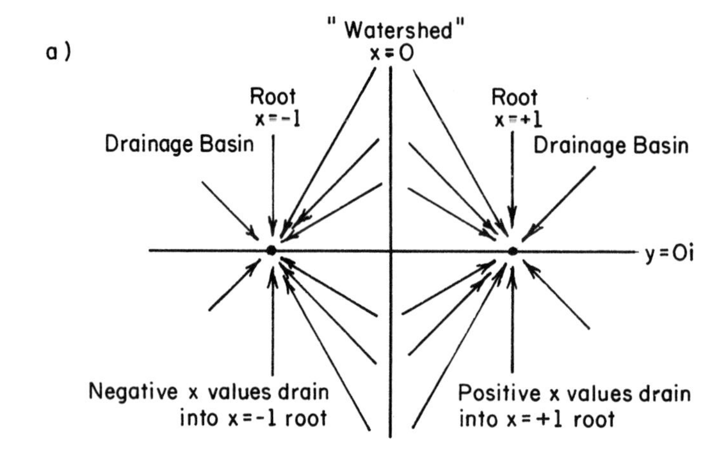

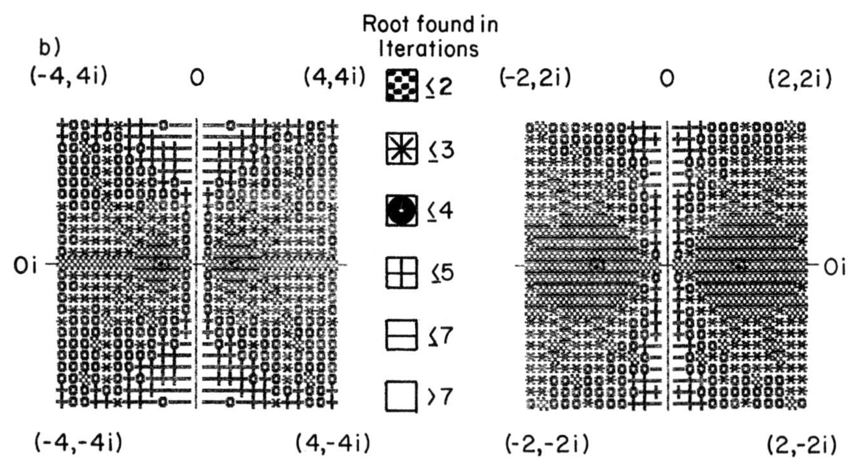

Figure 14.4. Drainage basins formed by the iteration of various values of z in the equation $z_{n+1} = (2z_n^2 + 1)/2z_n$ can be tracked using a pixel rainbow similar to that of Section 1.5.

discover that any positive estimate of z_0 homes in on the root 1 and that any negative guess homes in on the root -1. As in our drawing of a pixel rainbow in Section 1.5 we can create a linear space rainbow and we can color the speed at which an initial guess moves towards one or other of the two roots of the equations as illustrated in *Figure 14.4*. In the language of the mathematician, the zero point on line space becomes a watershed for the movement of the pixels and just as water will run downhill from a watershed so guesses run downhill to the drainage basin of the root of the equation.

From our experience in line space tackling the location of roots of the equation $z^2 - 1$, what would we anticipate as the behavior of a pixel representing a point in two-dimensional space as it moved toward the location of the roots of the equation $z^3 - 1$? Moving the pixel according to the iteration of the calculation for improved estimates from an initial guess using Newton's methods of approximation will help find the location of the roots. The roots of the equation $z^3 - 1$ are known to be $z = 1 + 0i$, $z = -0.5 + 0.866i$, $z = -0.5 - 0.866i$.

If we were to represent these roots in two-dimensional space as shown in *Figure 14.5*(a) it would seem reasonable to anticipate, that the iteration would converge to one of the three solutions (mapped in Figure 14.5) depending on the sector in which the initial value z_0 is located [4]. We would expect that three basins of attraction partition the complex plane into three 120° pie-shaped pieces.

To use Newton's method to track the movement of a pixel we write the calculation necessary for the iteration in the form:

$$z_{n+1} = z_n - \frac{f(z)}{f'(z)} = z_n - \frac{z_n^3 - 1}{3z_n^2}$$

Again this can be simplified to the form:

$$z_{n+1} = \frac{2z_n^3 + 1}{3z_n^2}$$

If one were to select the three points shown as P_1, P_2 and P_3 in Figure 14.5 then nothing particularly surprising would happen and the pixels located at these starting points would move into the drainage basins representing the roots of the equation. The result of the iterations for the three points shown in Figure 14.5(a) leading to the three roots labelled R_1, R_2 and R_3 are shown in *Table 14.2*.

The orthodox behavior of these three pixels as they move into the drainage basin of the expected roots does not prepare the student for the amazing behavior of pixels close to the suggested watersheds of the three drainage basins. The student can discover the surprising behavior of pixels in the vicinity of the watershed by looking at the behavior of pixels which start off at some point P_4 on the line joining R_2 and R_3 and moving it along the line joining the two points to generate the iterative calculations summarized in Figure 14.5(b).

This apparently crazy behavior of the iterative calculation in the vicinity of the postulated simple watersheds for the drainage basins of the roots of the equation $z^3 - 1$ had been noted by the British mathematician Arthur *Cayley* (1821–1895). In 1879 he had tried to extend the iterative process for exploring movement in the complex plane when seeking solutions to equations similar to $z^3 - 1$ and was fascinated by the unpredictable behavior of the iteration process in the watershed region. Since the power of modern computers was not available to Cayley he had to leave the problem for others to explore in detail.

The first modern mathematician to explore the behavior of pixels in the complex plane when iterated in the search for roots for equations was John Hubbard. Gleick

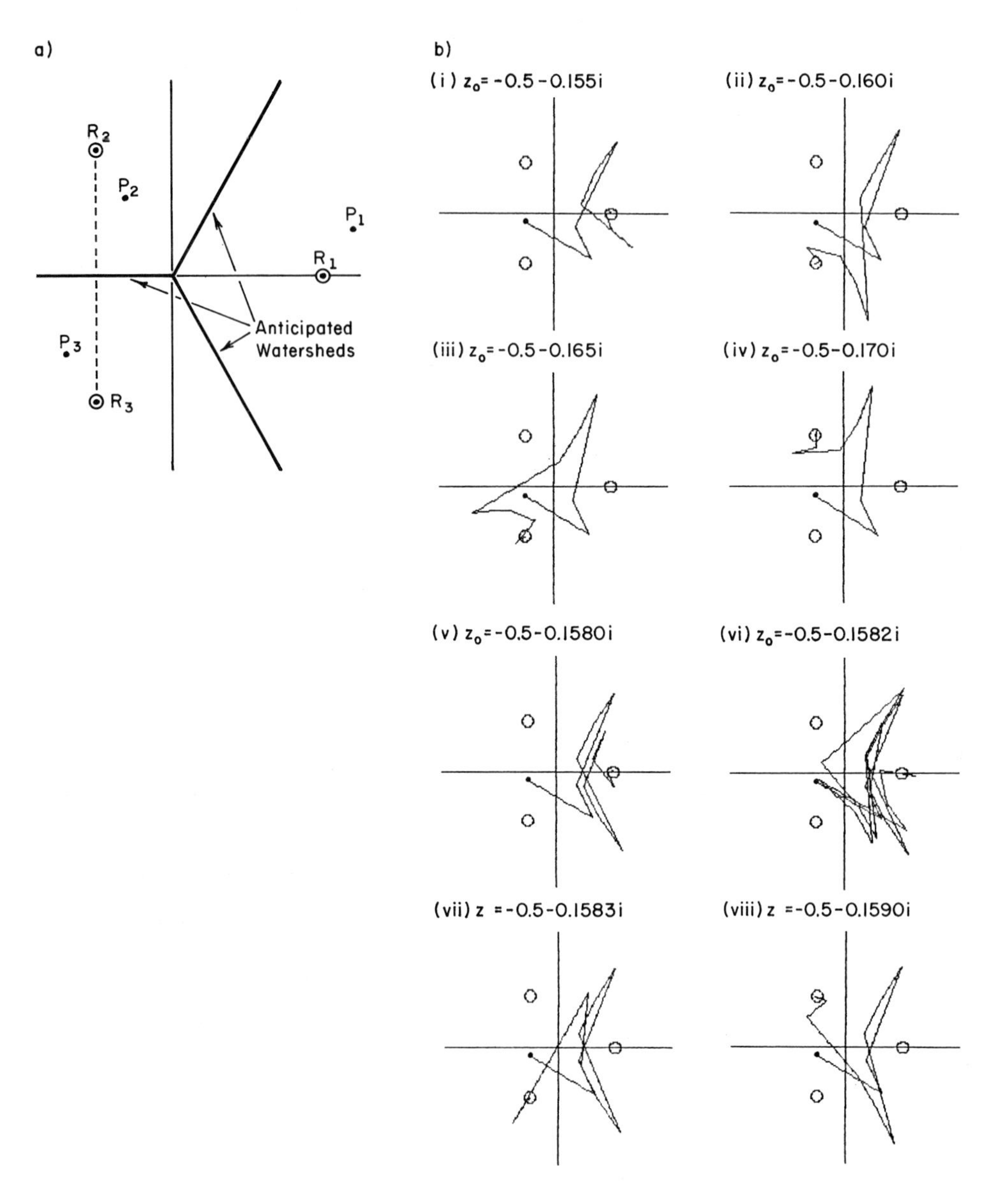

Figure 14.5. The complicated watersheds in the complex plane as one tries to locate the roots of the equation $z^3 - 1 = 0$ was a complete surprise to the early explorers of this problem. a) Anticipated simplicity of Newton's method iterated pixel movements when guesses are moved towards the basins of attraction by iterated estimates. b) Pixels located near the watershed between any two drainage basins exhibit complicated behavior.

Table 14.2. The iterative calculation of the movement of three points near the roots of the equation $z^3 - 1 = 0$ by Newton's methods of approximation results in rapid convergence of the value to their respective roots.

Iteration of the point $P_1 = z_0 = 1.2 + 0.3i$

Iteration	Real Part	Imaginary Part
1	0.99223	9.7475×10^{-2}
2	0.99041	2.7224×10^{-4}
3	1.00009	5.3211×10^{-6}
4	1.00000	9.8955×10^{-10}
5	1.00000	8.3122×10^{-18}
6	1.00000	0.0000
7	1.00000	0.0000

Iteration of the point $P_{n+1} = z_0 = -0.3 + 0.5i$

Iteration	Real Part	Imaginary Part
1	− 0.6614	1.1984
2	− 0.5358	0.9494
3	− 0.5022	0.8730
4	− 0.5000	0.8661
5	− 0.4999	0.8660
6	− 0.5000	0.8660
7	− 0.5000	0.8660

Iteration of the point $P_3 = z_0 = -0.7 - 0.5i$

Iteration	Real Part	Imaginary Part
1	− 0.3206	− 0.7594
2	− 0.5559	− 0.8578
3	− 0.5009	− 0.8631
4	− 0.4999	− 0.8660
5	− 0.5000	− 0.8660
6	− 0.5000	− 0.8660

has given us the graphic account of Hubbard's collision with the problem of iterative mapping of the type set out in Figure 14.5(b) [1]. At the time of his struggle with the iterative search for roots to the Newton Equation in 1976 Hubbard was teaching first year university students in Orsay, France. His students had asked him about the paths that were followed by iterative approximation of roots of the third degree equation. Hubbard told his students:

> "*The situation seems more complicated (than for second order equations). I will think of it and tell you next week*".

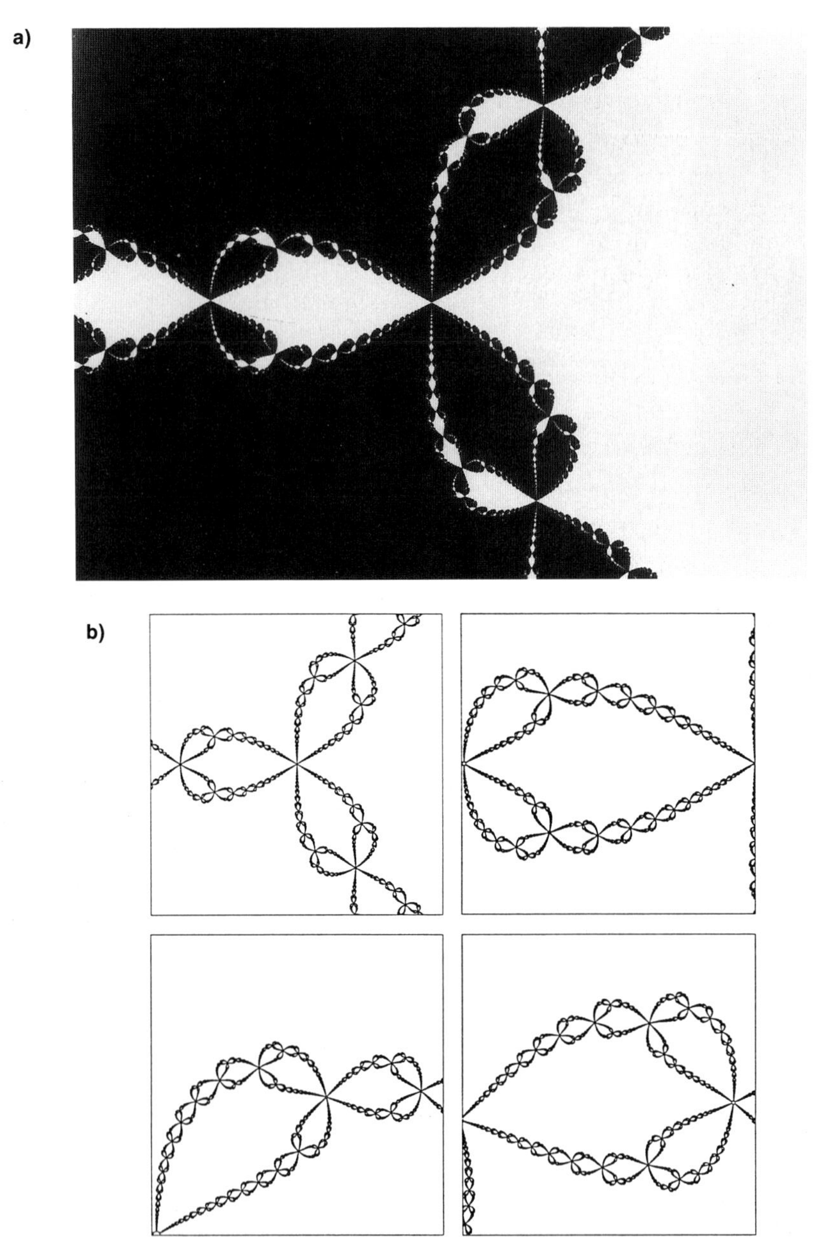

Figure 14.6. The watersheds between the regions of the three drainage basins of solutions for $z^3 - 1 = 0$ are fractally complicated and the network of interconnected boundaries separating the various regions of the pixel drainage region constitute a Julia set. a) Computer exploration of the watershed between the three regions of the drainage basins for solutions of the equations $z^3 - 1$ prove to be an intricate lace network of "three-petaled" flowers. b) Increasing resolution of the structure of the three-petaled basic pattern of the intricate watershed between the solution basins showed that each petal was bounded by an endless sequence of self-similar three-petaled structures and hence the boundary is a fractal and infinite in intricacy.
Peitgen, H.-O./Richter, P. H.: The Beauty of Fractals, 1986 © Springer-Verlag, Berlin-Heidelberg

However, the more he thought about it during the ensuing week, he realized, the less he knew about what constituted an intelligent guess or for that matter about what Newton's method really did. Gleick tells us that Hubbard was the type of mathematician who preferred to work with theorems and rigorous proofs but found himself in a situation where he had to use a computer to explore what was happening in the hope that the pattern of events that he observed would stimulate him into creating the necessary theorems and proofs. In other words, he began to use the techniques of experimental mathematics to discover patterns of events which he would then attempt to predict theoretically.

Hubbard started to explore the behavior of different starting points in the complex plane and he arranged for his computer to color a pixel that moved to R_1 red, points that led to R_2 blue, and green for those that moved to the third root. Initially his work showed that pixels starting near the expected solutions soon converged into the anticipated root. However in the words of Gleik:

> "*As Hubbard pushed his computer to explore the space in finer and finer detail, he and his students were bewildered by the picture that began to emerge. Instead of a neat ridge between the blue and red valleys, he saw blotches of green strung together like jewels. On even closer inspection, the line between the green blotch and the blue valley proved to have patches of red. The boundary finally revealed to Hubbard by his computer seemed bewildering. Wherever two colors tried to come together the third always inserted itself with a series of new self-similar intrusions*".

In *Figure 14.6* a simplified version of the type of map generated by Hubbard and other workers in exploring the behavior of pixels iterated into solutions of the equations $z^3 - 1$ is shown. This is a black and white version of the boundary and any pixel point located in a white part of the map would move to the real root 1, on the axis of the pattern. In fact, we now know that the watershed between the drainage basins is actually a Julia set [4] (see the brief discussion of Julia and Fatou sets given in Chapter 2).

It is also apparent that the pattern as shown in Figure 14.1 establishes that the fractal boundaries between the watersheds separating the four solutions of the equations $z^4 - 1$ is also a Julia set with four-petaled flowers. One would now anticipate that if one were to carry out the same exploration for the equation $z^5 - 1$ one would obtain fractal boundaries between the drainage basins which are a collage of intricately linked five-petaled flowers. Readers are encouraged to try various pixel explorations for themselves in iterating guesses at solutions of equations of various orders in the complex plane.

References

[1] J. Gleick, *Chaos, Making A New Science*, Viking Penguin Inc., New York 1987.

[2] For an introduction to Blissymbolics see C. K. Bliss, *Semantography*, third enlarged edition, P.O. box 222, Coogee, Sydney, Australia, 2034, 1978.

[3] B. H. Kaye, "Mathematical Dyslexia and Symbolic Bliss", presentation to the "SCORE" conference on educational development, May, 1992.

[4] M. Schroeder, *Fractals, Chaos Power Laws: Minutes from an Infinite Paradise*, W. H. Freeman, New York 1991.

Chapter 15

Fourier Analysis, Fractal Dimension and Formation Dynamics

The waves on the sea create fractals on the shore

Photo used with the permission of Mr. Bob Croxford Publisher, Atmosphere Postcards, Willis Vean, Mullion Helston, Cornwall TRI 27 DF

Chapter 15 Fourier Analysis, Fractal Dimension and Formation Dynamics 553

Section 15.1 Hot Rocks and Musical Notes 555
Section 15.2 Harmonious Rocks and Fractal Profiles 561
Section 15.3 Fourier Analysis, Fractal Structure and Fortune Hunting 569
References 574

Chapter 15

Fourier Analysis, Fractal Dimension and Formation Dynamics

Section 15.1. Hot Rocks and Musical Notes

Some scientists have been reluctant to accept the usefulness of the fractal dimension description of various systems since they believe that they already have adequate tools for describing such structures in a technique known as Fourier analysis. *Fourier analysis* takes its name from the French mathematician who created the technique for the mathematical analysis of complicated wave structures. Jean Baptist *Fourier* (1768–1830) lived through the French Revolution (Asimov tells us that Fourier came close to a terminal encounter with the guillotine during the Revolution but was saved by change of government). Fourier went to Egypt with Napoleon in 1798 and was a governor of part of French occupied Egypt. In 1808 Napoleon made him a baron for his mathematical discoveries.

When I was an undergraduate student of physics most textbooks contained a brief discussion of Fourier analysis based upon the study of musical notes. In particular I remember attending a lecture on how one could use Fourier analysis to tell the difference between middle C played on a violin and a piano. I remember wondering why anyone should want to do such a complicated treatment of such a simple problem [1]. In the 1950s, Fourier analysis was basically an academic novelty to the average student since the mathematical manipulations of data required to achieve Fourier analysis was too complicated and expensive for it to be used in most real scientific problems. However, today, modern computers of quite modest capacity have software enabling the student to carry out Fourier analysis of wave functions so that now what was once a novelty has become part of the mathematical skills equipment of the physics undergraduate student [1, 2].

To understand the basics of Fourier analysis, consider the information shown in *Figure 15.1*. Fourier, who was studying such complex waveforms, showed that a waveform no matter how complex can be broken down into a series of simple regular wave motions, the sum of which will be the original complex periodic variation. Fourier was a great student of the movement of heat through objects. I remember reading somewhere that one of the problems that he studied was the movement of heat through a wall (at the time, I never dreamed that I might be

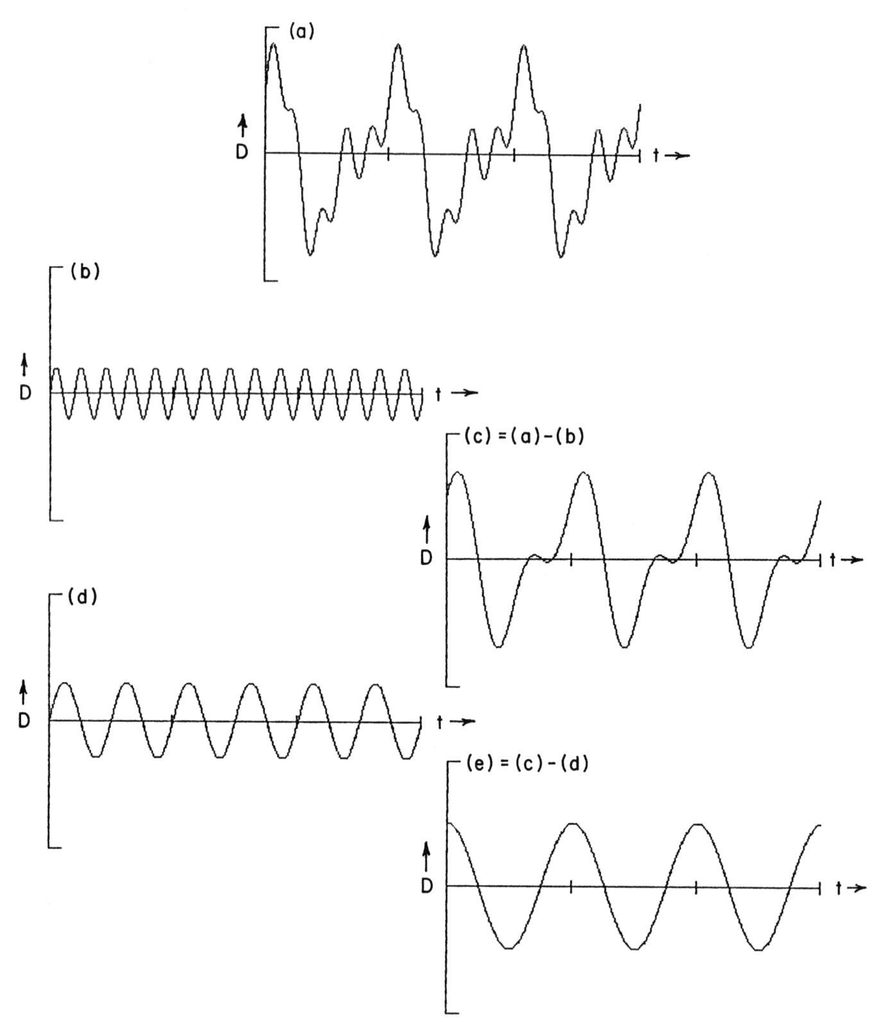

Figure 15.1. Fourier analysis breaks down a complex wave into uncomplicated components known as simple harmonic waves.

writing a book which involved a section on Fourier analysis and I didn't make a note of the article in which I read the information. Unfortunately, I cannot give a reference for the reader to be able to follow up on this information).

Again, I remember reading somewhere that by trial and error, the Europeans have discovered that if they made the stone walls of their houses of a certain thickness that the wall would heat up on the outside during the day and it would take about 12 hours for the heat to move through the stone to the inside of the house. Therefore the house was cool during the day and then in the evening the heat moving through the wall would warm up the inside of the house to fight off the chill of the night.

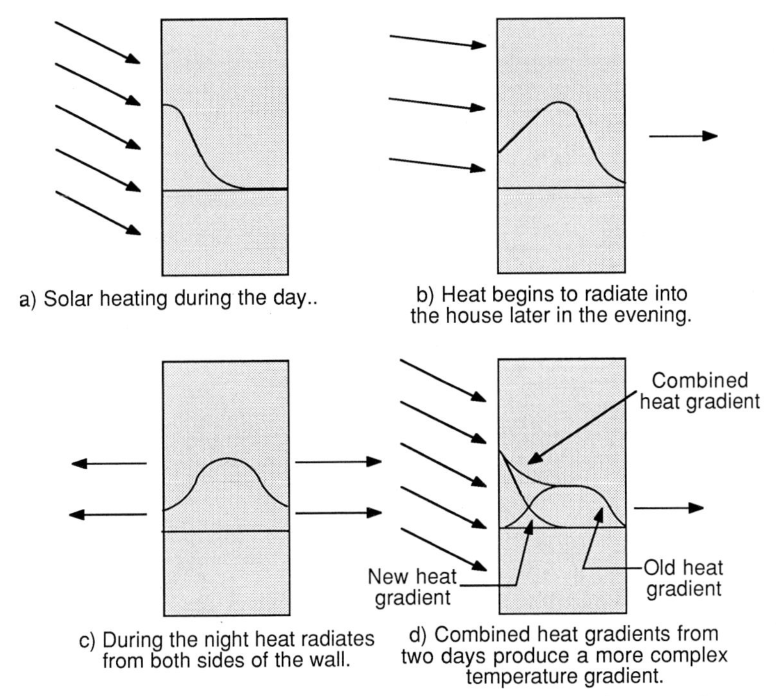

a) Solar heating during the day..
b) Heat begins to radiate into the house later in the evening.
c) During the night heat radiates from both sides of the wall.
d) Combined heat gradients from two days produce a more complex temperature gradient.
Combined heat gradient
New heat gradient
Old heat gradient

Figure 15.2. The build up of thermal energy in a wall constitutes a complex wave of heat energy pulses which can be studied by Fourier analysis.

If one were to study this movement of the heat pulse moving through a piece of stone then by the early evening the heat that had accumulated in the stone would start to move in both directions because the outside of the house would be cool during the night. This situation is illustrated in *Figure 15.2.* Over a period, the heating during the day and the cooling of the night would create a series of pulses in the thickness of the wall. The overall effect of the heating cycles of the day would be very complicated in a wall that was say six foot thick. If one was interested in studying this problem for the heat penetration of a castle wall, one would have a wave function describing the daily heating of the wall combined with another wave function representing the difference between heat of the summer and the winter. During a rain storm or on a cloudy day the shape and size of the heat pulse moving through the wall would vary so that one would end up with a very complicated wave in the wall if one could measure the thermal gradients in the wall.

I have often wondered if Fourier was inspired to create his method of analysis by realizing that the complicated heat waves in a rock wall were created by the interaction of what was essentially two simple wave functions, the daily temperature variation and the annual temperature variation. Fourier believed heat to be essential to health so he always kept his dwelling place overheated and swathed himself in layer and upon layer of clothes. Asimov comments on the personality of Fourier that "Even great scientists can have their irrational beliefs".

In the early 1990s, because of the rising costs of fuel, the thermal control of a building by natural solar heating became an important study of building design. Moreover, in many parts of the industrial world, electricity is best produced on a continuous basis and technologists have tried to devise new methods for storing electricity at night when more electricity is being produced than needed.

In Great Britain, one of the strategies adopted is to sell electricity at night at a low rate. This cheap electricity is used to heat up a stack of bricks which by the morning are hot enough on the outside to heat the houses during the day in winter (the stack of bricks are covered with a nice looking cover and the combination is called a night storage heater). People investing in such systems would not like a brick that was too thick so that the heat did not arrive at the outside of the brick until noon time neither would they like a thin brick which gave them heat by five o'clock in the morning.

The thickness of a brick which became hot enough in the morning to give gentle heat could be an interesting project for a physics class. They could place an electrical heater between two bricks and drill a series of holes in the bricks so that they could record the temperature gradient over a period of several hours when the electric heater was on for different amounts of time [3].

As we have suggested above, the basic physical principles of Fourier analysis can be appreciated from a study of how Fourier created his theory. We can gain further understanding of Fourier analysis by considering the reverse process of synthesizing waves from simple components to create complicated waves. Before studying the result of adding two waves together, it is necessary to have a clear understanding of the terms used to describe the structure of waves. In *Figure 15.3* the meaning of the terms *wavelength*, *amplitude* and *frequency* are summarized. An important property of two waves when they are added together is the phase relationship between the two waves (see discussion of the phase of a waveform in Chapter 1).

In *Figure 15.4* the result of adding two waves to each other when they have different phase angles is shown. In *Figure 15.5* the *harmonic notation* used to describe waves in terms of a reference low-frequency wave is shown. A wave which has five times the frequency of a basic wave is called the fifth harmonic of the basic wave. The use of harmonic notation to describe wave functions developed when scientists were breaking down the wave motions that created musical notes into the harmonies which contributed to that note. In *Figure 15.6*(a) the way in which a set of waves of different frequency, amplitude and phase angle can be combined to develop what is known as a *saw toothed wave* is illustrated graphically. In Figure 15.6(b), the early stages of the combination of various harmonic components to produce a triangular shaped wave is illustrated. In Figure 15.6(c), the way in which a square shaped wave of the type used in digital electronics can be synthesized from harmonic components is illustrated.

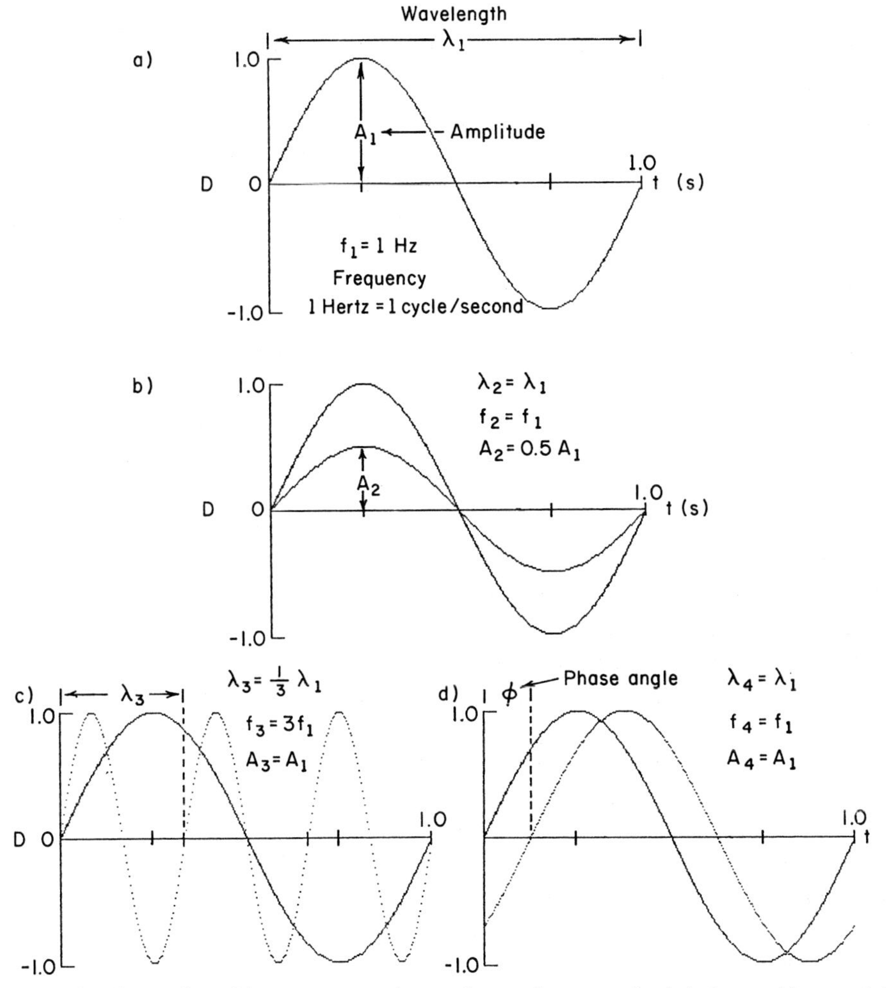

Figure 15.3. The number of times a wave motion reaches maximum amplitude in 1 second is termed the frequency of a wave motion. Wave motion is describable by four parameters, amplitude (energy) frequency, phase and wavelength.

In the science of *acoustics* (a dictionary definition of acoustics is that it is the science of sounds, from the Greek word akoustikos meaning "to hear") it is quite sufficient to record the amplitude of the contributory harmonics to the synthesized wave without registering the phase differences between the various components of the wave. In *Figure 15.7* the relative strength of the contributory harmonics to the three synthesized waves of Figure 15.6 are summarized in a bar graph which is called the *harmonic spectrum* of the wave. In a similar way, the result of electronic analysis of the waveforms present in a musical note played by a clarinet and an oboe are shown.

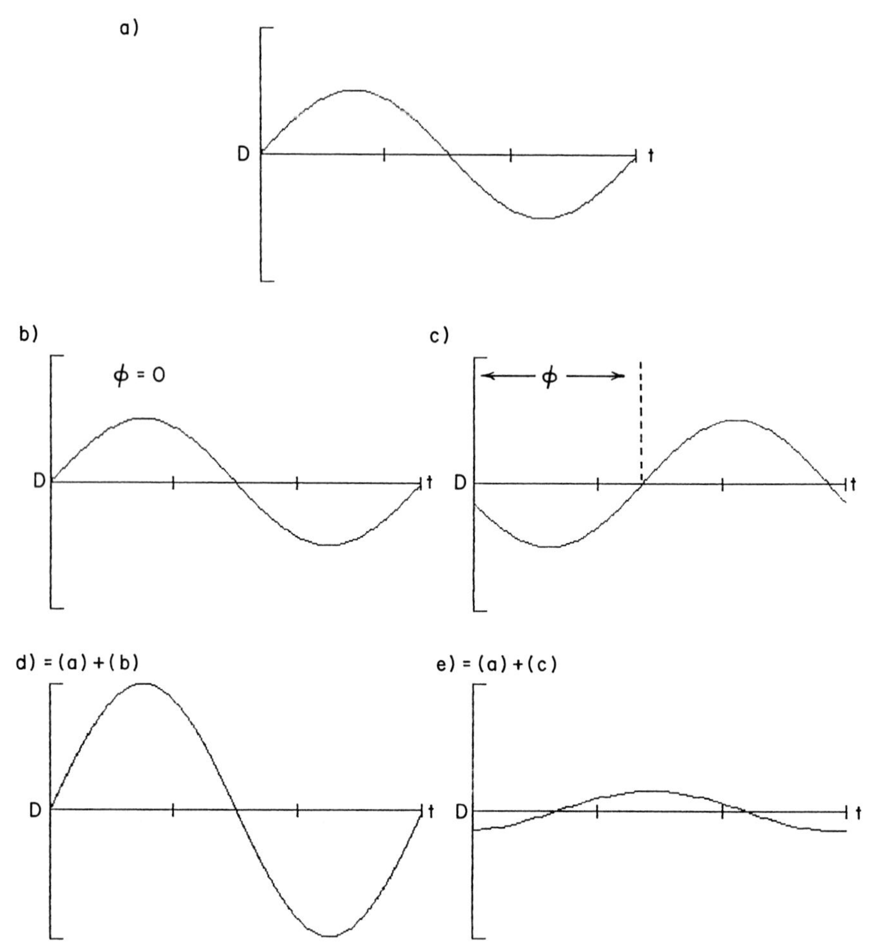

Figure 15.4. The resultant wave formed by adding two waves of the same amplitude and wavelength depends dramatically on the phase difference between the two waves. D = The displacement of the wave. ϕ = The phase difference between the two waves being added.

Electronic music is synthesized from simple oscillators using the known harmonic spectrum of a note [1]. A computer is given the harmonic spectrum of a note as played on a trumpet and the computer is able to synthesize that note by adding waves of different frequency and amplitude according to the recipe summarized in the harmonic spectrum. Fourier analysis of a musical note is probably a very appropriate form of analysis in that various parts of the musical instrument probably oscillate with the various frequencies present in the complex note so that the Fourier analysis is a search for the contributing causes to the complex note and is related to the formation dynamics of the note.

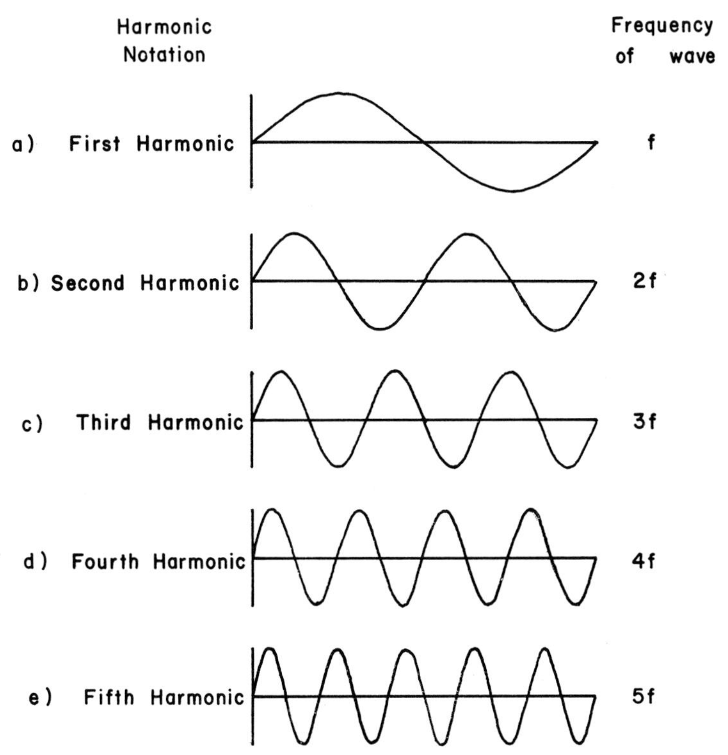

Figure 15.5. Two different notations are used to describe waves which differ with respect to each other by the frequency with which they repeat their displacement pattern. Sometimes the frequency is expressed as a multiple of the basic frequency known as the fundamental frequency. Because frequency analysis was developed in the study of musical tones an alternative notation known as harmonic notation is used to describe waves of different frequencies. The lowest frequency is known as the first harmonic and a wave of twice that frequency is known as the second harmonic etc. a shown in the diagram.

Section 15.2. Harmonious Rocks and Fractal Profiles

I took a whole new interest in the concept of Fourier analysis when the increasing power and falling costs of computers made it possible to start using Fourier analysis to describe and recognize the shape of fractured rock and other fine particles [4–11]. The first step in applying Fourier analysis to this technical problem is to take a fine particle profile in two-dimensional space and generate what is known as the *geometric signature waveform* (one can use this basic type of technique in three-dimensional space but the explanation of the procedures becomes quite complicated. For the purpose of this book we can restrict our discussion to the technology for manipulation of data in two-dimensional space).

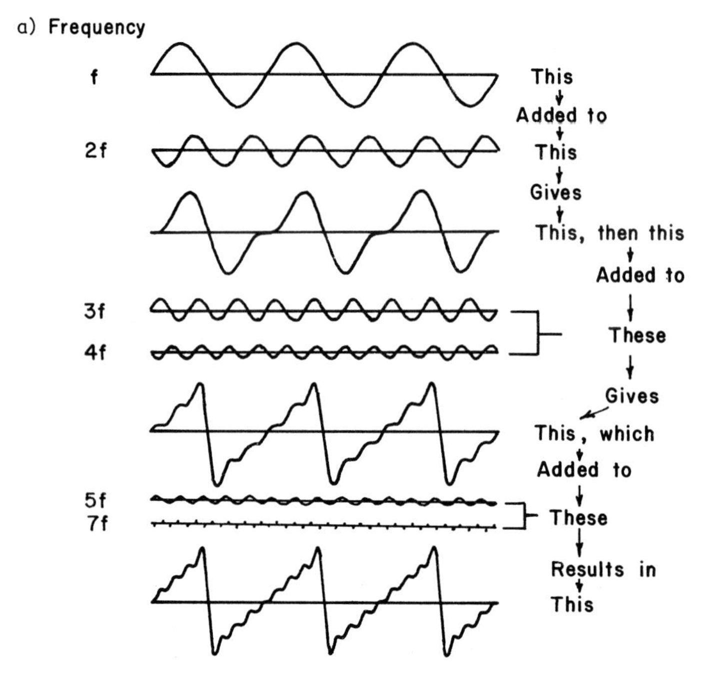

Figure 15.6. Complex waves can be built up from component waves. a) A saw toothed wave can be synthesized from simple harmonic components by adding waves of different frequency, amplitude and phase angle to each other. b) A triangular shaped wave can also be synthesized from a combination of simple harmonic waves as shown. c) A square shaped wave of the type used in digital signals can be synthesized from a series of simple harmonic waves as shown.

Several procedures have been developed for generating geometric signature waveforms from two-dimensional profiles. We will use the basic procedure developed by Beddows and co-workers [8]. In this technique, one treats the profile such as that shown in *Figure 15.8*(a) as if it were a thin piece of cardboard. One then calculates the centroid of such a piece of cardboard. Calculating the centroid of a thin shape is a relatively complex problem and for experiments at the high school level for the purposes of generating geometric signature waveforms one can often locate the centroid by eye or by making a cut-out of the shape and trying different balance points with a pin. When the centroid has been located, one then records the magnitude of the radial vector (the magnitude of the line drawn from the centroid to the boundary of the profile) as it moves around the shape as illustrated in Figure 15.8(a). One usually starts this process by choosing the longest radial vector within the profile as a reference direction. One has to assume that the radial vector is moving around the profile with a uniform steady angular velocity about the centroid.

For convenience, it is useful to normalize the magnitude of the vectors drawn such a R_2, R_3, R_4 etc. by dividing them by the longest vector R_1. When one plots the

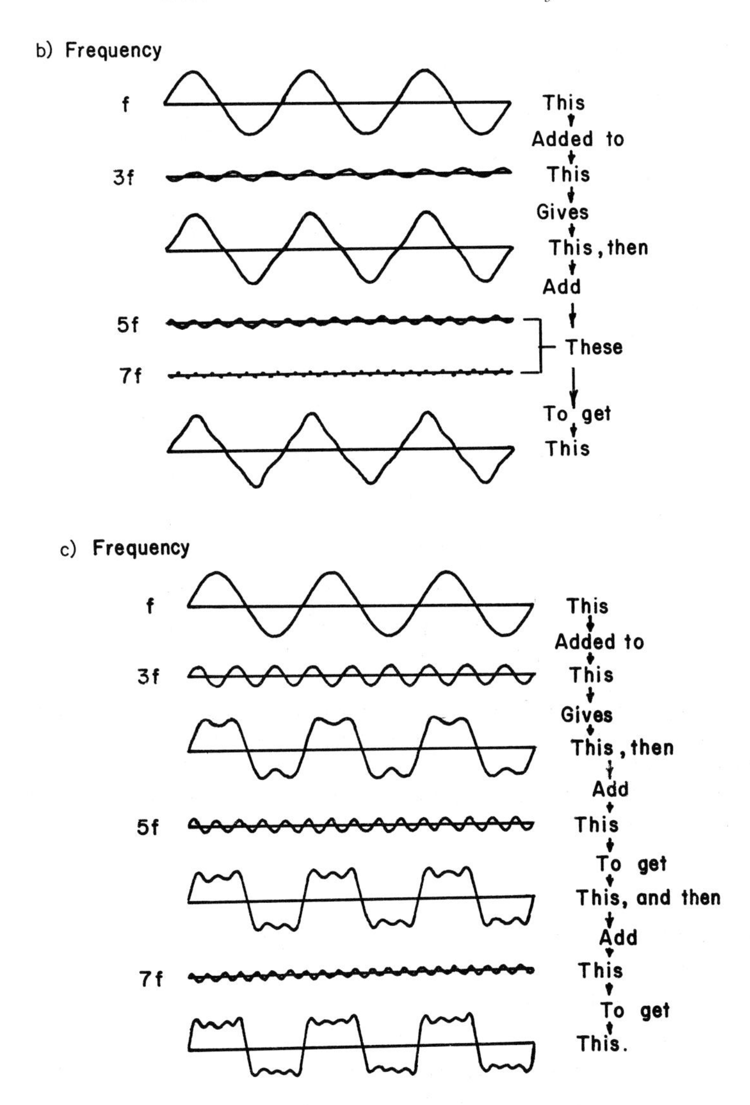

values of the vector against the angle using normalized values of the radial vector one generates the system shown in Figure 15.8(b) which is described as the geometric signature waveform of the profile. The structure of this waveform obviously depends upon how intensely one examines the boundary. In Figure 15.8(c) a higher resolution version of the signature waveform of (b) generated from 72 locations around the perimeter is presented.

When one is looking at the Fourier analysis of a waveform, theoretically, one has to consider very long waveforms consisting of many repeats of the same basic structure. However, for practical purposes, one can take the complex waveform such

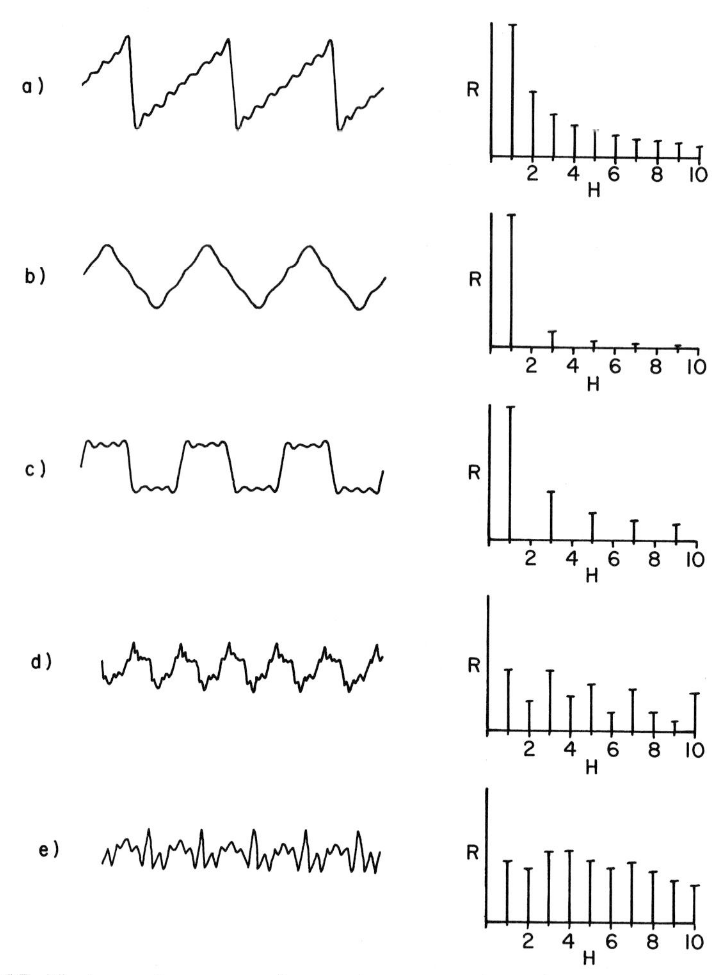

Figure 15.7. The harmonic spectrum of a complex wave can be used to summarize the various waves contributing to its structure. R represents the relative amplitude of the waves present in the complex wave. a) Harmonic spectrum of the saw tooth wave of Figure 15.6(a). b) Harmonic spectrum of the triangular wave of Figure 15.6(b). c) Harmonic spectrum of the square wave of Figure 15.6(c). d) Harmonic spectrum of a note played on a clarinet. e) Harmonic spectrum of a note played on a oboe. H represents the harmonic of the fundamental tone (first harmonic).

as that of Figure 15.8(c) and treat it as if it were one part of an infinite wavetrain and then subject it to Fourier analysis to discover the components contributing to the waveform. In Figure 15.8(d) the harmonic spectrum of the line shown in Figure 15.8(c) is shown. Note that in technical terms, the harmonic spectrum of the signature waveform is a dimensionless list of components which can be stored in the memory of a computer.

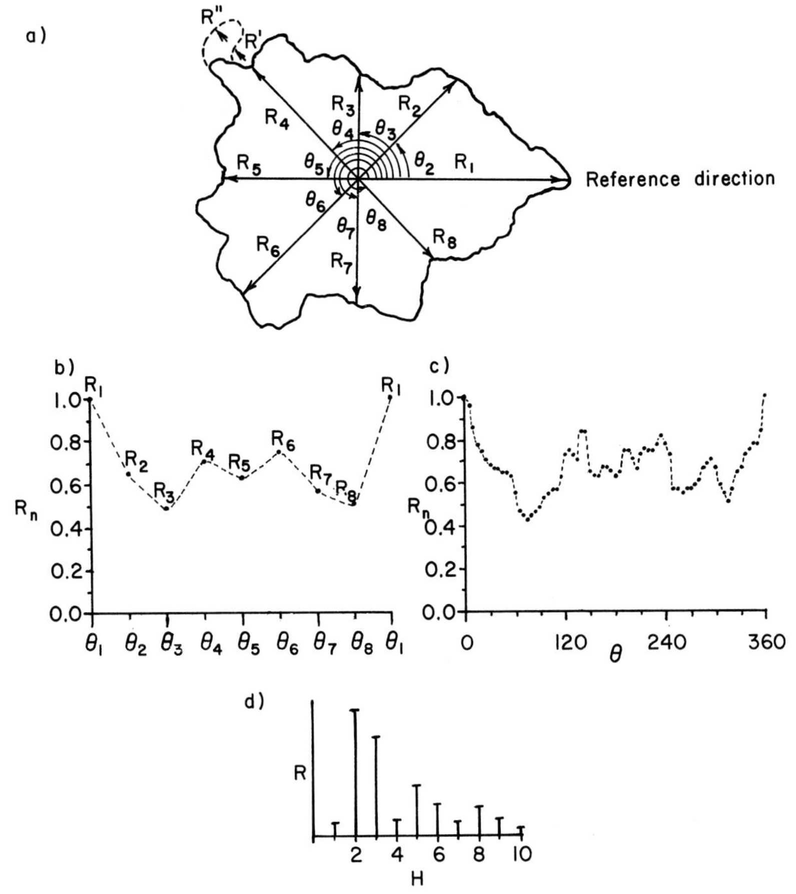

Figure 15.8. The geometric signature waveform of a fine particle profile can be subjected to Fourier analysis to summarize information on the structure of the profile. a) Profile showing the reference line, centroid and various radial vectors. b) Sketch showing how the geometric signature waveform is generated by scanning the radial vector through 360°. c) 72 point geometric signature waveform of profile in (a). d) Fourier spectrum of the complex wave of (c).

As we go from the profile to the waveform, we degenerate the dimensional description process from two to one-dimensional space. Then, as we go from the waveform to the harmonic spectrum, we go from one-dimensional to zero dimensions in data space. If one had a whole array of different shaped profiles, one could list the harmonic spectra of the various profiles in a computer. When an unknown profile was characterized by a computer-aided image analysis system, the computer could compare the harmonic spectra of the new profile with those stored in its memory and recognize the shape of the profile from its previous experience.

From a physics point of view, there are two difficulties encountered when attempting to use the geometric signature waveform to characterize fine particle

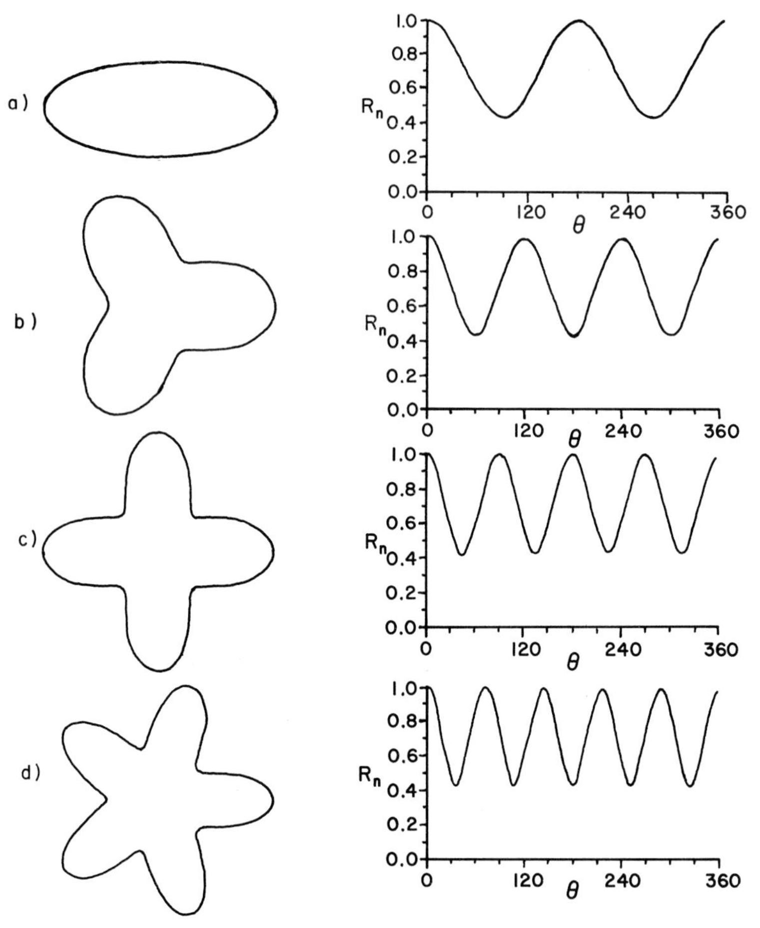

Figure 15.9. The first five harmonics of a waveform contain information on the basic structure of the profile. a) Second harmonic profile. b) Third harmonic profile. c) Fourth harmonic profile. d) Fifth harmonic profile.

profiles. One of them is that for some rugged boundaries the radial vector has two values as it crosses a convolution. If the profile of Figure 15.8(a) had a convolution such as that shown by the dotted line at the top left hand corner of the profile then the value of R_4 could have three different values depending on when one decided that the vector has touched the profile. One can overcome this problem by using what are known as two-dimensional Fourier transforms of the profile but a discussion of such technology is beyond the scope of this book. Interested readers will find a full discussion of this aspect of the problem in References 11, 12 and 13.

The other problem with the Fourier analysis of geometric signature waveforms arises when there are sharp edges on the profile. Theoretically, the Fourier transform

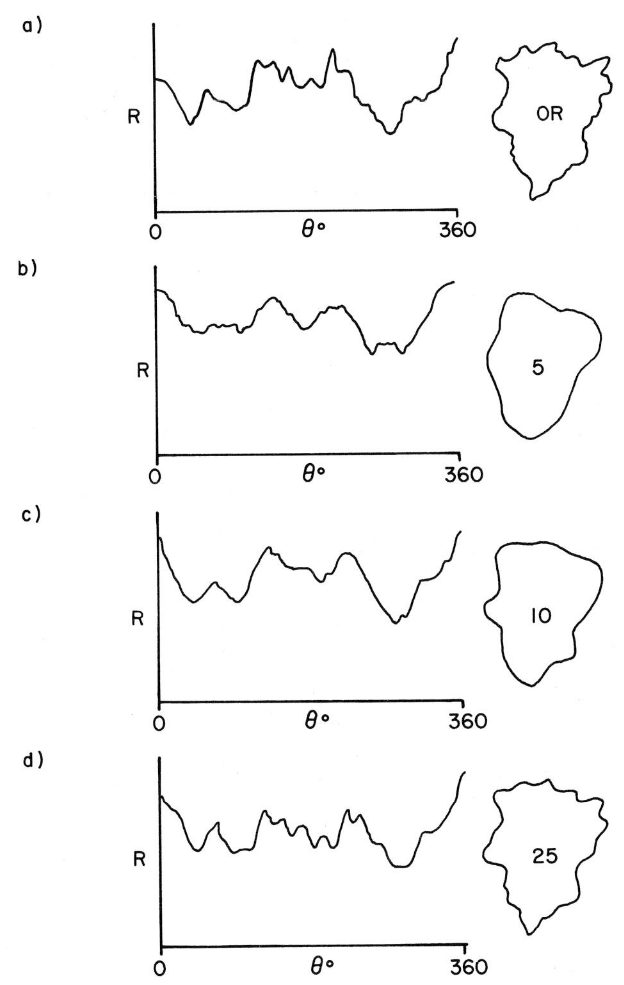

Figure 15.10. A. G. Flook demonstrated the physical interpretation of the components of a geometric signature waveform by reconstructing a profile from the various components of the harmonic spectrum of the original profile waveform. a) Original profile. b) Waveform and profile reconstructed from the first 5 harmonics. c) Waveform and profile reconstructed from the first 10 harmonics. d) Waveform and profile reconstructed from the first 25 harmonics.

of a very sharp edge contains an infinite number of components and the Fourier analysis of complex waveforms with sharp peaks upon them becomes a difficult problem mathematically and requires more computing power. Alternate methods are used to characterize the presence of sharp edges on a profile (see Reference 12).

The physical significance of the various harmonics present in the harmonic spectrum of a geometric signature waveform can be appreciated from the data summarized in *Figure 15.9*. Flook, who has made a study of the physical significance of the

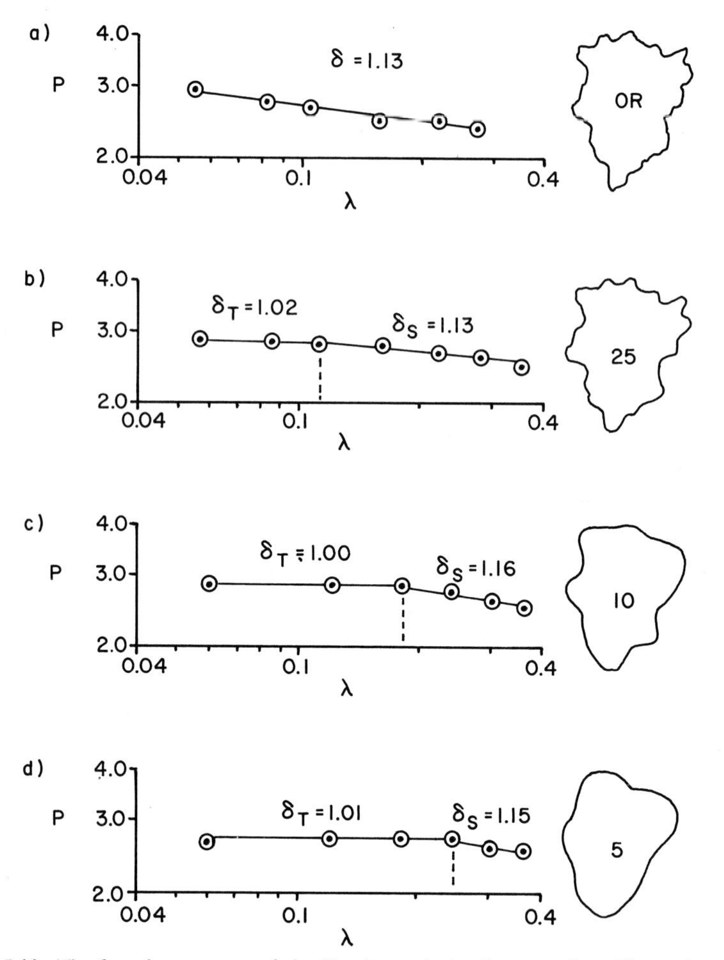

Figure 15.11. The fractal structure and the Fourier analysis of a rugged profile are interrelated.

geometric signature waveform, points out that the basic shape of a rock is contained in the first five harmonics and that beyond the fifth harmonic it is a philosophical debate whether the starfish of the fifth harmonic represents a deformed circle or a five lobed structure.

After the fifth harmonic, the various harmonics present in the harmonic spectrum actually contribute to the texture of the profile rather than the shape of the profile. Specialists in this area regard the first five harmonic components of the spectra as contributing to a description of the basic shape of a profile, whereas the higher harmonics are said to describe the texture. Flook illustrated this point with some interesting data on the shape of a profile marked "original" in *Figure 15.10.* Flook Fourier-analyzed the geometric signature waveform of this profile and then synthesized the profile information present in the first five harmonics. It can be seen that the profile reconstructed from the first five harmonics describes the basic structure of the profile. As one adds more and more harmonics to the synthesized profile, the basic shape of the profile stays the same and the texture changes. By the time one has reached the twenty-fifth harmonic, the original and the synthesized profile start to look very similar.

From the electronic engineers point of view, the texture of the profile represents noise on the basic waveform describing the structure of the profile. Although the Fourier analysis of the geometric signature waveform results in a useful nondimensional description of the profile, it is most unlikely that the various components of the analyzed wave represent any real description of the formation dynamics of the rock. In fact, the freshly fractured rock will start off with the highest ruggedness and then erosion forces would probably contribute to a loss of texture. A fractal description of the profile is more likely to be related to the formation dynamics (readers interested in this subject should read the discussion on fractals and fracture in Reference 14).

Consider the data shown in *Figure 15.11*. The original Flook profile has a fractal dimension of 1.13. If one now carries out explorations of the synthesized profiles of Figure 15.10 to determine their fractal dimensions one generates the set of Richardson plots summarized in Figure 15.11. This could represent a progressive polishing of a rock tumbling in sand with the progressive removal of the texture of the rock corresponding to the increasing dominance of the Euclidian portion of the Richardson plot.

Ever since I generated the data shown in Figure 15.10, I have toyed with the idea of presenting the harmonic spectra of the original rock and the synthesized rock shapes to an electronic synthesizer to make musical notes. The sound of such synthesized music should be related to the structure and the texture of the rocks and the creation of such musical notes would give a whole new meaning to the term "Rock Music"!

Section 15.3. Fourier Analysis, Fractal Structure and Fortune Hunting

In an exercise at the end of Chapter 8, we suggested that the fluctuations in the record of the stock market could be described by a fractal dimension. Many people are interested in predicting future movements of the stock market record in the hope that if they can predict the movements accurately then one can buy and sell stocks at a good profit. If one looks at a relatively short run of data from the stock markets, it appears to be fractal with the property of self-similarity in that the record of transactions in the short run look exactly the same as the transactions for the entire day. Readers can check this fact for themselves by looking at the records of transactions of the Dow Jones Average and/or the Toronto Stock Exchange in the newspaper such as the Wall Street Journal or the Toronto Globe and Mail (comparable European and world-wide data will be available in similar international newspapers).

As discussed earlier, it is very reasonable to expect that the structure of a detailed record of the sales and purchases of stocks would be a chaotic system since the record represents a summary of the multitude of motives and impulsive behavior of many different people buying and selling for different reasons. However from time to time, if one looks at the structure of the stock market there will appear to be cyclical events which can be analyzed by Fourier analysis.

In *Figure 15.12* two records of a run of several months on the Dow Jones Average are shown. Figure 15.12(a) is for the recovery of the stock market after what is known as the crash of October, 1987. On that day there was a catastrophic nose dive of the stocks and shares and then a gradual recovery of the confidence of investors in the stock market. To a physicist, the smoothed out variations of the stock market after the crash look very similar to the behavior of a damped oscillator which is given a very large displacement before being allowed to oscillate. It is not surprising that the application of Fourier analysis to such behavior would give some components which would suggest cyclical fluctuations in the market. The same comments apply to the displacement and recovery of the market before and after the attempt to overthrow Mikael Gorbachev as the president of the former Soviet Union in August, 1991, as shown in the data of Figure 15.12(b) (note however that the "spike" in the record, representing the panic in the stock market which dominated emotions for two or three days when the outcome of the attempted overthrow was unknown, causes problems for the Fourier Analysis. As in the case of the sharp-edged profile discussed earlier, the Gorbachev spike gives almost an infinite number of components in the Fourier spectrum).

Many people have attempted to apply Fourier analysis to the stock market and it is well established that there are several well-known cyclical patterns to be discovered. The first is the four-year cycle of the U.S. Presidential election. As a run up to an election, the President who is seeking re-election makes many promises and hands out many grants to people. This stimulates the economy. Then, immediately after the election reality has to be faced and controls and cut-backs to programs are

Figure 15.12. The stock market record of the United States known as the Dow Jones average manifests both fractal and harmonic oscillation structure. a) The smoothed out variations in the fluctuations of the Dow Jones average observed after the stock market crash of 1987 looks to the physicist like the behavior of a damped simple harmonic oscillator which had suffered a major displacement before it began to oscillate. i) Actual period. ii) "Smoothed out" variations of (i). b) The record of the stock market before and after the failed attempt to dislodge Gorbachow showed marked oscillations. i) Raw record. ii) Smoothed out record.

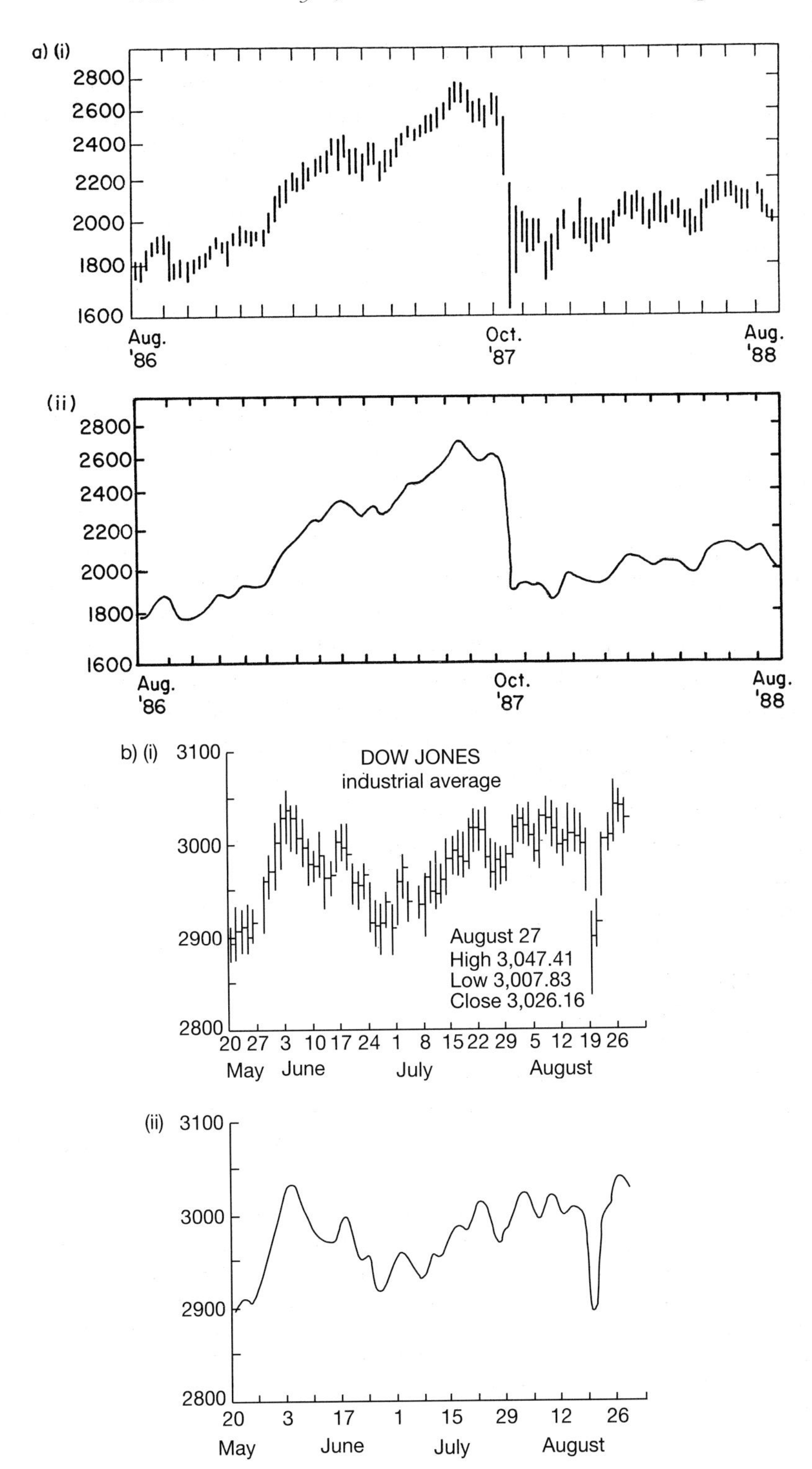

a) (i)
2800
2600
2400
2200
2000
1800
1600
Aug. '86
Oct. '87
Aug. '88
(ii)
2800
2600
2400
2200
2000
1800
1600
Aug. '86
Oct. '87
Aug. '88
b) (i) 3100
DOW JONES
industrial average
3000
2900
August 27
High 3,047.41
Low 3,007.83
Close 3,026.16
2800
20 27 3 10 17 24 1 8 15 22 29 5 12 19 26
May June July August
(ii) 3100
3000
2900
2800
20 3 17 1 15 29 12 26
May June July August

introduced with a delayed effect on the market of about a year. The Fourier cycle associated with Presidential elections in the United States lag the actual election date by about a year. However, sometimes in periods of deep recession, the expected peak fails to appear (see stock market records of the early 1980s).

An Austrian economist by the name of *Schumpeter* (1883–1950), studied the stock market for many years. He claimed from historical evidence that there were three types of trade cycles to which he gave the names Juglar, Kitchin and Kondratienff [15]. The first trade cycle which he called the Juglar was caused by the amount of stock that companies kept over an eighteen-month period. It is interesting to note that in modern industry very little stock is held in warehouses. The development of rapid communication systems such as the fax machine and fast freight systems such as the aircraft means that very often an article is not even made until it is ordered and it is directly moved from the factory to the customer. Regional warehouses as an intermediate storage system for manufactured goods such as auto parts have almost disappeared. One would expect that a modern study may well fail to show the Juglar cycle in modern business records.

The second cycle Schumpeter called the Kitchin cycle and claimed that is was caused by changes in investment and output over a three or four year period. I am not an economist. I am not too sure if the Kitchin cycle is the same as the Presidential cycle of the United States economy. The Kondratienff cycle, named after a Russian economist is a long-run 50 year pattern with 25 years of prosperity and then 25 years of slow growth and depression. Some economists explain this latter cycle as being due to the fact that after there has been a severe depression the generation of workers following that period are extremely cautious and hard working. Whereas twenty-five years later, the memory of the depression fades and the people become slack and soft in their work attitudes. When the older workers have all retired, those without memories of hard times, which disciplined the older workers, develop a sloppy attitude towards their work which in turn can lead to an economic crisis and the whole cycle begins again. These ideas are left for the reader to explore for themselves in the economics textbooks.

The presence of the four-year cycle, and other cycles, in the behavior of the stock market record indicates that this is a system which is both "fractal and Fourier" structured. This tells us that the formation dynamics of the records are subject to gross fluctuations from large-scale economic and political events but that the day to day structure is fractal because of the multitudinous interaction of many small randomly interacting causes.

The climate of the earth is probably very similar to the stock market in that day to day fluctuations and year to year fluctuations tend to be fractal but over the centuries there are long-term cycles which can be retrieved from a climate–time data record of the form shown in *Figure 15.13* for the occurrence of extinctions of species over the past several million years of the Earth's history. It is well known that the local climate in the northern hemisphere is subject to an 11 year cycle which can be

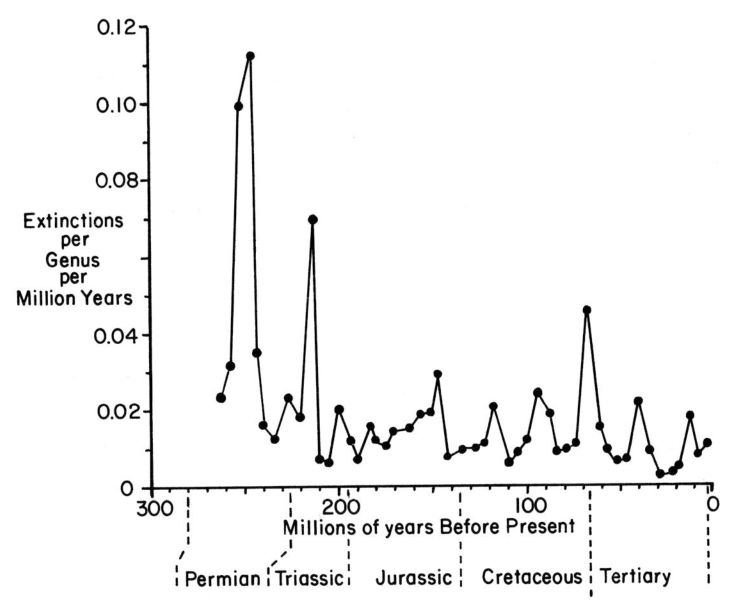

Figure 15.13. Scientists who have looked at the record of climatic changes over recorded periods of history have attempted to apply Fourier analysis to the pattern to detect various contributory cycles creating the Earth's climate. Such a search may be deluded since the pattern may be fractal with the variations caused by random interaction of different causes.

related to the sun spot cycle on the sun. The turbulence of the surface of the sun appears to fluctuate and when the sun's surface is active what are known as sun spots appear on the surface. The number and intensity of the sun spots vary according to an 11 year cycle however, over historic time, periods of quietness in the sun-spot activity have been recorded.

Other scientists have discovered a 120 year cycle and others claim to have found an 80 year cycle. There is no doubt that one can analyze a waveform such as that represented by the data of Figure 15.13. However, the situation is somewhat like the analysis of the waveform of the rock profile. The presence of the waveforms does not in any way prove that the formation of the observed structure was formed by the interaction of cyclical phenomena. Just as we were able to observe components in the waveform as a useful description, one has to be very careful when interpreting the Fourier analysis of the time series.

In summary, when one has a rugged structure, it can be analyzed both by Fourier and fractal structures and sometimes one technique is more useful than the other. At all times, one must be very careful not to build exotic theories on the analysis that one has formed unless the structure of the system can be related to the formation dynamics suggested by the analysis.

There is no doubt that there are many fascinating patterns in the records of various phenomena around us, I hope that the various techniques for describing such patterns which we have explored together in this book will help the reader gain a better understanding of the fascinating universe in which we live.

References

[1] H. E. White, D. H. White, *Physics and Music*, Saunders College, Philadelphia 1980.

[2] This topic is discussed in many general physics textbooks; see for example. D. H. Fender, *General Physics and Sound*, English University Press, London 1957.

[3] The measurement of heat gradients in an object such as a brick heated at one end and cooled at another is a classical type of experiment which is discussed in the older physics textbooks. The teacher wanting to do this kind of class study could find useful information in such books as S. G. Starling, A. G. Woodall, *Physics*, Longmans Press, London 1950.

[4] H. M. Sutton, N. Bundalli, "A Suggestion For The Exact Determination of Particle Size and Shape" in proceedings of the Society for Analytical Chemistry, January, 1973, Pgs. 13–17.

[5] T. P. Meloy in *Search For Signatures in Advanced Particulate Morphology*, J. K. Beddow, T. P. Meloy (Eds.), CTC Press, Boca Raton 1980.

[6] R. Ehrlich, B. Weinberg, "An Exact Method For Characterization of Grain Shape", *J. Sedimentary Petrol. 40* (1970) 205–212.

[7] H. P. Schwartz, K. C. Shane, "Measurement of Particle Shape by Fourier Analysis", *Sedimentology 13* (1969) 213–231.

[8] J. K. Beddow, G. C. Philip, A. F. Vetter, "On Relating Some Particle Profile Characteristics to the Profile Fourier Coefficients", *Powder Technology 18* (1977) 19–25.

[9] A. G. Flook, "Fourier Analysis of Particle Shape", *Proc. 4. Particle Size Analysis Conf.*, Loughborough University of Technology, September 21–24, 1981, in *Particle Size Analysis 1981–1982*, edited by N. G. Stanley Wood, T. Allen, Wiley Heyden Limited, London.

[10] A. G. Flook, "A Comparison of Quantitative Methods of Shape Characterization", *6. Int. Congress of Stereology*, *Acta Sterol. 3* (1984) 159–164.

[11] B. H. Kaye, "Shape Characterization of Fine Particles", in preparation, Elsevier Press.

[12] B. H. Kaye, G. G. Clark, Y, Liu, "Characterizing the Structure of Abrasive Fine Particles", *Particle and Particle Systems Characterization 9* (1992) 1–8.

[13] B. H. Kaye, "Dangers of Curve Fitting in the Deduction of Size Distribution from Diffraction Data", presented at the Bulk Powder Solids Conference, Rosemont, May 6–9, 1991. Proceedings published by Cahners Exposition Group, Cahners Plaza, 1350 E. Touhy Avenue, P.O. Box 5060, Des Plaines, Illinois, USA, 60019-9593.

[14] B. H. Kaye, *A Random Walk Through Fractal Dimensions*, VCH, Weinheim 1989.

[15] T. Congdon, D. McWilliams, *Basic Economics – A Dictionary of Terms, Concepts and Ideas*, Arrow Books, London 1976.

Author Index

Authors quoted in text with page number.
Pages in italics indicate references.

A

Aernoudt, E. *415*
Ahmed, A. K. *385*
Akhter, S. K. *386*
Allen, J. E. *265*
Allen, T. *175, 384, 462*
Amann, C. A. *386*
Appleby, L. *386*
Asimov, I *77, 436*
Avnir, D. 331, 334, *384*

B

Bak, P. 528, *533, 534*
Barnett, L. *77*
Barz, G. L. 280, *304*
Beckmann, P. 16, *78*
Beddow, J. K. *532*
Beddows, J. K. 322, 562, *384, 574*
Bell, E. T. *78*
Bell, R. C. 194, *206*
Bliss, C. K. 538, *552*
Bolsaitis, P. P. 344, 348, *385*
Bowen, R. *384*
Brancazio, P. J. *265*
Briggs, J. *78, 105*
Broad, J. 372
Brown, W. *78*
Bundalli, N. *574*

C

Carey, D. *78*
Carrol, L. (Dodson, Ludwig) 307, *383*
Chayes, F. *175*
Chen, K. *533, 534*
Cheng, Z. *304*
Chesters, S. *386*
Clark, G. G. 220, 336, *229, 304, 383, 384, 462, 533, 574*
Clarke, K. C. 421, *384, 436*
Colbeck, I. *386*
Colborne, R. *78*
Congdon, T. *574*
Cotton, W. R. *385*

D

Davidovits, P. *265*
Davies, R. *533*
Debus, K. H. *533*
Dekking, K. 9, *78*
DeMeester, P. *415*
Devaney, R. L. *105*
Dewdney, A. K. 85, *105*
Dhodapkar, S. *386*
Donan, D. F. *386*
Drake, F. *482*
Dreyer, K. *533*
Ducheyne, P. *415*

E

Eastman, J. *385*
Efros, A. L. *532*
Ehrlich, R. *574*
Elam, W. T. 280, *304*

Elings, V. B. *462*
Elliot, J. F. *385*
Emi, H. 282, *304*
Ensor, D. S. 503, *532*
Evans, B. *78, 175*
Evans, C. *384*
Exner, H. E. *383*

F

Fairstein, E. *532*
Fayed, M. *532*
Feder, J. *386*
Feller, W. 45, *78*
Fender *265*
Fender, D. H. *574*
Fink, D. G. 16, 168, *78, 175*
Flook, A. G. 567, *574*
Foukal, P. V. *386*

G

Gamow, G. 50, 54, 156, *78, 175*
Gleick, J. 7, 102, 233, 547, 551, *77, 105, 265, 386, 482, 533, 552*
Glen, W. *386*
Gratton J. 452, 521, 529, *533*
Greigo, R. J. *462*
Grinstein, G. *534*
Gubster, D. U. 280, *304*

H

Haig, W. J. *534*
Hale, J. K. 69, *77*
Halliday, D. *206*
Hardman, E. J. *386*
Hardy, G. *78*
Harrison, R. M. *386*
Hartwig, W. C. 359, *386*
Hayakawa, Y. 283, *304*
Hearsh, R. *462*
Held, G. A. 528, *534*
Helle-Shaw, H. S. 287, *304*
Herdan, G. *415*
Hickey, F. R. *533*
Hiragi, S. 282, *304*
Hiseieh, M. T. *385*
Hoffman, A. *304*
Honjo, H. 283, *304*
Horn, P. M. *534*
Huntley, H. E. *265*

I

Ipsen, D. C. *265*

J

Japuntnich, D. 348, *385*
Jha, A. K. 361, *386*

K

Kanoaka, C. 282, *304*
Kasahara, K. *78*
Kasper, G. *386*
Kaye, B. H. 220, 336, 503 *77, 78, 175, 206, 229, 265, 304, 383, 384, 385, 386, 415, 436, 462, 532, 533, 534, 552*
Keane, D. T. *534*
Kendal, M. J. 157, *175, 415*
Kendall, K. *385*
Kiely, T. *415*
Kittleson, D. B. *386*
Klingsing, G. E. *386*
Kotrappa, P. 345, *385*
Kutty, T. R. N. 349, *385*

L

LaSalle, J. P. 69, *77*
Lavenda, B. H. *462*
Lawless, B. A. *385*
Leblanc, J. 395
Lewis, R. 401

Liu, Y. 524, *533, 574*
Long, C. A. *386*
Lundin, M. *386*

M

Mandelbrot, B. B. 5, 45, 49, 60, 62, 86, 100, 236, 308, 310, 401, 428, 439, 440, 442, *78, 265, 383, 386, 415, 436, 462*
Matsushita, M. 283, *304*
McCarthy, J. F. *385*
McCracken, D. D. 59, *78*
McFarlane, B. 324, *384*
McGreevy, J. P. *415*
McIver, D. J. L. 324, *384*
McWilliams, D. *574*
Meakin, P. 280, *304*
Medley, J. B. *415*
Meloy, T. P. *574*
Mendes-France, M. *78*
Mercer, T. *385*
Middleton *533*
Mohiuddin, G. *385*
Montroll, E. W. 443, *462*
Moon, F. C. 7, *78*
Moran, P. A. P. 157, *175, 415*
Moroney, M. J. 37, 210, *78, 175, 206, 229*
Morrison, P. *78*
Morrow, F. *385*
Muller, J. 453, *462*
Mullins, M. E. 503, *532*
Myler, C. A. *386*
Myojo, T. 282, *304*

N

Nadis, S. *534*
Nelson, J. *385*
Newman, J. R. 157, *175, 206*
Nicoli, D. F. *462*

O

Olds, G. C. E. *385*
Ooms, M. 361, *386*
Orbach, R. *533*
Otten, L. *532*

P

Parratt, L. J. 110, *175*
Passoja, D. E. 425, *436*
Paterson, L. 289, *304*
Paully, A. J. 425, *436*
Pearce, F. *532*
Peat, F. D. 78, *105*
Peitgen, H. O. 7, 100, *78, 105*
Perera, F. P. *385*
Peterson, I. 104, *78, 105*
Pfeifer, H. J. 378, 503, *386, 532*
Philip G. C. *574*
Pierce, J. R. *462*
Pilliar, R. M. *415*
Pilpel, N. *533*
Pinnick, R. G. *385*
Plessis, P. *534*
Porter, M. C. *532*
Prasad, S. V. 361, *386*
Purcell, E. M. *265*
Purvis, A. *415*

R

Reist, P. C. 348, *385*
Resnick, R. *206*
Richter, P. H. 7, 100, *78, 105*
Richter, R. *304*
Roberts, A. W. 361, 362, *386*
Rogers, B. J. 324, *384*
Roy D. *415*
Russ, J. C. *385*

S

Sagan, C. *482*
Sander, L. M. 274, 277, *304*

Sander, L. M. *304*
Sano, M. 283, *304*
Sawada, Y. 283, *304*
Schaeffer, D. W. *385*
Schafer, D. W. 350, *385*
Schafer, H. J. 378, 503, *386, 532*
Schneider, H. J. *532*
Schroeder, M. *78, 552*
Schroeder, M. 513, *533*
Schuster, H. G. *78*
Schwartz, H. P. *574*
Schwarz, J. *383*
Scott, D. *384*
Seifert, W. W. *384*
Sem, G. J. *385*
Shafer, D. W. *385*
Shane, K. C. *574*
Shapiro, A. H. *265*
Shaw, R. *533*
Shlesinger, M. 275, 443, *175, 229, 304, 462*
Siegel, R. W. *385*
Siegla, D. C. *386*
Solina, D. H. *534*
Sprague, J. 280, *304*
Starling, S. G. *574*
Stauffer, D. *532*
Stenhouse, I. 503, *532*
Stewart, I. 12, 18, 82, 465, 479, 482, *78, 105, 265, 482*
Stober, W. 345, *385*
Studt, T. *533*
Summerfield, M. A. *415*
Sutton, H. M. *574*

T

Tang, C. *533*
Taylor, D. *532*
Trottier, R. A. 503, *304, 384, 385, 532*

U

Underwood, E. E. *415*
Upadhyaya, G. S. 361, 362, *386*

V

Van der Boorten, A. *78*
Van Nostrand, D. *78*
Van Vechten, D. 280, *304*
Vetter, A. F. *574*
Vilenken, N. Ya. 22, *78*
Vivekananden, R. 349, *385*
Vogel, S. *265*
Voss, R. F. *304*

W

Wagner, C. *385*
Walker, J *304*
Weinberg, B. *574*
Weiner, B. B. *462*
Weisenfeld, K. *533*
Wen, H. Y. *386*
West, B. J. 275, 443, *175, 229, 304, 462*
Westcott, V. C. *384*
Whalley, W. B. *415*
Wheat, T. A. *385*
White, D. C. S. *265*
White, D. H. *574*
White, H. E. *574*
Whitten, T. A. 274, 277, *304*
Wilkie, D. *532*
Winche, J. 361, *386*
Wolf, S. A. 280, *304*
Wolfram, S. *462*
Woodall, A. G. *574*

Z

Zaltash, A. *386*
Zeeman, E. C. *533*
Zimmer, C. *534*
Zipf, G. K. 439, *461*

Subject Index

A

acceleration/time/distance/velocity relation 540, 543
accident occurrence, simulation of 492
accidents, occurrence (log-normal distribution) 491
accidents, pattern of occurrence 487
accidents, clustering of 491
accuracy 134
address of the point 23
adsorption, gas 329
aerodynamic diameter 344
aerodynamic diameter, effect of fractal dimension 346
aggregate/agglomerate 274
allowable variations in content of a mixture 128–133
amalgamation, mosaic 331, 332
amplitude/frequency/phase/wavelength 558, 559
antithetic 160
antithetic variates 158, 160
applied mathematics 12
Archimedes principle 249
area estimate by dot count 140, 143
area estimate by square count 141
area, quantum of 22
asbestos, inhalation hazard of 345
aspect ratio 258
asymptote 75
attractor 93
attractor in phase space 70
attractor, strange 70, 468
Auebarch, Felix 443
avalanching 528
average 36
average displacement for a random walk 54
average displacement of a drunk 54
average value 117
Avnir, David 331

B

ball mill 347
bar, Cantorian 46, 50
bell curve 111, 113–115, 125
bell curve, skewed 192
Bequerel, Antoine Henri 184
berserk numbers 38
Bessel's correction 118
Bessel, Frederich Wilhelm 118
BET method 330
bimodal distribution 153
bimodal distribution of city size 443
biodegradable plastics 513
black-body radiation 19
Bliss, Charles 538
bone, structure of (Sierpinski fractal) 408, 409
bone, synthetic 408
boundary fractal dimension 311
boundary layer 255
bread characterizing 389, 390, 392
bread texture related to Sierpinski Carpet 400
bread, structure of 389, 390
Briggs, Henry 118
Brown, Robert 56
Brownian motion 56, 450

Brunauer, Emmett, Teller method (BET) 330
Buffon 180
Buffon's needle 155
Buffon's needle, determination of π 156
Buffon's needle, lines versus grids 162
Buffon's needle, needles versus crosses 160
Buffon, Georges 155
butterfly effect 11

C

Cabot Corporation 340
Cantor, Georg 46
Cantorian bar 46, 50
Cantorian dust 44, 46, 48
Cantorian set 392
capture trees of monosized aerosols 503, 504
capture trees on a fiber 503, 504
capture trees within a filter 281, 282
capture trees, simulation of 281, 282
carbon black, comparison of various grades 340
carbon black, distribution of fractal dimensions 342, 343
carbon black, growth of 342
Cardano, Girolamo 87
Carrol, Lewis (Dodson, Ludwig) 307
Cartesian coordinates 23
Cartesian product 84
catastrophe, violet 19
cats and mice 466
Cayley, Arthur 547
cell, electrolytic 283, 285
cell, Helle-Shaw 287
cell, Helle-Shaw, simple 287, 288
ceramics 348
chain, Markovian 41
chance 40
chaos, deterministic 3
chaotic 7
chaotic tumbling mixers 220–223
charge coupled device 331, 332
Chaye's dot-count technique 145
Chaye's method for estimating areas 390, 391
cherry cake 400
chocolate buttons 150
chocolate, consistency of 401
chromatography, fractal in fronts in paper 370
circle in quantized space 83, 85
circle, equation of 81, 82
climate, variations in 114
cluster size distribution, theoretical 513, 514
clustering of accidents 491
clustering of random events, simulation 498
clustering of rare events, epidemiology 496, 497
coastline of Great Britain 60
coastline, length of 62
cobweb diagram, May equation 473
coin flipping 109
coin flipping in sets 111
coin flipping, predicting the outcome of 109
coin tossing 37, 40, 181
coincidence errors, optical sizing 501
coincidence errors, density of coverage 501
coincidence errors, fine particle counting 500
coincidence of pores in a surface filter 508, 509
coincidence, density of coverage to avoid 505
common sense 20

compensating errors 137
complex equation, iteration of 91
complex number 87
complex number, modulus of 87, 98, 99
complex number, representation of 87
complex numbers, operations on 89
complexity 8
conclusions, jumping to 215
conic sections 429, 430
constants, values of various physical 264
content by Chaye's method 390, 391
content by Rosiwal intercept 390, 391
content fluctuations in a mixture 128–133
content of a mixture, allowable variations in 128–133
convex hull 256, 257
coordinates of a point 23
cosmology 3
cosmos 3
critical velocity 260
critical velocity, laminar to turbulent flow 252
cumulative oversize distribution 120
cumulative undersize distribution 120

D

damped vibration 65
data space, fractal dimension in, dripping tap 518, 516
data space, fractal dimension, avalanching powder 529
deMoivre, Abraham 115
decay, exponential 182
decay, exponential, equation 182
decay, logarithmic 181
decay, radioactive, simulation of 186
decorative stone, preservation of 410
DeMorgan, Augustus 156, 180
density fractal dimension 276
density of coverage for optical sizing 501
density of coverage to avoid coincidence 505
deposition of monosized dust, simulation of 214
deposition, electrolytic 282
depth (sponge) filter, manufacture of 511
derived dimensions 236
Descartes, René 22
determinism 3
deterministic chaos 3, 5
Devil's staircase 46, 48, 50
diameter, aerodynamic 344
diesel soot, fractal structure of 350, 351
diesel soot, respirable hazard of 350
Diet foods taken to the extreme 401
differential equations of motion 543
diffusion limited aggregate (DLA) 6, 273, 274
dimension, topological 239
dimensional analysis 234
dimensionality, operational 236
dimensions of various physical quantities 264
dimensions, classical 235
dimensions, derived 236
dimensions, fractional 49
dimensions, fundamental 236
disease, spread of 459
dispersion of drunks 51
displacement distribution for random walks 451, 452
displacement vector 55
distance/velocity/acceleration/time relation 540, 543
distribution of drunks 109

distribution of random walk displacements 451, 452
distribution of Smarties, color (Poisson) 215
distribution of the number of heads 111
distribution, bimodal 153
distribution, bimodal, of city size 443
distribution, cluster size, theoretical 513, 514
distribution, cumulative oversize 120
distribution, cumulative undersize 120
distribution, game length, Snakes and Ladders 199
distribution, Gaussian probability 115
distribution, Gaussian, tails of 443
distribution, hyperbolic, Levy flight 458
distribution, log-Gaussian 191, 451
distribution, log-normal 191, 451
distribution, log-normal, median value of 198, 200
distribution, log-normal, mode of 198, 200
distribution, log-normal, of accident occurrence 491
distribution, Poisson 209
distribution, Rosin-Rammler 446–450
distribution, tails of 226
DLA (diffusion limited aggregate) 6, 274
DLA simulation program 292–297
DLA simulation, flowchart 291
DLA, effect of sticking probability 278, 279
DLA, fractal dimension of 277
DLA, simple 273
DLA, simple, stages of growth 298
DLA, simulation of 277
Doppler effect 454
Doppler, Christian Johan 454
dot count area estimation 140, 143
dot count percent content estimation 145
double randomness in data collection 145
drainage basin 93, 546
dripping faucet/tap 516
dripping tap, fractal dimension in data space 518, 516
drunk, average displacement 54
drunks, dispersion of 51
drunks, distribution of 109
drunks, staggering 51
dust deposition, simulation of monosized 214
dust, Cantorian 44, 46, 48
dynamic viscosity 260

E

e, the exponential function 118
electrolysis, chemical process 285
electrolytic cell 283, 285
electrolytic deposition 282
elutriation of an ink in paper 370
energy, kinetic, equation for 263
energy, potential, equation for 263
equation of a circle 81, 82
equation of a line 81, 82
equation, behaviour of points near the roots of 548, 549
equation, roots of 543
equations of motion, differential 543
equations of motion, falling rock 540, 543
equipaced method of fractal characterization 314
equipaced method, chord length variation 317, 318
equipaced method, fractal dimension by 366, 367, 373, 375
ergodic sequence 41

error 133
errors, compensating 137
errors, self-canceling 137
Euclidean geometry 5
exponential decay 182
exponential growth 183
exponential notation 26

F

factorial operation, (!) 119
Fatou dust 103
Fatou, P. 90
faucet, dripping 516
Feigenbaum number 465
felt, simulating the structure of 403, 404
ferrography 322
fiber, definition of 258
fiber filter, simulating the structure of 403, 404
fiber, capture trees on 503, 504
filter paper, fractal ink fronts in 370
filter, capture trees in 503, 504
filter, capture trees within a 281, 282
filter, closing large holes in 407
filter, persistence of large holes in 406, 407
filter, simulating the structure of 403, 404
filter, smoking of 407
filter ability of metal fumes 348
fine particle counter, resistazone stream 499, 500
fine particle counting, coincidence errors 500
fine particle counting, primary count loss 500
fine particle counting, secondary count gain 501
fine particle sizing, resistazone stream counter 500
fingering, fluid 287, 288
fire, spread of 459
flipping, coin 109
flipping, coin, in sets 111
flipping, coin, predicting the outcome of 109
"flooding", creating islands on a map by 419, 420
flow, laminar 245
flowchart, DLA simulation 291
fluid fingering 287, 288
fluid, Newtonian 248
fluids, non-Newtonian 356
foundations, shaky 113
Fourier analysis of geometric signature waveform 563, 565
Fourier analysis, concept of 556
Fourier analysis, limitations 566
Fourier analysis, profile reconstruction 567
Fourier, Jean Baptist 555
fractal characterization, equipaced technique 314
fractal characterization, structured walk method 310
fractal determination by gas adsorption 330
fractal dimension 64, 233
fractal dimension by equipaced method 366, 367, 373, 375
fractal dimension by mosaic amalgamation 366, 368
fractal dimension by structured walk method 363, 364, 373, 374
fractal dimension in data space of avalanching 529
fractal dimension in data space, dripping tap 518, 516
fractal dimension of a DLA 277

fractal dimension of a Levy flight 458, 459
fractal dimension of a Sierpinski Carpet 394, 395
fractal dimension using synthetic islands 363, 364
fractal dimension, boundary 311
fractal dimension, density 276
fractal dimension, effect of topography on 369
fractal dimension, effect on aerodynamic diameter 346
fractal dimension, effect on the strange attractor 524
fractal dimension, effect on viscosity 357, 358
fractal dimension, Korcak, islands/fragments 429–433
fractal dimension, quadric island 72
fractal dimension, Sierpinski, data handling 396–399
fractal dimension, structural 313
fractal dimension, surface, calculation 422, 424
fractal dimension, surface, Passoja's method 425–427
fractal dimension, textural 313
fractal dimension, triadic island 72
fractal dimensions of carbon black, distribution 342, 343
fractal fronts, ink in paper (chromatography) 370
fractal geometry 5
fractal rabbits 334
fractal structure 275
fractal structure of diesel soot 350, 351
fractal structure, regions of various 314, 315, 335
fractal, Sierpinski, of paper 407
fractal, use of the term 311
fractalicious 5
fractional dimensions 49
fragmentation fractal (Korcak) 429–433
Fred versus Freda 37
frequency of words used in text 439, 440
frequency/phase/wavelength/amplitude 558, 559
frost 280
fume resulting from incineration of waste 344, 348, 349
fume, welding, filterability of 348
fumes, filterability of metal 348
fuming 347
fundamental dimensions 236

G

Galileo 12
Galileo, acceleration due to gravity experiment 251
Galton board 125
Galton, Sir Francis 126
"gamblers ruin" 179
gambling 179
Gamow, George 50
gas adsorption 329
gas adsorption, fractal determination by 330
Gauss, Karl Friedrich 87, 115
Gaussian distribution, tails of 443
Gaussian probability distribution 115
Gaussian probability graph paper 120, 123
Gaussian probability, equation for 116, 118
genus of a system 239
geometric signature waveform 561, 565
geometric signature waveform, construction of 562, 565

geometric signature waveform, Fourier analysis 563, 565
geometric signature waveform, harm. spectrum 563, 565
geometric signature waveform, limitations 565
geometry, Euclidean 5
geometry, fractal 5
geometry, quantum 22
graph paper, Gaussian probability 120, 123
graph paper, log-Gaussian 192
graph paper, log-normal 192
graph paper, log-probability 192
graph paper, logarithmic, simplified 76
graph paper, Poisson 211
graph paper, Poisson, simplified 212
graph paper, Rosin-Rammler 448
graph paper, triangular, use of 520, 522
Great Britain, coastline of 60
Great Britain, east versus west 315
growth, exponential 183

H

half-life, radioactive material 186–189
harmonic notation 558, 561
harmonic spectrum 559, 564
hazard, inhalation 344
heads, distribution of the number of 111
Heisenberg's Uncertainty Principle 4
Helle-Shaw cell 287
Helle-Shaw cell, simple 287, 288
Helle-Shaw, H. S. 287
Hubbard, John 547
hull, convex 256, 257
hyperbolic distribution function, Levy flight 458
hyperbolic function, Korcak relation 429, 440
hyperbolic function, Zipf's law 440

I

i, square root of −1 87
ideal fractal boundary 308
imaginary number 87
incineration of waste, fumes resulting from 344, 348, 349
infinity, operational definition of 30
infinity, ∞ 30
inhalation hazard 344
inhalation hazard of dust, assessment of 505
ink fronts in paper, fractal (chromatography) 370
intersection count 275, 276
irrational numbers 16, 53
island, quadric, construction of 72
island, triadic, construction of 72
islands created by "flooding" a map 419, 420
islands in a lake, number of 427
isoaerodynamic fine particles 345
iterate 26
iteration of a complex equation 91
iteration of numbers near 1.00 31

J

jelly beans 223
jelly beans, sampling 223
Julia set 103
Julia set, forth order (Newton's method for $z^4 - 1 = 0$) 538
Julia set, self-similarity within the third order 550
Julia set, third order (Newton's method for $z^3 - 1 = 0$) 550
Julia, G. 90
jumping to conclusions. 215

K

Karman Street 251
kinematic viscosity 260
kinetic energy, equation for 263
Koch islands 71
Korcak 429
Korcak fractal dimension for islands/fragments 429–433
Korcak relation 429

L

Lake Ramsey 141
Lake Ramsey, fractal determination by mosaic 331, 332
Lake Ramsey, ice melting data 170
laminar flow 245
Laplacian determinism 4
large holes in a filter, closing 407
large holes, persistence of in a filter 406, 407
laws of probability 11
leaching of minerals from an ore 513
Leibniz, Gottfried Wilhelm 539
length of a boundary, estimation of 308
length of a coastline 62
leukocytes, activated 323
Levy flight 453
Levy flight, fractal dimension of 458, 459
Levy flight, hyperbolic distribution function 458
Levy flight, pseudo Levy flight 459
Levy flight, simulation of 455
Levy flight, space filling ability 458
Levy, Paul 453
light trapping by soot 352
line, equation of 81, 82
log-Gaussian distribution 191, 451
log-Gaussian graph paper 192
log-log graph paper 29
log-normal distribution 191, 451
log-normal distribution of accident occurrence 491
log-normal distribution, median value of 198, 200
log-normal distribution, mode of 198, 200
log-normal graph paper 192
log-probability graph paper 192
logarithm 29
logarithmic graph paper 29
logarithmic graph paper, simplified 76
logarithms, base 10 118
logarithms, base e 118
logarithmic decay 181
long runs, simulation 184

M

Mach number 259
Mach, Ernst 259
Malthus 466
Mandelbrot set 5
Mandelbrot set, construction of 91
Mandelbrot set, dynamics of 92–99
Mandelbrot set, equation for 91
Mandelbrot set, simplified map 91
Mandelbrot set, zoom 101
Mandelbrot, Benoit B. 5
Manitoulin Island 334, 335
map area 429
Markovian chain 41
mathematics, applied 12
mathematics, pure 12
May equation for population growth 471
May equation, cobweb diagram 473
May equation, number of roots found 480
May, R. 471
mean 36

mean from probability graph paper 123
median of a log-normal distribution 198, 200
metal film, sputter deposited 280
metal fumes, filterability of 348
mine tailings 353
Minkowski sausage 324, 325
Minkowski, Herman 324
mixer, chaotic tumbling 220–223
mixing, monitoring mixing, tracking powder 220
mixture, allowable variations in content of 128–133
mixture, content fluctuations in 128–133
mixtures of powders, sampling 127
mixtures, simulating 128
modal analysis 145
mode of a log-normal distribution 198, 200
modulus of a complex number 87, 98, 99
moments of a strange attractor 522, 523
monosized aerosols, capture trees of 503, 504
Monte Carlo routine 59, 269
mosaic amalgamation 331, 332
mosaic amalgamation, fractal dimension by 366, 368
motion, differential equations of 543
motion, equation of 234, 237
motion, falling rock, equations of 540, 543
muses 14

N

Napier, John 29, 118
natural fractal boundary 308
Newton's method for complex numbers, example 545, 546
Newton's method for finding roots 537, 544
Newton's method for $Z^3 - 1 = 0$, typical iterations 547, 548
Newton's method for $Z^4 - 1 = 0$, results 538
Newton, Sir Issac 539
Newtonian fluid 248
non-Newtonian fluids 356
normalization 62
normalization factor 62
notation, exponential 26
"nothingbutism" 8
nuclear winter 352
nucleating center 270
number, complex 87
number, complex, modulus of 87, 98, 99
number, imaginary 87
numbers, berserk 38
numbers, complex, operations on 89
numbers, irrational 16, 53

O

occlusion, projection 372
odds 179
oil, movement through sandstone 409
operational definition of infinity 30
operational definition of zero 31
operational dimensionality 236
operations on complex numbers 89
optical sizing, density of coverage for 501
optical sizing, simulated dispersion, monosized 501
ore, leaching minerals from 513
ore, richness of 145
origami 9, 10
Ouch! 446
oversize, cumulative distribution 120

P

π', shape factor 258
π, determination of by Buffon's needle 156
π, pi 14
paper chromatography, fractal ink fronts 370
paper, Sierpinski fractal of 407
paper, simulating the structure of 403, 404
Passoja's method for surface fractals 425–427
pendulum 64, 65
pendulum, clock 69
pendulum, driven 69
pendulum, period of, derivation 242
percent content by Chaye's method 390, 391
percent content by Rosiwal intercept 390, 391
percent content estimation by dot count 145
percolating pathways through a matrix 511, 515
percolation 511, 512
percolation, critical point for (59.28%) 511, 514
period of a pendulum, derivation of 242
Perrin, J. 451, *462*
Petri dish 284, 287
Petri, Richard 284
phase angle 67
phase angle, effect of when adding waves 560
phase space diagram 469, 470
phase space representation 68
phase/wavelength/amplitude/frequency 558, 559
phases of the moon 66
photon 21
physical constants, values of various 264
physical quantities, dimensions of various 264
pin board 125
pixel 14
pixel rainbow 25
pixoid 22
Planck, Max 19
plastics, biodegradable 513
pneumoconiosis 345
point, address of 23
point, coordinates of 23
point, definition of 18
points on a circle, number of 21
Poise, unit of viscosity 255
Poiseuille formula for fluid flow 253, 254
Poiseuille formula, derivation of units for 253
Poiseuille, J. L. M. 253
Poisson distribution 209
Poisson distribution, equation for 209
Poisson graph paper 211
Poisson graph paper, simplified 212
Poisson probability distribution 209
Poisson, Simeon Denis 209
polygons, inscribed and excribed 169
population distribution of cities (Zipf's law) 443, 445
population ecology 466
population growth, May equation 471
potential energy, equation for 263
powder mixing, monitoring 220
powder mixing, tracking 220
powders, sample size 128
powders, sampling mixtures of 127
precipitation 347
precision 134
predator-prey cycle 468

preservation of decorative stone 410
primary count loss, fine particle counting 500
prisms, triangular, surface area calculation with 421–423
probability distribution, Gaussian 115
probability relationships 11
probability theory 110
probability, laws of 11
probability, Poisson 209
Procrustean thinking 446
product, Cartesian 84
profilometer 327, 328
profilometer, effect of needle size 328, 329
projection occlusion 372
pseudo Levy flight 459
pure mathematics 12
Pythagoras 319
Pythagoras' theorem 55

Q

quadric island, construction of 72
quadric island, perimeter estimate 309
quantized space 83
quantized space, circle in 83, 85
Quantum Theory 19
quantum geometry 22
quantum of area 22
quenching 342
Quincunx board 125

R

rabbits, fractal 334
radiation, black-body 19
radioactive decay, simulation of 186
radioactive material, half-life of 186–189
radioactivity 185
radioactivity, discovery of 184
rainbow, pixel 25
Ramsey Lake 62, 141
Ramsey Lake, fractal determination by mosaic 331, 332
Ramsey Lake, ice melting data 170
random 7, 38
random clustering of events 496
random events, clustering, simulation 498
random number table 190
random number table, numbers 1–6 only 197
random number table, numbers 1–4 only 272
random number table, simplified 39
random number table, track lengths on 189
randomized Sierpinski Carpet 394, 395
randomness, double, in data collection 145
random walk 51
random walk theory 59
random walk, average displacement 54
random walk, self-avoiding 57, 58
random walk, simulation of 51, 54
random walks, displacement distribution 451, 452
rare events, clustering of, epidemiology 496, 497
rate of change, differential calculus 541, 543
Rayleigh, Lord 243
reciprocal 27
reductionism 8
regions of various ruggedness 373
respirable dust, assessing the hazard of 505
respirable hazard of a dust, assessing 347

respirable hazard of diesel soot 350
respirable hazard, asbestos 345
Reynolds number 245
Reynolds number, equation for 259
Reynolds number, examples 246
Reynolds number, typical values 261
Reynolds, Sir Osbourne 259
Rice up to here! 183
Richardson plot 64, 310
Richardson, Lewis Fry 60
richness of an ore 145
Robot (game) 203
robot 203
Robot, examples 204
Robot, parts bin 203, 204
Robot, rotation 205
Robot, Rules of Play 203
roots of an equation 543
roots of an equation, behavior of points near 548, 549
Rosin–Rammler distribution 446–450
Rosin–Rammler graph paper 448
Rosiwal intercept method for estimating content 390, 391
rugged lines, synthetic island technique 363, 364
ruggedness, regions of various 314, 315
ruggedness, regions of various 335
Rules of Play, Robot 203
Rules of Play, Snakes and Ladders 195

S

sample size for powders 128
sample size required to reduce standard deviation 165
sample size, comparison of 128–133
sample size, effect on standard deviation 128–133
sampling jelly beans 223
sampling mixtures of powders 127
sandstone, movement of oil through 409
sandstone, Sierpinski fractal of 409, 410
sausage, Minkowski 324, 325
saw tooth wave, construction of 562
scalar 55
scaling function 44
secondary count gain, fine particle counting 501
Sierpinski, Waclaw 393
Sierpinski fractal of paper 407
self-avoiding random walk 57, 58
self-canceling errors 137
self-similarity in the stock market 569
self-similarity within the third order Julia set 550
self-similarity, statistical 47, 275
sense, common 20
sequence, ergodic 41
shaky foundations 113
shape factor 256
shape index 256
shear stress 248
Sierpinski carpet 393, 394
Sierpinski carpet related to bread texture 400
Sierpinski carpet, connectedness of 396
Sierpinski carpet, construction of 394
Sierpinski carpet, fractal dimension of 394, 395
Sierpinski carpet, graphical presentation of data 396
Sierpinski carpet, randomized 394, 395
Sierpinski carpet, statistically self-similar 394, 395
Sierpinski fractal dimension, data handling 396–399
Sierpinski fractal of bone 408, 409

Sierpinski fractal of sandstone 409, 410
sieve, aperture size distribution of 148, 150
sieve, woven wire 147
simplified random number table 39
simulate 38
simulated monosized dispersion for optical sizing 501
simulating mixtures 128
simulating the structure of a felt 403, 404
simulating the structure of a fiber filter 403, 404
simulating the structure of paper 403, 404
simulation of a Levy flight 455, 457
simulation of a random walk 51, 54
simulation of accident occurrences 492
simulation of DLA 277
simulation of monosized dust deposition 214
simulation of radioactive decay 186
simulation of the clustering of random events 498
simulation, DLA program 292–297
simulation, DLA, Flowchart 291
simulation, Smarties color distribution 216
sintering 350
sirens, stochastic 114
sizing of fine particles, resistazone stream counter 500
skewed bell curve 192
slurries, transportation and handling of 355
Smarties® 150
Smarties®, color distribution simulation 216
Smarties®, color distribution of (Poisson) 215
Smarties®, diameter and thickness 171
Smarties®, weight distribution of 152
Snakes and Ladders 193
Snakes and Ladders game board 194
Snakes and Ladders, game length distribution 199
Snakes and Ladders, rules of play 195
solid fumes 344
solids content, effect on viscosity of a suspension 357, 358
soot 277
soot as a smoke screen 352
soot, diesel, fractal structure of 350, 351
soot, light trapping by 352
space filling ability 49, 275, 310
space filling ability, Levy flight 458
space, quantized 83
specific gravity 258
sponge (depth) filter, manufacture of 511
sputter deposited metal film 280
square count area estimation 141
square root, concept of 52
square wave, construction of 563
staggering drunks 51
staircase, Devil's 46, 48, 50
standard deviation from probability graph paper 123
standard deviation, change in with sample size 128–133
standard deviation, reduction of 165
standard deviation, calculation of 117, 118
statistical self-similarity 47
statistically self-similar Sierpinski carpet 394, 395
stereology 389, 392
sticking probability, DLA, effect of 278, 279

Stober Disc Centrifuge 345
stochastic process 40
stochastic sirens 114
stock market, self-similarity in 569
Stokes law 245, 250, 251, 344
Stokes law, derivation of units for 250
Stokes, Sir George 251
strain 248
strange attractor 70, 468
strange attractor of tumbling rocks 518, 519
strange attractor, dripping faucet/tap 517, 516
strange attractor, effect of fractal dimension on 524
strange attractor, moments of 522, 523
stratified count logic 411
stream fineparticle counter, resistazone 499, 500
stress, shear 248
structural fractal dimension 313
structured walk method of fractal characterization 310
structured walk method, fractal dimension by 363, 364, 373, 374
surface area calculation with triangular prisms 421–423
surface filter 509
surface filter, manufacture of 508
surface fractal dimension by Passoja's method 425–427
surface fractal dimension calculation 422, 424
suspension viscosity, effect of fractal dimension 357, 358
suspension viscosity, effect of solids content 357, 358
suspension, viscosity of 356, 357
synthetic bone 408
lines 363, 364

T

tailings, mine 353
tails of the distribution 226
tails of the Gaussian distribution 443
tangent 75
tap, dripping 516
telephone lines, noise on 45
Teller, Edward 330
textural fractal dimension 313
thixotropy 355
thixotropy of food mixtures 356
thixotropy of slurries 355
thorium dioxide (Thoria) 347
time/distance/velocity/acceleration relation 540, 543
topographical map 419, 420
topography 240
topography, effect on river fractal dimension 369
topological dimension 239
topology 238
track lengths on a random number table 189
tracking powder mixing 220
tremas 393
triadic island, construction of 72
triadic island, perimeter estimate 309
triangular graph paper, use of 520, 522
triangular wave, construction of 563
tumbling rock, effect of rotational speed 526, 527
tumbling rock, strange attractor of 518, 519

U

uncertainty 133
undersize, cumulative distribution 120
sunspot cycle 572
uranium dioxide (Urania) 347

V

variance 118
vector 55
vector, displacement 55
velocity, critical, laminar to turbulent flow 252
velocity/acceleration/time/distance relation 540, 543
Vermilion River, fractal dimension of regions 369
violet catastrophe 19
viscometer 356, 357
viscosity of a fluid 247
viscosity of a suspension 356, 357
viscosity, derivation of units for coefficient of 247, 249
viscosity, dynamic 260
viscosity, kinematic 260
viscosity, suspension, effect of fractal dimension 357, 358
viscosity, suspension, effect of solids content 357, 358
viscosity, unit of (Poise) 255
von Karman, Theodor 252
von Koch, Helge 71
vulcans 340

W

wager 180
Wallis, John 30
watershed 546
wave-particle duality of light 20
wave form analysis, components of a complex wave 556
wavelength/amplitude/frequency/phase 558, 559
waves, terminology associated with 558
wear 322
wear debris 321
weight distribution of Smarties® 152
welding fume, filterability of 348
Whittaker, E. T. 120
Whitten-Sander aggregate 273, 274
word frequency in text 439, 440

Z

zero, operational definition of 31
Zipf's law 440
Zipf's law, population distribution of cities 443, 445
Zipf, George Kingsley 439